Lecture Notes in Com

Founding Editors

Gerhard Goos
Juris Hartmanis

Editorial Board Members

Elisa Bertino, *Purdue University, West Lafayette, IN, USA*
Wen Gao, *Peking University, Beijing, China*
Bernhard Steffen, *TU Dortmund University, Dortmund, Germany*
Moti Yung, *Columbia University, New York, NY, USA*

The series Lecture Notes in Computer Science (LNCS), including its subseries Lecture Notes in Artificial Intelligence (LNAI) and Lecture Notes in Bioinformatics (LNBI), has established itself as a medium for the publication of new developments in computer science and information technology research, teaching, and education.

LNCS enjoys close cooperation with the computer science R & D community, the series counts many renowned academics among its volume editors and paper authors, and collaborates with prestigious societies. Its mission is to serve this international community by providing an invaluable service, mainly focused on the publication of conference and workshop proceedings and postproceedings. LNCS commenced publication in 1973.

Aleš Leonardis · Elisa Ricci · Stefan Roth ·
Olga Russakovsky · Torsten Sattler · Gül Varol
Editors

Computer Vision – ECCV 2024

18th European Conference
Milan, Italy, September 29–October 4, 2024
Proceedings, Part LXXXIX

Editors
Aleš Leonardis
University of Birmingham
Birmingham, UK

Stefan Roth
Technical University of Darmstadt
Darmstadt, Germany

Torsten Sattler
Czech Technical University in Prague
Prague, Czech Republic

Elisa Ricci
University of Trento
Trento, Italy

Olga Russakovsky
Princeton University
Princeton, NJ, USA

Gül Varol
École des Ponts ParisTech
Marne-la-Vallée, France

ISSN 0302-9743　　　　　　　ISSN 1611-3349　(electronic)
Lecture Notes in Computer Science
ISBN 978-3-031-73023-8　　　ISBN 978-3-031-73024-5　(eBook)
https://doi.org/10.1007/978-3-031-73024-5

© The Editor(s) (if applicable) and The Author(s), under exclusive license to Springer Nature Switzerland AG 2025

This work is subject to copyright. All rights are solely and exclusively licensed by the Publisher, whether the whole or part of the material is concerned, specifically the rights of translation, reprinting, reuse of illustrations, recitation, broadcasting, reproduction on microfilms or in any other physical way, and transmission or information storage and retrieval, electronic adaptation, computer software, or by similar or dissimilar methodology now known or hereafter developed.
The use of general descriptive names, registered names, trademarks, service marks, etc. in this publication does not imply, even in the absence of a specific statement, that such names are exempt from the relevant protective laws and regulations and therefore free for general use.
The publisher, the authors and the editors are safe to assume that the advice and information in this book are believed to be true and accurate at the date of publication. Neither the publisher nor the authors or the editors give a warranty, expressed or implied, with respect to the material contained herein or for any errors or omissions that may have been made. The publisher remains neutral with regard to jurisdictional claims in published maps and institutional affiliations.

This Springer imprint is published by the registered company Springer Nature Switzerland AG
The registered company address is: Gewerbestrasse 11, 6330 Cham, Switzerland

If disposing of this product, please recycle the paper.

Foreword

Welcome to the proceedings of the European Conference on Computer Vision (ECCV) 2024, the 18th edition of the conference, which has been held biennially since its founding in France in 1990. We are holding the event in Italy for the third time, in a convention centre where we can accommodate nearly 7000 in-person attendees.

The journey to ECCV 2024 started at CVPR 2019, where the general chairs met at a sketchy bar in Long Beach to discuss ideas on how to organize an amazing future ECCV. After that, Covid came and changed conferences probably forever, but we maintained our excitement to organize an awesome in-person conference in 2024, particularly having seen the wonderful event that was ECCV 2022 in Tel Aviv. And here we are, 5 years later, delighted that we finally get to welcome all of you to Milan!

First and foremost, ECCV is about the exchange of scientific ideas, and hence the most important organizational task is the selection of the programme. To our Program Chairs, Aleš Leonardis, Elisa Ricci, Gül Varol, Olga Russakovsky, Stefan Roth, and Torsten Sattler, our entire community owe a huge debt of gratitude. As you will read in their Preface to the proceedings, they dealt with over 8,500 submissions, and their diligence, thoroughness, and sheer effort has been inspirational. From the early stages of designing the call for papers, through recruiting and selecting area chairs and reviewers, to ensuring every paper has a fair and thorough assessment, their professionalism, commitment, and attention to detail has been exemplary. From all the choices that we made to organize ECCV, choosing this amazing team of Program Chairs has been, without a doubt, our very best decision.

The Program Chairs were advised and supported by an Ethics Review Committee, Chloé Bakalar, Kate Saenko, Remi Denton, and Yisong Yue, who provided invaluable input on papers where reviewers or Area Chairs had raised ethical concerns.

The Publication Chairs, Jovita Lukasik, Michael Möller, François Brémond, and Mahmoud Ali, did great work in assembling the camera-ready papers into these Springer volumes, dealing with numerous issues that arose promptly and professionally.

With the ever-expanding importance of workshops, tutorials, and demos at our major conferences comes a corresponding increase in the challenge of organizing over 70 workshops and 9 tutorials attached to the main conference. The Workshop and Tutorial Chairs, Alessio Del Bue, Jordi Pont-Tuset, Cristian Canton, and Tatiana Tommasi, and Demo Chairs, Hyung Chang and Marco Cristani, worked tirelessly to coordinate the very varied demands and structures of these different events, yielding an extremely rich auxiliary program which will enhance the conference experience for everyone who attends.

The Industry Chairs, Shaogang Gong and Cees Snoek, working with Limor Urfaly and Lior Gelfand, managed to attract numerous companies and organizations from all around the world to the industrial Expo, making ECCV 2024 an international event that not only showcases excellent foundational research, but also presents how such research can be transformed in real products. It is notable that together with the big companies, several start-ups chose ECCV to present their business ideas and activities.

As usual in the recent main conference, we also organized Speed Mentoring and Doctoral Consortium events. While the latter is an institutional occasion allowing senior PhDs to show their work, the former is especially important to answer to the many doubts and uncertainties young scholars have while undertaking a career in Computer Vision and Artificial Intelligence. We warmly thank our Social Activities Chairs, Raffaella Lanzarotti, Simone Bianco and Giovanni Farinella, and the Doctoral Consortium Chairs, Cigdem Beyan and Or Litany, for their dedication to organize such events, so important for our young colleagues, as well as all the mentors who were available to share their experience.

As the conference becomes larger and larger, the overall budget increases to a level where uncertainties in estimates of attendance can result in significant losses, perhaps enough to significantly damage the prospects of running the conference in the future. This risk imposed the need to require authors to commit to full conference registrations, with the undesirable consequent possibility to cause hardship to authors, particularly student authors, from organizations where funding attendance is difficult. To mitigate this effect, we allocated travel grants to students who might otherwise have had difficulty in attending. Our Diversity Chairs, David Fouhey and Rita Cucchiara, put tremendous effort into a wonderfully careful and thoughtful programme, balancing diversity in all its forms with a changing budget landscape, allowing us to support the participation of many people who might not otherwise have been able to attend.

This year's conference also marks the move to a unified, multi-year conference website, based on that used for other leading computer vision and AI conferences. This was an effort that we decided to take on to make the ECCV website more familiar to the community and support the organization of the future ECCV editions. Our thanks go to Lee Campbell of Eventhosts who managed the entire design and development of the new website and provided responsive support and updates throughout the conference organization timeline.

Our Finance Chairs, Gérard Medioni and Nicole Finn, provided invaluable advice and assistance on all aspects of the financial planning of the conference, in conjunction with the professional conference organizers AIM Group, where Lavinia Ricci led a fantastic team that managed thousands of details from space planning to food to registrations. In particular, Elder Bromley and Mara Carletti were a tremendous support to the entire conference organization. The conference centre, Mico DMC, also provided an excellent service, showing considerable flexibility in adapting the venue and processes to the particular demands of an academic-focused conference.

Finally, we cannot forget to thank our Social Media Chair, Kosta Derpanis, supported by Abby Stylianou and Jia-Bin Huang, who, with their presence on social media, spread ECCV news and responded to small and big questions in a timely manner, helping the community to promptly receive information and fix their problems.

October 2024

Andrew Fitzgibbon
Laura Leal-Taixé
Vittorio Murino

Preface

ECCV 2024 saw a remarkable 48% increase in submissions, with 8585 valid papers submitted, compared to 5804 in ECCV 2022, 5150 in 2020, and 2439 in 2018. Of these, 2387 papers (27.9%) were accepted for publication, with 200 (2.3% overall) selected for oral presentations.

A total of 173 complete submissions were desk-rejected for various reasons. Common issues included revealing author identities in the paper or supplementary material (for example: by including the author names, referring to specific grants in the acknowledgments, including links to GitHub accounts with visible author information, etc.), exceeding the page limit or otherwise significantly altering the submission template, or posting an arXiv preprint explicitly mentioning the work as an ECCV 2024 submission. The desk-rejected submissions also include 14 papers identified as dual submissions to other concurrent conferences, such as ICML, NeurIPS, and MICCAI. Seven papers were rejected among the accepted papers due to plagiarism issues, after the Program Chairs conducted a thorough plagiarism check with the iThenticate software followed by manual inspection.

The double-blind review process was managed using the CMT system, ensuring anonymity between authors and Area Chairs/reviewers. Each paper received at least three reviews, amounting to over 26,000 reviews in total. To maintain the quality and the fairness of the reviews, we recruited 469 Area Chairs (ACs) and more than 8200 reviewers. Of the recruited reviewers, 7293 ultimately participated.

Area Chairs were chosen for their technical expertise and reputation; many of them had served in similar roles at other top conferences. We also recruited additional Area Chairs through an open call for self-nominations. Among the selected ACs, around 18% were women. Geographical distribution was taken into account during the selection process, with 31% of ACs working in Europe, 38% in the Americas, 27% in Asia, and 4% in the rest of the world. Despite our best efforts to consider diversity factors in the selection process, the gender diversity and geographic diversity of ACs is still below what we were hoping for.

To ensure a smooth decision-making process, we grouped the ACs into triplets. Upon accepting their role, ACs were requested to enter into the CMT system domain conflicts and subject areas, as well as information about their DBLP, TPMS (Toronto Paper Matching System), and OpenReview profiles. These were used to determine whether different subject areas were sufficiently covered, to detect conflicts, and to automatically form AC triplets. An AC triplet was a group of three ACs who worked together throughout the process, which helped to facilitate discussion, calibrate the decision-making process, and provide support for first-time ACs. AC triplets were manually refined by Program Chairs to ensure no triplet contained more than two major time zones to enable easier coordination for virtual AC meetings. Paper assignments were made such that no AC within the triplet was conflicted with any of the papers. Each AC was responsible for overseeing 18 papers on average.

In each triplet, we identified a Lead AC who was an experienced member of the computer vision community willing to take on additional administrative duties. The primary responsibilities of Lead ACs included coordinating the AC triplet to ensure deadlines were met and offering guidance to less experienced ACs. They also served as the main contact person with the Program Chairs and helped in handling papers if an original AC was unable to complete their tasks.

Reviewers were invited from reviewer pools of previous conferences and also selected from the pool of authors. We asked experienced ACs to recommend more potential reviewers. We further recruited reviewers through self-nominations, where researchers volunteered to review for ECCV 2024 via an open call. The self-nominations were then filtered, selecting those reviewers with at least two papers published in related conferences such as ECCV, ICCV, CVPR, ICLR, NeurIPS, etc.

Similarly to ACs, reviewers were also asked to update their CMT, Toronto Paper Matching System (TPMS), and OpenReview profiles. Reviewers had the option to classify themselves as either senior researchers or junior researchers and students, with the number of assigned papers varying accordingly (a quota of 6 for senior researchers with 3 or more times reviewing experience, and a quota of 4 otherwise). On average, each reviewer was assigned about 4 papers.

Conflicts of interest among authors, ACs, and reviewers were managed automatically via conflict domains and co-authorship information from DBLP. Paper assignments to ACs and reviewers were determined using an algorithm that combined subject-area affinity scores from CMT, TPMS, and OpenReview, along with additional criteria such as ensuring that every paper had at least one non-student reviewer. For AC assignments, we additionally incorporated a newly built affinity score based on embedding similarity with the papers on the Google Scholar profile of the ACs. Once the assignments were made, ACs and reviewers were asked to promptly report any undetected conflicts of interest or other concerns that would preclude them from handling the assigned papers to ensure immediate reassignment, if necessary.

Given the widespread use of Large Language Models (LLMs), ECCV 2024, like previous conferences, implemented a special policy regulating their use. Authors were permitted to use any tools, including LLMs, in preparing their papers, but they remained fully responsible for any misrepresentation, factual inaccuracies, or plagiarism. Reviewers were also allowed to use LLMs to refine the wording of their reviews; however, they were held accountable for the accuracy of their content and were strictly prohibited from inputting submissions into an LLM. One challenge we ran into was that tools that detect LLM-generated content were not able to distinguish between text polished vs. text written from scratch by an LLM. Thus, it became quite difficult to enforce the policy in practice.

Overall, the challenges encountered during the review process were consistent with those faced at previous computer vision conferences. For instance, a small number of reviewers ultimately did not submit their reviews or provided brief, uninformative reviews, necessitating last-minute reassignment to emergency reviewers. As the community expands and the number of submissions rises, securing enough qualified reviewers and performing a quality check of the received reviews becomes increasingly difficult.

The review process included a rebuttal phase where authors were given a week to address any concerns raised by the reviewers. Each response was limited to a single PDF page using a predefined template. Rebuttals with formatting or anonymity violations were removed by the Program Chairs. ACs then facilitated discussions with reviewers to evaluate the merits of each submission. While the goal was to reach a consensus, the final decision was left to the ACs in the triplet to ensure fairness. As is common, the ACs within each triplet were tasked with organizing meetings to discuss papers, especially borderline cases. The entire process was conducted online, with no in-person meetings. For each paper, a primary and a secondary AC were assigned from the AC triplet. The primary AC was responsible for entering the decisions and meta-reviews into the CMT system, which were then reviewed and approved by the secondary ACs. Besides making accept/reject decisions, the ACs were also required to provide recommendations regarding oral/poster presentations and award nominations.

Reviewers and ACs were asked to flag papers with potential ethical concerns. An ethics review committee, consisting of four leading researchers from both industry and academia, was appointed. The role of the committee was to evaluate papers escalated by the AC for further investigation. In total, 31 papers were flagged to the committee. The committee requested that the authors of 15 papers address specific concerns in their camera-ready submissions. The authors all complied with the requests.

Following the recent IEEE decision to ban the use of the photograph of Lena Forsén, we automatically identified all submissions that included this image. Although the ban did not officially apply to ECCV, we nevertheless shared the context with the authors of the affected papers and suggested that they remove the photograph in the camera-ready submission. All authors complied.

To help the ACs to track the overall progress of the reviewing and decision-making processes, we created AC triplet activity dashboards for each AC triplet (via Google sheets), which were updated on a regular basis. During the review process, the activity dashboard enabled ACs to track the progress of incoming reviews for each paper and promptly identify papers that necessitated attention, e.g., papers potentially requiring the invitation of emergency reviewers, or papers with suspiciously short reviews. During the decision-making process, the dashboard was also used by the Program Chairs to identify papers with incomplete review information, e.g., for which reviewers did not enter their final recommendation, the AC did not enter their final decision or meta-review, the secondary AC did not confirm the recommendation, etc.

At the end of the decision-making progress, the Program Chairs carefully examined the very small number of cases where ACs overruled a unanimous reviewer recommendation for acceptance or rejection. In such instances, it was crucial for the primary and secondary ACs to explain in detail the motivation behind the decision. To ensure a fair evaluation, an additional Area Chair, not originally part of the decision-making triplet, was appointed to provide an independent opinion and advise the Program Chairs in determining whether to uphold or reverse the overturn.

Inevitably, after decisions were released, some authors were dissatisfied. The authors were given the possibility to appeal the final decision in case of procedural issues encountered during the review process by filling out a form to directly contact the Program Chairs. Valid reasons for appealing a decision included policy errors (e.g., reviewers or ACs enforcing a non-existent policy), clerical errors (e.g., the meta-review clearly indicated an intention to accept a paper, but it was mistakenly rejected), and significant misunderstandings by the reviewers or ACs. In total, we received 59 appeals and upheld 6 of them. One was due to a clear policy error by an AC, and the paper decision changed from reject to accept following a successful appeal. The other 5 papers were already conditionally accepted – but the AC/reviewers wrongly flagged them as papers for which the associated dataset was required to be released prior to final acceptance; this condition was removed following the appeal.

After the camera-ready deadline, additional checks were conducted, particularly for papers contributing a new dataset and for those flagged for potential ethical concerns. Specifically, during the submission phase, authors were required to specify if the primary contribution of their papers was the release of a new dataset, and for these papers, a valid dataset link had to be provided by the camera-ready deadline. A total of 435 submissions included a dataset as part of their contribution, and all complied with the requirement to release the dataset by the camera-ready deadline.

We sincerely thank all the ACs and reviewers for their invaluable contributions to the review process. Their dedication and expertise were crucial in maintaining the high standards of the conference. We would also like to thank our Technical Program Chair, Sascha Hornauer, who did tremendous work behind the scenes, and to the entire CMT team for their prompt support with any technical issues.

October 2024

Aleš Leonardis
Elisa Ricci
Stefan Roth
Olga Russakovsky
Torsten Sattler
Gül Varol

Organization

General Chairs

Andrew Fitzgibbon Graphcore, UK
Laura Leal-Taixé NVIDIA, Italy
Vittorio Murino University of Verona & University of Genoa, Italy

Program Chairs

Aleš Leonardis University of Birmingham, UK
Elisa Ricci University of Trento, Italy
Stefan Roth Technical University of Darmstadt, Germany
Olga Russakovsky Princeton University, USA
Torsten Sattler Czech Technical University in Prague, Czech Republic
Gül Varol Ecole des Ponts Paris Tech, France

Technical Program Chair

Sascha Hornauer Mines Paris – PSL, France

Industrial Liaison Chairs

Cees Snoek University of Amsterdam, Netherlands
Shaogang Gong Queen Mary University of London, UK

Publication Chairs

Francois Bremond Inria, France
Mahmoud Ali Inria, France
Jovita Lukasik University of Siegen, Germany
Michael Moeller University of Siegen, Germany

Poster Chairs

Aljoša Ošep — Carnegie Mellon University, USA
Zuzana Kukelova — Czech Technical University in Prague, Czech Republic

Diversity Chairs

David Fouhey — New York University, USA
Rita Cucchiara — Università di Modena e Reggio Emilia, Italy

Local Chairs

Raffaella Lanzarotti — Università degli Studi di Milano, Italy
Simone Bianco — University of Milano-Bicocca, Italy

Conference Ombuds

Georgia Gkioxari — California Institute of Technology, USA
Greg Mori — Borealis AI & Simon Fraser University, Canada

Ethics Review Committee

Chloé Bakalar — Meta, USA
Kate Saenko — Boston University, USA
Remi Denton — Google Research, USA
Yisong Yue — Caltech, Asari AI & Latitude AI, USA

Workshop and Tutorial Chairs

Alessio Del Bue — Istituto Italiano di Tecnologia, Italy
Cristian Canton — Meta AI, USA
Jordi Pont-Tuset — Google DeepMind, Switzerland
Tatiana Tommasi — Politecnico di Torino, Italy

Finance Chairs

Gerard Medioni	Amazon, USA
Nicole Finn	c to c events, USA

Demo Chairs

Hyung Chang	University of Birmingham, UK
Marco Cristani	University of Verona, Italy

Publicity and Social Media Chairs

Konstantinos Derpanis	York University & Samsung AI Centre Toronto, Canada
Jia-Bin Huang	University of Maryland College Park, USA
Abby Stylianou	Saint Louis University, USA

Social Activities Chairs

Giovanni Maria Farinella	University of Catania, Italy
Raffaella Lanzarotti	Università degli Studi di Milano, Italy
Simone Bianco	University of Milano-Bicocca, Italy

Doctoral Consortium Chairs

Cigdem Beyan	University of Verona, Italy
Or Litany	NVIDIA & Technion, USA

Web Developer

Lee Campbell	Eventhosts, USA

Area Chairs

Ehsan Adeli	Stanford University, USA
Aishwarya Agrawal	University of Montreal & Mila & DeepMind, Canada
Naveed Akhtar	University of Western Australia, Australia
Yagiz Aksoy	Simon Fraser University, Canada
Karteek Alahari	Inria, France
Jose Alvarez	NVIDIA, USA
Djamila Aouada	Interdisciplinary Centre for Security, Reliability and Trust, University of Luxembourg, Luxembourg
Andre Araujo	Google, Brazil
Pablo Arbelaez	Universidad de los Andes, Colombia
Iro Armeni	Stanford University, USA
Yuki Asano	University of Amsterdam, Netherlands
Karl Åström	Lund University, Sweden
Mathieu Aubry	Ecole des Ponts ParisTech, France
Shai Bagon	Weizmann Institute of Science, Israel
Song Bai	ByteDance, Singapore
Xiang Bai	Huazhong University of Science and Technology, China
Yang Bai	Tencent, China
Lamberto Ballan	University of Padova, Italy
Linchao Bao	University of Birmingham, UK
Ronen Basri	Weizmann Institute of Science, Israel
Sara Beery	Massachusetts Institute of Technology, USA
Vasileios Belagiannis	University of Erlangen–Nuremberg, Germany
Serge Belongie	University of Copenhagen, Denmark
Sagie Benaim	Hebrew University of Jerusalem, Israel
Rodrigo Benenson	Google, Switzerland
Cigdem Beyan	University of Bergamo, Italy
Bharat Bhatnagar	Meta Reality Labs Research, Switzerland
Tolga Birdal	Imperial College London, UK
Matthew Blaschko	KU Leuven, Belgium
Federica Bogo	Meta Reality Labs Research, Switzerland
Timo Bolkart	Google, Switzerland
Katherine Bouman	California Institute of Technology, USA
Lubomir Bourdev	Primepoint, USA
Edmond Boyer	Inria, France
Yuri Boykov	University of Waterloo, Canada
Eric Brachmann	Niantic, Germany

Wieland Brendel	University of Tübingen, Germany
Gabriel Brostow	University College London, UK
Michael Brown	York University, Canada
Andrés Bruhn	University of Stuttgart, Germany
Andrei Bursuc	valeo.ai, France
Benjamin Busam	Technical University of Munich, Germany
Holger Caesar	TU Delft, Netherlands
Jianfei Cai	Monash University, Australia
Simone Calderara	University of Modena and Reggio Emilia, Italy
Octavia Camps	Northeastern University, Boston, USA
Joao Carreira	DeepMind, UK
Silvia Cascianelli	University of Modena and Reggio Emilia, Italy
Ayan Chakrabarti	Google Research, USA
Tat-Jen Cham	Nanyang Technological University, Singapore
Antoni Chan	City University of Hong Kong, China
Manmohan Chandraker	University of California, San Diego, USA
Xiaojun Chang	University of Technology Sydney, Australia
Devendra Singh Chaplot	Mistral AI, USA
Liang-Chieh Chen	TikTok, USA
Long Chen	Hong Kong University of Science and Technology, China
Mei Chen	Microsoft, USA
Shizhe Chen	Inria, France
Shuo Chen	RIKEN, Japan
Xilin Chen	Institute of Computing Technology, Chinese Academy of Sciences, China
Anoop Cherian	Mitsubishi Electric Research Labs, USA
Tat-Jun Chin	University of Adelaide, Australia
Minsu Cho	Pohang University of Science and Technology, Korea
Hisham Cholakkal	Mohamed bin Zayed University of Artificial Intelligence, United Arab Emirates
Yung-Yu Chuang	National Taiwan University, Taiwan
Ondrej Chum	Czech Technical University in Prague, Czech Republic
Marcella Cornia	University of Modena and Reggio Emilia, Italy
Marco Cristani	University of Verona, Italy
Cristian Canton	Meta AI, USA
Zhaopeng Cui	Zhejiang University, China
Angela Dai	Technical University of Munich, Germany
Jifeng Dai	Tsinghua University, China
Yuchao Dai	Northwestern Polytechnical University, China

Dima Damen	University of Bristol, UK
Kostas Daniilidis	University of Pennsylvania, USA
Antitza Dantcheva	Inria, France
Raoul de Charette	Inria, France
Shalini De Mello	NVIDIA Research, USA
Tali Dekel	Weizmann Institute of Science, Israel
Ilke Demir	Intel Corporation, USA
Joachim Denzler	Friedrich Schiller University Jena, Germany
Konstantinos Derpanis	York University, Canada
Jose Dolz	École de technologie supérieure Montreal, Canada
Jiangxin Dong	Nanjing University of Science and Technology, China
Hazel Doughty	Leiden University, Netherlands
Iddo Drori	Boston University and Columbia University, USA
Enrique Dunn	Stevens Institute of Technology, USA
Thibaut Durand	Borealis AI, Canada
Sayna Ebrahimi	Google, USA
Mohamed Elhoseiny	King Abdullah University of Science and Technology, Saudi Arabia
Bin Fan	University of Science and Technology Beijing, China
Yi Fang	New York University, USA
Giovanni Maria Farinella	University of Catania, Italy
Ryan Farrell	Brigham Young University, USA
Alireza Fathi	Google, USA
Christoph Feichtenhofer	Meta & FAIR, USA
Rogerio Feris	MIT-IBM Watson AI Lab & IBM Research, USA
Basura Fernando	Agency for Science, Technology and Research (A*STAR), Singapore
David Fouhey	New York University, USA
Katerina Fragkiadaki	Carnegie Mellon University, USA
Friedrich Fraundorfer	Graz University of Technology, Austria
Oren Freifeld	Ben-Gurion University, Israel
Huan Fu	University of Sydney, China
Yasutaka Furukawa	Simon Fraser University, Canada
Andrea Fusiello	University of Udine, Italy
Fabio Galasso	Sapienza University, Italy
Jürgen Gall	University of Bonn, Germany
Orazio Gallo	NVIDIA Research, USA
Chuang Gan	MIT-IBM Watson AI Lab, USA
Lianli Gao	University of Electronic Science and Technology of China, China

Shenghua Gao	University of Hong Kong, China
Sourav Garg	University of Adelaide, Australia
Efstratios Gavves	University of Amsterdam, Netherlands
Peter Gehler	Zalando, Germany
Theo Gevers	University of Amsterdam, Netherlands
Golnaz Ghiasi	Google DeepMind, USA
Rohit Girdhar	Carnegie Mellon University, USA
Xavier Giro-i-Nieto	Universitat Politecnica de Catalunya, Spain
Ross Girshick	Allen Institute for Artificial Intelligence, USA
Zan Gojcic	NVIDIA, Switzerland
Vladislav Golyanik	Max Planck Institute for Informatics, Germany
Stephen Gould	Australian National University, Australia
Liangyan Gui	University of Illinois Urbana-Champaign, USA
Fatma Guney	Koc University, Turkey
Yulan Guo	Sun Yat-sen University, China
Yunhui Guo	University of Texas at Dallas, USA
Mohit Gupta	University of Wisconsin-Madison, USA
Minh Ha Quang	RIKEN Center for Advanced Intelligence Project, Japan
Bumsub Ham	Yonsei University, Korea
Bohyung Han	Seoul National University, Korea
Kai Han	University of Hong Kong, China
Xiaoguang Han	Shenzhen Research Institute of Big Data & Chinese University of Hong Kong (Shenzhen), China
Tatsuya Harada	University of Tokyo & RIKEN, Japan
Tal Hassner	Weir AI, USA
Xuming He	ShanghaiTech University, China
Felix Heide	Princeton & Algolux, USA
Anthony Hoogs	Kitware, USA
Timothy Hospedales	Edinburgh University, UK
Mahdi Hosseini	Concordia University, Canada
Peng Hu	College of Computer Science, Sichuan University, China
Gang Hua	Wormpex AI Research, USA
Jia-Bin Huang	University of Maryland College Park, USA
Qixing Huang	University of Texas at Austin, USA
Sharon Xiaolei Huang	Pennsylvania State University, USA
Junhwa Hur	Google, USA
Nazli Ikizler-Cinbis	Hacettepe University, Turkey
Eddy Ilg	Saarland University, Germany
Ahmet Iscen	Google, France

Nathan Jacobs	Washington University in St. Louis, USA
Varun Jampani	Stability AI, USA
C. V. Jawahar	International Institute of Information Technology - Hyderabad, India
Laszlo Jeni	Carnegie Mellon University, USA
Jiaya Jia	Chinese University of Hong Kong, China
Qin Jin	Renmin University of China, China
Jungseock Joo	NVIDIA & University of California, Los Angeles, USA
Fredrik Kahl	Chalmers, Sweden
Yannis Kalantidis	NAVER LABS Europe, France
Vicky Kalogeiton	Ecole Polytechnique, IP Paris, France
Evangelos Kalogerakis	University of Massachusetts Amherst, USA
Hiroshi Kawasaki	Kyushu University, Japan
Margret Keuper	University of Mannheim & Max Planck Institute for Informatics, Germany
Salman Khan	Mohamed bin Zayed University of Artificial Intelligence, United Arab Emirates
Anna Khoreva	Bosch Center for Artificial Intelligence, Germany
Gunhee Kim	Seoul National University, Korea
Junmo Kim	Korea Advanced Institute of Science and Technology, Korea
Min H. Kim	Korea Advanced Institute of Science and Technology, Korea
Seon Joo Kim	Yonsei University, Korea
Tae-Kyun Kim	Korea Advanced Institute of Science and Technology & Imperial College London, Korea
Benjamin Kimia	Brown University, USA
Hedvig Kjellström	KTH Royal Institute of Technology, Sweden
Laurent Kneip	ShanghaiTech University, China
A. Sophia Koepke	University of Tübingen, Germany
Iasonas Kokkinos	University College London, UK
Wai-Kin Adams Kong	Nanyang Technological University, Singapore
Piotr Koniusz	Data61/CSIRO & Australian National University, Australia
Jana Kosecka	George Mason University, USA
Adriana Kovashka	University of Pittsburgh, USA
Hilde Kuehne	University of Bonn, Germany
Zuzana Kukelova	Czech Technical University in Prague, Czech Republic
Ajay Kumar	Hong Kong Polytechnic University, China
Kiriakos Kutulakos	University of Toronto, Canada

Suha Kwak	Pohang University of Science and Technology, Korea
Zorah Laehner	University of Bonn, Germany
Shang-Hong Lai	National Tsing Hua University, Taiwan
Iro Laina	University of Oxford, UK
Jean-Francois Lalonde	Université Laval, Canada
Loic Landrieu	École des Ponts ParisTech, France
Diane Larlus	Naver Labs Europe, France
Viktor Larsson	Lund University, Sweden
Stéphane Lathuilière	Telecom-Paris, France
Gim Hee Lee	National University of Singapore, Singapore
Yong Jae Lee	University of Wisconsin-Madison, USA
Bastian Leibe	RWTH Aachen University, Germany
Ales Leonardis	University of Birmingham, UK
Vincent Lepetit	Ecole des Ponts ParisTech, France
Fuxin Li	Oregon State University, USA
Hongdong Li	Australian National University, Australia
Xi Li	Zhejiang University, China
Yin Li	University of Wisconsin-Madison, USA
Yu Li	International Digital Economy Academy, Singapore
Xiaodan Liang	Sun Yat-sen University, China
Shengcai Liao	Inception Institute of Artificial Intelligence, United Arab Emirates
Jongwoo Lim	Seoul National University, Korea
Ser-Nam Lim	University of Central Florida, USA
Stephen Lin	Microsoft Research, China
Tsung-Yi Lin	NVIDIA Research, USA
Yen-Yu Lin	National Yang Ming Chiao Tung University, Taiwan
Yutian Lin	Wuhan University, China
Zhe Lin	Adobe Research, USA
Haibin Ling	Stony Brook University, USA
Or Litany	NVIDIA, USA
Jiaying Liu	Peking University, China
Jun Liu	Lancaster University, UK
Miaomiao Liu	Australian National University, Australia
Si Liu	Beihang University, China
Sifei Liu	NVIDIA, USA
Wei Liu	Tencent, China
Yanxi Liu	Pennsylvania State University, USA
Yaoyao Liu	Johns Hopkins University, USA

Yebin Liu	Tsinghua University, China
Zicheng Liu	Microsoft, USA
Ziwei Liu	Nanyang Technological University, Singapore
Ismini Lourentzou	University of Illinois, Urbana-Champaign, USA
Chen Change Loy	Nanyang Technological University, Singapore
Feng Lu	Beihang University, China
Le Lu	Alibaba Group, USA
Oisin Mac Aodha	University of Edinburgh, UK
Luca Magri	Politecnico di Milano, Italy
Michael Maire	University of Chicago, USA
Subhransu Maji	University of Massachusetts, Amherst, USA
Atsuto Maki	KTH Royal Institute of Technology, Sweden
Yasushi Makihara	Osaka University, Japan
Massimiliano Mancini	University of Trento, Italy
Kevis-Kokitsi Maninis	Google Research, Switzerland
Kenneth Marino	Google DeepMind, USA
Renaud Marlet	Ecole des Ponts ParisTech & Valeo, France
Niki Martinel	University of Udine, Italy
Jiri Matas	Czech Technical University in Pragu, Czech Republic
Yasuyuki Matsushita	Osaka University, Japan
Simone Melzi	University of Milano-Bicocca, Italy
Thomas Mensink	Google Research, Netherlands
Michele Merler	IBM Research, USA
Pascal Mettes	University of Amsterdam, Netherlands
Ajmal Mian	University of Western Australia, Australia
Krystian Mikolajczyk	Imperial College London, UK
Ishan Misra	Facebook AI Research, USA
Niloy Mitra	University College London, UK
Anurag Mittal	Indian Institute of Technology Madras, India
Davide Modolo	Amazon, USA
Davide Moltisanti	University of Bath, UK
Philippos Mordohai	Stevens Institute of Technology, USA
Francesc Moreno	Institut de Robotica i Informatica Industrial, Spain
Pietro Morerio	Istituto Italiano di Tecnologia, Italy
Greg Mori	Simon Fraser University & Borealis AI, Canada
Roozbeh Mottaghi	FAIR @ Meta, USA
Yadong Mu	Peking University, China
Vineeth N. Balasubramanian	Indian Institute of Technology, India
Hajime Nagahara	Osaka University, Japan
Vinay Namboodiri	University of Bath, UK
Liangliang Nan	Delft University of Technology, Netherlands

Michele Nappi	University of Salerno, Italy
Srinivasa Narasimhan	Carnegie Mellon University, USA
P. J. Narayanan	International Institute of Information Technology, Hyderabad, India
Nassir Navab	Technical University of Munich, Germany
Ram Nevatia	University of Southern California, USA
Natalia Neverova	Facebook AI Research, France
Juan Carlos Niebles	Salesforce & Stanford University, USA
Michael Niemeyer	Google, Germany
Simon Niklaus	Adobe Research, USA
Ko Nishino	Kyoto University, Japan
Nicoletta Noceti	MaLGa-DIBRIS & Università degli Studi di Genova, Italy
Matthew O'Toole	Carnegie Mellon University, USA
Jean-Marc Odobez	Idiap Research Institute, École polytechnique fédérale de Lausanne, Switzerland
Francesca Odone	Università di Genova, Italy
Takayuki Okatani	Tohoku University & RIKEN Center for Advanced Intelligence Project, Japan
Carl Olsson	Lund University, Sweden
Vicente Ordonez	Rice University, USA
Aljosa Osep	Technical University of Munich, Germany
Martin R. Oswald	University of Amsterdam, Netherlands
Andrew Owens	University of Michigan, USA
Tomas Pajdla	Czech Technical University in Prague, Czechia
Manohar Paluri	Meta, USA
Jinshan Pan	Nanjing University of Science and Technology, China
In Kyu Park	Inha University, Korea
Jaesik Park	Seoul National University, Korea
Despoina Paschalidou	Stanford, USA
Georgios Pavlakos	University of Texas at Austin, USA
Vladimir Pavlovic	Rutgers University, USA
Marcello Pelillo	University of Venice, Italy
David Picard	École des Ponts ParisTech, France
Hamed Pirsiavash	University of California, Davis, USA
Bryan Plummer	Boston University, USA
Thomas Pock	Graz University of Technology, Austria
Fabio Poiesi	Fondazione Bruno Kessler, Italy
Marc Pollefeys	ETH Zurich & Microsoft, Switzerland
Jean Ponce	Ecole normale supérieure - PSL, France
Gerard Pons-Moll	University of Tübingen, Germany

Jordi Pont-Tuset	Google, Switzerland
Fatih Porikli	Qualcomm R&D, USA
Brian Price	Adobe, USA
Guo-Jun Qi	Westlake University, USA
Long Quan	Hong Kong University of Science and Technology, China
Venkatesh Babu Radhakrishnan	Indian Institute of Science, India
Hossein Rahmani	Lancaster University, UK
Deva Ramanan	Carnegie Mellon University, USA
Vignesh Ramanathan	Facebook, USA
Vikram V. Ramaswamy	Princeton University, USA
Nalini Ratha	University at Buffalo, State University of New York, USA
Yogesh Rawat	University of Central Florida, USA
Hamid Rezatofighi	Monash University, Australia
Helge Rhodin	University of British Columbia, Canada
Stephan Richter	Apple, Germany
Anna Rohrbach	Technical University of Darmstadt, Germany
Marcus Rohrbach	Technical University of Darmstadt, Germany
Gemma Roig	Goethe University Frankfurt, Germany
Negar Rostamzadeh	Google, Canada
Paolo Rota	University of Trento, Italy
Stefan Roth	Technical University of Darmstadt, Germany
Carsten Rother	University of Heidelberg, Germany
Subhankar Roy	University of Trento, Italy
Michael Rubinstein	Google, USA
Christian Rupprecht	University of Oxford, UK
Olga Russakovsky	Princeton University, USA
Bryan Russell	Adobe Research, USA
Michael Ryoo	Stony Brook University & Google, USA
Reza Sabzevari	Delft University of Technology, Netherlands
Mathieu Salzmann	École polytechnique fédérale de Lausanne, Switzerland
Dimitris Samaras	Stony Brook University, USA
Aswin Sankaranarayanan	Carnegie Mellon University, USA
Sudeep Sarkar	University of South Florida, USA
Yoichi Sato	University of Tokyo, Japan
Manolis Savva	Simon Fraser University, Canada
Simone Schaub-Meyer	Technical University of Darmstadt, Germany
Bernt Schiele	Max Planck Institute for Informatics, Germany
Cordelia Schmid	Inria & Google, France
Nicu Sebe	University of Trento, Italy

Laura Sevilla-Lara	University of Edinburgh, UK
Fahad Shahbaz Khan	Mohamed bin Zayed University of Artificial Intelligence, United Arab Emirates
Qi Shan	Apple, USA
Shiguang Shan	Institute of Computing Technology, Chinese Academy of Sciences, China
Viktoriia Sharmanska	University of Sussex & Imperial College London, UK
Eli Shechtman	Adobe Research, USA
Evan Shelhamer	DeepMind, UK
Lu Sheng	Beihang University, China
Boxin Shi	Peking University, China
Jianbo Shi	University of Pennsylvania, USA
Mike Zheng Shou	National University of Singapore, Singapore
Abhinav Shrivastava	University of Maryland, USA
Aliaksandr Siarohin	Snap, USA
Kaleem Siddiqi	McGill University, Canada
Leonid Sigal	University of British Columbia, Canada
Oriane Siméoni	valeo.ai, France
Krishna Kumar Singh	Adobe Research, USA
Richa Singh	Indian Institute of Technology Jodhpur, India
Yale Song	Facebook AI Research, USA
Concetto Spampinato	University of Catania, Italy
Srinath Sridhar	Brown University, USA
Bjorn Stenger	Rakuten, Japan
Abby Stylianou	Saint Louis University, USA
Akihiro Sugimoto	National Institute of Informatics, Japan
Chen Sun	Brown University, USA
Deqing Sun	Google, USA
Jian Sun	Xi'an Jiaotong University, China
Min Sun	National Tsing Hua University, Taiwan
Qianru Sun	Singapore Management University, Singapore
Yu-Wing Tai	Dartmouth College, USA
Ayellet Tal	Technion, Israel
Ping Tan	Hong Kong University of Science and Technology, China
Robby Tan	National University of Singapore, Singapore
Chi-Keung Tang	Hong Kong University of Science and Technology, China
Makarand Tapaswi	International Institute of Information Technology, Hyderabad & Wadhwani AI, India
Justus Thies	Technical University of Darmstadt, Germany

Radu Timofte	University of Wurzburg & ETH Zurich, Germany
Giorgos Tolias	Czech Technical University in Prague, Czechia
Federico Tombari	Google & Technical University of Munich, Switzerland
Tatiana Tommasi	Politecnico di Torino, Italy
James Tompkin	Brown University, USA
Matthew Trager	AWS AI Labs, USA
Anh Tran	VinAI, Vietnam
Du Tran	Google, USA
Shubham Tulsiani	Carnegie Mellon University, USA
Sergey Tulyakov	Snap, USA
Dimitrios Tzionas	University of Amsterdam, Netherlands
Maria Vakalopoulou	CentraleSupelec, France
Joost van de Weijer	Computer Vision Center, Spain
Laurens van der Maaten	Meta, USA
Jan van Gemert	Delft University of Technology, Netherlands
Sebastiano Vascon	Ca' Foscari University of Venice & European Centre for Living Technology, Italy
Nuno Vasconcelos	University of California, San Diego, USA
Mayank Vatsa	Indian Institute of Technology Jodhpur, India
Javier Vazquez-Corral	Autonomous University of Barcelona, Spain
Andrea Vedaldi	Oxford University, UK
Olga Veksler	University of Waterloo, Canada
Luisa Verdoliva	University Federico II of Naples, Italy
Rene Vidal	University of Pennsylvania, USA
Bo Wang	CtrsVision, USA
Heng Wang	TikTok, USA
Jingdong Wang	Baidu, China
Jingya Wang	ShanghaiTech University, China
Ruiping Wang	Institute of Computing Technology, Chinese Academy of Sciences, China
Shenlong Wang	University of Illinois, Urbana-Champaign, USA
Ting-Chun Wang	NVIDIA, USA
Wenguan Wang	Zhejiang University, China
Xiaolong Wang	University of California, San Diego, USA
Xiaoqian Wang	Purdue University, USA
Xiaoyu Wang	Hong Kong University of Science and Technology, USA
Xin Wang	Microsoft Research, USA
Xinchao Wang	National University of Singapore, Singapore
Yang Wang	Concordia University, Canada
Yiming Wang	Fondazione Bruno Kessler, Italy

Yu-Chiang Frank Wang	National Taiwan University, Taiwan
Yu-Xiong Wang	University of Illinois, Urbana-Champaign, USA
Zhangyang Wang	University of Texas at Austin, USA
Olivia Wiles	DeepMind, UK
Christian Wolf	Naver Labs Europe, France
Michael Wray	University of Bristol, UK
Jiajun Wu	Stanford University, USA
Jianxin Wu	Nanjing University, China
Shangzhe Wu	Stanford University, USA
Ying Wu	Northwestern University, USA
Yu Wu	Wuhan University, China
Jianwen Xie	Baidu Research, USA
Saining Xie	New York University, USA
Weidi Xie	Shanghai Jiao Tong University, China
Dan Xu	Hong Kong University of Science and Technology, China
Huijuan Xu	Pennsylvania State University, USA
Xiangyu Xu	Xi'an Jiaotong University, China
Yan Yan	Illinois Institute of Technology, USA
Jiaolong Yang	Microsoft Research, China
Kaiyu Yang	California Institute of Technology, USA
Linjie Yang	ByteDance AI Lab, USA
Ming-Hsuan Yang	University of California, Merced, USA
Yanchao Yang	University of Hong Kong, China
Yi Yang	Zhejiang University, China
Angela Yao	National University of Singapore, Singapore
Mang Ye	Wuhan University, China
Kwang Moo Yi	University of British Columbia, Canada
Xi Yin	Facebook, USA
Chang D. Yoo	Korea Advanced Institute of Science and Technology, Korea
Shaodi You	University of Amsterdam, Netherlands
Jingyi Yu	Shanghai Tech University, China
Ning Yu	Netflix Eyeline Studios, USA
Qi Yu	Rochester Institute of Technology, USA
Stella Yu	University of Michigan, USA
Yizhou Yu	University of Hong Kong, China
Junsong Yuan	University at Buffalo, State University of New York, USA
Lu Yuan	Microsoft, USA
Sangdoo Yun	NAVER AI Lab, Korea
Hongbin Zha	Peking University, China

Jing Zhang	Australian National University, Australia
Lei Zhang	Hong Kong Polytechnic University, China
Ning Zhang	Facebook AI, USA
Richard Zhang	Adobe, USA
Tianzhu Zhang	University of Science and Technology of China, China
Hengshuang Zhao	University of Hong Kong, China
Qi Zhao	University of Minnesota, USA
Sicheng Zhao	Tsinghua University, China
Liang Zheng	Australian National University, Australia
Wei-Shi Zheng	Sun Yat-sen University, China
Zhun Zhong	University of Nottingham, UK
Bolei Zhou	University of California, Los Angeles, USA
Kaiyang Zhou	Hong Kong Baptist University, China
Luping Zhou	University of Sydney, Australia
S. Kevin Zhou	University of Science and Technology of China, China
Lei Zhu	Hong Kong University of Science and Technology (Guangzhou), China
Linchao Zhu	Zhejiang University, China
Andrew Zisserman	University of Oxford, UK
Daniel Zoran	DeepMind, UK
Wangmeng Zuo	Harbin Institute of Technology, China

Technical Program Committee

Sathyanarayanan N. Aakur
Andrea Abate
Wim Abbeloos
Amos L. Abbott
Rameen Abdal
Salma Abdel Magid
Rabab Abdelfattah
Ahmed Abdelkader
Eslam Mohamed Abdelrahman
Ahmed Abdelreheem
Akash Abdu Jyothi
Kfir Aberman
Shahira Abousamra
Yazan Abu Farha
Shady Abu-Hussein

Abulikemu Abuduweili
Hanno Ackermann
Chandranath Adak
Aikaterini Adam
Lukáš Adam
Kamil Adamczewski
Donald Adjeroh
Mohammed Adnan
Mahmoud Afifi
Triantafyllos Afouras
Akshay Agarwal
Madhav Agarwal
Nakul Agarwal
Sharat Agarwal
Shivang Agarwal
Vatsal Agarwal

Abhinav Agarwalla
Shivam Aggarwal
Sara Aghajanzadeh
Shashank Agnihotri
Harsh Agrawal
Antonio Agudo
Shai Aharon
Saba Ahmadi
Waqar Ahmed
Byeongjoo Ahn
Chanho Ahn
Jaewoo Ahn
Namhyuk Ahn
Seoyoung Ahn
Nilesh A. Ahuja
Yihao Ai

Abhishek Aich
Emanuele Aiello
Noam Aigerman
Kiyoharu Aizawa
Kenan Emir Ak
Reza Akbarian Bafghi
Sai Aparna Aketi
Derya Akkaynak
Ibrahim Alabdulmohsin
Yuval Alaluf
Md Ferdous Alam
Stephan Alaniz
Badour A. Sh AlBahar
Paul Albert
Isabela Albuquerque
Emanuel Aldea
Luca Alessandrini
Emma Alexander
Konstantinos P. Alexandridis
Stamatis Alexandropoulos
Saghir Alfasly
Mohsen Ali
Waqar Ali
Daniel Aliaga
Sadegh Aliakbarian
Garvita Allabadi
Thiemo Alldieck
Antonio Alliegro
Jurandy Almeida
Gal Alon
Hadi Alzayer
Ibtihel Amara
Giuseppe Amato
Sameer Ambekar
Niki Amini-Naieni
Shir Amir
Roberto Amoroso
Dongsheng An
Jie An
Junyi An
Liang An
Xiang An
Yuexuan An
Zhaochong An
Zhulin An
Saket Anand
Pavan Kumar Anasosalu Vasu
Codruta O. Ancuti
Cosmin Ancuti
Fernanda Andaló
Connor Anderson
Juan Andrade-Cetto
Bruno Andreis
Alexander Andreopoulos
Bjoern Andres
Shivangi Aneja
Dragomir Anguelov
Damith Kawshan Anhettigama
Aditya Annavajjala
Michel Antunes
Tejas Anvekar
Saeed Anwar
Evlampios Apostolidis
Srikar Appalaraju
Relja Arandjelović
Nikita Araslanov
Alexandre Araujo
Helder Araujo
Eric Arazo
Dawit Mureja Argaw
Anurag Arnab
Federica Arrigoni
Alexey Artemov
Aditya Arun
Nader Asadi
Kumar Ashutosh
Vishal Asnani
Mahmoud Assran
Amir Atapour-Abarghouei
Nikos Athanasiou
Ali Athar
ShahRukh Athar
Rowel O. Atienza
Benjamin Attal
Matan Atzmon
Nicolas Audebert
Romaric Audigier
Tristan Aumentado-Armstrong
Yannis Avrithis
Muhammad Awais
Md Rabiul Awal
Görkay Aydemir
Mehmet Aygün
Melika Ayoughi
Ali Ayub
Kumar Ayush
Mehdi Azabou
Basim Azam
Mohammad Farid Azampour
Hossein Azizpour
Vivek B. S.
Yunhao Ba
Mohammadreza Babaee
Deepika Bablani
Sudarshan Babu
Roman Bachmann
Inhwan Bae
Jeonghun Baek
Kyungjune Baek
Seungryul Baek
Piyush Nitin Bagad
Mohammadhossein Bahari
Gaetan Bahl
Sherwin Bahmani
Bing Bai
Haoyue Bai
Jiawang Bai
Jinbin Bai
Lei Bai
Qingyan Bai
Shutao Bai
Xuyang Bai
Yalong Bai
Yuanchao Bai
Yue Bai
Yunpeng Bai
Ziyi Bai
Daniele Baieri
Sungyong Baik
Ramon Baldrich

Coloma Ballester
Irene Ballester
Sonat Baltaci
Biplab Banerjee
Monami Banerjee
Sandipan Banerjee
Snehasis Banerjee
Soumya Banerjee
Sreya Banerjee
Jihwan Bang
Ankan Bansal
Nitin Bansal
Siddhant Bansal
Chong Bao
Hangbo Bao
Runxue Bao
Wentao Bao
Yiwei Bao
Zhipeng Bao
Zhongyun Bao
Amir Bar
Omer Bar Tal
Manel Baradad Jurjo
Lorenzo Baraldi
Lorenzo Baraldi
Tal Barami
Danny Barash
Daniel Barath
Adrian Barbu
Nick Barnes
Alina J. Barnett
German Barquero
Joao P. Barreto
Jonathan T. Barron
Rodrigo C. Barros
Luca Barsellotti
Myles Bartlett
Kristijan Bartol
Hritam Basak
Dina Bashkirova
Chaim Baskin
Lennart C. Bastian
Abhipsa Basu
Soumen Basu
Peyman Bateni

Anil Batra
David Bau
Zuria Bauer
Eduard Gabriel Bazavan
Federico Becattini
Nathan A. Beck
Jan Bednarik
Peter A. Beerel
Ardhendu Behera
Harkirat Singh Behl
Jens Behley
William Beksi
Soufiane Belharbi
Giovanni Bellitto
Ismail Ben Ayed
Abdessamad Ben Hamza
Yizhak Ben-Shabat
Guy Ben-Yosef
Robert Benavente
Assia Benbihi
Yasser Benigmim
Urs Bergmann
Amit H Bermano
Jesus Bermudez-Cameo
Florian Bernard
Stefano Berretti
Gabriele Berton
Petra Bevandić
Matthew Beveridge
Lalit Bhagat
Yash Bhalgat
Homanga Bharadhwaj
Skanda Bharadwaj
Ambika Saklani Bhardwaj
Goutam Bhat
Gaurav Bhatt
Neel P. Bhatt
Deblina Bhattacharjee
Anish Bhattacharya
Rajarshi Bhattacharya
Uttaran Bhattacharya
Apratim Bhattacharyya
Anand Bhattad
Binod Bhattarai
Snehal Bhayani

Brojeshwar Bhowmick
Aritra Bhowmik
Neelanjan Bhowmik
Amran Bhuiyan
Ankan Kumar Bhunia
Ayan Kumar Bhunia
Jing Bi
Qi Bi
Xiuli Bi
Michael Bi Mi
Hao Bian
Jia-Wang Bian
Shaojun Bian
Simone Bianco
Pia Bideau
Xiaoyu Bie
Roberto Bigazzi
Mario Bijelic
Bahri Batuhan Bilecen
Guillaume-Alexandre
  Bilodeau
Alexander Binder
Massimo Bini
Niccolò Bisagno
Carmen Bisogni
Anthony R. Bisulco
Sandika Biswas
Sanket Biswas
Ksenia Bittner
Kevin Blackburn-Matzen
Nathaniel Blanchard
Hunter Blanton
Emile Blettery
Hermann Blum
Deyu Bo
Janusz Bobulski
Marius Bock
Lea Bogensperger
Simion-Vlad Bogolin
Moritz Böhle
Piotr Bojanowski
Alexey Bokhovkin
Georg Bökman
Ladislau Boloni
Daniel Bolya

Lorenzo Bonicelli
Gianpaolo Bontempo
Vivek Boominathan
Adrian Bors
Damian Borth
Matteo Bortolon
Davide Boscaini
Laurie Nicholas Bose
Shirsha Bose
Mark Boss
Andrea Bottino
Larbi Boubchir
Alexandre Boulch
Terrance E. Boult
Amine Bourki
Walid Bousselham
Fadi Boutros
Nicolas C. Boutry
Thierry Bouwmans
Richard S. Bowen
Kevin W. Bowyer
Ivaylo Boyadzhiev
Aidan Boyd
Joseph Boyd
Aljaz Bozic
Behzad Bozorgtabar
Manuel Brack
Samarth Brahmbhatt
Nikolas Brasch
Cameron E. Braunstein
Toby P. Breckon
Mathieu Bredif
Romain Brégier
Carlo Bretti
Joel R. Brogan
Clifford Broni-Bediako
Andrew Brown
Ellis L. Brown
Thomas Brox
Qingwen Bu
Tong Bu
Shyamal Buch
Alvaro Budria
Ignas Budvytis
José M. Buenaposada

Kyle R. Buettner
Marcel C. Bühler
Nhat-Tan Bui
Adrian Bulat
Valay M. Bundele
Nathaniel J. Burgdorfer
Ryan D. Burgert
Evgeny Burnaev
Pietro Buzzega
Marco Buzzelli
Kwon Byung-Ki
Fabian Caba
Felipe Cadar
Martin Cadik
Bowen Cai
Changjiang Cai
Guanyu Cai
Haoming Cai
Hongrui Cai
Jiarui Cai
Jinyu Cai
Kaixin Cai
Lile Cai
Minjie Cai
Mu Cai
Pingping Cai
Rizhao Cai
Ruisi Cai
Ruojin Cai
Shen Cai
Shengqu Cai
Weidong Cai
Weiwei Cai
Wenjia Cai
Wenxiao Cai
Yancheng Cai
Yuanhao Cai
Yujun Cai
Yuxuan Cai
Zhaowei Cai
Zhipeng Cai
Zhixi Cai
Zhongang Cai
Zikui Cai
Salvatore Calcagno

Roberto Caldelli
Akin Caliskan
Lilian Calvet
Necati Cihan Camgoz
Tommaso Campari
Dylan Campbell
Guglielmo Camporese
Yigit Baran Can
Shaun Canavan
Gemma Canet Tarrés
Marco Cannici
Ang Cao
Anh-Quan Cao
Bing Cao
Bingyi Cao
Chenjie Cao
Dongliang Cao
Hu Cao
Jiahang Cao
Jiale Cao
Jie Cao
Jiezhang Cao
Jinkun Cao
Meng Cao
Mingdeng Cao
Peng Cao
Pu Cao
Qinglong Cao
Qingxing Cao
Shengcao Cao
Song Cao
Weiwei Cao
Xiangyong Cao
Xiao Cao
Xu Cao
Xu Cao
Yan-Pei Cao
Yang Cao
Yichao Cao
Ying Cao
Yuan Cao
Yuhang Cao
Yukang Cao
Yulong Cao
Yun-Hao Cao

Yunhao Cao
Yutong Cao
Zhangjie Cao
Zhiguo Cao
Zhiwen Cao
Ziang Cao
Giacomo Capitani
Luigi Capogrosso
Andrea Caraffa
Jaime S. Cardoso
Chris Careaga
Zachariah Carmichael
Giuseppe Cartella
Vincent Cartillier
Paola Cascante-Bonilla
Lucia Cascone
Pol Caselles Rico
Angela Castillo
Lluis Castrejon
Modesto
   Castrillón-Santana
Francisco M. Castro
Pedro Castro
Jacopo Cavazza
Oya Celiktutan
Luigi Celona
Jiazhong Cen
Fernando J. Cendra
Fabio Cermelli
Duygu Ceylan
Eunju Cha
Junbum Cha
Rohan Chabra
Rohan Chacko
Julia Chae
Nicolas Chahine
Junyi Chai
Lucy Chai
Tianrui Chai
Wenhao Chai
Zenghao Chai
Anish Chakrabarty
Goirik Chakrabarty
Anirban Chakraborty
Deep Chakraborty

Souradeep Chakraborty
Punarjay Chakravarty
Jacob Chalk
Loick Chambon
Chee Seng Chan
David Chan
Dorian Y. Chan
Kin Sun Chan
Matthew Chan
Adrien Chan-Hon-Tong
Sampath Chanda
Shivam Chandhok
Prashanth Chandran
Kenneth Chaney
Chirui Chang
Di Chang
Dongliang Chang
Hong Chang
Hyung Jin Chang
Ju Yong Chang
Matthew Chang
Ming-Ching Chang
MingFang Chang
Nadine Chang
Qi Chang
Seunggyu Chang
Wei-Yi Chang
Xiaojun Chang
Yakun Chang
Yi Chang
Zheng Chang
Sumohana S.
   Channappayya
Hanqing Chao
Pradyumna Chari
Korawat Charoenpitaks
Dibyadip Chatterjee
Moitreya Chatterjee
Pratik Chaudhari
Muawiz Chaudhary
Ritwick Chaudhry
Ruchika Chavhan
Sofian Chaybouti
Zhengping Che
Gullal Singh Cheema

Kunal Chelani
Rama Chellappa
Allison Y. Chen
Anpei Chen
Aozhu Chen
Bin Chen
Bin Chen
Bin Chen
Bohong Chen
Bor-Chun Chen
Bowei Chen
Brian Chen
Chang Wen Chen
Changan Chen
Changhao Chen
Changhuai Chen
Chaofan Chen
Chaofan Chen
Chaofeng Chen
Chen Chen
Cheng Chen
Chengkuan Chen
Chieh-Yun Chen
Chong Chen
Chun-Fu Richard Chen
Congliang Chen
Cuiqun Chen
Da Chen
Daoyuan Chen
Defang Chen
Dian Chen
Dian Chen
Ding-Jie Chen
Dingfan Chen
Dong Chen
Dongdong Chen
Fangyi Chen
Guang-Yong Chen
Guangyao Chen
Guangyi Chen
Guanying Chen
Guikun Chen
Guo Chen
Haibo Chen
Haiwei Chen

Hansheng Chen
Hanting Chen
Hanzhi Chen
Hao Chen
Haodong Chen
Haokun Chen
Haoran Chen
Haoxin Chen
Haoyu Chen
Haoyu Chen
Hong-You Chen
Honghua Chen
Honglin Chen
Hongming Chen
Huanran Chen
Hui Chen
Hwann-Tzong Chen
Jiacheng Chen
Jiajing Chen
Jianqiu Chen
Jiaqi Chen
Jiaxin Chen
Jiayi Chen
Jie Chen
Jie Chen
Jie Chen
Jieneng Chen
Jierun Chen
Jinghui Chen
Jingjing Chen
Jintai Chen
Jinyu Chen
Joya Chen
Jun Chen
Jun Chen
Jun Chen
Jun-Cheng Chen
Jun-Kun Chen
Junjie Chen
Kai Chen
Kai Chen
Kaifeng Chen
Kefan Chen
Kejiang Chen
Kuan-Wen Chen

Kunyao Chen
Li Chen
Liang Chen
Lianggangxu Chen
Liangyu Chen
Ling-Hao Chen
Linghao Chen
Liyan Chen
Min Chen
Min-Hung Chen
Minghao Chen
Minghui Chen
Nenglun Chen
Peihao Chen
Pengguang Chen
Ping Chen
Qi Chen
Qi Chen
Qi Chen
Qiang Chen
Qimin Chen
Qingcai Chen
Rongyu Chen
Rui Chen
Runjian Chen
Shang-Fu Chen
Shen Chen
Sherry X. Chen
Shi Chen
Shiming Chen
Shiyan Chen
Shizhe Chen
Shoufa Chen
Shu-Yu Chen
Shuai Chen
Si Chen
Siheng Chen
Simin Chen
Sixiang Chen
Sizhe Chen
Songcan Chen
Tao Chen
Tianshui Chen
Tianyi Chen
Tsai-Shien Chen

Wei Chen
Weidong Chen
Weihua Chen
Weijie Chen
Weikai Chen
Wuyang Chen
Xi Chen
Xi Chen
Xiang Chen
Xiangyu Chen
Xianyu Chen
Xiao Chen
Xiao Chen
Xiaohan Chen
Xiaokang Chen
Xiaotong Chen
Xin Chen
Xin Chen
Xin Chen
Xinghao Chen
Xingyu Chen
Xingyu Chen
Xinlei Chen
Xiuwei Chen
Xuanbai Chen
Xuanhong Chen
Xuesong Chen
Xuweiyi Chen
Xuxi Chen
Yanbei Chen
Yang Chen
Yaofo Chen
Yen-Chun Chen
Yi-Ting Chen
Yi-Wen Chen
Yilun Chen
Yinbo Chen
Yingcong Chen
Yingwen Chen
Yiting Chen
Yixin Chen
Yixin Chen
Yu Chen
Yuanhong Chen
Yuanqi Chen

Yuedong Chen	Ziyang Chen	Seunggeun Chi
Yueting Chen	An-Chieh Cheng	Zhixiang Chi
Yuhao Chen	Bowen Cheng	Naoya Chiba
Yuhao Chen	De Cheng	Julian Chibane
Yujin Chen	Erkang Cheng	Aditya Chinchure
Yukang Chen	Feng Cheng	Eleni Chiou
Yun-Chun Chen	Guangliang Cheng	Mia Chiquier
Yunjin Chen	Hao Cheng	Chiranjeev Chiranjeev
Yunlu Chen	Hao Cheng	Kashyap Chitta
Yuntao Chen	Harry Cheng	Hsu-kuang Chiu
Yushuo Chen	Ho Kei Cheng	Wei-Chen Chiu
Yuxiao Chen	Jingchun Cheng	Chee Kheng Chng
Zehui Chen	Jun Cheng	Shin-Fang Chng
Zerui Chen	Ka Leong Cheng	Donghyeon Cho
Zewen Chen	Kelvin B. Cheng	Gyusang Cho
Zeyuan Chen	Kevin Ho Man Cheng	Hanbyel Cho
Zeyuan Chen	Lechao Cheng	Hoonhee Cho
Zhang Chen	Shuo Cheng	Jae Won Cho
Zhao-Min Chen	Silin Cheng	Jang Hyun Cho
Zhaoliang Chen	Ta-Ying Cheng	Jungchan Cho
Zhaoxi Chen	Tianhang Cheng	Junhyeong Cho
Zhaoyu Chen	Weihao Cheng	Nam Ik Cho
Zhaozheng Chen	Wencan Cheng	Seokju Cho
Zhaozheng Chen	Wensheng Cheng	Seungju Cho
Zhe Chen	Xi Cheng	Suhwan Cho
Zhen Chen	Xize Cheng	Sunghyun Cho
Zhen Chen	Xuxin Cheng	Sungjun Cho
Zheng Chen	Yen-Chi Cheng	Sungmin Cho
Zhenghao Chen	Yi Cheng	Wonwoong Cho
Zhengyu Chen	Yihua Cheng	Nathaniel E. Chodosh
Zhenyu Chen	Yu Cheng	Jaesung Choe
Zhi Chen	Zesen Cheng	Junsuk Choe
Zhibo Chen	Zezhou Cheng	Changwoon Choi
Zhimin Chen	Zhanzhan Cheng	Chiho Choi
Zhineng Chen	Zhen Cheng	Ching Lam Choi
Zhiwei Chen	Zhi-Qi Cheng	Dongyoung Choi
Zhixiang Chen	Ziang Cheng	Dooseop Choi
Zhiyang Chen	Ziheng Cheng	Hongsuk Choi
Zhiyuan Chen	Soon Yau Cheong	Hyesong Choi
Zhuangzhuang Chen	Anoop Cherian	Jaewoong Choi
Zhuo Chen	Daniil Cherniavskii	Janghoon Choi
Zhuoxiao Chen	Ajad Chhatkuli	Jin Young Choi
Zihan Chen	Cheng Chi	Jinwoo Choi
Zijiao Chen	Haoang Chi	Jinyoung Choi
Ziwen Chen	Hyung-gun Chi	Jonghyun Choi

Jongwon Choi
Jooyoung Choi
Kanghyun Choi
Myungsub Choi
Seokeon Choi
Wonhyeok Choi
Bhavin Choksi
Jaegul Choo
Baptiste Chopin
Ayush Chopra
Gene Chou
Shih-Han Chou
Divya Choudhary
Siddharth Choudhary
Sunav Choudhary
Rohan Choudhury
Vasileios Choutas
Ka-Ho Chow
Pinaki Nath Chowdhury
Sanjoy Chowdhury
Sammy Christen
Fu-Jen Chu
Qi Chu
Ruihang Chu
Sanghyeok Chu
Wei-Ta Chu
Wen-Hsuan Chu
Wen-Hsuan Chu
Wei Qin Chuah
Andrei Chubarau
Ilya Chugunov
Sanghyuk Chun
Se Young Chun
Hyungjin Chung
Inseop Chung
Jaeyoung Chung
Jihoon Chung
Jiwan Chung
Joon Son Chung
Thomas A. Ciarfuglia
Umur A. Ciftci
Antonio E. Cinà
Ramazan Gokberk Cinbis
Anthony Cioppa
Javier Civera

Christopher A. Clark
James J. Clark
Ronald Clark
Brian S. Clipp
Mark Coates
Federico Cocchi
Felipe Codevilla
Cecília Coelho
Danny Cohen-Or
Forrester Cole
Julien Colin
John Collomosse
Marc Comino-Trinidad
Nicola Conci
Marcos V. Conde
Alexandru Paul
   Condurache
Runmin Cong
Tianshuo Cong
Wenyan Cong
Yuren Cong
Alessandro Conti
Andrea Conti
Enric Corona
John R. Corring
Jaime Corsetti
Jason J. Corso
Theo W. Costain
Guillaume Couairon
Robin Courant
Davide Cozzolino
Donato Crisostomi
Francesco Croce
Florinel Alin Croitoru
Ioana Croitoru
James L. Crowley
Steve Cruz
Gabriela Csurka
Alfredo Cuesta Infante
Aiyu Cui
Hainan Cui
Jiali Cui
Jiequan Cui
Justin Cui
Qiongjie Cui

Ruikai Cui
Yawen Cui
Ying Cui
Yuning Cui
Zhen Cui
Zhenyu Cui
Ziteng Cui
Benjamin J. Culpepper
Xiaodong Cun
Federico Cunico
Claudio Cusano
Sebastian Cygert
Guido Maria D'Amely di
   Melendugno
Moreno D'Incà
Feipeng Da
Mosam Dabhi
Rishabh Dabral
Konstantinos M. Dafnis
Matthew L. Daggitt
Anders B. Dahl
Bo Dai
Bo Dai
Haixing Dai
Mengyu Dai
Peng Dai
Peng Dai
Pingyang Dai
Qi Dai
Qiyu Dai
Wenrui Dai
Wenxun Dai
Zuozhuo Dai
Nicola Dall'Asen
Naser Damer
Bo Dang
Meghal Dani
Duolikun Danier
Donald G. Dansereau
Quan Dao
Son Duy Dao
Trung Tuan Dao
Mohamed Daoudi
Ahmad Darkhalil
Rangel Daroya

Abhijit Das
Abir Das
Anurag Das
Ayan Das
Nilotpal Das
Partha Das
Soumi Das
Srijan Das
Sukhendu Das
Swagatam Das
Ananyananda Dasari
Avijit Dasgupta
Gourav Datta
Achal Dave
Ishan Rajendrakumar Dave
Andrew Davison
Aram Davtyan
Axel Davy
Youssef Dawoud
Laura Daza
Francesca De Benetti
Daan de Geus
Alfredo S. De Goyeneche
Riccardo de Lutio
Maria De Marsico
Christophe De
 Vleeschouwer
Tonmoay Deb
Sayan Deb Sarkar
Dale Decatur
Thanos Delatolas
Fabien Delattre
Mauricio Delbracio
Ginger D. Delmas
Andong Deng
Bailin Deng
Bin Deng
Boyang Deng
Chaorui Deng
Cheng Deng
Congyue Deng
Jiacheng Deng
Jiajun Deng
Jiankang Deng
Jingjing Deng

Jinhong Deng
Kangle Deng
Kenan Deng
Lei Deng
Liang-Jian Deng
Shijian Deng
Weijian Deng
Xiaoming Deng
Xueqing Deng
Ye Deng
Yongjian Deng
Youming Deng
Yu Deng
Zhijie Deng
Zhongying Deng
Matteo Denitto
Mohammad Mahdi
 Derakhshani
Alakh Desai
Kevin Desai
Nishq P. Desai
Sai Vikas Desai
Jean-Emmanuel Deschaud
Frédéric Devernay
Rahul Dey
Arturo Deza
Helisa Dhamo
Amaya Dharmasiri
Ankit Dhiman
Vikas Dhiman
Yan Di
Zonglin Di
Ousmane A. Dia
Renwei Dian
Haiwen Diao
Xiaolei Diao
Gonçalo José Dias Pais
Abdallah Dib
Juan Carlos Dibene
 Simental
Sebastian Dille
Christian Diller
Mariella Dimiccoli
Anastasios Dimou
Or Dinari

Caiwen Ding
Changxing Ding
Choubo Ding
Fangqiang Ding
Guiguang Ding
Guodong Ding
Henghui Ding
Hui Ding
Jian Ding
Jian-Jiun Ding
Jianhao Ding
Li Ding
Liang Ding
Lihe Ding
Ming Ding
Runyu Ding
Shouhong Ding
Shuangrui Ding
Shuxiao Ding
Tianjiao Ding
Tianyu Ding
Wenhao Ding
Xiaohan Ding
Xinpeng Ding
Yaqing Ding
Yong Ding
Yuhang Ding
Yuqi Ding
Zheng Ding
Zhengming Ding
Zhipeng Ding
Zhuangzhuang Ding
Zihan Ding
Zihan Ding
Ziluo Ding
Anh-Dung Dinh
Markos Diomataris
Christos Diou
Alish Dipani
Alara Dirik
Ajay Divakaran
Abdelaziz Djelouah
Nemanja Djuric
Thanh-Toan Do
Tien Do

Anh-Dzung Doan
Bao Gia Doan
Khoa D. Doan
Carl Doersch
Andreea Dogaru
Nehal Doiphode
Csaba Domokos
Bowen Dong
Haoye Dong
Jiahua Dong
Jianfeng Dong
Jinpeng Dong
Junhao Dong
Junting Dong
Li Dong
Minjing Dong
Nanqing Dong
Peijie Dong
Qiaole Dong
Qihua Dong
Qiulei Dong
Runpei Dong
Shichao Dong
Shuting Dong
Sixun Dong
Siyan Dong
Tian Dong
Wei Dong
Wei Dong
Weisheng Dong
Xiaoyi Dong
Xiaoyu Dong
Xingping Dong
Yinpeng Dong
Zhenxing Dong
Jonathan C. Donnelly
Naren Doraiswamy
Gianfranco Doretto
Michael Dorkenwald
Muskan Dosi
Huanzhang Dou
Shuguang Dou
Yiming Dou
Zhiyang Frank Dou
Zi-Yi Dou

Arthur Douillard
Michail C. Doukas
Matthijs Douze
Amil Dravid
Vincent Drouard
Dawei Du
Jia-Run Du
Jinhao Du
Ruoyi Du
Weiyu Du
Xiaodan Du
Ye Du
Yingjun Du
Yong Du
Yuanqi Du
Yuntao Du
Zexing Du
Zhuoran Du
Radhika Dua
Haodong Duan
Haoran Duan
Huiyu Duan
Jiafei Duan
Jiali Duan
Peiqi Duan
Ye Duan
Yuchen Duan
Yueqi Duan
Zhihao Duan
Amanda Cardoso Duarte
Shiv Ram Dubey
Anastasia Dubrovina
Daniel Duckworth
Timothy Duff
Nicolas Dufour
Rahul Duggal
Shivam Duggal
Bardienus P. Duisterhof
Felix Dülmer
Matteo Dunnhofer
Chi Nhan Duong
Elona Dupont
Nikita Durasov
Zoran Duric
Mihai Dusmanu

Titir Dutta
Ujjal Kr Dutta
Isht Dwivedi
Sai Kumar Dwivedi
Sebastian Dziadzio
Thomas Eboli
Ramin Ebrahim Nakhli
Benjamin Eckart
Takeharu Eda
Johan Edstedt
Alexei A. Efros
Nikos Efthymiadis
Bernhard Egger
Max Ehrlich
Iván Eichhardt
Farshad Einabadi
Martin Eisemann
Peter Eisert
Mohamed El Banani
Mostafa El-Khamy
Alaa El-Nouby
Randa I. M. Elanwar
James Elder
Abdelrahman Eldesokey
Ismail Elezi
Ehsan Elhamifar
Moshe Eliasof
Sara Elkerdawy
Qing En
Ian Endres
Andreas Engelhardt
Francis Engelmann
Martin Engilberge
Kenji Enomoto
Chanho Eom
Guy Erez
Linus Ericsson
Maria C. Escobar
Marcos Escudero-Viñolo
Sedigheh Eslami
Valter Estevam
Ali Etemad
Martin Nicolas Everaert
Ivan Evtimov
Ralph Ewerth

Fevziye Irem Eyiokur
Cristobal Eyzaguirre
Gabriele Facciolo
Mohammad Fahes
Masud ANI Fahim
Fabrizio Falchi
Alex Falcon
Aoxiang Fan
Baojie Fan
Bin Fan
Chao Fan
David Fan
Haoqiang Fan
Hehe Fan
Heng Fan
Hongyi Fan
Huijie Fan
Junsong Fan
Lei Fan
Lei Fan
Lifeng Fan
Lue Fan
Mingyuan Fan
Qi Fan
Rui R. Fan
Xiran Fan
Yifei Fan
Yue Fan
Zejia Fan
Zhaoxin Fan
Zhenfeng Fan
Zhentao Fan
Zhipeng Fan
Zhiwen Fan
Zicong Fan
Sean Fanello
Chaowei Fang
Chuan Fang
Gongfan Fang
Guian Fang
Han Fang
Jiansheng Fang
Jiemin Fang
Jin Fang
Jun Fang

Kun Fang
Pengfei Fang
Rui Fang
Sheng Fang
Xiang Fang
Xiaolin Fang
Xiuwen Fang
Zhaoyuan Fang
Zhiyuan Fang
Eros Fanì
Matteo Farina
Farzan Farnia
Aiman Farooq
Azade Farshad
Mohsen Fayyaz
Arash Fayyazi
Ben Fei
Jingjing Fei
Xiaohan Fei
Zhengcong Fei
Michael Felsberg
Berthy T. Feng
Bin Feng
Brandon Y. Feng
Chao Feng
Chen Feng
Chengjian Feng
Chun-Mei Feng
Fan Feng
Guorui Feng
Hao Feng
Jianjiang Feng
Lan Feng
Lei Feng
Litong Feng
Mingtao Feng
Qianyu Feng
Ruicheng Feng
Ruili Feng
Ruoyu Feng
Ryan T. Feng
Shiwei Feng
Songhe Feng
Tuo Feng
Wei Feng

Weitao Feng
Weixi Feng
Xuelu Feng
Yanglin Feng
Yao Feng
Yue Feng
Zhanxiang Feng
Zhenhua Feng
Zunlei Feng
Clara Fernandez
Victoria Fernandez
    Abrevaya
Miguel-Ángel
    Fernández-Torres
Sira Ferradans
Claudio Ferrari
Ethan Fetaya
Mustansar Fiaz
Guénolé Fiche
Panagiotis P. Filntisis
Marco Fiorucci
Kai Fischer
Tobias Fischer
Tom Fischer
Volker Fischer
Robert B. Fisher
Alessandro Flaborea
Corneliu O. Florea
Georgios Floros
Alejandro Fontan
Tomaso Fontanini
Lin Geng Foo
Wolfgang Förstner
Niki M. Foteinopoulou
Simone Foti
Simone Foti
Victor Fragoso
Gianni Franchi
Jean-Sebastien Franco
Emanuele Frascaroli
Moti Freiman
Rafail Fridman
Sara Fridovich-Keil
Felix Friedrich
Stanislav Frolov

Andre Fu
Bin Fu
Changcheng Fu
Changqing Fu
Jianlong Fu
Jie Fu
Jingjing Fu
Keren Fu
Lan Fu
Lele Fu
Minghao Fu
Qichen Fu
Rao Fu
Taimeng Fu
Tianwen Fu
Yanwei Fu
Ying Fu
Yonggan Fu
Yun Fu
Yuqian Fu
Zehua Fu
Zhenqi Fu
Zipeng Fu
Wolfgang Fuhl
Yasuhisa Fujii
Yuki Fujimura
Katsuki Fujisawa
Kent Fujiwara
Takuya Funatomi
Christopher Funk
Antonino Furnari
Ryo Furukawa
Ryosuke Furuta
David Futschik
Akshay Gadi Patil
Siddhartha Gairola
Rinon Gal
Adrian Galdran
Silvio Galesso
Meirav Galun
Rofida Gamal
Weijie Gan
Yiming Gan
Yulu Gan
Vineet Gandhi

Kanchana Vaishnavi Gandikota
Aditya Ganeshan
Aditya Ganeshan
Swetava Ganguli
Sujoy Ganguly
Hanan Gani
Alireza Ganjdanesh
Harald Ganster
Roy Ganz
Angela F. Gao
Bin-Bin Gao
Changxin Gao
Chen Gao
Chongyang Gao
Daiheng Gao
Daoyi Gao
Difei Gao
Hang Gao
Hongchang Gao
Huiyu Gao
Junyu Gao
Junyu Gao
Kaifeng Gao
Katelyn Gao
Kuofeng Gao
Lin Gao
Ling Gao
Maolin Gao
Quankai Gao
Ruopeng Gao
Shanghua Gao
Shangqi Gao
Shangqian Gao
Shaobing Gao
Shenyuan Gao
Xiang Gao
Xiangjun Gao
Yang Gao
Yipeng Gao
Yuan Gao
Yue Gao
Yunhe Gao
Zhanning Gao
Zhi Gao

Zhihan Gao
Zhitong Gao
Zhongpai Gao
Ziteng Gao
Nicola Garau
Elena Garces
Noa Garcia
Nuno Cruz Garcia
Ricardo Garcia Pinel
Guillermo Garcia-Hernando
Saurabh Garg
Mathieu Garon
Risheek Garrepalli
Pablo Garrido
Quentin Garrido
Stefano Gasperini
Vincent Gaudilliere
Chandan Gautam
Shivam Gautam
Paul Gavrikov
Jonathan Z. Gazak
Chongjian Ge
Liming Ge
Runzhou Ge
Shiming Ge
Songwei Ge
Weifeng Ge
Wenhang Ge
Yanhao Ge
Yixiao Ge
Yunhao Ge
Yuying Ge
Zhiqi Ge
Zongyuan Ge
James Gee
Chen Geng
Daniel Geng
Xue Geng
Zhengyang Geng
Zichen Geng
Kyle Genova
Georgios Georgakis
Mariana-Iuliana Georgescu

Yiangos Georgiou
Markos Georgopoulos
Stamatios Georgoulis
Michal Geyer
Mina Ghadimi Atigh
Abhijay Ghildyal
Reza Ghoddoosian
Mohsen Gholami
Peyman Gholami
Enjie Ghorbel
Soumya Suvra Ghosal
Anindita Ghosh
Anurag Ghosh
Arnab Ghosh
Arthita Ghosh
Aurobrata Ghosh
Shreya Ghosh
Soham Ghosh
Sreyan Ghosh
Andrea Giachetti
Paris Giampouras
Alexander Gielisse
Andrew Gilbert
Guy Gilboa
Nate Gillman
Jhony H. Giraldo
Roger Girgis
Sharath Girish
Mario Valerio Giuffrida
Francesco Giuliari
Nikolaos Gkanatsios
Ioannis Gkioulekas
Alexi Gladstone
Derek Gloudemans
Aurele T. Gnanha
Hyojun Go
Akos Godo
Danilo D. Goede
Arushi Goel
Kratarth Goel
Rahul Goel
Shubham Goel
Vidit Goel
Tejas Gokhale
Vignesh Gokul

Aditya Golatkar
Micah Goldblum
S. Alireza Golestaneh
Gabriele Goletto
Lluis Gomez
Guillermo Gomez-Trenado
Alex Gomez-Villa
Nuno Gonçalves
Biao Gong
Boqing Gong
Chen Gong
Chengyue Gong
Dayoung Gong
Dong Gong
Jingyu Gong
Kaixiong Gong
Kehong Gong
Liyu Gong
Mingming Gong
Ran Gong
Rui Gong
Tao Gong
Xiaojin Gong
Xuan Gong
Xun Gong
Yifan Gong
Yu Gong
Yuanhao Gong
Yuning Gong
Yunye Gong
Cristina I. González
Adam Goodge
Nithin Gopalakrishnan
   Nair
Cameron Gordon
Dipam Goswami
Gaurav Goswami
Hanno Gottschalk
Chenhui Gou
Jianping Gou
Shreyank N. Gowda
Mohit Goyal
Julia Grabinski
Helmut Grabner
Patrick L. Grady

Alexandros Graikos
Eric Granger
Douglas R. Gray
Michael Alan Greenspan
Connor Greenwell
David Griffiths
Artur Grigorev
Alexey A. Gritsenko
Ivan Grubišić
Geonmo Gu
Jianyang Gu
Jiaqi Gu
Jiayuan Gu
Jinwei Gu
Peiyan Gu
Qiao Gu
Tianpei Gu
Xianfan Gu
Xiang Gu
Xiangming Gu
Xiao Gu
Xiuye Gu
Yanan Gu
Yiwen Gu
Yuchao Gu
Yuming Gu
Yun Gu
Zeqi Gu
Zhangxuan Gu
Banglei Guan
Junfeng Guan
Qingji Guan
Shanyan Guan
Tianrui Guan
Tongkun Guan
Ziqiao Guan
Ziyu Guan
Denis A. Gudovskiy
Antoine Guédon
Paul Guerrero
Ricardo Guerrero
Liangke Gui
Sadaf Gulshad
Manuel Günther
Chen Guo

Chenqi Guo
Chuan Guo
Guodong Guo
Haiyun Guo
Heng Guo
Hengtao Guo
Hongji Guo
Jia Guo
Jianwei Guo
Jianyuan Guo
Jiayi Guo
Jie Guo
Jie Guo
Jingcai Guo
Jingyuan Guo
Jinyang Guo
Lanqing Guo
Meng-Hao Guo
Mengxi China Guo
Minghao Guo
Pengfei Guo
Pengsheng Guo
Pinxue Guo
Qi Guo
Qin Guo
Qing Guo
Qingpei Guo
Saidi Guo
Shi Guo
Shuxuan Guo
Song Guo
Taian Guo
Tiantong Guo
Tianyu Guo
Wen Guo
Wenzhong Guo
Xianda Guo
Xiaobao Guo
Xiefan Guo
Yangyang Guo
Yanhui Guo
Yao Guo
Yichen Guo
Yong Guo
Yuan-Chen Guo
Yuliang Guo
Yuxiang Guo
Yuyu Guo
Zixian Guo
Ziyu Guo
Zujin Guo
Aarush Gupta
Akash Gupta
Akshita Gupta
Ankush Gupta
Anshul Gupta
Anubhav Gupta
Anup Kumar Gupta
Honey Gupta
Kunal Gupta
pravir singh gupta
Rohit Gupta
Saumya Gupta
Corina Gurau
Yeti Z. Gurbuz
Saket Gurukar
Siddharth Gururani
Vladimir Guzov
Matthew A. Gwilliam
Hyunho Ha
Ryo Hachiuma
Isma Hadji
Armin Hadzic
Daniel Haehn
Nicolai Haeni
Ronny Haensch
Meera Hahn
Oliver Hahn
Nguyen Hai
Yuval Haitman
Levente Hajder
Moayed Haji-Ali
Sina Hajimiri
Alexandros Haliassos
Peter M. Hall
Bumsub Ham
Ryuhei Hamaguchi
Abdullah J. Hamdi
Max Hamilton
Lars Hammarstrand
Hasan Abed Al Kader Hammoud
Beining Han
Boran Han
Dongyoon Han
Feng Han
Guangxing Han
Guangzeng Han
Hu Han
Jiaming Han
Jin Han
Jungong Han
Junlin Han
Kai Han
Kang Han
Kun Han
Ligong Han
Longfei Han
Mei Han
Mingfei Han
Muzhi Han
Sungwon Han
Tengda Han
Tengda Han
Wencheng Han
Xinran Han
Xintong Han
Yizeng Han
Yufei Han
Zhenjun Han
Zhi Han
Zhongyi Han
Zongbo Han
Zongyan Han
Ankur Handa
Tiankai Hang
Yucheng Hang
Asif Hanif
Param Hanji
Joëlle Hanna
Niklas Hanselmann
Nicklas A. Hansen
Fusheng Hao
Shaozhe Hao
Shengyu Hao

Shijie Hao
Tianxiang Hao
Yanbin Hao
Yu Hao
Zekun Hao
Takayuki Hara
Mehrtash Harandi
Sanjay Haresh
Haripriya Harikumar
Adam Harley
David M. Hart
Md Yousuf Harun
Md Rakibul Hasan
Connor Hashemi
Atsushi Hashimoto
Khurram Azeem Hashmi
Nozomi Hata
Ali Hatamizadeh
Ryuichiro Hataya
Timm Haucke
Joakim Bruslund Haurum
Stephen Hausler
Mohammad Havaei
Hideaki Hayashi
Zeeshan Hayder
Chengan He
Conghui He
Dailan He
Fan He
Fazhi He
Gaoqi He
Haoyu He
Hongliang He
Jiangpeng He
Jianzhong He
Jiawei He
Ju He
Jun-Yan He
Junfeng He
Lihuo He
Liqiang He
Nanjun He
Ruian He
Runze He
Ruozhen He

Shengfeng He
Shuting He
Songtao He
Tao He
Tianyu He
Tong He
Wei He
Xiang He
Xiangteng He
Xiangyu He
Xiaoxiao He
Xin He
Xingyi He
Xingzhe He
Xinlei He
Xinwei He
Yang He
Yang He
Yannan He
Yeting He
Yinan He
Ying He
Yisheng He
Yuhang He
Yuhang He
Zecheng He
Zewei He
Zexin He
Zhenyu He
Ziwen He
Ramya S. Hebbalaguppe
Peter Hedman
Deepti B. Hegde
Sindhu B. Hegde
Matthias Hein
Mattias Paul Heinrich
Aral Hekimoglu
Philipp Henzler
Byeongho Heo
Jae-Pil Heo
Miran Heo
Stephane Herbin
Alexander Hermans
Pedro Hermosilla Casajus
Jefferson E. Hernandez

Monica Hernandez
Charles Herrmann
Roei Herzig
Fabian Herzog
Georg Hess
Mauricio Hess-Flores
Robin Hesse
Chamin P. Hewa
   Koneputugodage
Masatoshi Hidaka
Richard E. L. Higgins
Mitchell K. Hill
Carlos Hinojosa
Tobias Hinz
Yusuke Hirota
Elad Hirsch
Shinsaku Hiura
Chih-Hui Ho
Man M. Ho
Tuan N. A. Hoang
Jennifer Hobbs
Jiun Tian Hoe
David T. Hoffmann
Derek Hoiem
Yannick Hold-Geoffroy
Lukas Höllein
Gregory I. Holste
Hiroto Honda
Cheeun Hong
Cheng-Yao Hong
Danfeng Hong
Deokki Hong
Fa-Ting Hong
Fangzhou Hong
Feng Hong
Guan Zhe Hong
Je Hyeong Hong
Ji Woo Hong
Jie Hong
Joanna Hong
Lanqing Hong
Lingyi Hong
Sangwoo Hong
Sungeun Hong
Sunghwan Hong

Susung Hong
Weixiang Hong
Xiaopeng Hong
Yan Hong
Yi Hong
Yuchen Hong
Julia Hornauer
Maxwell C. Horton
Eliahu Horwitz
Mir Rayat Imtiaz Hossain
Tonmoy Hossain
Mehdi Hosseinzadeh
Kazuhiro Hotta
Andrew Z. Hou
Fei Hou
Hao Hou
Ji Hou
Jian Hou
Jinghua Hou
Qibin Hou
Qiqi Hou
Ruibing Hou
Saihui Hou
Tingbo Hou
Xinhai Hou
Yuenan Hou
Yunzhong Hou
Zhi Hou
Lukas Hoyer
Petr Hruby
Jun-Wei Hsieh
Chih-Fan Hsu
Gee-Sern Hsu
Hung-Min Hsu
Yen-Chi Hsu
Anthony Hu
Benran Hu
Bo Hu
Dapeng Hu
Dongting Hu
Guosheng Hu
Haigen Hu
Hanjiang Hu
Hanzhe Hu
Hengtong Hu

Hezhen Hu
Jian Hu
Jian-Fang Hu
Jie Hu
Juan Hu
Junjie Hu
Kai Hu
Lianyu Hu
Lily Hu
MengShun Hu
Minghui Hu
Mu Hu
Panwen Hu
Panwen Hu
Ruizhen Hu
Shengshan Hu
Shiyu Hu
Shizhe Hu
Shu Hu
Tao Hu
Tao Hu
Vincent Tao Hu
Wenbo Hu
Xiaodan Hu
Xiaolin Hu
Xiaoling Hu
Xiaotao Hu
Xiaowei Hu
Xinting Hu
Xinting Hu
Xixi Hu
Xuefeng Hu
Yang Hu
Yaosi Hu
Yinlin Hu
Yueyu Hu
Zeyu Hu
Zhanhao Hu
zhengyu hu
Zhenzhen Hu
Zhiming Hu
Zhiming Hu
Zhongyun Hu
Zijian Hu
Andong Hua

Hang Hua
Miao Hua
Wei Hua
Mengdi Huai
Binbin Huang
Bo Huang
Buzhen Huang
Chao Huang
Chao-Tsung Huang
Chaoqin Huang
Ching-Chun Huang
Cong Huang
Danqing Huang
Di Huang
Feihu Huang
Haibin Huang
Haiyang Huang
Hao Huang
Hsin-Ping Huang
Huaibo Huang
Ian Y. Huang
Jiabo Huang
Jiahui Huang
Jiancheng Huang
Jiangyong Huang
Jiaxing Huang
Jin Huang
Jinfa Huang
Jing Huang
Jonathan Huang
Junkai Huang
Junwen Huang
Kejie Huang
Kuan-Chih Huang
Lei Huang
Libo Huang
Lin Huang
Linjiang Huang
Linyan Huang
linzhi Huang
Lun Huang
Luojie Huang
Mingzhen Huang
Qidong Huang
Qiusheng Huang

Runhui Huang
Ruqi Huang
Shaofei Huang
Sheng Huang
Sheng-Yu Huang
Shiyuan Huang
Shuaiyi Huang
Shuangping Huang
Siteng Huang
Siyu Huang
Siyuan Huang
Tao Huang
Thomas E. Huang
Tianyu Huang
Wenjian Huang
Wenke Huang
Xiaohu Huang
Xiaohua Huang
Xiaoke Huang
Xiaoshui Huang
Xiaoyang Huang
Xijie Huang
Xinyu Huang
Xuhua Huang
Yan Huang
Yaping Huang
Ye Huang
Yi Huang
Yi Huang
Yi Huang
Yi-Hua Huang
Yifei Huang
Yihao Huang
Ying Huang
Yizhan Huang
You Huang
Yue Huang
Yufei Huang
Yuge Huang
Yukun Huang
Yuyao Huang
Zaiyu Huang
Zehao Huang
Zeyi Huang
Zhe Huang

Zhewei Huang
Zhida Huang
Zhiqi Huang
Zhiyu Huang
Zhongzhan Huang
Ziling Huang
Zilong Huang
Ziqi Huang
Zixuan Huang
Ziyuan Huang
Jaesung Huh
Ka-Hei Hui
Le Hui
Tianrui Hui
Xiaofei Hui
Ahmed Imtiaz Humayun
Thomas Hummel
Jing Huo
Shuwei Huo
Yuankai Huo
Julio Hurtado
Rukhshanda Hussain
Mohamed Hussein
Noureldien Hussein
Chuong Minh Huynh
Tran Ngoc Huynh
Muhammad Huzaifa
Inwoo Hwang
Jaehui Hwang
Jenq-Neng Hwang
Sehyun Hwang
Seunghyun Hwang
Sukjun Hwang
Sunhee Hwang
Wonjun Hwang
Jeongseok Hyun
Sangeek Hyun
Ekaterina Iakovleva
Tomoki Ichikawa
A. S. M. Iftekhar
Masaaki Iiyama
Satoshi Ikehata
Sunghoon Im
Woobin Im
Nevrez Imamoglu

Abdullah Al Zubaer Imran
Tooba Imtiaz
Nakamasa Inoue
Naoto Inoue
Eldar Insafutdinov
Catalin Ionescu
Radu Tudor Ionescu
Nasar Iqbal
Umar Iqbal
Go Irie
Muhammad Zubair Irshad
Francesco Isgrò
Yasunori Ishii
Berivan Isik
Syed M. S. Islam
Mariko Isogawa
Vamsi Krishna K. Ithapu
Koichi Ito
Leonardo Iurada
Shun Iwase
Sergio Izquierdo
Sarah Jabbour
David Jacobs
David E. Jacobs
Yasamin Jafarian
Azin Jahedi
Achin Jain
Ayush Jain
Jitesh Jain
Kanishk Jain
Yash Jain
Nikita Jaipuria
Tomas Jakab
Muhammad Abdullah
 Jamal
Hadi Jamali-Rad
Stuart James
Nataraj Jammalamadaka
Donggon Jang
Sujin Jang
Young Kyun Jang
Youngjoon Jang
Steeven Janny
Paul Janson
Maximilian Jaritz

Ronnachai Jaroensri
Guillaume Jaume
Sajid Javed
Saqib Javed
Bhavin Jawade
Hirunima Jayasekara
Senthilnath Jayavelu
Guillaume Jeanneret
Pranav Jeevan
Rohit Jena
Tomas Jenicek
Simon Jenni
Hae-Gon Jeon
Jeimin Jeon
Seogkyu Jeon
Boseung Jeong
Jongheon Jeong
Yonghyun Jeong
Yoonwoo Jeong
Koteswar Rao Jerripothula
Artur Jesslen
Nikolay Jetchev
Ankit Jha
Sumit K. Jha
I-Hong Jhuo
Deyi Ji
Ge-Peng Ji
Hui Ji
Jianmin Ji
Jiayi Ji
Jingwei Ji
Naye Ji
Pengliang Ji
Shengpeng Ji
Wei Ji
Xiang Ji
Xiaopeng Ji
Xiaozhong Ji
Xinya Ji
Yu Ji
Yuanfeng Ji
Zhanghexuan Ji
Baoxiong Jia
Ding Jia
Jinrang Jia
Menglin Jia
Shan Jia
Shuai Jia
Tong Jia
Wenqi Jia
Xiaojun Jia
Xiaosong Jia
Xu Jia
Yiren Jian
Bo Jiang
Borui Jiang
Chen Jiang
Guangyuan Jiang
Hai Jiang
Haiyang Jiang
Haiyong Jiang
Hanwen Jiang
Hanxiao Jiang
Hao Jiang
Haobo Jiang
Huaizu Jiang
Huajie Jiang
Jianmin Jiang
Jianwen Jiang
Jiaxi Jiang
Jin Jiang
Jing Jiang
Kaixun Jiang
Kui Jiang
Li Jiang
Liming Jiang
Meirui Jiang
Ming Jiang
Ming Jiang
Peng Jiang
Peng-Tao Jiang
Runqing Jiang
Siyang Jiang
Tianjian Jiang
Tingting Jiang
Weisen Jiang
Wen Jiang
Xingyu Jiang
Xueying Jiang
Xuhao Jiang
Yanru Jiang
Yi Jiang
Yifan Jiang
Yingying Jiang
Yue Jiang
Yuming Jiang
Zeren Jiang
Zheheng Jiang
Zhenyu Jiang
Zhongyu Jiang
Zi-Hang Jiang
Zixuan Jiang
Ziyu Jiang
Zutao Jiang
Zutao Jiang
Jianbin Jiao
Jianbo Jiao
Licheng Jiao
Ruochen Jiao
Shuming Jiao
Wenpei Jiao
Zequn Jie
Dongkwon Jin
Gaojie Jin
Haian Jin
Haibo Jin
Kyong Hwan Jin
Lei Jin
Lianwen Jin
Linyi Jin
Peng Jin
Peng Jin
Sheng Jin
SouYoung Jin
Weiyang Jin
Xiao Jin
Xiaojie Jin
Xin Jin
Xin Jin
Yeying Jin
Yi Jin
Ying Jin
Zhenchao Jin
Chenchen Jing
Junpeng Jing

Mengmeng Jing
Taotao Jing
Jeonghee Jo
Yeonsik Jo
Younghyun Jo
Cameron A. Johnson
Faith M. Johnson
Maxwell Jones
Michael J. Jones
R. Kenny Jones
Ameya Joshi
Shantanu H. Joshi
Brendan Jou
Chen Ju
Wei Ju
Xuan Ju
Yan Ju
Felix Juefei-Xu
Florian Jug
Yohan Jun
Kim Jun-Seong
Masum Shah Junayed
Chanyong Jung
Claudio R. Jung
Hoin Jung
Hyunyoung Jung
Sangwon Jung
Steffen Jung
Sacha Jungerman
Hari Chandana K.
Prajwal K. R.
Berna Kabadayi
Anis Kacem
Anil Kag
Kumara Kahatapitiya
Bernhard Kainz
Ivana Kajic
Ioannis Kakogeorgiou
Niveditha Kalavakonda
Mahdi M. Kalayeh
Ajinkya Kale
Anmol Kalia
Sinan Kalkan
Jayateja Kalla
Tarun Kalluri

Uday Kamal
Sandesh Kamath
Chandra Kambhamettu
Meina Kan
Sai Srinivas Kancheti
Takuhiro Kaneko
Cuicui Kang
Dahyun Kang
Guoliang Kang
Hyolim Kang
Jaeyeon Kang
Juwon Kang
Li-Wei Kang
Mingon Kang
MinGuk Kang
Minjun Kang
Weitai Kang
Zhao Kang
Yash Mukund Kant
Yueying Kao
Saarthak Kapse
Aupendu Kar
Oğuzhan Fatih Kar
Ozgur Kara
Tamás Karácsony
Srikrishna Karanam
Neerav Karani
Mert Asim Karaoglu
Laurynas Karazija
Navid Kardan
Amirhossein Kardoost
Nour Karessli
Michelle Karg
Mohammad Reza Karimi
  Dastjerdi
Animesh Karnewar
Arjun M. Karpur
Shyamgopal Karthik
Korrawe Karunratanakul
Tejaswi Kasarla
Satyananda Kashyap
Yoni Kasten
Marc A. Kastner
Hirokatsu Kataoka
Isinsu Katircioglu

Kai Katsumata
Ilya Kaufman
Manuel Kaufmann
Chaitanya Kaul
Prannay Kaul
Prakhar Kaushik
Isaak Kavasidis
Ryo Kawahara
Yuki Kawana
Justin Kay
Evangelos Kazakos
Jingcheng Ke
Lei Ke
Tsung-Wei Ke
Wei Ke
Zhanghan Ke
Nikhil V. Keetha
Thomas Kehrenberg
Marilyn Keller
Rohit Keshari
Janis Keuper
Daniel Keysers
Hrant Khachatrian
Seyran Khademi
Wesley A. Khademi
Taras Khakhulin
Samir Khaki
Hasam Khalid
Umar Khalid
Amr Khalifa
Asif Hussain Khan
Faizan Farooq Khan
Muhammad Haris Khan
Naimul Khan
Qadeer Khan
Zeeshan Khan
Bishesh Khanal
Pulkit Khandelwal
Ishan Khatri
Muhammad Uzair Khattak
Vahid Reza Khazaie
Vaishnavi M. Khindkar
Rawal Khirodkar
Pirazh Khorramshahi
Sahil S. Khose

Kourosh Khoshelham
Sena Kiciroglu
Benjamin Kiefer
Kotaro Kikuchi
Mert Kilickaya
Benjamin Killeen
Beomyoung Kim
Boah Kim
Bumsoo Kim
Byeonghwi Kim
Byoungjip Kim
Changhoon Kim
Changick Kim
Chanho Kim
Chanyoung Kim
Dahun Kim
Dahyun Kim
Diana S. Kim
Dong-Jin Kim
Donggun Kim
Donghyun Kim
Dongkeun Kim
Dongwan Kim
Dongyoung Kim
Eun-Sol Kim
Eunji Kim
Euyoung Kim
Geeho Kim
Giseop Kim
Guisik Kim
Gwanghyun Kim
Hakyeong Kim
Hanjae Kim
Hanjung Kim
Heewon Kim
Howon Kim
Hyeongwoo Kim
Hyeongwoo Kim
Hyo Jin Kim
Hyung-Il Kim
Hyunwoo J. Kim
Insoo Kim
Jae Myung Kim
Jeongsol Kim
Jihwan Kim

Jinkyu Kim
Jinwoo Kim
Jongyoo Kim
Joonsoo Kim
Jung Uk Kim
Junho Kim
Junho Kim
Junsik Kim
Kangyeol Kim
Kunhee Kim
Kwang In Kim
Manjin Kim
Minchul Kim
Minji Kim
Minjung Kim
Namil Kim
Namyup Kim
Nayeong Kim
Sanghyun Kim
Seong Tae Kim
Seungbae Kim
Seungryong Kim
Seungwook Kim
Soo Ye Kim
Soohwan Kim
Sungnyun Kim
Sungyeon Kim
Sunnie S. Y. Kim
Tae Hyun Kim
Tae Hyung Kim
Taehoon Kim
Taehun Kim
Taehwan Kim
Taekyung Kim
Taeoh Kim
Taewoo Kim
Won Hwa Kim
Wonjae Kim
Woo Jae Kim
Young Min Kim
YoungBin Kim
Youngeun Kim
Youngseok Kim
Youngwook Kim
Akisato Kimura

Andreas Kirsch
Nikita Kister
Furkan Osman Kınlı
Marcus Klasson
Florian Kleber
Tzofi M. Klinghoffer
Jan P. Klopp
Florian Kluger
Hannah Kniesel
David M. Knigge
Byungsoo Ko
Dohwan Ko
Jongwoo Ko
Takumi Kobayashi
Muhammed Kocabas
Yeong Jun Koh
Kathlén Kohn
Subhadeep Koley
Nick Kolkin
Soheil Kolouri
Jacek Komorowski
Deying Kong
Fanjie Kong
Hanyang Kong
Hui Kong
Kyeongbo Kong
Lecheng Kong
Lingdong Kong
Linghe Kong
Lingshun Kong
Naejin Kong
Quan Kong
Shu Kong
Xianghao Kong
Xiangtao Kong
Xiangwei Kong
Xiaoyu Kong
Xin Kong
Youyong Kong
Yu Kong
Zhenglun Kong
Aishik Konwer
Nicholas C. Konz
Gwanhyeong Koo
Juil Koo

Julian F. P. Kooij
George Kopanas
Sanjeev J. Koppal
Bruno Korbar
Giorgos Kordopatis-Zilos
Dimitri Korsch
Adam Kortylewski
Divya Kothandaraman
Suraj Kothawade
Iuliia Kotseruba
Sasikanth Kotti
Alankar Kotwal
Shashank Kotyan
Alexandros Kouris
Petros Koutras
Rama Kovvuri
Dilip Krishnan
Praveen Krishnan
Ranganath Krishnan
Rohan M. Krishnan
Georg Krispel
Alexander Krull
Tianshu Kuai
Haowei Kuang
Zhengfei Kuang
Andrey Kuehlkamp
David Kügler
Arjan Kuijper
Anna Kukleva
Jonas Kulhanek
Peter Kulits
Akshay R. Kulkarni
Ashutosh C. Kulkarni
Kuldeep Kulkarni
Nilesh Kulkarni
Abhinav Kumar
Abhishek Kumar
Akash Kumar
Avinash Kumar
B. V. K. Vijaya Kumar
Chandan Kumar
Pulkit Kumar
Ratnesh Kumar
Sateesh Kumar
Satish Kumar

Suryansh Kumar
Yogesh Kumar
Nupur Kumari
Sudhakar Kumawat
Nilakshan Kunananthaseelan
Rohit Kundu
Souvik Kundu
Meng-Yu Jennifer Kuo
Weicheng Kuo
Shuhei Kurita
Yusuke Kurose
Takahiro Kushida
Uday Kusupati
Alina Kuznetsova
Jobin K. V.
Henry Kvinge
Ho Man Kwan
Hyeokjun Kweon
Donghyeon Kwon
Gihyun Kwon
Heeseung Kwon
Hyoukjun Kwon
Myung-Joon Kwon
Taein Kwon
YoungJoong Kwon
Cameron Kyle-Davidson
Christos Kyrkou
Jorma Laaksonen
Patrick Labatut
Yann Labbé
Manuel Ladron de Guevara
Florent Lafarge
Jean Lahoud
Bolin Lai
Farley Lai
Jian-Huang Lai
Shenqi Lai
Xin Lai
Yu-Kun Lai
Yung-Hsuan Lai
Zeqiang Lai
Zhengfeng Lai
Barath Lakshmanan
Rohit Lal

Rodney LaLonde
Hala Lamdouar
Meng Lan
Yushi Lan
Federico Landi
George V. Landon
Chunbo Lang
Jochen Lang
Nico Lang
Georg Langs
Raffaella Lanzarotti
Dong Lao
Yixing Lao
Yizhen Lao
Zakaria Laskar
Alexandros Lattas
Chun Pong Lau
Shlomi Laufer
Justin Lazarow
Svetlana Lazebnik
Duy Tho Le
Hieu Le
Hoang Le
Hoang Le
Thi-Thu-Huong Le
Trung Le
Trung-Nghia Le
Tung Thanh Le
Hoàng-Ân Lê
Herve Le Borgne
Guillaume Le Moing
Erik Learned-Miller
Tim Lebailly
Byeong-Uk Lee
Byung-Kwan Lee
Cheng-Han Lee
Chul Lee
Daeun Lee
Dogyoon Lee
Dong Hoon Lee
Eugene Eu Tzuan Lee
Eung-Joo Lee
Gyuseong Lee
Hsin-Ying Lee
Hwee Kuan Lee

Hyeongmin Lee
Hyungtae Lee
Jae Yong Lee
Jaeho Lee
Jaeseong Lee
Jaewon Lee
Jangho Lee
Jangwon Lee
Jihyun Lee
Jiyoung Lee
Jong-Seok Lee
Jongho Lee
Jongmin Lee
Joo-Ho Lee
Joon-Young Lee
Joonseok Lee
Jungbeom Lee
Jungho Lee
Jungwoo Lee
Junha Lee
Junhyun Lee
Junyong Lee
Kibok Lee
Kuan-Ying Lee
Kwonjoon Lee
Kwot Sin Lee
Kyungmin Lee
Kyungmoon Lee
Minhyeok Lee
Minsik Lee
Pilhyeon Lee
Saehyung Lee
Sangho Lee
Sanghyeok Lee
Sangmin Lee
Sehun Lee
Sehyung Lee
Seon-Ho Lee
Seong Hun Lee
Seongwon Lee
Seung Hyun Lee
Seung-Ik Lee
Seungho Lee
Seunghun Lee
Seungmin Lee

Seungyong Lee
Sohyun Lee
Suhyeon Lee
Sungho Lee
Sungmin Lee
Suyoung Lee
Taehyun Lee
Wooseok Lee
Yao-Chih Lee
Yi-Lun Lee
Yonghyeon Lee
Youngwan Lee
Leonidas Lefakis
Bowen Lei
Chenyang Lei
Chenyi Lei
Jiahui Lei
Na Lei
Qinqian Lei
Yinjie Lei
Thomas Leimkuehler
Abe Leite
Abdelhak Lemkhenter
Jiaxu Leng
Luziwei Leng
Zhiying Leng
Hendrik P. A. Lensch
Jan E. Lenssen
Ted Lentsch
Simon Lepage
Stefan Leutenegger
Filippo Leveni
Axel Levy
Ailin Li
Aixuan Li
Baiang Li
Baoxin Li
Bin Li
Bing Li
Bing Li
Bo Li
Bowen Li
Boying Li
Changlin Li
Changlin Li

Chao Li
Chenghong Li
Chenglin Li
Chenglong Li
Chengze Li
Chun-Guang Li
Daiqing Li
Dasong Li
Dian Li
Dong Li
Fangda Li
Feiran Li
Fenghai Li
Gen Li
Guanbin Li
Guangrui Li
Guihong Li
Guorong Li
Haifeng Li
Han Li
Hang Li
Hangyu Li
Hanhui Li
Hao Li
Hao Li
Haoang Li
Haoran Li
Haoxiang Li
Haoxin Li
He Li
Heng Li
Hengduo Li
Hongshan Li
Hongwei Bran Li
Hongxiang Li
Hongyang Li
Hongyu Li
Huafeng Li
Huan Li
Hui Li
Jiacheng Li
Jiahao Li
Jialu Li
Jiaman Li
Jiangmeng Li

Jiangtong Li
Jiangyuan Li
Jianing Li
Jianwei Li
Jianwu Li
Jiaqi Li
Jiaqi Li
Jiatong Li
Jiaxuan Li
Jiazhi Li
Jichang Li
Jie Li
Jin Li
Jinglun Li
Jingzhi Li
Jingzong Li
Jinlong Li
Jinlong Li
Jinpeng Li
Jinxing Li
Jun Li
Jun Li
Junbo Li
Juncheng Li
Junxuan Li
Junyi Li
Kai Li
Kaican Li
Kailin Li
Ke Li
Kehan Li
Keyu Li
Kun Li
Kunchang Li
Kunpeng Li
Lei Li
Lei Li
Li Li
Li Erran Li
Liang Li
Lin Li
Lincheng Li
Liulei Li
Liunian Harold Li
Lujun Li

Manyi Li
Maomao Li
Meng Li
Mengke Li
Mengtian Li
Mengtian Li
Ming Li
Ming Li
Minghan Li
Mingjie Li
Nannan Li
Nianyi Li
Peike Li
Peizhao Li
Peng Li
Pengpeng Li
Pengyu Li
Ping Li
Puhao Li
Qiang Li
Qing Li
Qingyong Li
Qiufu Li
Qizhang Li
Ren Li
Rong Li
Rongjie Li
Ru Li
Rui Li
Ruibo Li
Ruihui Li
Ruilong Li
Ruining Li
Ruixuan Li
Runze Li
Ruoteng Li
Shaohua Li
Shasha Li
Shigang Li
Shijie Li
Shikun Li
Shile Li
Shuai Li
Shuai Li
Shuang Li

Shuwei Li
Si Li
Siyao Li
Siyuan Li
Siyuan Li
Taihui Li
Tianye Li
Wanhua Li
Wanqing Li
Wei Li
Wei Li
Wei Li
Wei-Hong Li
Weihao Li
Weijia Li
Weiming Li
Wenbin Li
Wenbo Li
Wenhao Li
Wenjie Li
Wenshuo Li
Wentong Li
Wenxi Li
Xiang Li
Xiang Li
Xiang Li
Xiang Li
Xiang Li
Xiangtai Li
Xiangyang Li
Xianzhi Li
Xiao Li
Xiao Li
Xiaoguang Li
Xiaomeng Li
Xiaoming Li
Xiaoqi Li
Xiaoqiang Li
Xiaotian Li
Xiaoyu Li
Xin Li
Xin Li
Xin Li
Xinghui Li
Xingyi Li

Xingyu Li
Xinjie Li
Xinyu Li
Xiu Li
Xiujun Li
Xuan Li
Xuanlin Li
Xuelong Li
Xuelu Li
Xueqian Li
Ya-Li Li
Yanan Li
Yang Li
Yang Li
Yangyan Li
Yanjing Li
Yansheng Li
Yanwei Li
Yanyu Li
Yaohui Li
Yaowei Li
Yawei Li
Yi Li
Yi Li
Yicong Li
Yicong Li
Yifei Li
Yijin Li
Yijun Li
Yijun Li
Yikang Li
Yimeng Li
Yiming Li
Yiming Li
Yingwei Li
Yiting Li
Yixuan Li
Yize Li
Yizhuo Li
Yong Li
Yong-Lu Li
Yongjie Li
Yuanman Li
Yuanming Li
Yuelong Li

Yuexiang Li
Yuezun Li
Yuhang Li
Yuheng Li
Yulin Li
Yumeng Li
Yunfan Li
Yunheng Li
Yunqiang Li
Yunsheng Li
Yuyan Li
Yuyang Li
Zejian Li
Zekun Li
Zekun Li
Zhangheng Li
Zhangzikang Li
Zhaoshuo Li
Zhaowen Li
Zhe Li
Zhe Li
Zhen Li
Zhen Li
Zhen Li
Zheng Li
Zhengqin Li
Zhengyuan Li
Zhenyu Li
Zhichao Li
Zhihao Li
Zhihao Li
Zhiheng Li
Zhiqi Li
Zhixuan Li
Zhong Li
Zhuoling Li
Zhuowan Li
Zhuowei Li
Zhuoxiao Li
Zihan Li
Ziqiang Li
Wen Qiao Li
Dongze Lian
Long Lian
Qing Lian

Ruyi Lian
Zhouhui Lian
Jin Lianbao
Chao Liang
Chia-Kai Liang
Dingkang Liang
Feng Liang
Gongbo Liang
Hanxue Liang
Hao Liang
Hui Liang
Jiadong Liang
Jiajun Liang
Jian Liang
Jingyun Liang
Jinxiu S. Liang
Junwei Liang
Kaiqu Liang
Ke Liang
Kevin J. Liang
Luming Liang
Mingfu Liang
Pengpeng Liang
Siyuan Liang
Xiaoxiao Liang
Xinran Liang
Xiwen Liang
Yang Liang
Yixun Liang
Yongqing Liang
Youwei Liang
Yuanzhi Liang
Zhexin Liang
Zhihao Liang
Zhixuan Liang
Kang Liao
Liang Liao
Minghui Liao
Ting-Hsuan Liao
Wei Liao
Xin Liao
Yinghong Liao
Yue Liao
Zhibin Liao
Ziwei Liao

Benedetta Liberatori
Daniel J. Lichy
Maiko Lie
Qin Likun
Isaak Lim
Teck Yian Lim
Bannapol Limanond
Baijiong Lin
Beibei Lin
Cheng Lin
Chenhao Lin
Chia-Wen Lin
Chieh Hubert Lin
Chuang Lin
Chung-Ching Lin
Chunyu Lin
Ci-Siang Lin
Di Lin
Fanqing Lin
Feng Lin
Fudong Lin
Guangfeng Lin
Haotong Lin
Haozhe Lin
Hubert Lin
Hui Lin
Jason Lin
Jianxin Lin
Jiaqi Lin
Jiaying Lin
Jiehong Lin
Jierui Lin
Jintao Lin
Kai-En Lin
Ke Lin
Kevin Lin
Kevin Qinghong Lin
Kuan Heng Lin
Kun-Yu Lin
Kunyang Lin
Kwan-Yee Lin
Lijian Lin
Liqiang Lin
Liting Lin
Luojun Lin

Mingyuan Lin
Qiuxia Lin
Shaohui Lin
Shih-Yao Lin
Sihao Lin
Siyou Lin
Tiancheng Lin
Tsung-Yu Lin
Wanyu Lin
Wei Lin
Wei Lin
Wen-Yan Lin
Wenbin Lin
Xiangbo Lin
Xianhui Lin
Xiaofan Lin
Xiaofeng Lin
Xin Lin
Xudong Lin
Xue Lin
Xuxin Lin
Ya-Wei Eileen Lin
Yan-Bo Lin
Yancong Lin
Yi Lin
Yijie Lin
Yiming Lin
Yiqi Lin
Yiqun Lin
Yongliang Lin
Yu Lin
Yuanze Lin
Yuewei Lin
Zhi-Hao Lin
Zhiqiu Lin
ZhiWei Lin
Zinan Lin
Ziyi Lin
David B. Lindell
Philipp Lindenberger
Jingwang Ling
Jun Ling
Yongguo Ling
Zhan Ling
Alexander Liniger

Stefan P. Lionar
Phillip Lippe
Lahav O. Lipson
Joey Litalien
Ron Litman
Mattia Litrico
Dor Litvak
Aishan Liu
Ajian Liu
Akide L. Y. Liu
Andrew Liu
Ao Liu
Bang Liu
Benlin Liu
Bin Liu
Bin Liu
Bing Liu
Binghao Liu
Bingyuan Liu
Bo Liu
Bo Liu
Bo Liu
Boning Liu
Bowen Liu
Boxiao Liu
Chang Liu
Chang Liu
Chang Liu
Chang Liu
Chao Liu
Chengxin Liu
Chengxu Liu
Chih-Ting Liu
Chuanjian Liu
Chun-Hao Liu
Daizong Liu
Decheng Liu
Di Liu
Difan Liu
Dong Liu
Dongnan Liu
Fang Liu
Fang Liu
Fangyi Liu
Feng Liu

Fengbei Liu
Fenglin Liu
Fengqi Liu
Furui Liu
Fuxiao Liu
Haisong Liu
Han Liu
Hanwen Liu
Hanyuan Liu
Hao Liu
Haolin Liu
Haotian Liu
Haozhe Liu
Heshan Liu
Hong Liu
Hongbin Liu
Hongfu Liu
Hongyu Liu
Hsueh-Ti Derek Liu
Huidong Liu
Isabella Liu
Ji Liu
Jia-Wei Liu
Jiachen Liu
Jiaheng Liu
Jiahui Liu
Jiaming Liu
Jiancheng Liu
Jiang Liu
Jianmeng Liu
Jiashuo Liu
Jiawei Liu
Jiawei Liu
Jiayang Liu
Jiayi Liu
Jie Liu
Jie Liu
Jie Liu
Jihao Liu
Jing Liu
Jing Liu
Jing Liu
Jingyuan Liu
Jingyuan Liu
Jiuming Liu

Jiyuan Liu
Jun Liu
Kang-Jun Liu
Kangning Liu
Kenkun Liu
Kunhao Liu
Li Liu
Lijuan Liu
Lingbo Liu
Lingqiao Liu
Liu Liu
Liyang Liu
Meng Liu
Mengchen Liu
Miao Liu
Ming Liu
Minghao Liu
Minghua Liu
Mingxuan Liu
Mingyuan Liu
Nan Liu
Nian Liu
Ning Liu
Peidong Liu
Peirong Liu
Peiye Liu
Pengju Liu
Ping Liu
Qi Liu
Qiankun Liu
Qihao Liu
Qing Liu
Qingjie Liu
Richard Liu
Risheng Liu
Rui Liu
Ruicong Liu
Ruoshi Liu
Ruyu Liu
Shaohui Liu
Shaoteng Liu
Shaowei Liu
Sheng Liu
Shenglan Liu
Shikun Liu

Shilong Liu
Shuaicheng Liu
Shuaicheng Liu
Shuming Liu
Songhua Liu
Tao Liu
Tian Yu Liu
Tianci Liu
Tianshan Liu
Tongliang Liu
Tyng-Luh Liu
Wei Liu
Weifeng Liu
Weixiao Liu
Weiyu Liu
Wen Liu
Wenxi Liu
Wenyu Liu
Wenyu Liu
Wu Liu
Xian Liu
Xianglong Liu
Xianpeng Liu
Xiao Liu
Xiaohong Liu
Xiaoyu Liu
Xiaoyu Liu
Xihui Liu
Xin Liu
Xin Liu
Xinchen Liu
Xingtong Liu
Xingyu Liu
Xinhang Liu
Xinhui Liu
Xinpeng Liu
Xinwei Liu
Xinyu Liu
Xiulong Liu
Xiyao Liu
Xu Liu
Xubo Liu
Xudong Liu
Xueting Liu
Xueyi Liu

Yan Liu
Yanbin Liu
Yang Liu
Yang Liu
Yang Liu
Yang Liu
Yang Liu
Yanwei Liu
Yaojie Liu
Ye Liu
Yi Liu
Yihao Liu
Yingcheng Liu
Yingfei Liu
Yipeng Liu
Yipeng Liu
Yixin Liu
Yizhang Liu
Yong Liu
Yong Liu
Yonghuai Liu
Yongtuo Liu
Yu Liu
Yu-Lun Liu
Yu-Shen Liu
Yuan Liu
Yuang Liu
Yuanpei Liu
Yuanpeng Liu
Yuanwei Liu
Yuchen Liu
Yuchen Liu
Yuchi Liu
Yueh-Cheng Liu
Yufan Liu
Yuhao Liu
Yuliang Liu
Yun Liu
Yun Liu
Yun Liu
Yunfan Liu
Yunfei Liu
Yunze Liu
Yupei Liu
Yuqi Liu

Yuyang Liu
Yuyuan Liu
Zhaoqiang Liu
Zhe Liu
Zhe Liu
Zhen Liu
Zheng Liu
Zhenguang Liu
Zhi Liu
Zhihua Liu
Zhijian Liu
Zhili Liu
Zhuoran Liu
Ziquan Liu
Ziyi Liu
Zuxin Liu
Zuyan Liu
Josep Llados
Ling Lo
Shao-Yuan Lo
Liliana Lo Presti
Sylvain Lobry
Yaroslava Lochman
Fotios Logothetis
Suhas Lohit
Marios Loizou
Vishnu Suresh Lokhande
Cheng Long
Chengjiang Long
Fuchen Long
Guodong Long
Rujiao Long
Shangbang Long
Teng Long
Xiaoxiao Long
Zijun Long
Ivan Lopes
Vasco Lopes
Adrian Lopez-Rodriguez
Javier Lorenzo-Navarro
Yujing Lou
Brian C. Lovell
Weng Fei Low
Changsheng Lu
Chun-Shien Lu

Daohan Lu
Dongming Lu
Erika Lu
Fan Lu
Guangming Lu
Guo Lu
Hao Lu
Hao Lu
Hongtao Lu
Jiachen Lu
Jiaxin Lu
Jiwen Lu
Lewei Lu
Liying Lu
Quanfeng Lu
Shenyu Lu
Shun Lu
Tao Lu
Xiangyong Lu
Xiankai Lu
Xin Lu
Xuanchen Lu
Xuequan Lu
Yan Lu
Yang Lu
Yanye Lu
Yawen Lu
Yifan Lu
Yongchun Lu
Yongxi Lu
Yu Lu
Yu Lu
Yuzhe Lu
Zhichao Lu
Zhihe Lu
Zijia Lu
Tianyu Luan
Pauline Luc
Simon Lucey
Timo Lüddecke
Jonathon Luiten
Jovita Lukasik
Ao Luo
Cheng Luo
Chuanchen Luo

Donghao Luo
Fangzhou Luo
Gen Luo
Gongning Luo
Hongchen Luo
Jiahao Luo
Jiebo Luo
Jinqi Luo
Jinqi Luo
Jun Luo
Katie Z. Luo
Kunming Luo
Lei Luo
Mandi Luo
Mi Luo
Ruotian Luo
Sihui Luo
Tiange Luo
Wenhan Luo
Xiao Luo
Xiaotong Luo
Xiongbiao Luo
Xu Luo
Yadan Luo
Yawei Luo
Ye Luo
Yisi Luo
Yong Luo
You-Wei Luo
Yuanjing Luo
Zelun Luo
Zhengxiong Luo
Zhengyi Luo
Zhiming Luo
Zhipeng Luo
Zhongjin Luo
Zilin Luo
Ziyang Luo
Tung M. Luu
Diogo C. Luvizon
Jun Lv
Pei Lv
Yunqiu Lv
Zhaoyang Lv
Gengyu Lyu

Jiancheng Lyu
Jipeng Lyu
Junfeng Lyu
Mengyao Lyu
Mingzhi Lyu
Weimin Lyu
Xiaoyang Lyu
Xinyu Lyu
Yiwei Lyu
Youwei Lyu
Ailong Ma
Andy J. Ma
Benteng Ma
Bingpeng Ma
Chao Ma
Chuofan Ma
Cong Ma
Cuixia Ma
Fan Ma
Fangchang Ma
Fei Ma
Guozheng Ma
Haoyu Ma
Hengbo Ma
Huimin Ma
Jiahao Ma
Jianqi Ma
Jiawei Ma
Jiayi Ma
Kai Ma
Kede Ma
Lei Ma
Li Ma
Lin Ma
Liqian Ma
Lizhuang Ma
Mengmeng Ma
Ning Ma
Qianli Ma
Rui Ma
Shijie Ma
Shiqiang Ma
Shiqing Ma
Shuailei Ma
Sizhuo Ma

Tao Ma
Teli Ma
Wenxuan Ma
Wufei Ma
Xianzheng Ma
Xiaoxuan Ma
Xinyin Ma
Xinzhu Ma
Xu Ma
Yeyao Ma
Yifeng Ma
Yuexiao Ma
Yuexin Ma
Yunsheng Ma
Zhan Ma
Zhanyu Ma
Ziping Ma
Ziqiao Ma
Muhammad Maaz
Anish Madan
Neelu Madan
Spandan Madan
Sai Advaith Maddipatla
Rishi Madhok
Filippo Maggioli
Simone Magistri
Marcus Magnor
Sabarinath Mahadevan
Shweta Mahajan
Aniruddha Mahapatra
Sarthak Kumar Maharana
Behrooz Mahasseni
Upal Mahbub
Arif Mahmood
Kaleel Mahmood
Mohammed Mahmoud
Tanvir Mahmud
Jinjie Mai
Helena de Almeida Maia
Josef Maier
Shishira R. Maiya
Snehashis Majhi
Orchid Majumder
Sagnik Majumder
Ilya Makarov

Sina Malakouti
Hashmat Shadab Malik
Mateusz Malinowski
Utkarsh Mall
Srikanth Malla
Clement Mallet
Dimitrios Mallis
Abed Malti
Yunze Man
Oscar Mañas
Karttikeya Mangalam
Fabian Manhardt
Ioannis Maniadis Metaxas
Fahim Mannan
Rafal Mantiuk
Dongxing Mao
Jiageng Mao
Wei Mao
Weian Mao
Weixin Mao
Ye Mao
Yongsen Mao
Yunyao Mao
Yuxin Mao
Zhiyuan Mao
Emanuela Marasco
Matthew Marchellus
Alberto Marchisio
Diego Marcos
Alina E. Marcu
Riccardo Marin
Manuel J. Marín-Jiménez
Octave Mariotti
Dejan Markovic
Imad Eddine Marouf
Valerio Marsocci
Diego Martin Arroyo
Ricardo Martin-Brualla
Brais Martinez
Renato Martins
Damien Martins Gomes
Tetiana Martyniuk
Pierre Marza
David Masip
Carlo Masone

Timothée Masquelier
André G. Mateus
Minesh Mathew
Yusuke Matsui
Bruce A. Maxwell
Christoph Mayer
Prasanna Mayilvahanan
Amir Mazaheri
Amrita Mazumdar
Pratik Mazumder
Alessio Mazzucchelli
Amarachi B. Mbakwe
Scott McCloskey
Naga Venkata Kartheek
    Medathati
Henry Medeiros
Guofeng Mei
Haiyang Mei
Jie Mei
Jieru Mei
Kangfu Mei
Lingjie Mei
Xiaoguang Mei
Dennis Melamed
Luke Melas-Kyriazi
Iaroslav Melekhov
Yifang Men
Ricardo A. Mendoza-León
Depu Meng
Fanqing Meng
Jingke Meng
Lingchen Meng
Qier Meng
Qingjie Meng
Quan Meng
Yanda Meng
Zibo Meng
Otniel-Bogdan Mercea
Pablo Mesejo
Safa Messaoud
Nico Messikommer
Nando Metzger
Christopher Metzler
Vasileios Mezaris
Liang Mi

Zhenxing Mi
S. Mahdi H. Miangoleh
Bo Miao
Changtao Miao
Jiaxu Miao
Zichen Miao
Bjoern Michele
Christian Micheloni
Marko Mihajlovic
Zoltán Á. Milacski
Simone Milani
Leo Milecki
Roy Miles
Christen Millerdurai
Monica Millunzi
Chaerin Min
Cheol-Hui Min
Dongbo Min
Hyun-Seok Min
Jie Min
Juhong Min
Kyle Min
Yifei Min
Yuecong Min
Zhixiang Min
Matthias Minderer
Di Ming
Qi Ming
Xiang Ming
Riccardo Miotto
Aymen Mir
Pedro Miraldo
Parsa Mirdehghan
Seyed Ehsan Mirsadeghi
Muhammad Jehanzeb
    Mirza
Ashkan Mirzaei
Dmytro Mishkin
Anand Mishra
Ashish Mishra
Samarth Mishra
Shlok K. Mishra
Diganta Misra
Abhay Mittal
Gaurav Mittal

Surbhi Mittal
Trisha Mittal
Taiki Miyanishi
Daisuke Miyazaki
Hong Mo
Kaichun Mo
Sangwoo Mo
Sicheng Mo
Sicheng Mo
Zhipeng Mo
Michael Moeller
Peyman Moghadam
Hadi Mohaghegh
    Dolatabadi
Salman Mohamadi
Mirgahney H. Mohamed
Deen Dayal Mohan
Fnu Mohbat
Satyam Mohla
Tony C. W. Mok
Liliane Momeni
Pascal Monasse
Ajoy Mondal
Anindya Mondal
Mathew Monfort
Tom Monnier
Yusuke Monno
Eduardo F. Montesuma
Gyeongsik Moon
Taesup Moon
WonJun Moon
Dror Moran
Julie R. C. Mordacq
Deeptej S. More
Arthur Moreau
Davide Morelli
Luca Morelli
Pedro Morgado
Alexandre Morgand
Henrique Morimitsu
Matteo Moro
Lia Morra
Matteo Mosconi
Ali Mosleh

Sayed Mohammad
    Mostafavi Isfahani
Saman Motamed
Chong Mou
Dana Moukheiber
Pierre Moulon
Ramy A. Mounir
Théo Moutakanni
Fangzhou Mu
Jiteng Mu
Yao Mark Mu
Manasi Muglikar
Yasuhiro Mukaigawa
Amitangshu Mukherjee
Avideep Mukherjee
Prerana Mukherjee
Tanmoy Mukherjee
Anirban Mukhopadhyay
Soumik Mukhopadhyay
Yusuke Mukuta
Ravi Teja Mullapudi
Lea Müller
Norman Müller
Chaithanya Kumar
    Mummadi
Muhammad Akhtar Munir
Subrahmanyam Murala
Sanjeev Muralikrishnan
Ana C. Murillo
Nils Murrugarra-Llerena
Mohamed Adel Musallam
Damien Muselet
Josh David Myers-Dean
Byeonghu Na
Taeyoung Na
Muhammad Ferjad Naeem
Sauradip Nag
Pravin Nagar
Rajendra Nagar
Varun Nagaraja
Tushar Nagarajan
Seungjun Nah
Shu Nakamura
Gaku Nakano
Yuta Nakashima

Kiyohiro Nakayama
Mitsuru Nakazawa
Krishna Kanth Nakka
Yuesong Nan
Karthik Nandakumar
Paolo Napoletano
Syed S. Naqvi
Dinesh Reddy
    Narapureddy
Supreeth
    Narasimhaswamy
Kartik Narayan
Sriram Narayanan
Fabio Narducci
Erickson R. Nascimento
Muzammal Naseer
Kamal Nasrollahi
Lakshmanan Nataraj
Vishwesh Nath
Avisek Naug
Alexander Naumann
K. L. Navaneet
Pablo Navarrete Michelini
Shah Nawaz
Nazir Nayal
Niv Nayman
Amin Nejatbakhsh
Negar Nejatishahidin
Reyhaneh Neshatavar
Pedro C. Neto
Lukáš Neumann
Richard Newcombe
Alejandro Newell
Evonne Ng
Kam Woh Ng
Trung T. Ngo
Tuan Duc Ngo
Anh Nguyen
Anh Duy Nguyen
Cuong Cao Nguyen
Duc Anh Nguyen
Hoang Chuong Nguyen
Huy Hong Nguyen
Khai Nguyen
Khanh-Binh Nguyen

Khanh-Duy Nguyen
Khoi Nguyen
Khoi D. Nguyen
Kiet A. Nguyen
Ngoc Cuong Nguyen
Pha Nguyen
Phi Le Nguyen
Phong Ha Nguyen
Rang Nguyen
Tam V. Nguyen
Thao Nguyen
Thuan Hoang Nguyen
Toan Tien Nguyen
Trong-Tung Nguyen
Van Nguyen Nguyen
Van-Quang Nguyen
Thuong Nguyen Canh
Thien Trang Nguyen Vu
Haomiao Ni
Jiangqun Ni
Minheng Ni
Yao Ni
Zhen-Liang Ni
Zixuan Ni
Dong Nie
Hui Nie
Jiahao Nie
Lang Nie
Liqiang Nie
Qiang Nie
Ying Nie
Yinyu Nie
Yongwei Nie
Aditya Nigam
Kshitij N. Nikhal
Nick Nikzad
Jifeng Ning
Rui Ning
Xuefei Ning
Li Niu
Muyao Niu
Shuaicheng Niu
Wei Niu
Xuesong Niu
Yi Niu

Yulei Niu
Zhenxing Niu
Zhong-Han Niu
Shohei Nobuhara
Jongyoun Noh
Junhyug Noh
Nadhira Noor
Parsa Nooralinejad
Sotiris Nousias
Tiago Novello
Gal Novich
David Novotny
Slawomir Nowaczyk
Ewa M. Nowara
Evangelos Ntavelis
Valsamis Ntouskos
Leonardo Nunes
Oren Nuriel
Zhakshylyk Nurlanov
Simbarashe Nyatsanga
Lawrence O'Gorman
Anton Obukhov
Michael Oechsle
Ferda Ofli
Changjae Oh
Dongkeun Oh
Junghun Oh
Seoung Wug Oh
Youngtaek Oh
Hiroki Ohashi
Takehiko Ohkawa
Takeshi Oishi
Takahiro Okabe
Fumio Okura
Daniel Olmeda Reino
Suguru Onda
Trevine S. J. Oorloff
Michael Opitz
Roy Or-El
Jose Oramas
Jordi Orbay
Tribhuvanesh Orekondy
Evin Pınar Örnek
Alessandro Ortis
Magnus Oskarsson

Julian Ost
Daniil Ostashev
Mayu Otani
Naima Otberdout
Hatef Otroshi Shahreza
Yassine Ouali
Amine Ouasfi
Cheng Ouyang
Wanli Ouyang
Wenqi Ouyang
Xu Ouyang
Poojan B. Oza
Milind G. Padalkar
Johannes C. Paetzold
Gautam Pai
Anwesan Pal
Simone Palazzo
Avinash Paliwal
Cristina Palmero
Chengwei Pan
Fei Pan
Hao Pan
Jianhong Pan
Junting Pan
Liang Pan
Lili Pan
Linfei Pan
Liyuan Pan
Tai-Yu Pan
Xichen Pan
Xingjia Pan
Xinyu Pan
Yingwei Pan
Zhaoying Pan
Zhihong Pan
Zixuan Pan
Zizheng Pan
Rohit Pandey
Saurabh Pandey
Bo Pang
Guansong Pang
Lu Pang
Meng Pang
Tianyu Pang
Youwei Pang

Ziqi Pang
Omiros Pantazis
Juan J. Pantrigo
Hsing-Kuo Kenneth Pao
Marina Paolanti
Joao P. Papa
Samuele S. Papa
Dim P. Papadopoulos
Symeon Papadopoulos
George Papandreou
Toufiq Parag
Chethan Parameshwara
Foivos Paraperas
 Papantoniou
Shaifali Parashar
Alejandro Pardo
Jason R. Parham
Kranti K. Parida
Rishubh Parihar
Chunghyun Park
Daehee Park
Dongmin Park
Dongwon Park
Eunbyung Park
Eunhyeok Park
Eunil Park
Geon Yeong Park
Gyeong-Moon Park
Hyoungseob Park
Jae Sung Park
JaeYoo Park
Jin-Hwi Park
Jinhyung Park
Jinyoung Park
Jongwoo Park
JoonKyu Park
JungIn Park
Junheum Park
Kiru Park
Kwanyong Park
Seongsik Park
Seulki Park
Song Park
Sungho Park
Sungjune Park

Taesung Park
Yeachan Park
Gaurav Parmar
Paritosh Parmar
Maurizio Parton
Magdalini Paschali
Vito Paolo Pastore
Or Patashnik
Gaurav Patel
Maitreya Patel
Diego Patino
Suvam Patra
Viorica Patraucean
Badri Narayana Patro
Danda Pani Paudel
Angshuman Paul
Sneha Paul
Soumava Paul
Sudipta Paul
Sujoy Paul
Rémi Pautrat
Ioannis Pavlidis
Svetlana Pavlitska
Raju Pavuluri
Kim Steenstrup Pedersen
Marco Pedersoli
Adithya Pediredla
Pieter Peers
Jiju Peethambaran
Sen Pei
Wenjie Pei
Yuru Pei
Simone Alberto Peirone
Chantal Pellegrini
Latha Pemula
Abhirama Subramanyam
 V. B. Penamakuri
Adrian Penate-Sanchez
Baoyun Peng
Bo Peng
Can Peng
Cheng Peng
Chi-Han Peng
Chunlei Peng
Jie Peng

Jingliang Peng
Kebin Peng
Kunyu Peng
Liang Peng
Liangzu Peng
Pai Peng
Peixi Peng
Sida Peng
Songyou Peng
Wei Peng
Wen-Hsiao Peng
Xi Peng
Xiaojiang Peng
Yi-Xing Peng
Yuxin Peng
Zhiliang Peng
Ziqiao Peng
Matteo Pennisi
Or Perel
Gabriel Perez
Gustavo Perez
Juan C. Perez
Andres Felipe Perez
 Murcia
Eduardo Pérez-Pellitero
Neehar Peri
Skand Peri
Gabriel J. Perin
Federico Pernici
Chiara Pero
Elia Peruzzo
Marco Pesavento
Dmitry M. Petrov
Ilya A. Petrov
Mathis Petrovich
Vitali Petsiuk
Tomas Pevny
Shubham Milind Phal
Chau Pham
Hai X. Pham
Khoi Pham
Long Hoang Pham
Trong Thang Pham
Trung X. Pham
Tung Pham

Hoang Phan
Huy Phan
Minh Hieu Phan
Julien Philip
Stephen Phillips
Cheng Perng Phoo
Hao Phung
Shruti S. Phutke
Weiguo Pian
Yongri Piao
Luigi Piccinelli
A. J. Piergiovanni
Sara Pieri
Vipin Pillai
Wu Pingyu
Silvia L. Pintea
Francesco Pinto
Maura Pintor
Giovanni Pintore
Vittorio Pippi
Robinson Piramuthu
Fiora Pirri
Leonid Pishchulin
Francesca Pistilli
Francesco Pittaluga
Fabio Pizzati
Edward Pizzi
Benjamin Planche
Iuliia Pliushch
Chiara Plizzari
Ryan Po
GIovanni Poggi
Matteo Poggi
Kilian Pohl
Chandradeep Pokhariya
Ashwini Pokle
Matteo Polsinelli
Adrian Popescu
Teodora Popordanoska
Nikola Popović
Ronald Poppe
Samuele Poppi
Andrea Porfiri Dal Cin
Angelo Porrello
Pedro Porto Buarque de Gusmão
Rudra P. K. Poudel
Kossar Pourahmadi Meibodi
Hadi Pouransari
Ali Pourramezan Fard
Omid Poursaeed
Anish J. Prabhu
Mihir Prabhudesai
Aayush Prakash
Aditya Prakash
Shraman Pramanick
Mantini Pranav
B. H. Pawan Prasad
Meghshyam Prasad
Prateek Prasanna
Ekta Prashnani
Bardh Prenkaj
Derek S. Prijatelj
Véronique Prinet
Malte Prinzler
Victor Adrian Prisacariu
Federica Proietto Salanitri
Sergey Prokudin
Bill Psomas
Dongqi Pu
Mengyang Pu
Nan Pu
Shi Pu
Rita Pucci
Kuldeep Purohit
Senthil Purushwalkam
Waqas A. Qazi
Charles R. Qi
Chenyang Qi
Haozhi Qi
Jiaxin Qi
Lei Qi
Mengshi Qi
Peng Qi
Xianbiao Qi
Xiangyu Qi
Yuankai Qi
Zhangyang Qi
Guocheng Qian
Hangwei Qian
Jianing Qian
Qi Qian
Rui Qian
Shengsheng Qian
Shengyi Qian
Shenhan Qian
Wen Qian
Xuelin Qian
Yaguan Qian
Yijun Qian
Yiming Qian
Zhenxing Qian
Wenwen Qiang
Feng Qiao
Fengchun Qiao
Xiaotian Qiao
Yanyuan Qiao
Yi-Ling Qiao
Yu Qiao
Hangyu Qin
Haotong Qin
Jie Qin
Peiwu Qin
Siyang Qin
Wenda Qin
Xuebin Qin
Xugong Qin
Yang Qin
Yipeng Qin
Yongqiang Qin
Yuzhe Qin
Zequn Qin
Zeyu Qin
Zheng Qin
Zhenyue Qin
Ziheng Qin
Jiaxin Qing
Congpei Qiu
Haibo Qiu
Hang Qiu
Heqian Qiu
Jiayan Qiu
Jielin Qiu

Longtian Qiu
Mufan Qiu
Ri-Zhao Qiu
Weichao Qiu
Xuchong Qiu
Xuerui Qiu
Yuda Qiu
Yuheng Qiu
Zhongxi Qiu
Maan Qraitem
Chao Qu
Linhao Qu
Yanyun Qu
Kha Gia Quach
Ruijie Quan
Fabio Quattrini
Yvain Queau
Faisal Z. Qureshi
Rizwan Qureshi
Hamid R. Rabiee
Paolo Rabino
Ryan L. Rabinowitz
Petia Radeva
Bhaktipriya Radharapu
Krystian Radlak
Bodgan Raducanu
M. Usman Rafique
Francesco Ragusa
Sahar Rahimi Malakshan
Tanzila Rahman
Aashish Rai
Arushi Rai
Shyam Nandan Rai
Zobeir Raisi
Amit Raj
Kiran Raja
Sachin Raja
Deepu Rajan
Jathushan Rajasegaran
Gnana Praveen Rajasekhar
Ramanathan Rajendiran
Marie-Julie Rakotosaona
Gorthi Rama Krishna Sai
    Subrahmanyam
Sai Niranjan
    Ramachandran
Santhosh Kumar
    Ramakrishnan
Srikumar Ramalingam
Michaël Ramamonjisoa
Ravi Ramamoorthi
Shanmuganathan Raman
Mani Ramanagopal
Ashish Ramayee Asokan
Andrea Ramazzina
Jason Rambach
Sai Saketh Rambhatla
Sai Saketh Rambhatla
Clément Rambour
Francois Bernard Julien
    Rameau
Visvanathan Ramesh
Adín Ramírez Rivera
Haoxi Ran
Xuming Ran
Aakanksha Rana
Srinivas Rana
Kanchana N. Ranasinghe
Poorva G. Rane
Aneesh Rangnekar
Harsh Rangwani
Viresh Ranjan
Anyi Rao
Sukrut Rao
Yongming Rao
ZhiBo Rao
Carolina Raposo
Hanoona Abdul Rasheed
Amir Rasouli
Deevashwer Rathee
Christian Rathgeb
Avinash Ravichandran
Bharadwaj Ravichandran
Arijit Ray
Dripta S. Raychaudhuri
Sonia Raychaudhuri
Haziq Razali
Daniel Rebain
William T. Redman
Albert W. Reed
Aniket Rege
Christoph Reich
Christian Reimers
Simon Reiß
Konstantinos Rematas
Tal Remez
Davis Rempe
Bin Ren
Chao Ren
Chuan-Xian Ren
Dayong Ren
Dongwei Ren
Jiawei Ren
Jiaxiang Ren
Jing Ren
Mengwei Ren
Pengfei Ren
Pengzhen Ren
Qibing Ren
Shuhuai Ren
Sucheng Ren
Tianhe Ren
Weihong Ren
Wenqi Ren
Xuanchi Ren
Yanli Ren
Yihui Ren
Yixuan Ren
Yufan Ren
Zhenwen Ren
Zhihang Ren
Zhiyuan Ren
Zhongzheng Ren
Jose Restom
George Retsinas
Ambareesh Revanur
Ferdinand Rewicki
Manuel Rey Area
Md Alimoor Reza
Farnoush Rezaei Jafari
Hamed Rezazadegan
    Tavakoli
Rafael S. Rezende
Wonjong Rhee

Anthony D. Rhodes
Daniel Riccio
Alexander Richard
Christian Richardt
Luca Rigazio
Benjamin Risse
Dominik Rivoir
Luigi Riz
Mamshad Nayeem Rizve
Antonino M. Rizzo
Wes J. Robbins
Damien Robert
Jonathan Roberts
Joseph Robinson
Antonio Robles-Kelly
Mrigank Rochan
Chris Rockwell
Chris Rockwell
Ivan Rodin
Erik Rodner
Ranga Rodrigo
Andres C. Rodriguez
Cristian Rodriguez
Carlos Rodriguez-Pardo
Antonio J.
   Rodriguez-Sanchez
Barbara Roessle
Paul Roetzer
Alina Roitberg
Javier Romero
Meitar Ronen
Keran Rong
Xuejian Rong
Yu Rong
Marco Rosano
Bodo Rosenhahn
Gabriele Rosi
Candace Ross
Andreas Rössler
Giulio Rossolini
Mohammad Rostami
Edward Rosten
Daniel Roth
Karsten Roth
Mark S. Rothermel

Matthias Rottmann
Anastasios Roussos
Aniket Roy
Anirban Roy
Debaditya Roy
Shuvendu Roy
Sudipta Roy
Ahana Roy Choudhury
Amit Roy-Chowdhury
Aruni RoyChowdhury
Dávid Rozenberszki
Denys Rozumnyi
Lixiang Ru
Lingyan Ruan
Shulan Ruan
Viktor Rudnev
Daniel Rueckert
Nataniel Ruiz
Ewelina Rupnik
Evgenia Rusak
Chris Russell
Marc Rußwurm
Fiona Ryan
Dawid Damian Rymarczyk
DongHun Ryu
Sari Saba-Sadiya
Robert Sablatnig
Mohammad Sabokrou
Ragav Sachdeva
Ali Sadeghian
Arka Sadhu
Sadra Safadoust
Bardia Safaei
Ryusuke Sagawa
Avishkar Saha
Gobinda Saha
Oindrila Saha
Aditya Sahdev
Lakshmi Babu Saheer
Aadarsh Sahoo
Pritish Sahu
Aneeshan Sain
Nirat Saini
Saurabh Saini
Kuniaki Saito

Shunsuke Saito
Rahul Sajnani
Fumihiko Sakaue
Parikshit V. Sakurikar
Riccardo Salami
Soorena Salari
Mohammadreza Salehi
Leonard Salewski
Driton Salihu
Benjamin Salmon
Cristiano Saltori
Joel Saltz
Tim Salzmann
Sina Samangooei
Babak Samari
Nermin Samet
Fawaz Sammani
Leo Sampaio Ferraz
   Ribeiro
Shailaja Keyur Sampat
Alessio Sampieri
Jorge Sanchez
Pedro Sandoval-Segura
Nong Sang
Shengtian Sang
Patsorn Sangkloy
Depanshu Sani
Juan C. Sanmiguel
Hiroaki Santo
Joshua Santoso
Bikash Santra
Soubhik Sanyal
Hitesh Sapkota
Ayush Saraf
Nikolaos Sarafianos
István Sárándi
Kyle Sargent
Andranik Sargsyan
Josip Šarić
Mert Bulent Sariyildiz
Abhijit Sarkar
Anirban Sarkar
Chayan Sarkar
Michel Sarkis
Paul-Edouard Sarlin

Sara Sarto
Josua Sassen
Srikumar Sastry
Imari Sato
Takami Sato
Shin'ichi Satoh
Ravi Kumar Satzoda
Jack Saunders
Corentin Sautier
Mattia Savardi
Bogdan Savchynskyy
Mohamed Sayed
Marin Scalbert
Gianluca Scarpellini
Gerald Schaefer
Guilherme G. Schardong
David Schinagl
Phillip Schniter
Patrick Schramowski
Matthias Schubert
Peter Schüffler
Samuel Schulter
René Schuster
Klamer Schutte
Luca Scofano
Jesse Scott
Marcel Seelbach Benkner
Karthik Seemakurthy
Mattia Segù
Santi Seguí
Sinisa Segvic
Constantin Marc Seibold
Roman Seidel
Lorenzo Seidenari
Taiki Sekii
Yusuke Sekikawa
Matan Sela
Pratheba Selvaraju
Agniva Sengupta
Ahyun Seo
Jinhwan Seo
Junyoung Seo
Kwanggyoon Seo
Seonguk Seo
Seunghyeon Seo

Jinseok Seol
Hongje Seong
Ana F. Sequeira
Dario Serez
Dario Serez
David Serrano-Lozano
Pratinav Seth
Francesco Setti
Giorgos Sfikas
Mohammad Amin Shabani
Faisal Shafait
Anshul Shah
Chintan Shah
Jay Shah
Ketul Shah
Mubarak Shah
Viraj Shah
Mohamad Shahbazi
Muhammad Bilal B. Shaikh
Abdelrahman M. Shaker
Greg Shakhnarovich
Md Salman Shamil
Fahad Shamshad
Caifeng Shan
Dandan Shan
Hongming Shan
Xiaojun Shan
Chong Shang
Fanhua Shang
Jinghuan Shang
Lei Shang
Sifeng Shang
Wei Shang
Yuzhang Shang
Yuzhang Shang
Sukrit Shankar
Dian Shao
Mingwen Shao
Rui Shao
Ruizhi Shao
Shuai Shao
Shuwei Shao
Ron A. Shapira Weber
S. M. A. Sharif

Aashish Sharma
Avinash Sharma
Charu Sharma
Prafull Sharma
Prasen Kumar Sharma
Allam Shehata
Mark Sheinin
Sumit Shekhar
Oleksandr Shekhovtsov
Chuanfu Shen
Fei Shen
Fengyi Shen
Furao Shen
Hui-liang Shen
Jiajun Shen
Jianghao Shen
Jiangrong Shen
Jiayi Shen
Li Shen
Li-Yong Shen
Linlin Shen
Maying Shen
Qiu Shen
Qiuhong Shen
Shuai Shen
Shuhan Shen
Siqi Shen
Tianwei Shen
Tong Shen
Xiaolong Shen
Xiaoqian Shen
Yan Shen
Yanqing Shen
Yilin Shen
Ying Shen
Yiqing Shen
Yuan Shen
Yucong Shen
Yuhan Shen
Yunhang Shen
Zehong Shen
Zengming Shen
Zhijie Shen
Zhiqiang Shen
Hualian Sheng

Tao Sheng
Yichen Sheng
Zehua Sheng
Shivanand Venkanna Sheshappanavar
Ivaxi Sheth
Baoguang Shi
Botian Shi
Dachuan Shi
Daqian Shi
Haizhou Shi
Hengcan Shi
Jia Shi
Jing Shi
Jingang Shi
QingHongYa Shi
Ruoxi Shi
Tianyang Shi
Weishi Shi
Wu Shi
Wuxuan Shi
Xiaodan Shi
Xiaoshuang Shi
Xiaoyu Shi
Xingjian Shi
Xinyu Shi
Xuepeng Shi
Yichun Shi
Yujiao Shi
Zhenbo Shi
Zheng Shi
Zhensheng Shi
Zhenwei Shi
Zhihao Shi
Zifan Shi
Takashi Shibata
Meng-Li Shih
Yichang Shih
Dongseok Shim
Wataru Shimoda
Ilan Shimshoni
Changha Shin
Gyungin Shin
Hyungseob Shin
Inkyu Shin

Seungjoo Shin
Ukcheol Shin
Yooju Shin
Young Min Shin
Koichi Shinoda
Kaede Shiohara
Suprosanna Shit
Palaiahnakote Shivakumara
Sindi Shkodrani
Michal Shlapentokh-Rothman
Debaditya Shome
Hyounguk Shon
Sulabh Shrestha
Aman Shrivastava
Ayush Shrivastava
Gaurav Shrivastava
Aleksandar Shtedritski
Dong Wook Shu
Han Shu
Jun Shu
Xiangbo Shu
Xiujun Shu
Yang Shu
Bing Shuai
Hong-Han Shuai
Qing Shuai
Changjian Shui
Pushkar Shukla
Mustafa Shukor
Hubert P. H. Shum
Nina Shvetsova
Chenyang Si
Jianlou Si
Zilin Si
Mennatullah Siam
Sven Sickert
Désiré Sidibé
Ioannis Siglidis
Alberto Signoroni
Karan Sikka
Pedro Silva
Julio Silva-Rodríguez
Hyeonjun Sim

Jae-Young Sim
Chonghao Sima
Christian Simon
Martin Simon
Alessandro Simoni
Enis Simsar
Abhishek Singh
Apoorv Singh
Ashish Singh
Bharat Singh
Jasdeep Singh
Jaskirat Singh
Krishnakant Singh
Manish Kumar Singh
Mannat Singh
Nikhil Singh
Pravendra Singh
Rajat Vikram Singh
Simranjit Singh
Darshan Singh S.
Utkarsh Singhal
Dipika Singhania
Vasu Singla
Abhishek Kumar Sinha
Animesh Sinha
Sanjana Sinha
Saptarshi Sinha
Sudipta Sinha
Sophia A. Sirko-Galouchenko
Josef Sivic
Elena Sizikova
Geri Skenderi
Gregory Slabaugh
Habib Slim
Dmitriy Smirnov
James S. Smith
William Smith
Noah Snavely
Kihyuk Sohn
Bolivar E. Solarte
Mattia Soldan
Sobhan Soleymani
Samik Some
Nagabhushan Somraj

Jeany Son
Seung Woo Son
Byung Cheol Song
Chen Song
Guanglu Song
Jie Song
Jifei Song
Li Song
Liangchen Song
Lin Song
Luchuan Song
Mingli Song
Ran Song
Sibo Song
Sifan Song
Siyang Song
Weilian Song
Weinan Song
Wenfeng Song
Xiangchen Song
Xibin Song
Xinhang Song
Yafei Song
Yang Song
Yi-Yang Song
Yizhi Song
Yue Song
Zeen Song
Zhenbo Song
Zikai Song
Ekta Sood
Tomáš Souček
Rajiv Soundararajan
Albin Soutif-Cormerais
Jeremy Speth
Indro Spinelli
Jon Sporring
Manogna Sreenivas
Arvind Krishna Sridhar
Deepak Sridhar
Balaji Vasan Srinivasan
Pratul Srinivasan
Anuj Srivastava
Astitva Srivastava
Dhruv Srivastava
Koushik Srivatsan
Pierre-Luc St-Charles
Ioannis Stamos
Anastasis Stathopoulos
Colton Stearns
Jan Steinbrener
Jan-Martin O. Steitz
Sinisa Stekovic
Federico Stella
Michael Stengel
Alexandros Stergiou
Gleb Sterkin
Rainer Stiefelhagen
Noah Stier
Timo N. Stoffregen
Vladan Stojnić
Nick O. Stracke
Ombretta Strafforello
Julian Straub
Nicola Strisciuglio
Vitomir Struc
Yannick Strümpler
Joerg Stueckler
Chi Su
Hang Su
Hang Su
Kun Su
Rui Su
Shaolin Su
Sitong Su
Xingzhe Su
Xiu Su
Yao Su
Yiyang Su
Yongyi Su
Zhaoqi Su
Zhixun Su
Zhuo Su
Zhuo Su
Iago Suárez
Arulkumar Subramaniam
Sanjay Subramanian
A. Subramanyam
Swathikiran Sudhakaran
Yusuke Sugano
Masanori Suganuma
Yumin Suh
Mohammed Suhail
Xiuchao Sui
Yang Sui
Yao Sui
Heung-Il Suk
Pavel Suma
Baigui Sun
Baochen Sun
Bin Sun
Bo Sun
Changchang Sun
Che Sun
Cheng Sun
Chong Sun
Chunyi Sun
Gan Sun
Guofei Sun
Guoxing Sun
Haifeng Sun
Hanqing Sun
Haoliang Sun
He Sun
Heming Sun
Hongbin Sun
Huiming Sun
Jennifer J. Sun
Jian Sun
Jiande Sun
Jianhua Sun
Jiankai Sun
Jipeng Sun
Keqiang Sun
Lei Sun
Lichao Sun
Long Sun
Mingjie Sun
Peize Sun
Pengzhan Sun
Qiyue Sun
Shangquan Sun
Shanlin Sun
Shuyang Sun
Tao Sun

Tiancheng Sun
Wei Sun
Weiwei Sun
Weixuan Sun
Xianfang Sun
Xiaohang Sun
Xiaoshuai Sun
Xiaoxiao Sun
Ximeng Sun
Xuxiang Sun
Yanan Sun
Yasheng Sun
Yihong Sun
Ying Sun
Yixuan Sun
Yu Sun
Yuan Sun
Yuchong Sun
Zeren Sun
Zhanghao Sun
Zhaodong Sun
Zhaohui H. Sun
Zhicheng Sun
Zhicheng Sun
Haomiao Sun
Varun Sundar
Shobhita Sundaram
Minhyuk Sung
Kalyan Sunkavalli
Yucheng Suo
Indranil Sur
Saksham Suri
Naufal Suryanto
Vadim Sushko
David Suter
Roman Suvorov
Fnu Suya
Teppei Suzuki
Kunal Swami
Archana Swaminathan
Gurumurthy Swaminathan
Robin Swanson
Eran Swears
Alexander Swerdlow
Sirnam Swetha
Tabish A. Syed
Tanveer Syeda-Mahmood
Stanislaw K. Szymanowicz
Sethuraman T. V.
Calvin-Khang T. Ta
The-Anh Ta
Babak Taati
Samy Tafasca
Andrea Tagliasacchi
Haowei Tai
Yuan Tai
Francesco Taioli
Peng Taiying
Keita Takahashi
Naoya Takahashi
Jun Takamatsu
Nicolas Talabot
Hugues G. Talbot
Hossein Talebi
Davide Talon
Gary Tam
Toru Tamaki
Dipesh Tamboli
Andong Tan
Bin Tan
Cheng Tan
David Joseph New Tan
Fuwen Tan
Guang Tan
Jianchao Tan
Jing Tan
Jingru Tan
Lei Tan
Mingkui Tan
Mingxing Tan
Shuhan Tan
Shunquan Tan
Weimin Tan
Xin Tan
Zhentao Tan
Zhentao Tan
Masayuki Tanaka
Chen Tang
Chengzhou Tang
Chenwei Tang
Fan Tang
Feng Tang
Hao Tang
Haoran Tang
Jiajun Tang
Jiapeng Tang
Jiaxiang Tang
Jie Tang
Junshu Tang
Keke Tang
Luming Tang
Luyao Tang
Lv Tang
Ming Tang
Quan Tang
Shengji Tang
Sheyang Tang
Shitao Tang
Shixiang Tang
Tao Tang
Weixuan Tang
Xu Tang
Yang Tang
Yansong Tang
Yehui Tang
Yu-Ming Tang
Zheng Tang
Zhipeng Tang
Zitian Tang
Md Mehrab Tanjim
Julian Tanke
An Tao
Chaofan Tao
Chenxin Tao
Jiale Tao
Junli Tao
Keda Tao
Ming Tao
Ran Tao
Wenbing Tao
Xinhao Tao
Jean-Philippe G. Tarel
Laia Tarres
Laia Tarrés
Enzo Tartaglione

Keisuke Tateno
SaiKiran K. Tedla
Antonio Tejero-de-Pablos
Bugra Tekin
Purva Tendulkar
Minggui Teng
Ruwan Tennakoon
Andrew Beng Jin Teoh
Konstantinos Tertikas
Piotr Teterwak
Piotr Teterwak
Anh Thai
Kartik Thakral
Nupur Thakur
Sadbhawna Thakur
Balamurugan Thambiraja
Vikas Thamizharasan
Kevin Thandiackal
Sushil Thapa
Daksh Thapar
Jonas Theiner
Christian Theobalt
Spyridon Thermos
Fida Mohammad Thoker
Christopher L. Thomas
Diego Thomas
William Thong
Mamatha Thota
Mukund Varma
  Thottankara
Changyao Tian
Chunwei Tian
Jinyu Tian
Kai Tian
Lin Tian
Tai-Peng Tian
Xin Tian
Xinyu Tian
Yapeng Tian
Yu Tian
Yuan Tian
Yuesong Tian
Yunjie Tian
Yuxin Tian
Zhuotao Tian

Mert Tiftikci
Javier Tirado-Garín
Garvita Tiwari
Lokender Tiwari
Anastasia Tkach
Andrea Toaiari
Sinisa Todorovic
Pavel Tokmakov
Tri Ton
Adam Tonderski
Jinguang Tong
Peter Tong
Xin Tong
Zhan Tong
Francesco Tonini
Alessio Tonioni
Alessandro Torcinovich
Marwan Torki
Lorenzo Torresani
Fabio Tosi
Matteo Toso
Anh T. Tran
Hung Tran
Linh-Tam Tran
Minh-Triet Tran
Ngoc-Trung Tran
Phong Tran
Jonathan Tremblay
Alex Trevithick
Aditay Tripathi
Subarna Tripathi
Felix Tristram
Gabriele Trivigno
Emanuele Trucco
Prune Truong
Thanh-Dat Truong
Tomasz Trzcinski
Fu-Jen Tsai
Yu-Ju Tsai
Michael Tschannen
Tze Ho Elden Tse
Ethan Tseng
Yu-Chee Tseng
Shahar Tsiper
Hanzhang Tu

Rong-Cheng Tu
Yuanpeng Tu
Zhengzhong Tu
Zhigang Tu
Narek Tumanyan
Anil Osman Tur
Haithem Turki
Mehmet Ozgur Turkoglu
Daniyar Turmukhambetov
Victor G. Turrisi da Costa
Tinne Tuytelaars
Bartlomiej Twardowski
Radim Tylecek
Christos Tzelepis
Seiichi Uchida
Hideaki Uchiyama
Vishaal Udandarao
Mostofa Rafid Uddin
Kohei Uehara
Tatsumi Uezato
Nicolas Ugrinovic
Youngjung Uh
Norimichi Ukita
Amin Ullah
Markus Ulrich
Ardian Umam
Mesut Erhan Unal
Mathias Unberath
Devesh Upadhyay
Paul Upchurch
Shagun Uppal
Yoshitaka Ushiku
Anil Usumezbas
Yuzuko Utsumi
Roy Uziel
Anil Vadathya
Sharvaree Vadgama
Pratik Vaishnavi
Gregory Vaksman
Matias A. Valdenegro Toro
Lucas Valença
Eduardo Valle
Ernest Valveny
Laurens van der Maaten
Wouter Van Gansbeke

Nanne van Noord
Max W. F. van Spengler
Lorenzo Vaquero
Farshid Varno
Cristina Vasconcelos
Francisco Vasconcelos
Igor Vasiljevic
Florin-Alexandru
 Vasluianu
Subeesh Vasu
Arun Balajee Vasudevan
Vaibhav S. Vavilala
Kyle Vedder
Vijay Veerabadran
Ronny Xavier Velastegui
 Sandoval
Senem Velipasalar
Andreas Velten
Raviteja Vemulapalli
Deepika Vemuri
Edward Vendrow
Jonathan Ventura
Lucas Ventura
Jakob Verbeek
Dor Verbin
Eshan Verma
Manisha Verma
Monu Verma
Sahil Verma
Constantin Vertan
Eli Verwimp
Noranart Vesdapunt
Jordan J. Vice
Sara Vicente
Kavisha Vidanapathirana
Dat Viet Thanh Nguyen
Sudheendra
 Vijayanarasimhan
Sujal T. Vijayaraghavan
Deepak Vijaykeerthy
Elliot Vincent
Yael Vinker
Duc Minh Vo
Huy V. Vo
Khoa H. V. Vo

Romain Vo
Antonin Vobecky
Michele Volpi
Riccardo Volpi
Igor Vozniak
Nicholas Vretos
Vibashan V. S.
Ngoc-Son Vu
Tuan-Anh Vu
Khiem Vuong
Mårten Wadenbäck
Neal Wadhwa
Sophia J. Wagner
Muntasir Wahed
Nobuhiko Wakai
Devesh Walawalkar
Jacob Walker
Matthew Walmer
Matthew R. Walter
Bo Wan
Guancheng Wan
Jia Wan
Jin Wan
Jun Wan
Qiyang Wan
Renjie Wan
Wei Wan
Xingchen Wan
Yecong Wan
Zhexiong Wan
Ziyu Wan
Karan Wanchoo
Alex Jinpeng Wang
Angtian Wang
Baoyuan Wang
Benyou Wang
Biao Wang
Bin Wang
Bing Wang
Binghui Wang
Binglu Wang
Can Wang
Ce Wang
Changwei Wang
Chao Wang

Chaoyang Wang
Chen Wang
Chen Wang
Chen Wang
Chengrui Wang
Chien-Yao Wang
Chu Wang
Chuan Wang
Congli Wang
Dadong Wang
Di Wang
Dong Wang
Dong Wang
Dongdong Wang
Dongkai Wang
Dongqing Wang
Dongsheng Wang
X. Wang
Fan Wang
Fangfang Wang
Fangjinhua Wang
Fei Wang
Feng Wang
Feng Wang
Fu-Yun Wang
Gaoang Wang
Guangcong Wang
Guangming Wang
Guangrun Wang
Guangzhi Wang
Guanshuo Wang
Guo-Hua Wang
Guoqing Wang
Guoqing Wang
Haixin Wang
Haiyan Wang
Han Wang
Hanjing Wang
Hanyu Wang
Hao Wang
Hao Wang
Haobo Wang
Haochen Wang
Haochen Wang
Haohan Wang

Haoqi Wang
Haoran Wang
Haotao Wang
Haoxuan Wang
HaoYu Wang
Hengkang Wang
Hengli Wang
Hengyi Wang
Hesheng Wang
Hong Wang
Hongjun Wang
Hongxiao Wang
Hongyu Wang
Hongzhi Wang
Hua Wang
Huafeng Wang
Huan Wang
Huijie Wang
Huiyu Wang
Jiadong Wang
Jiahao Wang
Jiahao Wang
Jiahao Wang
Jiakai Wang
Jialiang Wang
Jiamian Wang
Jian Wang
Jiang Wang
Jiangliu Wang
Jianjia Wang
Jianyi Wang
Jianyuan Wang
Jiaqi Wang
Jiashun Wang
Jiayi Wang
Jiaze Wang
Jin Wang
Jinfeng Wang
Jingbo Wang
Jinghua Wang
Jingkang Wang
Jinglong Wang
Jinglu Wang
Jinpeng Wang
Jinqiao Wang

Jue Wang
Jun Wang
Jun Wang
Junjue Wang
Junke Wang
Junxiao Wang
Kai Wang
Kai Wang
Kai Wang
Kai Wang
Kaihong Wang
Kewei Wang
Keyan Wang
Kun Wang
Kup Wang
Lan Wang
Lanjun Wang
Le Wang
Lei Wang
Lei Wang
Lezi Wang
Liansheng Wang
Liao Wang
Lijuan Wang
Lijun Wang
Limin Wang
Lin Wang
Linwei Wang
Lishun Wang
Lixu Wang
Liyuan Wang
Lizhen Wang
Lizhi Wang
Longguang Wang
Luozhou Wang
Luting Wang
Mang Wang
Manning Wang
Mei Wang
Mengjiao Wang
Mengmeng Wang
Miaohui Wang
Min Wang
Naiyan Wang
Nannan Wang

Ning-Hsu Wang
Pei Wang
Peihao Wang
Peiqi Wang
Peng Wang
Pengfei Wang
Pengkun Wang
Pichao Wang
Pu Wang
Qi Wang
Qian Wang
Qiang Wang
Qiang Wang
Qiangchang Wang
Qianqian Wang
Qifei Wang
Qilong Wang
Qin Wang
Qing Wang
Qingzhong Wang
Qitong Wang
Qiufeng Wang
Ronggang Wang
Rui Wang
Rui Wang
Rui Wang
Ruibin Wang
Ruisheng Wang
Ruoyu Wang
Sai Wang
Sen Wang
Sen Wang
Shan Wang
Shaoru Wang
Sheng Wang
Sheng-Yu Wang
Shengze Wang
Shida Wang
Shijie Wang
Shipeng Wang
Shiping Wang
Shiyu Wang
Shizun Wang
Shuhui Wang
Shujun Wang

Shunli Wang
Shunxin Wang
Shuo Wang
Shuo Wang
Shuo Wang
Shuzhe Wang
Siqi Wang
Siwei Wang
Song Wang
Song Wang
Su Wang
Tan Wang
Tao Wang
Taoyue Wang
Teng Wang
Tengfei Wang
Tiancai Wang
Tianqi Wang
Tianyang Wang
Tianyu Wang
Tong Wang
Tsun-Hsuan Wang
Tuanfeng Wang
Tuanfeng Y. Wang
Wei Wang
Weihan Wang
Weikang Wang
Weimin Wang
Weiqiang Wang
Weixi Wang
Weiyao Wang
Weiyun Wang
Wen Wang
Wenbin Wang
Wenhao Wang
Wenjing Wang
Wenqian Wang
Wentao Wang
Wenxiao Wang
Wenxuan Wang
Wenzhe Wang
Xi Wang
Xi Wang
Xiang Wang
Xiao Wang

Xiao Wang
Xiao Wang
Xiaobing Wang
Xiaofeng Wang
Xiaohan Wang
Xiaosen Wang
Xiaosong Wang
Xiaoxing Wang
Xiaoyang Wang
Xijun Wang
Xijun Wang
Xinggang Wang
Xinghan Wang
Xinjiang Wang
Xinshao Wang
Xintong Wang
Xizi Wang
Xu Wang
Xuan Wang
Xuanhan Wang
Xue Wang
Xueping Wang
Xuyang Wang
Yali Wang
Yan Wang
Yan Wang
Yang Wang
Yangang Wang
Yangtao Wang
Yaohui Wang
Yaoming Wang
Yaxing Wang
Yaxiong Wang
Yi Wang
Yi Ru Wang
Yidong Wang
Yifan Wang
Yifeng Wang
Yifu Wang
Yikai Wang
Yilin Wang
Yilun Wang
Yin Wang
Yinggui Wang
Yingheng Wang

Yingqian Wang
Yipei Wang
Yiqun Wang
Yiran Wang
Yiwei Wang
Yixu Wang
Yizhi Wang
Yizhou Wang
Yizhou Wang
Yong Wang
Yu Wang
Yu-Shuen Wang
Yuan-Gen Wang
Yuchen Wang
Yude Wang
Yue Wang
Yuesong Wang
Yufei Wang
Yufu Wang
Yuguang Wang
Yuhan Wang
Yujia Wang
Yulin Wang
Yunke Wang
Yuting Wang
Yuxi Wang
YuXin Wang
Yuzheng Wang
Ze Wang
Zedong Wang
Zehan Wang
Zengmao Wang
Zeyu Wang
Zeyu Wang
Zhao Wang
Zhaokai Wang
Zhaowen Wang
Zhe Wang
Zhen Wang
Zhen Wang
Zhendong Wang
Zheng Wang
Zheng Wang
Zheng Wang
Zhengyi Wang

Zhennan Wang
Zhenting Wang
Zhenyi Wang
Zhenyu Wang
Zhenzhi Wang
Zhepeng Wang
Zhi Wang
Zhibo Wang
Zhihao Wang
Zhihui Wang
Zhijie Wang
Zhikang Wang
Zhixiang Wang
Zhiyong Wang
Zhongdao Wang
Zhonghao Wang
Zhouxia Wang
Zhu Wang
Zian Wang
Zifu Wang
Zihao Wang
Zijian Wang
Ziqiang Wang
Ziqin Wang
Ziqing Wang
Zirui Wang
Zirui Wang
Ziwei Wang
Ziyan Wang
Ziyang Wang
Ziyi Wang
Ziyun Wang
Frederik Warburg
Syed Talal Wasim
Daniel Watson
Jamie Watson
Ethan Weber
Silvan Weder
Jan Dirk Wegner
Chen Wei
Donglai Wei
Fangyin Wei
Fangyun Wei
Guoqiang Wei
Jia Wei
Jiacheng Wei
Kaixuan Wei
Kun Wei
Longhui Wei
Megan Wei
Mian Wei
Mingqiang Wei
Pengxu Wei
Ping Wei
Qiuhong Anna Wei
Shikui Wei
Tianyi Wei
Wei Wei
Wenqi Wei
Xian Wei
Xin Wei
Xing Wei
Xinyue Wei
Xiu-Shen Wei
Yi Wei
Yixuan Wei
Yunchao Wei
Yuxiang Wei
Yuxiang Wei
Zeming Wei
Zhipeng Wei
Zihao Wei
Zimian Wei
Jean-Baptiste Weibel
Luca Weihs
Martin Weinmann
Michael Weinmann
Bihan Wen
Bowen Wen
Chao Wen
Chenglu Wen
Chuan Wen
Congcong Wen
Jie Wen
Jing Wen
Qiang Wen
Rui Wen
Sijia Wen
Song Wen
Xiang Wen
Xin Wen
Yilin Wen
Youpeng Wen
Yuanbo Wen
Yuxin Wen
Chung-Yi Weng
Junwu Weng
Shuchen Weng
Wenming Weng
Yijia Weng
Zhenzhen Weng
Thomas Westfechtel
Christopher Johannes Wewer
Spencer Whitehead
Tobias Jan Wieczorek
Thaddäus Wiedemer
Julian Wiederer
Kevin Tirta Wijaya
Asiri Wijesinghe
Kimberly Wilber
Jeffrey R. Willette
Bryan M. Williams
Williem Williem
Christian Wilms
Benjamin Wilson
Richard Wilson
Felix Wimbauer
Vanessa Wirth
Scott Wisdom
Calden Wloka
Alex Wong
Chau-Wai Wong
Chi-Chong Wong
Ka Wai Wong
Kelvin Wong
Kok-Seng Wong
Kwan-Yee K. Wong
Yongkang Wong
Sangmin Woo
Simon S. Woo
Markus Worchel
Scott Workman
Marcel Worring
Safwan Wshah

Aming Wu
Bo Wu
Bojian Wu
Boxi Wu
Changguang Wu
Chaoyi Wu
Chen Henry Wu
Cheng-En Wu
Chenming Wu
Chenyan Wu
Chenyun Wu
Cho-Ying Wu
Chongruo Wu
Cong Wu
Dayan Wu
Di Wu
Dongming Wu
Fangzhao Wu
Fuxiang Wu
Gaojie Wu
Guanyao Wu
Guile Wu
Haiping Wu
Haiwei Wu
Haiyu Wu
Han Wu
Haoning Wu
Haoning Wu
Haotian Wu
Hefeng Wu
Huisi Wu
Jane Wu
Jay Zhangjie Wu
Jhih-Ciang Wu
Ji-Jia Wu
Jialian Wu
Jiaye Wu
Jimmy Wu
Jing Wu
Jing Wu
Jinjian Wu
Jiqing Wu
Jun Wu
Junfeng Wu
Junlin Wu
Junru Wu
Junyang Wu
Junyi Wu
Letian Wu
Lifang Wu
Lin Yuanbo Wu
Liwen Wu
Min Wu
Minye wu
Peng Wu
Penghao Wu
Qian Wu
Qiangqiang Wu
Qianyi Wu
Qingbo Wu
Rongliang Wu
Rui Wu
Rundi Wu
Shuang Wu
Shuzhe Wu
Tao Wu
Tao Wu
Tao Wu
Te-Lin Wu
Tianfu Wu
Tianhao Wu
Tianhao Wu
Ting-Wei Wu
Tong Wu
Tong Wu
Tsung-Han Wu
Tz-Ying Wu
Weibin Wu
Weijia Wu
Xian Wu
Xiao Wu
Xiaodong Wu
Xiaohe Wu
Xiaoqian Wu
Xiaoyang Wu
Xindi Wu
Xingjiao Wu
Xinxiao Wu
Xiuzhe Wu
Yang Wu
Yangzheng Wu
Yanze Wu
Yanzhao Wu
Yawen Wu
Yicheng Wu
Ying Nian Wu
Yingwen Wu
Yong Wu
Yuanwei Wu
Yue Wu
Yue Wu
Yuqun Wu
Yushu Wu
Yushuang Wu
Zhe Wu
Zheng Wu
Zhi-Fan Wu
Zhihao Wu
Zhijie Wu
Zhiliang Wu
Zhonghua Wu
Zijie Wu
Ziyi Wu
Zizhao Wu
Zongwei Wu
Zongyu Wu
Zongze Wu
Stefanie Wuhrer
Jamie M. Wynn
Monika Wysoczańska
Jianing Xi
Teng Xi
Bin Xia
Changqun Xia
Haifeng Xia
Jiaer Xia
Jiahao Xia
Kun Xia
Mingxuan Xia
Shihong Xia
Weihao Xia
Xiaobo Xia
Yan Xia
Ye Xia
Yifei Xia

Zhaoyang Xia
Zhihao Xia
Zhihua Xia
Zhuofan Xia
Zimin Xia
Chuhua Xian
Wenqi Xian
Yongqin Xian
Donglai Xiang
Jinhai Xiang
Liuyu Xiang
Tian-Zhu Xiang
Tiange Xiang
Wangmeng Xiang
Xiaoyu Xiang
Yuanbo Xiangli
Anqi Xiao
Aoran Xiao
Bei Xiao
Chunxia Xiao
Fanyi Xiao
Han Xiao
Jiancong Xiao
Jimin Xiao
Jing Xiao
Jing Xiao
Jun Xiao
Junbin Xiao
Junfei Xiao
Mingqing Xiao
Qingyang Xiao
Ruixuan Xiao
Taihong Xiao
Yang Xiao
Yanru Xiao
Yao Xiao
Yijun Xiao
Yuting Xiao
Zehao Xiao
Zeyu Xiao
Zihao Xiao
Binhui Xie
Chaohao Xie
Chi Xie
Christopher Xie

Chuanlong Xie
Fei Xie
Guo-Sen Xie
Haozhe Xie
Hongtao Xie
Jiahao Xie
Jiaxin Xie
Jin Xie
Jinheng Xie
Jiu-Cheng Xie
Jiyang Xie
Junyu Xie
Liuyue Xie
Ming-Kun Xie
Mingyang Xie
Qian Xie
Tingting Xie
Weicheng Xie
Xianghui Xie
Xiaohua Xie
Xudong Xie
Yichen Xie
Yiming Xie
You Xie
Yuan Xie
Yusheng Xie
Yutong Xie
Zeke Xie
Zhenda Xie
Zhenyu Xie
ZiYang Xie
Chaoyue Xing
Fuyong Xing
Jinbo Xing
Xiaoyan Xing
Xiaoying Xing
XiMing Xing
Xin Xing
Yazhou Xing
Yifan Xing
Yun Xing
Zhen Xing
Hongkai Xiong
Jingjing Xiong
Jinhui Xiong

Junwen Xiong
Peixi Xiong
Wei Xiong
Weihua Xiong
Yu Xiong
Yuanhao Xiong
Yuanjun Xiong
Yuwen Xiong
Zhexiao Xiong
Zhiwei Xiong
Yuliang Xiu
Alessio Xompero
An Xu
Angchi Xu
Baixin Xu
Bicheng Xu
Bo Xu
Chao Xu
Chenfeng Xu
Chenshu Xu
Chenxin Xu
Chi Xu
Dejia Xu
Dongli Xu
Feng Xu
Gangwei Xu
Haiming Xu
Haiyang Xu
Han Xu
Haofei Xu
Haohang Xu
Haoran Xu
Hongbin Xu
Hongmin Xu
Jiale Xu
Jianjin Xu
Jiaqi Xu
Jie Xu
Jilan Xu
Jinglin Xu
Jingyi Xu
Jun Xu
Kai Xu
Katherine Xu
Ke Xu

Kele Xu
Lan Xu
Lian Xu
Liang Xu
Linning Xu
Lumin Xu
Manjie Xu
Mengde Xu
Mengdi Xu
Mengmeng Frost Xu
Min Xu
Ming Xu
Mutian Xu
Peiran Xu
Peng Xu
Qi Xu
Qiang Xu
Qiangeng Xu
Qingshan Xu
Qingyang Xu
Qiuling Xu
Ran Xu
Renzhe Xu
Ruikang Xu
Runsen Xu
Runsheng Xu
Shichao Xu
Sirui Xu
Tongda Xu
Wanting Xu
Wei Xu
Weiwei Xu
Wenjia Xu
Wenju Xu
Wenqiang Xu
Xiang Xu
Xianghao Xu
Xiangyu Xu
Xiangyu Xu
Xiaogang Xu
Xiaohao Xu
Xin Xu
Xin Xu
Xin-Shun Xu
Xing Xu
Xinli Xu
Xinyu Xu
Xiuwei Xu
Xiyan Xu
Xudong Xu
Xuemiao Xu
Xun Xu
Yan Xu
Yan Xu
Yan Xu
Yangyang Xu
Yanwu Xu
Yating Xu
Yi Xu
Yi Xu
Yi Xu
Yihong Xu
YiKun Xu
Yinghao Xu
Yingyan Xu
Yinshuang Xu
Yiran Xu
Yixing Xu
Yongchao Xu
Yue Xu
Yufei Xu
Yunqiu Xu
Zexiang Xu
Zhan Xu
Zhe Xu
Zhengqin Xu
Zhenlin Xu
Zhiqiu Xu
Zhiyuan Xu
Zhongcong Xu
Zhuoer Xu
Zipeng Xu
Ziyue Xu
Zongyi Xu
Ziwei Xuan
Danna Xue
Fanglei Xue
Fei Xue
Feng Xue
Han Xue
Jianru Xue
Le Xue
Lixin Xue
Mingfu Xue
Nan Xue
Qinghan Xue
Shangjie Xue
Xiangyang Xue
Zihui Xue
Abhay Yadav
Amit Kumar Singh Yadav
Takuma Yagi
Tomas F Yago Vicente
I. Zeki Yalniz
Kota Yamaguchi
Shin'ya Yamaguchi
Burhaneddin Yaman
Toshihiko Yamasaki
Kohei Yamashita
Lee Juliette Yamin
Chaochao Yan
Hongyu Yan
Jiexi Yan
Kai Yan
Pei Yan
Qingan Yan
Qingsen Yan
Qingsong Yan
Rui Yan
Shaoqi Yan
Shi Yan
Siming Yan
Siming Yan
Siyuan Yan
Weilong Yan
Wending Yan
Xiangyi Yan
Xinchen Yan
Xingguang Yan
Xueting Yan
Yan Yan
Yichao Yan
Zhaoyi Yan
Zhiqiang Yan
Zhiyuan Yan

Zike Yan
Zizheng Yan
Keiji Yanai
Pinar Yanardag
Anqi Yang
Anqi Joyce Yang
Bangbang Yang
Baoyao Yang
Bin Yang
Binwei Yang
Bo Yang
Bo Yang
Boyu Yang
Changdi Yang
Chao Yang
Charig Yang
Cheng-Fu Yang
Cheng-Yen Yang
Chenhongyi Yang
Chuanguang Yang
De-Nian Yang
Dingcheng Yang
Dingkang Yang
Dong Yang
Erkun Yang
Fan Yang
Fan Yang
Fan Yang
Fan Yang
Fan Yang
Feng Yang
Fengting Yang
Fengxiang Yang
Fengyuan Yang
Fu-En Yang
Gang Yang
Gengshan Yang
Guandao Yang
Guanglei Yang
Haitao Yang
Hanqing Yang
Heran Yang
Honghui Yang
Huanrui Yang
Huiyuan Yang
Huizong Yang
Hunmin Yang
Jiange Yang
Jiaqi Yang
Jiawei Yang
Jiayu Yang
Jiazhi Yang
Jie Yang
Jie Yang
Jiewen Yang
Jihan Yang
Jing Yang
Jingkang Yang
Jinhui Yang
Jinlong Yang
Jinrong Yang
Jinyu Yang
Kaicheng Yang
Kailun Yang
Lan Yang
Le Yang
Lehan Yang
Lei Yang
Lei Yang
Lei Yang
Li Yang
Lihe Yang
Ling Yang
Lingxiao Yang
Linlin Yang
Lixin Yang
Longrong Yang
Lu Yang
Luwei Yang
Michael Ying Yang
Min Yang
Ming Yang
MingKun Yang
Mouxing Yang
Muli Yang
Peiyu Yang
Qi Yang
Qian Yang
Qiushi Yang
Ren Yang
Rui Yang
Ruihan Yang
Sejong Yang
Shan Yang
Shangrong Yang
Shiqi Yang
Shuai Yang
Shuai Yang
Shuang Yang
Shuo Yang
Shusheng Yang
Sibei Yang
Siwei Yang
Siyuan Yang
Siyuan Yang
Song Yang
Songlin Yang
Tianyu Yang
Tong Yang
Wankou Yang
Wenhan Yang
Wenhan Yang
Wenjie Yang
Wenqi Yang
William Yang
Xi Yang
Xi Yang
Xiangpeng Yang
Xiao Yang
Xiaofeng Yang
Xiaoshan Yang
Xin Jeremy Yang
Xingyi Yang
Xinlong Yang
Xitong Yang
Xiulong Yang
Xu Yang
Xuan Yang
Xue Yang
Xuelin Yang
Xun Yang
Yan Yang
Yan Yang
Yang Yang
Yaokun Yang

Yezhou Yang
Yiding Yang
Yijun Yang
Yijun Yang
Yin Yang
Yinfei Yang
Yixin Yang
Yongqi Yang
Yongqi Yang
Yue Yang
Yuewei Yang
Yuezhi Yang
Yujiu Yang
Yung-Hsu Yang
Yuwei Yang
Ze Yang
Ze Yang
Zetong Yang
Zhangsihao Yang
Zhaoyuan Yang
Zhen Yang
Zhenpei Yang
Zhibo Yang
Zhiwei Yang
Zhiwen Yang
Zhiyuan Yang
Zhuoqian Yang
Ziyan Yang
Ziyun Yang
Zongxin Yang
Zuhao Yang
Chengtang Yao
Cong Yao
Hantao Yao
Jiawen Yao
Lina Yao
Mingde Yao
Mingshuai Yao
Qingsong Yao
Shunyu Yao
Taiping Yao
Ting Yao
Xincheng Yao
Xinwei Yao
Xu Yao

Xufeng Yao
Yao Yao
Yazhou Yao
Yue Yao
Ziwei Yao
Sudhir Yarram
Rajeev Yasarla
Mohsen Yavartanoo
Botao Ye
Dengpan Ye
Fei Ye
Hanrong Ye
Jianglong Ye
Jiarong Ye
Jin Ye
Jingwen Ye
Jinwei Ye
Junjie Ye
Keren Ye
Maosheng Ye
Meng Ye
Meng Ye
Muchao Ye
Nanyang Ye
Peng Ye
Qi Ye
Qian Ye
Qinghao Ye
Qixiang Ye
Ruolin Ye
Shuquan Ye
Tian Ye
Vickie Ye
Wenqian Ye
Xinchen Ye
Yufei Ye
Moon Ye-Bin
Yousef Yeganeh
Chun-Hsiao Yeh
Raymond Yeh
Yu-Ying Yeh
Florence Yellin
Sriram Yenamandra
Tarun Yenamandra
Promod Yenigalla

Chandan Yeshwanth
Dong Yi
Hongwei Yi
Kai Yi
Ran Yi
Renjiao Yi
Xinyu Yi
Alper Yilmaz
Jonghwa Yim
Aoxiong Yin
Fei Yin
Fukun Yin
Jia-Li Yin
Ming Yin
Nan Yin
Ruihong Yin
Tianwei Yin
Wenzhe Yin
Xiaoqi Yin
Yingda Yin
Yu Yin
Yufeng Yin
Zhenfei Yin
Xianghua Ying
Xiaowen Ying
Naoto Yokoya
Chen YongCan
ByungIn Yoo
Innfarn Yoo
Jinsu Yoo
Sungjoo Yoo
Hee Suk Yoon
Jae Shin Yoon
Jihun Yoon
Sangwoong Yoon
Sejong Yoon
Sung Whan Yoon
Sung-Hoon Yoon
Sunjae Yoon
Youngho Yoon
Youngseok Yoon
Youngseok Yoon
Yuichi Yoshida
Ryota Yoshihashi
Yusuke Yoshiyasu

Chenyu You
Haoran You
Haoxuan You
Shan You
Yang You
Yingxuan You
Yurong You
Chan-Hyun Youn
Kim Youwang
Nikolaos-Antonios Ypsilantis
Baosheng Yu
Bei Yu
Bruce X. B. Yu
Chaohui Yu
Chunlin Yu
Cunjun Yu
Dahai Yu
En Yu
En Yu
Fenggen Yu
Gang Yu
Haibao Yu
Hanchao Yu
Hang Yu
Hao Yu
Hao Yu
Haojun Yu
Heng Yu
Hong-Xing Yu
Houjian Yu
Jianhui Yu
Jiashuo Yu
Jing Yu
Jiwen Yu
Jiyang Yu
Kaicheng Yu
Lei Yu
Lidong Yu
Lijun Yu
Mulin Yu
Peilin Yu
Qian Yu
Qihang Yu
Qing Yu
Rui Yu
Ruixuan Yu
Runpeng Yu
Shaozuo Yu
Shuzhi Yu
Sihyun Yu
Tan Yu
Tao Yu
Tianjiao Yu
Wei Yu
Weihao Yu
Wenwen Yu
Xi Yu
Xiaohan Yu
Xin Yu
Xin Yu
Xuehui Yu
Yingchen Yu
Yongsheng Yu
Yunlong Yu
Zehao Yu
Zhaofei Yu
Zhengdi Yu
Zhengdi Yu
Zhixuan Yu
Zhongzhi Yu
Zhuoran Yu
Zitong Yu
Chun Yuan
Chunfeng Yuan
Hangjie Yuan
Haobo Yuan
Jiakang Yuan
Jiangbo Yuan
Liangzhe Yuan
Maoxun Yuan
Shanxin Yuan
Shengming Yuan
Shuai Yuan
Shuaihang Yuan
Wentao Yuan
Xiaoding Yuan
Xiaoyun Yuan
Xin Yuan
Xin Yuan
Yixuan Yuan
Yu-Jie Yuan
Yuan Yuan
Yuan Yuan
Yuhui Yuan
Zheng Yuan
Zhuoning Yuan
Mehmet Kerim Yücel
Dongxu Yue
Haixiao Yue
Kaiyu Yue
Tao Yue
Xiangyu Yue
Zihao Yue
Zongsheng Yue
Heeseung Yun
Jooyeol Yun
Juseung Yun
Kimin Yun
Se-Young Yun
Sukwon Yun
Tian Yun
Raza Yunus
Ekim Yurtsever
Eloi Zablocki
Riccardo Zaccone
Martin Zach
Muhammad Zaigham Zaheer
Ilya Zakharkin
Egor Zakharov
Abhaysinh S. Zala
Pierluigi Zama Ramirez
Eduard Sebastian Zamfir
Amir Zamir
Luca Zancato
Yuan Zang
Yuhang Zang
Zelin Zang
Pietro Zanuttigh
Giacomo Zara
Samira Zare
Olga Zatsarynna
Denis Zavadski
Vitjan Zavrtanik

Jan Zdenek
Yanjie Ze
Bernhard Zeisl
John Zelek
Oliver Zendel
Ailing Zeng
Chong Zeng
Dan Zeng
Fangao Zeng
Haijin Zeng
Huimin Zeng
Jia Zeng
Jiabei Zeng
Kuo-Hao Zeng
Libing Zeng
Ling-An Zeng
Ming Zeng
Pengpeng Zeng
Runhao Zeng
Tieyong Zeng
Wei Zeng
Yan Zeng
Yanhong Zeng
Yawen Zeng
Yuyuan Zeng
Zilai Zeng
Ziyao Zeng
Kaiwen Zha
Ruyi Zha
Yaohua Zha
Bohan Zhai
Qiang Zhai
Runtian Zhai
Wei Zhai
YiKui Zhai
Yuanhao Zhai
Yunpeng Zhai
De-Chuan Zhan
Fangneng Zhan
Guanqi Zhan
Huangying Zhan
Huijing Zhan
Kun Zhan
Xueying Zhan
Aidong Zhang

Baochang Zhang
Baoheng Zhang
Baoming Zhang
Biao Zhang
Bingfeng Zhang
Binjie Zhang
Bo Zhang
Bo Zhang
Borui Zhang
Bowen Zhang
Can Zhang
Ce Zhang
Chang-Bin Zhang
Chao Zhang
Chao Zhang
Chen-Lin Zhang
Cheng Zhang
Cheng Zhang
Chenghao Zhang
Chenyangguang Zhang
Chi Zhang
Chongyang Zhang
Chris Zhang
Chuhan Zhang
Chunhui Zhang
Chuyu Zhang
Congyi Zhang
Daichi Zhang
Dan Zhang
Daoan Zhang
Daoqiang Zhang
David Junhao Zhang
Dexuan Zhang
Dingwen Zhang
Dingyuan Zhang
Dongsu Zhang
Fan Zhang
Fan Zhang
Fang-Lue Zhang
Feilong Zhang
Frederic Z. Zhang
Fuyang Zhang
Gang Zhang
Gengwei Zhang
Gengyu Zhang

Gengyuan Zhang
Gongjie Zhang
GuiXuan Zhang
Guofeng Zhang
Guozhen Zhang
Hang Zhang
Hang Zhang
Hanwang Zhang
Hao Zhang
Hao Zhang
Hao Zhang
Haokui Zhang
Haonan Zhang
Haotian Zhang
Hengrui Zhang
Hongguang Zhang
Hongrun Zhang
Hongyuan Zhang
Howard Zhang
Huaidong Zhang
Huaiwen Zhang
Hui Zhang
Hui Zhang
Jason Y. Zhang
Ji Zhang
Jiahui Zhang
Jiakai Zhang
Jiaming Zhang
Jian Zhang
Jianfu Zhang
Jiangning Zhang
Jianhua Zhang
Jianming Zhang
Jianpeng Zhang
Jianping Zhang
Jianrong Zhang
Jichao Zhang
Jie Zhang
Jie Zhang
Jie Zhang
Jimuyang Zhang
Jing Zhang
Jing Zhang
Jinghao Zhang
Jingyi Zhang

Jinlu Zhang
Jiqing Zhang
Jiyuan Zhang
Junbo Zhang
Junge Zhang
Junyi Zhang
Juyong Zhang
Kai Zhang
Kai Zhang
Kaidong Zhang
Kaihao Zhang
Kaipeng Zhang
Kaiyi Zhang
Ke Zhang
Ke Zhang
Kui Zhang
Le Zhang
Le Zhang
Lefei Zhang
Lei Zhang
Leo Yu Zhang
Li Zhang
Lianbo Zhang
Liang Zhang
Liangpei Zhang
Lin Zhang
Linfeng Zhang
Liqing Zhang
Lu Zhang
Malu Zhang
Manyuan Zhang
Mengmi Zhang
Mengqi Zhang
Mi Zhang
Min Zhang
Min-Ling Zhang
Mingda Zhang
Mingfang Zhang
Minghui Zhang
Mingjin Zhang
Mingyuan Zhang
Minjia Zhang
Ni Zhang
Pan Zhang
Peiyan Zhang
Pengze Zhang
Pingping Zhang
Qi Zhang
Qi Zhang
Qian Zhang
Qiang Zhang
Qijian Zhang
Qiming Zhang
Qing Zhang
Qing Zhang
Renrui Zhang
Rongyu Zhang
Ruida Zhang
Ruimao Zhang
Ruixin Zhang
Runze Zhang
Sanyi Zhang
Shan Zhang
Shanghang Zhang
Shaofeng Zhang
Sheng Zhang
Shengping Zhang
Shengyu Zhang
Shimian Zhang
Shiwei Zhang
Shizhou Zhang
Shu Zhang
Shuo Zhang
Siwei Zhang
Song-Hai Zhang
Tao Zhang
Tianyun Zhang
Ting Zhang
Tong Zhang
Weixia Zhang
Wendong Zhang
Wenlong Zhang
Wenqiang Zhang
Wentao Zhang
Wentian Zhang
Wenxiao Zhang
Wenxuan Zhang
Xi Zhang
Xiang Zhang
Xiang Zhang
Xianling Zhang
Xiao Zhang
Xiaohan Zhang
Xiaoming Zhang
Xiaoran Zhang
Xiaowei Zhang
Xiaoyun Zhang
Xikun Zhang
Xin Zhang
Xinfeng Zhang
Xingchen Zhang
Xingguang Zhang
Xingxuan Zhang
Xiong Zhang
Xiuming Zhang
Xu Zhang
Xuanyang Zhang
Xucong Zhang
Xuying Zhang
Yabin Zhang
Yabo Zhang
Yachao Zhang
Yahui Zhang
Yan Zhang
Yan Zhang
Yanan Zhang
Yang Zhang
Yanghao Zhang
Yawen Zhang
Yechao Zhang
Yi Zhang
Yi Zhang
Yi Zhang
Yi-Fan Zhang
Yifan Zhang
Yifei Zhang
Yifeng Zhang
Yihao Zhang
Yihua Zhang
Yimeng Zhang
Yiming Zhang
Yin Zhang
Yinan Zhang
Yinda Zhang
Ying Zhang

Yingliang Zhang	Zhong Zhang	Qingyu Zhao
Yitian Zhang	Zhongping Zhang	Qinyu Zhao
Yixin Zhang	Zhongqun Zhang	Rongchang Zhao
Yiyuan Zhang	Zicheng Zhang	Rui Zhao
Yongfei Zhang	Zicheng Zhang	Rui Zhao
Yonggang Zhang	Zihao Zhang	Rui Zhao
Yonghua Zhang	Ziming Zhang	Ruiqi Zhao
Youjian Zhang	Ziqi Zhang	Shanshan Zhao
Youmin Zhang	Qilong Zhangli	Shihao Zhao
Youshan Zhang	Bin Zhao	Shiyu Zhao
Yu Zhang	Bingchen Zhao	Shizhen Zhao
Yu Zhang	Bingyin Zhao	Shuai Zhao
Yuan Zhang	Cairong Zhao	Siheng Zhao
Yuechen Zhang	Can Zhao	Tianchen Zhao
Yuexi Zhang	Chen Zhao	TianHao Zhao
Yufei Zhang	Dong Zhao	Tiesong Zhao
Yuhan Zhang	Dongxu Zhao	Wang Zhao
Yuhang Zhang	Fang Zhao	Wangbo Zhao
Yunchao Zhang	Feng Zhao	Weichao Zhao
Yunhe Zhang	Fuqiang Zhao	Weiyue Zhao
Yunhua Zhang	Gangming Zhao	Wenda Zhao
Yunpeng Zhang	Ganlong Zhao	Wenliang Zhao
Yunzhi Zhang	Guiyu Zhao	Xiangyun Zhao
Yuxin Zhang	Haimei Zhao	Xiaoming Zhao
Yuyao Zhang	Hanbin Zhao	Xiaonan Zhao
Zaixi Zhang	Handong Zhao	Xiaoqi Zhao
Zeliang Zhang	Jian Zhao	Xin Zhao
Zewei Zhang	Jiaqi Zhao	Xin Zhao
Zeyu Zhang	Jie Zhao	Xingyu Zhao
Zhang Zhang	Kai Zhao	Xu Zhao
Zhao Zhang	Kaifeng Zhao	Yajie Zhao
Zhaoxiang Zhang	Lei Zhao	Yang Zhao
Zhen Zhang	Liang Zhao	Yifan Zhao
Zheng Zhang	Lirui Zhao	Ying Zhao
Zheng Zhang	Long Zhao	Yiqun Zhao
Zhenyu Zhang	Luxi Zhao	Yiqun Zhao
Zhenyu Zhang	Mingyang Zhao	Yizhou Zhao
Zheyuan Zhang	Minyi Zhao	Yizhou Zhao
Zhicheng Zhang	Na Zhao	Yucheng Zhao
Zhilu Zhang	Nanxuan Zhao	Yue Zhao
Zhishuai Zhang	Pu Zhao	Yunhan Zhao
Zhitian Zhang	Qi Zhao	Yuyang Zhao
Zhiwei Zhang	Qian Zhao	Yuzhi Zhao
Zhixing Zhang	Qibin Zhao	Zelin Zhao
Zhiyuan Zhang	Qingsong Zhao	Zengqun Zhao

Zhen Zhao
Zhenghao Zhao
Zhengyu Zhao
Zhou Zhao
Zibo Zhao
Zimeng Zhao
Ziwei Zhao
Ziwei Zhao
Zixiang Zhao
Jin Zhe
Anling Zheng
Ce Zheng
Chaoda Zheng
Chengwei Zheng
Chuanxia Zheng
Duo Zheng
Ervine Zheng
Guangcong Zheng
Haitian Zheng
Haiyong Zheng
Hao Zheng
Haotian Zheng
Haoxin Zheng
Huan Zheng
Huan Zheng
Jia Zheng
Jian Zheng
Jian-Qing Zheng
Jianqiao Zheng
Jianwei Zheng
Jin Zheng
Jingxiao Zheng
Kaiwen Zheng
Kecheng Zheng
Léon Zheng
Meng Zheng
Naishan Zheng
Peng Zheng
Qi Zheng
Qian Zheng
Rongkun Zheng
Shen Zheng
Shuai Zheng
Shuhong Zheng
Shunyuan Zheng

Siming Zheng
Tianhang Zheng
Wenting Zheng
Wenzhao Zheng
Xiaozheng Zheng
Xiawu Zheng
Xu Zheng
Yajing Zheng
Yalin Zheng
Yang Zheng
Ye Zheng
Yinglin Zheng
Yinqiang Zheng
Yu Zheng
Zangwei Zheng
Zehan Zheng
Zerong Zheng
Zhaoheng Zheng
Zhedong Zheng
Zhuo Zheng
Zilong Zheng
Shuaifeng Zhi
Tiancheng Zhi
Bineng Zhong
Fangcheng Zhong
Fangwei Zhong
Guoqiang Zhong
Nan Zhong
Yaoyao Zhong
Yijie Zhong
Yiqi Zhong
Yiran Zhong
Yiwu Zhong
Yunshan Zhong
Zichun Zhong
Ziming Zhong
Brady Zhou
Chong Zhou
Chu Zhou
Chunluan Zhou
Da-Wei Zhou
Dawei Zhou
Dewei Zhou
Dingfu Zhou
Donghao Zhou

Dongzhan Zhou
Fan Zhou
Hang Zhou
Hang Zhou
Hao Zhou
Hao Zhou
Haoyi Zhou
Hong-Yu Zhou
Honglu Zhou
Huayi Zhou
Jiahuan Zhou
Jiaming Zhou
Jian Zhou
Jianan Zhou
Jiantao Zhou
Jianxiong Zhou
JIngkai Zhou
Junsheng Zhou
Kailai Zhou
Kaiyang Zhou
Keyang Zhou
Kun Zhou
Lei Zhou
Mingyang Zhou
Mingyi Zhou
Mingyuan Zhou
Mo Zhou
Pan Zhou
Peng Zhou
Peng Zhou
Qianyi Zhou
Qianyu Zhou
Qihua Zhou
Qin Zhou
Qinqin Zhou
Qunjie Zhou
Sheng Zhou
Shenglong Zhou
Shijie Zhou
Shuchang Zhou
Sihang Zhou
Tao Zhou
Tianfei Zhou
Xiangdong Zhou
Xiaoqiang Zhou

Xin Zhou
Xingyi Zhou
Yan-Jie Zhou
Yang Zhou
Yang Zhou
Yanqi Zhou
Yi Zhou
Yi Zhou
Yichao Zhou
Yichen Zhou
Yin Zhou
Yiyi Zhou
Yu Zhou
Yucheng Zhou
Yufan Zhou
Yunsong Zhou
Yuqian Zhou
Yuxiao Zhou
Yuxuan Zhou
Zhenyu Zhou
Zijian Zhou
Zikun Zhou
Ziqi Zhou
Zixiang Zhou
Zongwei Zhou
Alex Z. Zhu
Benjin Zhu
Bin Zhu
Bin Zhu
Chenming Zhu
Chenyang Zhu
Deyao Zhu
Dongxiao Zhu
Fangrui Zhu
Fei Zhu
Feida Zhu
Fengqing Maggie Zhu
Guibo Zhu
Haidong Zhu
Hanwei Zhu
Hao Zhu
Hao Zhu
Heming Zhu
Jiachen Zhu
Jianke Zhu

Jiawen Zhu
Jiayin Zhu
Jinjing Zhu
Junyi Zhu
Kai Zhu
Ke Zhu
Lanyun Zhu
Lin Zhu
Linchao Zhu
Liyuan Zhu
Meilu Zhu
Muzhi Zhu
Qingtian Zhu
Ronghang Zhu
Rui Zhu
Rui Zhu
Rui-Jie Zhu
Ruizhao Zhu
Shengjie Zhu
Sijie Zhu
Siyu Zhu
Tyler Zhu
Wang Zhu
Weicheng Zhu
Wenwu Zhu
Xiangyu Zhu
Xiaofeng Zhu
Xiaoguang Zhu
Xiaosu Zhu
Xiaoyu Zhu
Xingkui Zhu
Xinxin Zhu
Xiyue Zhu
Yangguang Zhu
Yanjun Zhu
Yao Zhu
Ye Zhu
Feida Zhu
Fengqing Maggie Zhu
Guibo Zhu
Haidong Zhu
Hanwei Zhu
Hao Zhu
Hao Zhu
Heming Zhu

Jiachen Zhu
Jianke Zhu
Jiawen Zhu
Jiayin Zhu
Jinjing Zhu
Junyi Zhu
Kai Zhu
Ke Zhu
Lanyun Zhu
Lin Zhu
Linchao Zhu
Liyuan Zhu
Meilu Zhu
Muzhi Zhu
Qingtian Zhu
Ronghang Zhu
Rui Zhu
Rui Zhu
Rui-Jie Zhu
Ruizhao Zhu
Shengjie Zhu
Sijie Zhu
Siyu Zhu
Tyler Zhu
Wang Zhu
Weicheng Zhu
Wenwu Zhu
Xiangyu Zhu
Xiaofeng Zhu
Xiaoguang Zhu
Xiaosu Zhu
Xiaoyu Zhu
Xingkui zhu
Xinxin Zhu
Xiyue Zhu
Yangguang Zhu
Yanjun Zhu
Yao Zhu
Ye Zhu
Ye Zhu
Yichen Zhu
Yingying Zhu
Yousong Zhu
Yuansheng Zhu
Yurui Zhu

Zhen Zhu
Zhenwei Zhu
Zhenyao Zhu
Zhifan Zhu
Zhigang Zhu
Zhihong Zhu
Zihan Zhu
Zixin Zhu
Zunjie Zhu
Bingbing Zhuang
Jia-Xin Zhuang
Jiafan Zhuang
Peiye Zhuang
Wanyi Zhuang
Weiming Zhuang
Yihong Zhuang
Yixin Zhuang
Mingchen Zhuge
Tao Zhuo
Wei Zhuo

Yaoxin Zhuo
Bartosz Zieliński
Wojciech Zielonka
Filippo Ziliotto
Karel Zimmermann
Primo Zingaretti
Nikolaos Zioulis
Liu Ziyin
Mohammad Zohaib
Yongshuo Zong
Zhuofan Zong
Maria Zontak
Gaspard Zoss
Changqing Zou
Chuhang Zou
Danping Zou
Dongqing Zou
Haoming Zou
Longkun Zou
Shihao Zou

Xingxing Zou
Xueyan Zou
Yang Zou
Yuexian Zou
Yuli Zou
Yuliang Zou
Yunhao Zou
Zhiming Zou
Zihang Zou
Silvia Zuffi
Idil Esen Zulfikar
Maria A. Zuluaga
Ronglai Zuo
Xingxing Zuo
Xinxin Zuo
Yifan Zuo
Yiming Zuo
Reyer Zwiggelaar
Vlas Zyrianov

Contents – Part LXXXIX

How to Train the Teacher Model for Effective Knowledge Distillation 1
   *Shayan Mohajer Hamidi, Xizhen Deng, Renhao Tan, Linfeng Ye,
and Ahmed Hussein Salamah*

Tight and Efficient Upper Bound on Spectral Norm of Convolutional Layers . . . 19
   Ekaterina Grishina, Mikhail Gorbunov, and Maxim Rakhuba

Deciphering the Role of Representation Disentanglement: Investigating
Compositional Generalization in CLIP Models 35
   *Reza Abbasi, Mohammad Hossein Rohban,
and Mahdieh Soleymani Baghshah*

Modality Translation for Object Detection Adaptation Without Forgetting
Prior Knowledge 51
   *Heitor Rapela Medeiros, Masih Aminbeidokhti,
Fidel Alejandro Guerrero Peña, David Latortue, Eric Granger,
and Marco Pedersoli*

FroSSL: Frobenius Norm Minimization for Efficient Multiview
Self-supervised Learning 69
   *Oscar Skean, Aayush Dhakal, Nathan Jacobs,
and Luis Gonzalo Sanchez Giraldo*

Learning Multimodal Latent Generative Models with Energy-Based Prior 86
   Shiyu Yuan, Jiali Cui, Hanao Li, and Tian Han

On Learning Discriminative Features from Synthesized Data
for Self-supervised Fine-Grained Visual Recognition 101
   *Zihu Wang, Lingqiao Liu, Scott Ricardo Figueroa Weston, Samuel Tian,
and Peng Li*

LaWa: Using Latent Space for In-Generation Image Watermarking 118
   *Ahmad Rezaei, Mohammad Akbari, Saeed Ranjbar Alvar,
Arezou Fatemi, and Yong Zhang*

Hierarchical Conditioning of Diffusion Models Using Tree-of-Life for Studying Species Evolution 137
   Mridul Khurana, Arka Daw, M. Maruf, Josef C. Uyeda, Wasila Dahdul, Caleb Charpentier, Yasin Bakış, Henry L. Bart Jr., Paula M. Mabee, Hilmar Lapp, James P. Balhoff, Wei-Lun Chao, Charles Stewart, Tanya Berger-Wolf, and Anuj Karpatne

Markov Knowledge Distillation: Make Nasty Teachers Trained by Self-undermining Knowledge Distillation Fully Distillable 154
   En-hui Yang and Linfeng Ye

Co-speech Gesture Video Generation with 3D Human Meshes 172
   Aniruddha Mahapatra, Richa Mishra, Renda Li, Ziyi Chen, Boyang Ding, Shoulei Wang, Jun-Yan Zhu, Peng Chang, Mei Han, and Jing Xiao

Understanding the Impact of Negative Prompts: When and How Do They Take Effect? 190
   Yuanhao Ban, Ruochen Wang, Tianyi Zhou, Minhao Cheng, Boqing Gong, and Cho-Jui Hsieh

GS2Mesh: Surface Reconstruction from Gaussian Splatting via Novel Stereo Views 207
   Yaniv Wolf, Amit Bracha, and Ron Kimmel

CARFF: Conditional Auto-Encoded Radiance Field for 3D Scene Forecasting 225
   Jiezhi Yang, Khushi Desai, Charles Packer, Harshil Bhatia, Nicholas Rhinehart, Rowan McAllister, and Joseph E. Gonzalez

Snuffy: Efficient Whole Slide Image Classifier 243
   Hossein Jafarinia, Alireza Alipanah, Saeed Razavi, Nahal Mirzaie, and Mohammad Hossein Rohban

Learning to Build by Building Your Own Instructions 261
   Aaron Walsman, Muru Zhang, Adam Fishman, Ali Farhadi, and Dieter Fox

Exploring Active Learning in Meta-learning: Enhancing Context Set Labeling 279
   Wonho Bae, Jing Wang, and Danica J. Sutherland

BlenderAlchemy: Editing 3D Graphics with Vision-Language Models 297
   Ian Huang, Guandao Yang, and Leonidas Guibas

DϵpS: Delayed ϵ-Shrinking for Faster Once-for-All Training 315
Aditya Annavajjala, Alind Khare, Animesh Agrawal, Igor Fedorov,
Hugo Latapie, Myungjin Lee, and Alexey Tumanov

Customize-A-Video: One-Shot Motion Customization of Text-to-Video
Diffusion Models 332
Yixuan Ren, Yang Zhou, Jimei Yang, Jing Shi, Difan Liu, Feng Liu,
Mingi Kwon, and Abhinav Shrivastava

Author Index 351

How to Train the Teacher Model for Effective Knowledge Distillation

Shayan Mohajer Hamidi[1](✉), Xizhen Deng[2], Renhao Tan[1], Linfeng Ye[1], and Ahmed Hussein Salamah[1]

[1] University of Waterloo, Waterloo, ON N2L 3G1, Canada
{smohajer,cameron.tan,l44ye,ahamsalamah}@uwaterloo.ca
[2] University of Michigan, Ann Arbor, MI 48109, USA
xizhen@umich.edu

Abstract. Recently, it was shown that the role of the teacher in knowledge distillation (KD) is to provide the student with an estimate of the true Bayes conditional probability density (BCPD). Notably, the new findings propose that the student's error rate can be upper-bounded by the mean squared error (MSE) between the teacher's output and BCPD. Consequently, to enhance KD efficacy, the teacher should be trained such that its output is close to BCPD in MSE sense. This paper elucidates that training the teacher model with MSE loss equates to minimizing the MSE between its output and BCPD, aligning with its core responsibility of providing the student with a BCPD estimate closely resembling it in MSE terms. In this respect, through a comprehensive set of experiments, we demonstrate that substituting the conventional teacher trained with cross-entropy loss with one trained using MSE loss in state-of-the-art KD methods consistently boosts the student's accuracy, resulting in improvements of up to 2.6%. The code for this paper is publicly available at: https://github.com/ECCV2024MSE/ECCV_MSE_Teacher.

Keywords: Knowledge distillation · Bayes conditional probability density · Mean squared error

1 Introduction

Knowledge distillation (KD), as introduced by [5] and popularized by [17], has emerged as a highly effective model compression technique, and has received significant attention from both academia and industry in recent years. At its core, KD entails the process of transferring the knowledge of a cumbersome model (teacher) into a lightweight counterpart (student). After the pioneering work by [17], numerous researchers have attempted to improve the performance of KD [3,30,36], and to understand why distillation works [2,14,16,26,28,34,52].

S. M. Hamidi and X. Deng—Authors contributed equally.

Supplementary Information The online version contains supplementary material available at https://doi.org/10.1007/978-3-031-73024-5_1.

An aspect of KD that has received relatively limited attention is the training of the teacher model. In most of the existing KD methods, the teacher is typically trained to optimize its own performance, despite the fact that such optimization does not necessarily translate into enhanced student performance [13,27]. Hence, to effectively train a student to attain high performance, it is crucial to align the training of the teacher accordingly.

Recently, [26] showed that the teacher's soft predictions can act as a proxy for the unknown true Bayes conditional probability distribution (BCPD) of label y given an input x. Specifically, a teacher model trained using conventional cross-entropy (CE) loss function approximates the true BCPD of the underlying dataset [21], and then it passes this estimate to the student model. As such, the enhancement in the student's accuracy within KD stems from the fact that the student utilizes the teacher's BCPD approximation to train its own model. Additionally, [26] noted that as the teacher's predictions approach the BCPD, the generalization error of the student model decreases. More importantly, [14] showed that the classification error rate of the student is directly bounded by the MSE between the teacher's output and the BCPD, as established through the Rademacher analysis. This fact was further empirically confirmed by [35] where they treated the teacher's output as a supervisory signal for the student model. Hence, based on these findings, it is imperative for effective KD that the teacher is trained such that its output is close to the BCPD in the MSE sense.

Now the question is that how the teacher model should be trained to ensure that its prediction is close to the BCPD in the MSE sense? To answer to this question, in this paper, we prove that training a DNN via MSE (resp. CE) loss is equivalent to training it to minimize the expected MSE (resp. CE) between its output and the true BCPD. Hence, based on the discussions above, for an effective KD, the teacher should be trained using MSE loss function. We shall note that proximity in terms of MSE does not necessarily equate to proximity in terms of CE, and vice versa. Therefore, while the output of a teacher trained using CE loss serves as an estimate of the true BCPD, it may not necessarily be close to the BCPD in terms of MSE, which is essential for effective KD. In fact, we empirically show that although the student's accuracy is almost inversely proportional to the MSE between BCPD and teacher's output, such relationship does not exist between the student's accuracy and CE between BCPD and teacher's output.

Based on the discussions above, we claim that for an effective KD, the teacher should be trained using MSE loss. To demonstrate the effectiveness of the teacher trained via MSE loss in KD, we conduct a thorough set of experiments over CIFAR-100 and ImageNet datasets, and show that by **solely** replacing a conventional teacher trained with CE loss by one trained with MSE loss in the existing state-of-the-art KD methods, the student's accuracy consistently increases. We shall emphasize the fact that such gain is obtained without making any changes over the underlying KD methods such as its distillation loss function or any hyper-parameters. Additionally, we observe a slight decrease in the teacher's performance when trained using MSE loss, confirming that optimizing the teacher's performance is not necessary for an effective knowledge distillation process. To summarize, the contributions of this paper are as follows

- We introduce a theorem to show that training a DNN to minimize CE/MSE loss is equivalent to training it to minimize the CE/MSE of its output to the true BCPD.
- We show that for an effective KD, the student should be provided with an estimate of BCPD that is close to it in MSE sense; thus, as per our theorem, the teacher as a BCPD estimator should be trained via MSE loss.
- We conduct a thorough set of experiments over CIFAR-100 and ImageNet datasets, and show that by replacing the teacher trained by CE with the one trained by MSE in the existing state-of-the-art KD methods, the student's accuracy consistently increases.

2 Related Works

2.1 Knowledge Distillation

The concept of knowledge transfer, as a means of compression, was first introduced by [5]. Then, [17] popularized this concept by softening the teacher's and student's logits using temperature technique where the student mimics the soft probabilities of the teacher, and referred to it as KD. To improve the effectiveness of distillation, various forms of knowledge transfer methods have been introduced which could be mainly categorized into three types: (i) logit-based [4,6,23,39,55,57], (ii) representation-based [36,49,53,54], and (iii) relationship-based [25,30,33,50].

2.2 Training a Customized Teacher for KD

In the literature, only a few works trained teachers specifically tailored for KD. [48] attempted to train a tolerant teacher which provides more secondary information to the student. They realized this via adding an extra term to facilitating a few secondary classes to emerge to complement the primary class. [13] regularized the teacher utilizing early-stopping during the training. Nevertheless, achieving optimal results may necessitate a thorough hyperparameter search, as the epoch number for identifying the best checkpoint can be particularly sensitive to various training settings, such as the learning rate schedule. Additionally, it is feasible to save multiple early teacher checkpoints, allowing the student to be distilled from them sequentially [20].

In addition, [45] stated that a checkpoint in the middle of the training procedure, often serves as a better teacher compared to the fully converged model. The authors in [15] used Lipschitz regularization so that the teacher can can learn the label distribution of the underlying dataset. Also [52] trained the teacher to have high conditional mutual information so that it can better predict the true BCPD. [40] trained the teachers to have more dispersed soft probabilities.

2.3 Training DNNs Using MSE

Mean squared error (MSE) loss serves as a prevalent choice for training DNNs, particularly in the context of regression tasks. This loss function is widely embraced when the objective is to predict continuous values, and its formulation involves calculating the average of the squared differences between predicted and actual values. Despite its established efficacy in regression scenarios, the landscape of loss functions in the realm of DNNs is vast and dynamic.

In contemporary practices, DNNs dedicated to classification tasks predominantly leverage the CE loss function. This method has gained substantial empirical favor, often surpassing MSE in the context of classification-oriented objectives. However, the empirical superiority of CE over MSE remains a topic of ongoing exploration. Notably, the existing body of literature does not uniformly advocate for a distinct advantage of CE in all scenarios.

Recent insights, as highlighted by the study conducted by [18], challenge the prevailing notion by showcasing that models trained with MSE not only hold their ground against their CE-trained counterparts across a diverse spectrum of tasks and settings but, intriguingly, exhibit superior classification performance in the majority of experimental conditions. These findings prompt a reevaluation of the perceived hierarchy between MSE and CE, underscoring the need for nuanced considerations when selecting the most suitable loss function based on the specificities of the task at hand. In light of such empirical observations, the applicability and performance of MSE in DNN training extend beyond the traditional confines of regression, warranting a more comprehensive exploration of its utility across various domains and applications.

3 Notation and Preliminaries

3.1 Notation

For a positive integer C, let $[C] \triangleq \{1, \ldots, C\}$. We use bold lowercase letters (e.g., $\boldsymbol{p}$) to represent vectors. Denote by $\boldsymbol{p}[i]$ the i-th element of vector $\boldsymbol{p}$. Also, $\{\boldsymbol{p}[c]\}_{c \in \mathcal{C}}$ is the set of all components of $\boldsymbol{p}$ with indices from the set $\mathcal{C}$. For two vectors $\boldsymbol{u}$ and $\boldsymbol{v}$, denote by $\boldsymbol{u} \cdot \boldsymbol{v}$ their inner product. We use $|\mathcal{C}|$ to denote the cardinality of a set $\mathcal{C}$. The transpose operation is denoted by $(\cdot)^\mathrm{T}$.

The cross-entropy of two probability distributions $\boldsymbol{p}_1$ and $\boldsymbol{p}_2$ is defined as $H(\boldsymbol{p}_1, \boldsymbol{p}_2) = \sum_{c=1}^{C} -\boldsymbol{p}_1[c] \log \boldsymbol{p}_2[c]$. For a random variable x, denote by $\mathbb{P}_x$ its probability distribution, and by $\mathbb{E}_x[\cdot]$ the expected value operation w.r.t. x. For two random variables x and y, denote by $\mathbb{P}_{(x,y)}$ their joint distribution.

3.2 True Risk vs. Empirical Risk

In a classification task with C classes, a DNN could be regarded as a mapping $f_{\boldsymbol{\theta}} : \boldsymbol{x} \to \boldsymbol{p}_{\boldsymbol{x}}$, where $\boldsymbol{\theta}$ represents all the model parameters, $\boldsymbol{x} \in \mathbb{R}^d$ is an input image, and $\boldsymbol{p}_{\boldsymbol{x}} \in \Delta^C$, where Δ^C is the C dimensional probability simplex. Then, the classifier predicts the correct label of $\boldsymbol{x}$, denoted by y, as $\hat{y} = \arg\max_{c \in [C]} \boldsymbol{p}_{\boldsymbol{x}}[c]$. As such, the error rate of f is defined as $\epsilon = \Pr\{\hat{y} \neq y\}$, and its accuracy is equal to $1 - \epsilon$. One may learn such a classifier by minimizing the *true* risk

$$\begin{aligned} R(f_{\boldsymbol{\theta}}, \ell) &\triangleq \mathbb{E}_{(\boldsymbol{x},y)}\left[\ell\left(y, \boldsymbol{p}_{\boldsymbol{x}}\right)\right] = \mathbb{E}_{\boldsymbol{x}}\left[\mathbb{E}_{y|\boldsymbol{x}}\left[\ell\left(y, \boldsymbol{p}_{\boldsymbol{x}}\right)\right]\right] \\ &= \mathbb{E}_{\boldsymbol{x}}\left[(\boldsymbol{p}_{\boldsymbol{x}}^*)^\mathrm{T} \cdot \boldsymbol{\ell}(\boldsymbol{p}_{\boldsymbol{x}})\right], \end{aligned} \quad (1)$$

where $\ell(\cdot)$ is the loss function and $\boldsymbol{\ell}(\cdot) \triangleq [\ell(1,\cdot),\ldots,\ell(C,\cdot)]$ is the vector of loss function, and $\boldsymbol{p}_{\boldsymbol{x}}^* \triangleq [\Pr(y|\boldsymbol{x})]_{y\in[C]}$ is Bayes class probability distribution over the labels, i.e., the BCPD.

However, in a typical deep learning algorithm, both the probability density function of $\boldsymbol{x}$, namely $\mathbb{P}_{\boldsymbol{x}}$, and also $\boldsymbol{p}_{\boldsymbol{x}}^*$ are unknown. Hence, one may learn such a classifier by instead minimizing the *empirical* risk on a training sample $\mathcal{D} \triangleq \{(\boldsymbol{x}_n, y_n)\}_{n=1}^N$ defined as:

$$R_{\mathrm{emp}}(f_{\boldsymbol{\theta}}, \ell) \triangleq \frac{1}{N} \sum_{n\in[N]} \mathbf{y}_n^{\mathrm{T}} \cdot \boldsymbol{\ell}(\boldsymbol{p}_{\boldsymbol{x}_n}), \qquad (2)$$

where $\mathbf{y}_n^{\mathrm{T}}$ is the one-hot vector with its y_n-th entry set to one and all other entries set to zero. By comparing Eq. (1) and Eq. (2), we see that in Eq. (2): (i) $\mathbb{P}_{\boldsymbol{x}}$ is approximated by $\frac{1}{N}$, and (ii) $\boldsymbol{p}_{\boldsymbol{x}}^*$ is approximated by $\mathbf{y}$ which is an unbiased estimation of $\boldsymbol{p}_{\boldsymbol{x}}^*$. The former assumption is reasonable, however, the latter results in a significant loss in granularity. To delve deeper into this matter, it is crucial to recognize that images inherently carry a wealth of information, and the practice of assigning a one-hot vector to $\boldsymbol{p}_{\boldsymbol{x}}^*$ tends to lead to a significant loss of this information. As we will explore further in the subsequent subsection, KD emerges as a strategy that, to some extent, alleviates this issue, providing a mechanism to better transfer the nuanced information embedded within images.

3.3 Estimating BCPD by the Teacher in KD

In KD, the role of the teacher is to provide the student with a better estimate of $\boldsymbol{p}_{\boldsymbol{x}}^*$ compared to one-hot vectors. To elucidate, denote by $\boldsymbol{p}_{\boldsymbol{x}}^t$ and $\boldsymbol{p}_{\boldsymbol{x}}^s$ the pretrained teacher's and student's outputs to sample $\boldsymbol{x}$, respectively. Then, the student uses the teacher's estimate of $\boldsymbol{p}_{\boldsymbol{x}}^*$, and minimizes

$$R_{\mathrm{kd}}(f_{\boldsymbol{\theta}}, \ell) \triangleq \frac{1}{N} \sum_{n\in[N]} (\boldsymbol{p}_{\boldsymbol{x}_n}^t)^{\mathrm{T}} \cdot \boldsymbol{\ell}(\boldsymbol{p}_{\boldsymbol{x}_n}^s). \qquad (3)$$

Note that the one-hot vector $\mathbf{y}_n$ in Eq. (2) is now replaced by the teacher's output probability $\boldsymbol{p}_{\boldsymbol{x}_n}^t$ in Eq. (3). In fact, the effectiveness of KD lies in the fact that $\boldsymbol{p}_{\boldsymbol{x}_n}^t$ serves as a better estimate of $\boldsymbol{p}_{\boldsymbol{x}}^*$ compared to the one-hot vector $\mathbf{y}_n^{\mathrm{T}}$.

3.4 Student's Generalization Error and Accuracy

In this subsection, our objective is to identify the key characteristics that the estimated BCPD should possess in order to enhance the accuracy of the student. As shown by [26], the student's generalization error is upper-bounded as

$$\mathbb{E}\left[(R_{\mathrm{kd}}(f_{\boldsymbol{\theta}}, \ell) - R(f_{\boldsymbol{\theta}}, \ell))^2\right] \leq \frac{1}{N}\mathrm{Var}\left[(\boldsymbol{p}_{\boldsymbol{x}}^t)^{\mathrm{T}} \cdot \boldsymbol{\ell}(\boldsymbol{p}_{\boldsymbol{x}})\right] + \kappa \left(\mathbb{E}\left[\|\boldsymbol{p}_{\boldsymbol{x}}^t - \boldsymbol{p}_{\boldsymbol{x}}^*\|\right]\right)^2, \qquad (4)$$

where κ is a positive constant number. When N is large, which is commonly the case for datasets in existing literature, the second term will dominate the right-hand side of Eq. (4). This implies that smaller average $\|\bm{p}_x^t - \bm{p}_x^*\|_2$ will lead to $R_{\text{emp}}(f_{\bm{\theta}})$ being a better approximation of the true risk $R(f_{\bm{\theta}})$, minimizing it should then lead to a better learned model.

On the other hand, [35] empirically showed that the accuracy of the student is almost inversely proportional to $\|\bm{p}_x^t - \bm{p}_x^*\|$. In addition, [14] showed that the accuracy of the student is directly bounded by the MSE between teacher's prediction and BCPD through the Rademacher analysis.

Therefore, the quality of the estimates provided by the teacher to the student can be measured by the MSE between its output and to the true $\bm{p}_x^*$. Based on this, in the next section, we show that for an effective KD, the teacher should be indeed trained via MSE loss, and not CE loss.

4 Methodology

In this section, first in Sect. 4.1, we introduce a theorem demonstrating that training a DNN model with MSE (CE) loss function results in minimizing the MSE (CE) between its output and the BCPD. Then, in Sect. 4.2, we use a synthetic dataset to empirically validate the introduced theorem, and to show that (i) closeness in MSE sense does not necessarily mean closeness in CE sense, and (ii) for an effective KD the teacher's output should be close to the true BCPD in MSE sense. Then, we conclude that the teacher should be trained via MSE loss function.

4.1 MSE Loss vs. CE Loss

In a classification task with C classes, it has been shown that the risk in Eq. (1), for $\ell = \{CE, MSE\}$, is minimized when $\bm{p}_x = \bm{p}_x^*$ [21]. However, since the underlying $\mathbb{P}_{(x,y)}$ is unknown, the DNNs are trained to minimize the empirical risk in Eq. (2), and consequently they can only approximate the true BCPD[1].

Hence, a teacher trained by either CE or MSE can approximate the true BCPD. However, it is crucial to note that these two estimates differ, as precisely established in the following theorem.

Theorem 1. *For $\ell = \{CE, MSE\}$*

$$\min_{\theta} \mathbb{E}_{(x,y)} \left[\ell\left(y, \bm{p}_x\right) \right] \equiv \min_{\theta} \mathbb{E}_{(x,y)} \left[\ell\left(\bm{p}_x^*, \bm{p}_x\right) \right]. \tag{5}$$

Proof. Please refer to the *Supplementary materials*.

Theorem 1 implies that when minimizing the expected loss, the resulting model attempts to generate outputs that closely approximate those of an "ideal" model. More importantly, this degree of closeness is quantified by the loss function. Specifically, (i) if $\ell = $ MSE then a model trained to minimize the expected

---

[1] Aside from the fact that $\mathbb{P}_{(x,y)}$ is unknown, training cannot typically find the global minimum hindering the DNNs to give accurate BCPD.

squared error between y and $\boldsymbol{p}_x$ will generate outputs that minimize the expected squared error between $\boldsymbol{p}_x^*$ and $\boldsymbol{p}_x$; and (ii) if $\ell = $ CE then a model trained to minimize the expected cross-entropy between y and $\boldsymbol{p}_x$ will generate outputs that minimize the expected cross-entropy between $\boldsymbol{p}_x^*$ and $\boldsymbol{p}_x$.

Thus, the BCPD estimates provided by the teachers trained by CE loss and MSE loss are different; in that, the former estimate is close to the true BCPD in CE sense, also the latter estimate is close to the true BCPD in MSE sense. In the next subsection, we empirically show that for the student to have high accuracy, the teacher's estimate should be close to the true BCPD in MSE sense, rather than in CE sense.

4.2 MSE Proximity vs. CE Proximity

In this subsection, our intention is two-fold: (i) we aim to empirically show that although the student's accuracy is almost inversely proportional to the MSE between BCPD and teacher's output (as also demonstrated by [14,35]), such relationship does not exist between the student's accuracy and CE between BCPD and teacher's output; and (ii) to empirically validate Theorem 1. Toward this aim, since the true BCPD is unknown for the popular datasets in the literature, we generate a synthetic dataset whose BCPD is known.

- **Generating Dataset:** Inspired from [35], we generate a 3-class toy Gaussian dataset with 10^5 data points. The dataset is divided into training, validation, and test sets with a split ratio $[0.9, 0.05, 0.05]$.
  The sampling process is implemented as follows: we first choose the label y using a uniform distribution across all the 3 classes. Next, we sample $x|_{y=k} \sim \mathcal{N}(\mu_k, \sigma^2 I)$ as the input signal. Here, μ_k is a 30-dim vector with entries randomly selected from $\{-\delta_\mu, 0, \delta_\mu\}$. Then, we calculate the BCPD of the samples using the fact that $p^*(y|x) \propto p(x|y)p(y)$. Particularly, as y follows uniform distribution, we have $\boldsymbol{p}^*(y|x) = \frac{p(x|y=k)}{\sum_{j\neq k} p(x|y=j)}$. Following $p(x|y=k) \sim \mathcal{N}(\mu_k, \sigma^2 I)$, we find $\boldsymbol{p}^*(y|x)$ should have a Softmax form as

$$\boldsymbol{p}^*(y=k|x) = \frac{\exp(s_k)}{\sum_{j\neq k} \exp(s_j)}; \quad s_i = -\frac{1}{2\sigma^2}\|x - \mu_i\|_2^2. \tag{6}$$

- **Training Setup:** The underlying model (for both teacher and student) is a 3-layer MLP with ReLU activation function, and the hidden size is 128 for each layer. We set the learning rate as 5×10^{-4}, the batch size as 32, and the number of training epochs is 100. In our experiments, we set $\sigma = 4$ and $\delta_\mu = 1$.

Now, we conduct two sets of experiments as elaborated in the sequel.

Set 1: The Student's Accuracy as a Function of MSE($\boldsymbol{p}_x^*$, $\boldsymbol{p}_x$) and CE($\boldsymbol{p}_x^*$, $\boldsymbol{p}_x$). Here, we conduct some experiments to understand which type of

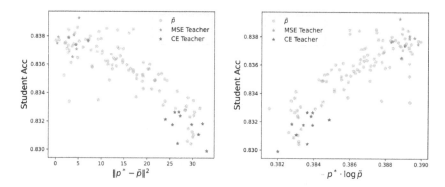

Fig. 1. Student accuracy as a function of (left) the MSE between p_x and $\tilde{p}_x$; and (right) CE between p_x and $\tilde{p}_x$. The gray dots are noisy versions of the true p_x. Also, the red and blue dots represent the points corresponding to the estimates provided by MSE and CE teachers, respectively.

proximity, i.e., whether proximity in CE sense or MSE sense, is better for the sake of the student accuracy. To this end, we generate 100 perturbed/noisy versions of p_x^* by adding some random noise to it, and denote any such perturbed version by $\tilde{p}_x^*$. Then, we train the student 100 times, each time using one of the noisy $\tilde{p}_x^*$, i.e., each time the student is trained via loss Eq. (3) with $p_{x_n}^t$ replaced by $\tilde{p}_x^*$. The resulting student's accuracy corresponding to these 100 noisy BCPD is depicted in Fig. 1, where in the left and right figure we set x-axis to the MSE and CE between the p_x^* and $\tilde{p}_x^*$, respectively (the points corresponding to noisy p_x^* are denoted by gray dots).

As seen in Fig. 1, the student's accuracy is almost inversely proportional to the MSE between p_x^* and $\tilde{p}_x^*$. Consequently, the teacher can enhance the student's performance by providing it with an estimate of p_x^* that is closer to it in MSE sense. On the other hand, as seen in the right figure depicted in Fig. 1, such relationship does not exist; in that, minimizing the CE between p_x^* and $\tilde{p}_x^*$ does not necessarily increase the student's performance.

Set 2: Empirically Validating Theorem 1. The previous set of experiments showed that the student's accuracy can be improved if it is given a BCPD estimate which is close to p_x^* in MSE sense. Here, we aim to empirically validate the Theorem 1, and therefore based on our discussions above, the teacher should be trained via MSE loss.

Hence, we train the teacher 10 times using MSE loss, and 10 times using CE loss. We denote the teacher's estimate of p_x^* by $\tilde{p}_{x,\text{MSE}}^*$ and $\tilde{p}_{x,\text{CE}}^*$ when it is trained by CE and MSE losses, respectively.

Now, once again we train the student using these two types of estimate (10 estimates for each type) and record the student's accuracy. The results are depicted in Fig. 1, where we used red and blue dots to show the points corresponding to $\tilde{p}_{x,\text{MSE}}^*$ and $\tilde{p}_{x,\text{CE}}^*$ estimates, respectively.

Observing the two figures we can conclude that (i) training the teacher using MSE/CE loss yields the BCPD estimate which is close to p_x^* in MSE/CE sense; (ii) the closeness in CE and MSE sense is rather different; and (iii) the teacher trained via MSE loss results in a better performance for the student model.

Lastly, we provide a toy example to understand why proximity in MSE sense is different from that in CE sense. Assume that the true BCPD is $p^* = [0.3, 0.7]$. Consider the following two estimates of the BCPD.

- Estimate 1, $E_1 = [0.29, 0.71]$. Here $\text{MSE}(p^*, E_1) = 0.01$ and $\text{CE}(p^*, E_1) = 0.610$.
- Estimate 2, $E_2 = [0.2, 0.8]$. Here $\text{MSE}(p^*, E_2) = 0.10$ and $\text{CE}(p^*, E_2) = 0.51$.

In the CE sense, E_2 is a superior estimator of p^*. However, from the standpoint of MSE, E_1 is a more accurate representation of p^*. Consequently, if a DNN is trained using the CE loss, it tends to produce an output akin to E_2 which is at a relatively greater distance from the true distribution p^* in the MSE sense.

5 Experiments

We conclude the paper with extensive experiments demonstrating the superior effectiveness of MSE teacher compared to CE teacher in different state-of-the-art KD methods. The outcomes of these experiments collectively contribute to a compelling argument for the preferential utilization of MSE teachers in the KD variants.

- **Terminology:** Hereafter, we refer to teachers trained by the CE and MSE losses as the "CE teacher" and "MSE teacher", respectively.
- **Organization:** The experiments are organized as follows: in Sect. 5.1 and Sect. 5.2, we conduct experiments on CIFAR-100 dataset; also Sect. 5.3 provides experiments on ImageNet dataset. In Sect. 5.4, we compare MSE and CE teachers in the semi-supervised distillation task. Lastly, we evaluate the performance of MSE teacher in binary classification knowledge distillation task in Sect. 5.5.
- **Plug-and-play nature of MSE teacher:** In all the experiments conducted in this section, when evaluating the performance of the MSE teacher, we refrain from tuning any hyper-parameters in the underlying knowledge transfer methods, i.e., all hyper-parameters remain the same as those used in the corresponding benchmark methods. We present the classification accuracies on the test set for both the teacher model and its distilled student model.

In addition, we should note that for training the MSE teacher, we use the same training setup as that used for the CE loss. One may further improve our results by further tuning the training hyper-parameters. In addition, similarly to the conventional KD methods, we use the same teacher for different students, without adjusting the teacher based on the specifics of each student model.

Table 1. The test accuracy (%) of student networks on CIFAR-100 (averaged over 5 runs), with teacher-student pairs of the same/different architectures. The subscript denotes the improvement achieved by replacing CE teacher with MSE teacher. We use **bold** numbers and asterisk (*) to denote the best results and to identify the results reproduced on our local machines, respectively.

	\multicolumn{12}{c}{Teachers and students with the **same** architectures.}											
Teacher	ResNet-56		ResNet-110		ResNet-110		WRN-40-2		WRN-40-2		VGG-13	
	CE	MSE	CE	MSE	CE	MSE	CE	MSE	CE	MSE	CE	MSE
Accuracy	72.34	72.14	74.31	73.43	74.31	73.43	75.61	74.99	75.61	74.99	74.64	73.20
Student	ResNet-20		ResNet-20		ResNet-32		WRN-16-2		WRN-40-1		VGG-8	
Accuracy	69.06		69.06		71.14		73.26		71.98		70.36	
AT	70.55	70.80 $_{+0.25}$	70.22	70.58 $_{+0.36}$	72.31	73.50 $_{+1.19}$	74.08	74.30 $_{+0.22}$	72.77	73.05 $_{+0.28}$	71.43	71.72 $_{+0.29}$
PKT	70.34	70.84 $_{+0.50}$	70.25	70.55 $_{+0.30}$	72.61	72.90 $_{+0.29}$	74.54	74.85 $_{+0.31}$	73.45	74.10 $_{+0.65}$	72.88	73.10 $_{+0.47}$
SP	69.67	70.77 $_{+1.10}$	70.04	70.75 $_{+0.71}$	72.69	73.34 $_{+0.65}$	73.83	74.60 $_{+0.77}$	72.43	73.30 $_{+0.87}$	72.68	73.19 $_{+0.61}$
CC	69.63	69.99 $_{+0.36}$	69.48	69.89 $_{+0.41}$	71.48	71.75 $_{+0.27}$	73.56	73.87 $_{+0.31}$	72.21	72.50 $_{+0.29}$	70.71	71.00 $_{+0.29}$
RKD	69.61	70.50 $_{+0.89}$	69.25	70.20 $_{+0.95}$	71.82	72.62 $_{+0.80}$	73.35	73.66 $_{+0.31}$	72.22	72.67 $_{+0.45}$	71.48	71.88 $_{+0.40}$
VID	70.38	70.61 $_{+0.23}$	70.16	70.49 $_{+0.33}$	72.61	73.05 $_{+0.44}$	74.11	74.44 $_{+0.33}$	73.30	73.58 $_{+0.28}$	71.23	71.57 $_{+0.34}$
CRD	71.16	71.43 $_{+0.27}$	71.46	71.87 $_{+0.41}$	73.48	74.03 $_{+0.55}$	75.48	75.85 $_{+0.37}$	74.14	74.86 $_{+0.72}$	73.94	74.25 $_{+0.31}$
REVIEW	71.89	72.15 $_{+0.26}$	71.65*	72.04 $_{+0.39}$	73.89	74.02 $_{+0.13}$	76.12	76.29 $_{+0.17}$	75.09	75.33 $_{+0.24}$	74.84	74.90 $_{+0.06}$
DKD	71.97	72.27 $_{+0.30}$	71.51*	71.88 $_{+0.37}$	74.11	74.32 $_{+0.21}$	76.24	76.76 $_{+0.52}$	74.81	75.53 $_{+0.72}$	74.68	74.92 $_{+0.24}$
HSAKD	72.58	**72.74** $_{+0.16}$	72.64*	**73.07** $_{+0.43}$	74.97*	**75.52** $_{+0.55}$	77.20	**77.55** $_{+0.35}$	77.00	**77.32** $_{+0.32}$	75.42*	**75.76** $_{+0.34}$

	\multicolumn{12}{c}{Teachers and students with **different** architectures.}											
Teacher	ResNet-50		ResNet-50		ResNet-32×4		ResNet-32×4		WRN-40-2		VGG-13	
	CE	MSE	CE	MSE	CE	MSE	CE	MSE	CE	MSE	CE	MSE
Accuracy	79.34	74.54	79.34	74.54	79.41	75.24	79.41	75.24	75.61	74.99	74.64	73.20
Student	MobileNetV2		VGG-8		ShuffleNetV1		ShuffleNetV2		ShuffleNetV1		MobileNetV2	
Accuracy	64.60		70.36		70.50		71.82		70.50		64.60	
AT	58.58	59.63 $_{+1.05}$	71.84	72.12 $_{+0.28}$	71.73	72.06 $_{+0.33}$	72.73	74.11 $_{+1.38}$	73.32	74.33 $_{+1.01}$	59.40	62.07 $_{+2.67}$
PKT	66.52	67.02 $_{+0.50}$	73.10	73.45 $_{+0.35}$	74.10	74.81 $_{+0.71}$	74.69	76.34 $_{+1.65}$	73.89	75.39 $_{+1.50}$	67.13	68.08 $_{+0.95}$
SP	68.08	69.00 $_{+0.92}$	73.34	74.04 $_{+0.70}$	73.48	74.57 $_{+1.09}$	74.56	75.70 $_{+1.14}$	74.52	75.72 $_{+1.20}$	66.30	67.03 $_{+0.73}$
CC	65.43	65.90 $_{+0.47}$	70.25	70.90 $_{+0.65}$	71.14	71.77 $_{+0.63}$	71.29	73.02 $_{+1.73}$	71.38	71.80 $_{+0.42}$	64.86	65.05 $_{+0.19}$
RKD	64.43	64.88 $_{+0.45}$	71.50	72.05 $_{+0.55}$	72.28	73.19 $_{+0.91}$	73.21	73.62 $_{+0.41}$	72.21	73.25 $_{+1.04}$	64.52	65.32 $_{+0.80}$
VID	67.57	67.77 $_{+0.20}$	70.30	70.55 $_{+0.25}$	73.38	73.89 $_{+0.51}$	73.40	74.67 $_{+1.27}$	73.61	75.03 $_{+1.42}$	65.56	65.82 $_{+0.26}$
CRD	69.11	69.15 $_{+0.04}$	74.30	74.55 $_{+0.25}$	75.11	75.81 $_{+0.70}$	75.65	76.54 $_{+0.89}$	76.05	76.43 $_{+0.38}$	69.70	69.77 $_{+0.07}$
REVIEW	69.89	70.03 $_{+0.14}$	73.43*	73.90 $_{+0.47}$	77.45	77.78 $_{+0.33}$	77.78	78.81 $_{+0.03}$	77.14	77.25 $_{+0.11}$	70.37	70.81 $_{+0.44}$
DKD	70.35	71.40 $_{+1.05}$	73.94*	75.14 $_{+1.17}$	76.45	77.15 $_{+0.70}$	77.07	77.52 $_{+0.45}$	76.70	77.30 $_{+0.60}$	69.71	70.22 $_{+0.51}$
HSAKD	71.83*	**72.67** $_{+0.84}$	75.87*	**76.34** $_{+0.47}$	79.51*	**79.81** $_{+0.30}$	79.93	**80.09** $_{+0.17}$	78.51	**78.82** $_{+0.31}$	71.09*	**72.40** $_{+1.31}$

5.1 CIFAR-100

This dataset comprises 50,000 training and 10,000 test color images, each of size 32×32, and is annotated for 100 classes [22].

- **Teacher-student pairs:** In line with the configurations of CRD [41], we use a couple of teacher-student pairs with identical and different network architectures for our experiments (see Table 1). We conduct each experiment over 5 independent runs and report the average accuracy (for the accuracy variances, refer to the *Supplementary materials*).
- **KD variants:** For comprehensive comparisons, we compare using a MSE teacher Vs. CE teacher in the existing state-of-the-art distillation methods, including KD [17], AT [54], PKT [32], SP [42], CC [33], RKD [31], VID [1], CRD [41], DKD [55], REVIEWKD [7], and HSAKD [49].
- **Training setup:** For all variants of knowledge distillation and settings in this paper, SGD is applied as the optimizer. We train the student for 240 epochs for all experiments with an initial learning rate of 0.05 by default, which will be decayed by factor of 0.1 at epoch 150, 180, 210. For MobileNetV2, ShuffleNetV1, and ShuffleNetV2, a smaller initial learning rate of 0.01 is used. We adopt batch size of 64. In addition, we report the hyper-parameters used for underlying KD variants in *Supplementary materials*.

Table 2. Additional experiments on CIFAR-100, when the feature based distillation methods are combined with conventional KD method. The test accuracy (%) of student networks on CIFAR-100 (averaged over 5 runs), with teacher-student pairs of the same/different architectures. The subscript denotes the improvement achieved by replacing CE teacher with MSE teacher.

Teacher	ResNet-110 CE	ResNet-110 MSE	WRN-40-2 CE	WRN-40-2 MSE	VGG-13 CE	VGG-13 MSE	ResNet-50 CE	ResNet-50 MSE	ResNet-32×4 CE	ResNet-32×4 MSE	VGG-13 CE	VGG-13 MSE
Accuracy	74.31	73.43	75.61	74.99	74.64	73.20	79.34	74.54	79.41	75.24	74.64	73.20
Student	ResNet-20		WRN-16-2		VGG-8		VGG-8		ShuffleNetV2		MobileNetV2	
Accuracy	69.06		73.26		70.36		71.14		73.26		64.60	
AT+KD	70.97	71.34 +0.37	75.32	75.65 +0.33	73.48	73.89 +0.41	74.01	74.22 +0.21	75.39	77.63 +2.24	65.13	66.76 +1.63
PKT+KD	70.72	71.15 +0.43	75.33	75.73 +0.40	73.25	73.50 +0.25	73.61	73.80 +0.19	74.66	76.13 +1.47	68.13	68.89 +0.76
SP+KD	71.02	71.40 +0.38	74.98	75.63 +0.65	73.49	73.80 +0.31	73.52	73.97 +0.45	74.88	76.86 +1.98	68.41	69.37 +0.96
CC+KD	70.88	71.31 +0.43	75.09	75.48 +0.39	73.04	73.46 +0.42	73.48	70.79	74.71	76.29 +1.58	68.02	68.39 +0.37
RKD+KD	70.77	71.48 +0.71	74.89	75.61 +0.52	72.97	73.30 +0.33	73.51	73.86 +0.35	74.55	75.82 +1.27	67.87	68.38 +0.51
VID+KD	71.10	71.53 +0.43	75.14	75.62 +0.48	73.19	73.47 +0.28	73.46	73.88 +0.42	74.85	75.97 +1.12	68.27	68.75 +0.48
CRD+KD	71.56	71.95 +0.39	75.64	75.85 +0.21	74.29	74.60 +0.31	74.58	74.98 +0.40	76.05	76.93 +0.88	69.94	70.23 +0.29
REVIEW+KD	71.78*	72.01 +0.23	76.22*	76.35 +0.13	74.96*	75.07 +0.11	73.89*	74.44 +0.55	77.89*	78.10 +0.21	71.05*	71.86 +0.81

- **Results:** The results are reported in Table 1, where the upper-part of the table comprises the feature-based methods that are combined with KD, and the lower-part contains three logit-based KD variants.

  By noting the results in Table 1, the following observations could be made:

  – Substituting the CE teacher with the MSE teacher in the benchmark KD methods consistently leads to an enhancement in the student's performance. This improvement can reach up to 2.67%.
  – The improvement achieved in teacher-student pairs with different architectures is notably more substantial (these pairs are the last three columns of Table 1).
  – The accuracy of the teacher experiences a slight decline when employing the MSE loss. This observation underscores the distinction between training the teacher model for KD and training it solely for optimizing its individual performance. It affirms that these are distinct tasks, each with its own set of considerations and trade-offs.

5.2 Additional Experiments on CIFAR-100

In this subsection, we follow the CRD paper [41], and combine the feature based distillation methods with conventional KD for achieving a higher performance. The results are listed in Table 2. As seen, the MSE teacher consistently yield a better accuracy compared to its CE counterpart.

Table 3. Top-1 and Top-5 student's test accuracy (%) on ImageNet validation set for 4 different KD methods using CE and MSE teachers (RN and MN stand for ResNet and MobileNet, respectively). **Bold** numbers and asterisk (*) to denote the best results and to identify the results reproduced on our local machines, respectively.

Teacher-Student		Teacher Performance		KD		DKD		REVIEW + KD		CRD +KD	
		Top1	Top5	Top1	Top5	Top1	Top5	Top1	Top5	Top1	Top5
RN34-R18	CE	73.31	91.42	71.03	90.05	71.70	90.41	71.84*	90.77*	71.38	90.49
	MSE	71.66	90.83	71.58	90.77	71.93	91.23	**72.16**	90.93	71.57	90.36
RN50-MNV2	CE	76.13	92.86	70.50	89.80	72.05	91.05	72.56*	91.00*	71.37*	90.41*
	MSE	73.88	90.97	70.92	90.08	72.34	91.17	**72.91**	92.38	71.58	90.62

5.3 ImageNet

ImageNet [37] is a large-scale dataset used in visual recognition tasks, containing around 1.2 million training and 50K validation images.

- **Teacher-student pairs:** Following the settings of [41, 51], we use 2 popular teacher-student pairs for our experiments (see Table 3).

- **Training setup:** We set the initial learning rate to 0.1 and divide the learning rate by 10 at 30, 60, and 90 epochs. We follow the standard training process but train for 20 more epochs (*i.e.*, 120 epochs in total). Weight decay is set to 0.0001.
- **Results:** We note that across all the knowledge transfer methods reported in Table 3, replacing the CE teacher by MSE teacher consistently leads to an increase in the student accuracy. For example, when considering teacher-student pairs ResNet34-ResNet18 and ResNet50-MobileNetV2, the increase in the student's Top-1 accuracy in KD is 0.55% and 0.42%, respectively.

It is important to highlight that achieving such gains over the ImageNet dataset is considered substantial. Taken together, the results obtained across both CIFAR-100 and ImageNet datasets underscore the effectiveness of opting for MSE teachers over their CE counterparts.

5.4 MSE Teacher in Semi-Supervised Distillation

Semi-supervised learning [8,9, 11,12,19,24,46,56] is a popular technique due to its ability to generate pseudo-labels for larger unlabeled dataset. In the context of KD, the teacher model has the responsibility of generating pseudo-labels for new, unlabeled examples (beyond its traditional function of providing soft targets during training).

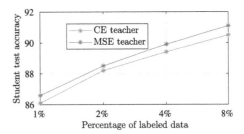

Fig. 2. The student's accuracy in semi supervised distillation for CE and MSE teachers.

To assess the MSE teacher's performance in a semi-supervised learning scenario, we conducted experiments using the CIFAR-10 dataset, following the settings outlined in [8], with the student model being ResNet18. In this experimental setup, although the dataset consists of 50,000 training images, only a small percentage of them are labeled—specifically, 1% (500), 2% (1000), 4% (2000), and 8% (4000). The outcomes are visualized in Fig. 2, where the student accuracy is plotted against the number of labeled samples. The results are averaged over three independent runs. As observed, the results demonstrate that not only can the MSE teacher effectively perform in a semi-supervised distillation scenario, but it also surpasses the performance of the CE teacher.

5.5 Binary Classification on Customized CIFAR-{10, 100}

The common understanding is that the enhancement in the accuracy of a student model in binary classification tends to be more modest compared to that observed in multi-class classification scenarios. This discrepancy arises due to the inherent limitations on the amount of information transferred from the teacher to the student network in binary classification settings, as documented in several studies [10,29,38,43,44,47].

In this section, we want to empirically verify the effectiveness of MSE teacher in binary classification tasks. To this end, we create three binary classification datasets from CIFAR-$\{10, 100\}$ datasets as explained in the sequel.

- **Dataset 1:** Following a similar approach to that described in [29], we construct the CIFAR $- 2 \times 5$ dataset, wherein the input distribution exhibits a "sub-classes" structure. Specifically, we merge the first 5 classes of CIFAR-10 to form class one, while the remaining 5 classes constitute class two.
- **Dataset 2:** We keep the training/testing samples from only two classes in the CIFAR-100 dataset: those belonging to class 20 and class 40, creating a dataset referred to as CIFAR-20-40.
- **Dataset 3:** Similar to dataset 2, but we keep class 50 and 70 from CIFAR-100, which we refer to as CIFAR-50-70.

Now, we use VGG-13 and MobileNetv2-0.1 as the teacher and student models, respectively; and use conventional KD, AT, CC and VID to distill knowledge from CE and MSE teachers to the student.

The results for all three tasks are summarized in Table 4, with the reported values representing the average across five distinct runs. As seen, the MSE teacher yields a better student's accuracy all the cases. Also, it is worth noting that in some cases, for instance, AT over CIFAR $- 2 \times 5$ dataset, the distillation method hurts the accuracy of the student.

Table 4. The student's accuracy (%) in binary classification knowledge distillation on variants of CIFAR dataset. The results are averaged over five runs.

Dataset	CIFAR $- 2 \times 5$		CIFAR-26-45		CIFAR-50-74	
Student Acc.	76.94		65.50		66.30	
Teacher	CE	MSE	CE	MSE	CE	MSE
KD	77.00	77.34	69.70	70.15	71.40	71.23
AT	67.39	68.21	64.70	65.33	67.10	69.12
CC	77.19	77.64	70.30	70.83	69.70	71.38
VID	77.16	77.47	69.25	69.88	69.80	69.97

6 Conclusion

This paper elucidated the significance of training the teacher model with MSE loss, which effectively minimizes the MSE between its output and BCPD. This approach aligns with the core responsibility of the teacher, namely, providing the student with a BCPD estimate that closely resembles it in terms of MSE.

Through a comprehensive series of experiments, we demonstrated the efficacy of substituting the conventional teacher trained with CE loss with one trained using MSE loss in state-of-the-art KD methods. Notably, this substitution consistently enhanced the student's accuracy, leading to improvements of up to 2.6%. In addition, we empirically showed the superior performance of MSE teacher in semi-supervised distillation task.

References

1. Ahn, S., Hu, S., Damianou, A., Lawrence, N., Dai, Z.: Variational information distillation for knowledge transfer, pp. 9155–9163 (2019). https://doi.org/10.1109/CVPR.2019.00938
2. Allen-Zhu, Z., Li, Y.: Towards understanding ensemble, knowledge distillation and self-distillation in deep learning. arXiv preprint arXiv:2012.09816 (2020)
3. Anil, R., Pereyra, G., Passos, A., Ormandi, R., Dahl, G.E., Hinton, G.E.: Large scale distributed neural network training through online distillation. arXiv preprint arXiv:1804.03235 (2018)
4. Beyer, L., Zhai, X., Royer, A., Markeeva, L., Anil, R., Kolesnikov, A.: Knowledge distillation: a good teacher is patient and consistent. In: Proceedings of the IEEE/CVF Conference on Computer Vision and Pattern Recognition, pp. 10925–10934 (2022)
5. Buciluǎ, C., Caruana, R., Niculescu-Mizil, A.: Model compression. In: Proceedings of the 12th ACM SIGKDD International Conference on Knowledge Discovery and Data Mining, pp. 535–541 (2006)
6. Chen, D., Mei, J.P., Wang, C., Feng, Y., Chen, C.: Online knowledge distillation with diverse peers. In: Proceedings of the AAAI Conference on Artificial Intelligence, pp. 3430–3437 (2020)
7. Chen, P., Liu, S., Zhao, H., Jia, J.: Distilling knowledge via knowledge review. In: Proceedings of the IEEE/CVF Conference on Computer Vision and Pattern Recognition, pp. 5008–5017 (2021)
8. Chen, T., Kornblith, S., Swersky, K., Norouzi, M., Hinton, G.E.: Big self-supervised models are strong semi-supervised learners. Adv. Neural. Inf. Process. Syst. **33**, 22243–22255 (2020)
9. Chi, Z., Gu, L., Liu, H., Wang, Y., Yu, Y., Tang, J.: MetaFSCIL: a meta-learning approach for few-shot class incremental learning. In: Proceedings of the IEEE/CVF Conference on Computer Vision and Pattern Recognition, pp. 14166–14175 (2022)
10. Chi, Z., et al.: Adapting to distribution shift by visual domain prompt generation. In: The Twelfth International Conference on Learning Representations (2024)
11. Chi, Z., Mohammadi Nasiri, R., Liu, Z., Lu, J., Tang, J., Plataniotis, K.N.: All at once: temporally adaptive multi-frame interpolation with advanced motion modeling. In: Vedaldi, A., Bischof, H., Brox, T., Frahm, J.-M. (eds.) ECCV 2020, Part XXVII. LNCS, vol. 12372, pp. 107–123. Springer, Cham (2020). https://doi.org/10.1007/978-3-030-58583-9_7
12. Chi, Z., Wang, Y., Yu, Y., Tang, J.: Test-time fast adaptation for dynamic scene deblurring via meta-auxiliary learning. In: Proceedings of the IEEE/CVF Conference on Computer Vision and Pattern Recognition, pp. 9137–9146 (2021)
13. Cho, J.H., Hariharan, B.: On the efficacy of knowledge distillation. In: Proceedings of the IEEE/CVF International Conference on Computer Vision, pp. 4794–4802 (2019)

14. Dao, T., Kamath, G.M., Syrgkanis, V., Mackey, L.: Knowledge distillation as semi-parametric inference. In: International Conference on Learning Representations (2020)
15. Dong, C., Liu, L., Shang, J.: Toward student-oriented teacher network training for knowledge distillation. In: The Twelfth International Conference on Learning Representations (2023)
16. Hamidi, S.M.: Training neural networks on remote edge devices for unseen class classification. IEEE Signal Process. Lett. **31**, 1004–1008 (2024). https://doi.org/10.1109/LSP.2024.3383948
17. Hinton, G., Vinyals, O., Dean, J.: Distilling the knowledge in a neural network. arXiv preprint arXiv:1503.02531 (2015)
18. Hui, L., Belkin, M.: Evaluation of neural architectures trained with square loss vs cross-entropy in classification tasks. In: International Conference on Learning Representations (2020)
19. Iliopoulos, F., Kontonis, V., Baykal, C., Menghani, G., Trinh, K., Vee, E.: Weighted distillation with unlabeled examples. Adv. Neural. Inf. Process. Syst. **35**, 7024–7037 (2022)
20. Jin, X., et al.: Knowledge distillation via route constrained optimization. In: Proceedings of the IEEE/CVF International Conference on Computer Vision, pp. 1345–1354 (2019)
21. Kanaya, F., Miyake, S.: Bayes statistical behavior and valid generalization of pattern classifying neural networks. IEEE Trans. Neural Netw. **2**(4), 471–475 (1991)
22. Krizhevsky, A., Hinton, G., et al.: Learning multiple layers of features from tiny images (2009)
23. Li, Z., Huang, Y., Chen, D., Luo, T., Cai, N., Pan, Z.: Online knowledge distillation via multi-branch diversity enhancement. In: Proceedings of the Asian Conference on Computer Vision (2020)
24. Liu, H., et al.: Few-shot class-incremental learning via entropy-regularized data-free replay. In: Avidan, S., Brostow, G., Cissé, M., Farinella, G.M., Hassner, T. (eds.) ECCV 2022. LNCS, vol. 13684, pp. 146–162. Springer, Cham (2022). https://doi.org/10.1007/978-3-031-20053-3_9
25. Liu, Y., et al.: Knowledge distillation via instance relationship graph. In: Proceedings of the IEEE/CVF Conference on Computer Vision and Pattern Recognition, pp. 7096–7104 (2019)
26. Menon, A.K., Rawat, A.S., Reddi, S., Kim, S., Kumar, S.: A statistical perspective on distillation. In: International Conference on Machine Learning, pp. 7632–7642. PMLR (2021)
27. Mirzadeh, S.I., Farajtabar, M., Li, A., Levine, N., Matsukawa, A., Ghasemzadeh, H.: Improved knowledge distillation via teacher assistant. In: Proceedings of the AAAI Conference on Artificial Intelligence, vol. 34, pp. 5191–5198 (2020)
28. Mobahi, H., Farajtabar, M., Bartlett, P.: Self-distillation amplifies regularization in Hilbert space. Adv. Neural. Inf. Process. Syst. **33**, 3351–3361 (2020)
29. Müller, R., Kornblith, S., Hinton, G.: Subclass distillation. arXiv preprint arXiv:2002.03936 (2020)
30. Park, W., Kim, D., Lu, Y., Cho, M.: Relational knowledge distillation. In: Proceedings of the IEEE Conference on Computer Vision and Pattern Recognition, pp. 3967–3976 (2019)
31. Park, W., Kim, D., Lu, Y., Cho, M.: Relational knowledge distillation, pp. 3967–3976 (2019)
32. Passalis, N., Tefas, A.: Learning deep representations with probabilistic knowledge transfer (2018)

33. Peng, B., et al.: Correlation congruence for knowledge distillation, pp. 5006–5015 (2019). https://doi.org/10.1109/ICCV.2019.00511
34. Phuong, M., Lampert, C.: Towards understanding knowledge distillation. In: International Conference on Machine Learning, pp. 5142–5151. PMLR (2019)
35. Ren, Y., Guo, S., Sutherland, D.J.: Better supervisory signals by observing learning paths. In: International Conference on Learning Representations (2021)
36. Romero, A., Ballas, N., Kahou, S.E., Chassang, A., Gatta, C., Bengio, Y.: FitNets: hints for thin deep nets. arXiv preprint arXiv:1412.6550 (2014)
37. Russakovsky, O., et al.: ImageNet large scale visual recognition challenge. Int. J. Comput. Vision **115**, 211–252 (2015)
38. Sajedi, A., Plataniotis, K.N.: On the efficiency of subclass knowledge distillation in classification tasks. arXiv preprint arXiv:2109.05587 (2021)
39. Stanton, S., Izmailov, P., Kirichenko, P., Alemi, A.A., Wilson, A.G.: Does knowledge distillation really work? Adv. Neural. Inf. Process. Syst. **34**, 6906–6919 (2021)
40. Tan, C., Liu, J.: Improving knowledge distillation with a customized teacher. IEEE Trans. Neural Netw. Learn. Syst. (2022)
41. Tian, Y., Krishnan, D., Isola, P.: Contrastive representation distillation (2020)
42. Tung, F., Mori, G.: Similarity-preserving knowledge distillation. In: 2019 IEEE/CVF International Conference on Computer Vision (ICCV), pp. 1365–1374 (2019). https://api.semanticscholar.org/CorpusID:198179476
43. Tzelepi, M., Passalis, N., Tefas, A.: Efficient online subclass knowledge distillation for image classification. In: 2020 25th International Conference on Pattern Recognition (ICPR), pp. 1007–1014 (2021). 10.1109/ICPR48806.2021.9411995
44. Tzelepi, M., Passalis, N., Tefas, A.: Online subclass knowledge distillation. Expert Syst. Appl. **181**, 115132 (2021)
45. Wang, C., Yang, Q., Huang, R., Song, S., Huang, G.: Efficient knowledge distillation from model checkpoints. Adv. Neural. Inf. Process. Syst. **35**, 607–619 (2022)
46. Wu, Y., Chi, Z., Wang, Y., Feng, S.: MetaGCD: learning to continually learn in generalized category discovery. In: Proceedings of the IEEE/CVF International Conference on Computer Vision, pp. 1655–1665 (2023)
47. Wu, Y., Chi, Z., Wang, Y., Plataniotis, K.N., Feng, S.: Test-time domain adaptation by learning domain-aware batch normalization. In: Proceedings of the AAAI Conference on Artificial Intelligence, vol. 38, pp. 15961–15969 (2024)
48. Yang, C., Xie, L., Qiao, S., Yuille, A.L.: Training deep neural networks in generations: a more tolerant teacher educates better students. In: Proceedings of the AAAI Conference on Artificial Intelligence, vol. 33, pp. 5628–5635 (2019)
49. Yang, C., An, Z., Cai, L., Xu, Y.: Hierarchical self-supervised augmented knowledge distillation. arXiv preprint arXiv:2107.13715 (2021)
50. Yang, C., Zhou, H., An, Z., Jiang, X., Xu, Y., Zhang, Q.: Cross-image relational knowledge distillation for semantic segmentation. In: Proceedings of the IEEE/CVF Conference on Computer Vision and Pattern Recognition, pp. 12319–12328 (2022)
51. Yang, J., Martinez, B., Bulat, A., Tzimiropoulos, G.: Knowledge distillation via softmax regression representation learning. In: International Conference on Learning Representations (2020)
52. Ye, L., Hamidi, S.M., Tan, R., Yang, E.H.: Bayes conditional distribution estimation for knowledge distillation based on conditional mutual information. In: The Twelfth International Conference on Learning Representations (2024). https://openreview.net/forum?id=yV6wwEbtkR

53. Yim, J., Joo, D., Bae, J., Kim, J.: A gift from knowledge distillation: fast optimization, network minimization and transfer learning. In: Proceedings of the IEEE Conference on Computer Vision and Pattern Recognition, pp. 4133–4141 (2017)
54. Zagoruyko, S., Komodakis, N.: Paying more attention to attention: improving the performance of convolutional neural networks via attention transfer (2017). https://arxiv.org/abs/1612.03928
55. Zhao, B., Cui, Q., Song, R., Qiu, Y., Liang, J.: Decoupled knowledge distillation. arXiv preprint arXiv:2203.08679 (2022)
56. Zhong, T., Chi, Z., Gu, L., Wang, Y., Yu, Y., Tang, J.: Meta-DMoE: adapting to domain shift by meta-distillation from mixture-of-experts. Adv. Neural. Inf. Process. Syst. **35**, 22243–22257 (2022)
57. Zhu, X., Gong, S., et al.: Knowledge distillation by on-the-fly native ensemble. Adv. Neural. Inf. Process. Syst. **31** (2018)

Tight and Efficient Upper Bound on Spectral Norm of Convolutional Layers

Ekaterina Grishina[✉][iD], Mikhail Gorbunov[iD], and Maxim Rakhuba[iD]

HSE University, Moscow, Russia
ergrishina@edu.hse.ru

Abstract. Controlling the spectral norm of the Jacobian matrix, which is related to the convolution operation, has been shown to improve generalization, training stability and robustness in CNNs. Existing methods for computing the norm either tend to overestimate it or their performance may deteriorate quickly with increasing the input and kernel sizes. In this paper, we demonstrate that the tensor version of the spectral norm of a four-dimensional convolution kernel, up to a constant factor, serves as an upper bound for the spectral norm of the Jacobian matrix associated with the convolution operation. This new upper bound is independent of the input image resolution, differentiable and can be efficiently calculated during training. Through experiments, we demonstrate how this new bound can be used to improve the performance of convolutional architectures.

Keywords: Spectral norm · Convolutional layer · Lipschitz constant

1 Introduction

Controlling spectral norm of convolutional layers' Jacobians is a way to make models more robust to input perturbations [2,20,22], increase generalization performance [16], prevent explosion of the gradients during back propagation [28]. In addition, bounds on spectral norm have been used to construct orthogonal convolutional layers [14,21].

Despite the advantages offered by controlling the spectral norm, its efficient computation for convolutional layers remains a challenging task. Finding the spectral norm is equivalent to finding the largest singular value of a matrix and the straightforward computation of the singular value decomposition (SVD) is not feasible due to the size of the convolution Jacobian T. Most of the other existing techniques rely on the input sizes of images, which can result in a significant computational load for high-resolution images or images with more than two spatial dimensions.

Supplementary Information The online version contains supplementary material available at https://doi.org/10.1007/978-3-031-73024-5_2.

In this work, we derive a new accurate upper bound for the spectral norm of the Jacobian $\|T\|_2$. For a kernel tensor $K \in \mathbb{R}^{c_{in} \times c_{out} \times h \times w}$, this bound can be computed with $\mathcal{O}(c_{in}c_{out}hw)$ operations, where c_{in}, c_{out} are the number of input and output channels and h, w are the filter sizes. Note that this complexity does not depend on the input or output image resolution. As a result, controlling the spectral norm during training comes at almost no additional cost.

To be more precise, our bound is based on a norm denoted as $\|K\|_\sigma$, which is induced by the kernel tensor K as a multilinear functional. Specifically, we show that $\|K\|_\sigma \leq \|T\|_2 \leq \sqrt{hw}\|K\|_\sigma$. The tensor norm $\|K\|_\sigma$ is well-known in applied multilinear algebra and can be efficiently computed by a higher-order generalization of a power method (HOPM) [3]. We will refer to our tensor norm upper bound $\sqrt{hw}\|K\|_\sigma$ as TN and will use it for several new regularization strategies. Compared to the existing approach [20], our bound is guaranteed to be more accurate. It is also applicable for zero-padded convolutions and can be modified to account for strided convolutions, which are both widely used in convolutional architectures. Moreover, it straightforwardly generalizes to convolutions with more than 2 spatial dimensions, for which efficient estimation of spectral properties is much more time-consuming.

2 Related Work

The work [18] was the first to develop an algorithm for the exact computation of singular values for the circular convolution. In this method, the authors obtain n^2 matrices of the shape $c_{out} \times c_{in}$ after applying the Fourier transform to a zero-padded kernel. Then the SVD of n^2 matrices is computed. The complexity (see Table 1) grows polynomially with n and can be computationally demanding for datasets with higher resolutions. Recently [7] extended the method of [18] for non-circular convolutions. The authors of [19] generalized the formula proposed in [18] to convolutions with more than two dimensions and arbitrary strides.

Several works [8,17] adapted power iteration to approximate the spectral norm of the convolution map. In this case, time complexity also depends on the input size n. The authors of [1] proposed an upper bound for the singular value of a doubly Toeplitz matrix, which can be efficiently computed, yet may be not very accurate for all filter sizes (see [6]). Works [5,6] introduced the so-called Gram iteration with superlinear convergence as opposed to the power iteration. However, it is applied to a padded kernel and, hence, [6] has a complexity similar to [18]. Both algorithms [5,6] can also be memory consuming, for example, in the algorithm [5] the spatial size of the kernel increases almost twice every iteration.

A differentiable upper bound independent of input size was introduced in [20] for the spectral norm of circular convolutions. It is computed as the minimum spectral norm of four unfoldings of convolution kernel multiplied by a constant factor $\sqrt{hw}$. Besides being a computationally efficient upper bound, it also provides new insights into the spectral normalization of GANs [15] and the regularization of CNNs from [30]. Using the Toeplitz matrix theory, [29] proved that the bound also holds in the non-circular case (zero-padded convolutions).

In this paper, we provide a new upper bound that combines the benefits of [20], while being both theoretically more precise and demonstrating noticeably higher accuracy in experiments. This new bound is based on the observation that the norm of a tensor can be bounded from above by the spectral norms of its unfoldings.

Table 1. Comparison of existing methods for computing the spectral norm of a convolution Jacobian. "Acc." stands for "accuracy", "Mem." for memory, "Diff." for "differentiable", "Ind. of n" for "independence of n". Accuracy and speed are measured experimentally. Our bound is differentiable, fast, accurate and independent of the input size n.

Method	Acc.	Mem.	Diff.	Ind. of n	Padding	Complexity ($\mathcal{O}$)
Power method [8,17]	+	+	+	−	zero	$n^2 c_{out} c_{in} hw$
Sedghi et al. [18]	+	−	−	−	circular	$n^2 c_{out} c_{in} (\log n + c_{in})$
F4 (Fantastic four) [20,29]	−	+	+	+	any	$c_{out} c_{in} hw$
LipBound [1]	−	+	+	+	zero	$c_{out} c_{in} hw$
Gram iteration [6]	+	−	+	−	circular	$n^2 c_{out} c_{in} (\log n + c_{in})$
PowerQR [7]	+	+	−	−	zero	$n^2 c_{out} c_{in} hw$
Tensor Norm (ours)	+	+	+	+	any	$c_{out} c_{in} hw$

3 Preliminaries and Notation

3.1 Background on Convolutions

For simplicity, let us first consider a one-dimensional convolution of a vector $X \in \mathbb{R}^n$ and a kernel vector $K \in \mathbb{R}^w$. In this paper we consider two types of convolution – with zero and circular paddings. Thanks to the linearity of convolution, it may be expressed as a matrix-vector product $Y = TX$. In the case of zero padding with an integer parameter $p \geq 0$, we have $Y_j = \sum_i K_{p+i} X_{j+i}$ and T is a matrix with constant values along diagonals – a Toeplitz matrix. In the case of circular padding, we have $Y_j = \sum_i K_{\lfloor \frac{w}{2} \rfloor + i} X_{(j+i) \bmod n}$ and T is a circulant, which is a special case of a Toeplitz matrix, where each column is a cyclic permutation of a previous one. Below we give examples of T for $n = 4, w = 3, p = 1$:

$$T = \begin{bmatrix} K_1 & K_2 & 0 & 0 \\ K_0 & K_1 & K_2 & 0 \\ 0 & K_0 & K_1 & K_2 \\ 0 & 0 & K_0 & K_1 \end{bmatrix} \begin{pmatrix} \text{zero} \\ \text{padding} \end{pmatrix}, \quad T = \begin{bmatrix} K_1 & K_2 & 0 & K_0 \\ K_0 & K_1 & K_2 & 0 \\ 0 & K_0 & K_1 & K_2 \\ K_2 & 0 & K_0 & K_1 \end{bmatrix} \begin{pmatrix} \text{circular} \\ \text{padding} \end{pmatrix}.$$

Let us now consider multi-channel two-dimensional convolution with stride 1 for ease of presentation. Larger stride values will be discussed separately. Let the convolution operation be given by a kernel tensor $K \in \mathbb{R}^{c_{out} \times c_{in} \times h \times w}$ applied to an input $X \in \mathbb{R}^{c_{in} \times n \times n}$ resulting in an output $Y \in \mathbb{R}^{c_{out} \times n \times n}$. The Jacobian of the convolution can be represented as a matrix $T \in \mathbb{R}^{c_{out} n^2 \times c_{in} n^2}$ satisfying

$$vec(Y) = T vec(X), \qquad (1)$$

where $vec(X)$ represents reshaping into a vector with $c_{in} n^2$ entries using colexicographic order (column-major).

Similarly to the one-dimensional case, zero padding that preserves spatial size leads to a matrix T that can be represented as a doubly block Toeplitz matrix. In more detail, each of its blocks B_k is a block Toeplitz matrix with unstructured dense blocks $T_{k,l} = K_{:,:,k+h_1,l+w_1} \in \mathbb{R}^{c_{out} \times c_{in}}$.

$$T = \begin{bmatrix} B_0 & \cdots & B_{h_2} & 0 & \cdots & 0 \\ \vdots & B_0 & \ddots & \ddots & \ddots & \vdots \\ B_{-h_1} & \ddots & \ddots & \ddots & \ddots & 0 \\ 0 & \ddots & \ddots & \ddots & \ddots & B_{h_2} \\ \vdots & \ddots & \ddots & \ddots & B_0 & \vdots \\ 0 & \cdots & 0 & B_{-h_1} & \cdots & B_0 \end{bmatrix}, \quad B_k = \begin{bmatrix} T_{k,0} & \cdots & T_{k,w_2} & 0 & \cdots & 0 \\ \vdots & T_{k,0} & \ddots & \ddots & \ddots & \vdots \\ T_{k,-w_1} & \ddots & \ddots & \ddots & \ddots & 0 \\ 0 & \ddots & \ddots & \ddots & \ddots & T_{k,w_2} \\ \vdots & \ddots & \ddots & \ddots & T_{k,0} & \vdots \\ 0 & \cdots & 0 & T_{k,-w_1} & \cdots & T_{k,0} \end{bmatrix}$$

where h_1, h_2 and w_1, w_2 subject to $h = h_1 + h_2 + 1, w = w_1 + w_2 + 1$ depend on the padding size in height and width. Depending on the order of vectorization, T may be either doubly block Toeplitz (as described above) or multi-block doubly Toeplitz. Nevertheless, singular values of the Jacobian remain the same, see, e.g. [29, Lemma 1].

Another way to maintain the input size is by using circular padding. In this case the layer "wraps around" and takes pixels from the opposite side of the image when the kernel gets beyond the edges. The corresponding Jacobian is a doubly block circulant matrix. It is a special case of a doubly block Toeplitz matrix with the entries satisfying $C_{k,l} = C_{(n-l) \bmod n, (n-k) \bmod n}$.

3.2 Matrix and Tensor Norms

Let us introduce several definitions from multilinear algebra. First, we define the Frobenius inner product for the d-dimensional tensors $A, B \in \mathbb{C}^{n_1 \times \cdots \times n_d}$ and the Frobenius norm:

$$\langle A, B \rangle_F = \sum_{i_1, \ldots i_d} \overline{A}_{i_1, \ldots i_d} B_{i_1, \ldots i_d}, \quad \|A\|_F = \sqrt{\langle A, A \rangle_F}.$$

Given also vectors $u_i \in \mathbb{C}^{n_i}$, we introduce the following multilinear functional:

$$[\![A; u_1, u_2 \ldots u_d]\!] = \sum_{i_1, \ldots i_d} A_{i_1, \ldots i_d} u_{1_{i_1}} u_{2_{i_2}} \cdots u_{d_{i_d}}. \qquad (2)$$

Spectral (second) norm of a matrix $A \in \mathbb{C}^{n_1 \times n_2}$ can be defined using this notation [11]:

$$\|A\|_2 = \sup_{\substack{u_i \in \mathbb{C}^{n_i}:\, \|u_i\|_2=1 \\ i=1,2}} |u_1^T A u_2| = \sup_{\substack{u_i \in \mathbb{C}^{n_i}:\, \|u_i\|_2=1 \\ i=1,2}} |[\![A; u_1, u_2]\!]|. \quad (3)$$

It is also well-known that $\|A\|_2$ equals to the largest singular value $\sigma_1(A)$ and the vectors u_1, u_2 on which the equality is attained are respectively the left and right singular vectors of A. Both the largest singular value and the respective singular vectors can be computed using, e.g., the power method.

Spectral norm (3) naturally extends from matrices to tensors with more than two dimensions [13]. For a d-dimensional tensor $A \in \mathbb{C}^{n_1 \times \cdots \times n_d}$, it is defined as a norm of a multilinear functional (2):

$$\|A\|_\sigma = \sup_{\substack{u_i \in \mathbb{C}^{n_i}:\, \|u_i\|_2=1 \\ i=1,\ldots,d}} |[\![A; u_1, u_2 \ldots u_d]\!]|. \quad (4)$$

We will further mean that the supremum above is taken over complex vectors u_i even for real A, which may be different from supremum over real u_i, see Appendix B for details. We use the notation $\|A\|_2$ for matrices and $\|A\|_\sigma$ for d-dimensional tensors when $d > 2$.

Our main result will be formulated in terms of the norm (4). We note that this expression also defines the largest singular value $\sigma_1(A)$ of a tensor A and is associated with the best rank-1 approximation problem [13]:

$$\inf_{\substack{\sigma \in \mathbb{R}_+,\, \|u_i\|_2=1 \\ i=1,\ldots,d}} \|A - \sigma\, u_1 \circ \cdots \circ u_d\|_F, \quad (5)$$

which admits the solution $\sigma = \sigma_1(A)$ ($\circ$ denotes tensor product). As we will discuss in more detail later, there is an analog of the power method called HOPM (Algorithm 1) that can be used to solve the best rank-1 approximation problem and calculate the spectral norm of a tensor.

3.3 Tensor Unfoldings

We call a matrix

$$A_{(i_1\ldots i_k;\, j_1\ldots j_{d-k})} \in \mathbb{R}^{(n_{i_1}\cdot\ldots\cdot n_{i_k}) \times (n_{i_k}\cdot\ldots\cdot n_{j_{d-k}})}$$

an unfolding of a tensor $A \in \mathbb{R}^{n_1 \times \cdots \times n_d}$, if it is obtained by first permuting indices of the tensor and then by reshaping the tensor into a matrix in a column-major order. In a Python numpy-like notation it reads as:

$$A = A.\texttt{transpose}(i_1, \ldots, i_k, j_1, \ldots, j_{d-k}),$$
$$A_{(i_1, i_2 \ldots i_k;\, j_1,\ldots j_{d-k})} = A.\texttt{reshape}(n_{i_1} \ldots n_{i_k}, n_{j_1} \ldots n_{j_{d-k}},\, \texttt{order='f'}).$$

The following lemma establishes a connection between the spectral norm of a tensor and the spectral norm of its unfoldings. We will later use this result to prove the lower bound on $\|T\|_2$ and to demonstrate how our result relates to [20].

Lemma 1 (Prop. 4.1, [27]). $\|K\|_\sigma \leq \|R\|_2$ for any unfolding matrix R of the tensor K.

3.4 Spectral Density Matrix

The approach we use to prove the main result of this paper is based on the following techniques. Following [29, Theorem 1], for a doubly block Toeplitz matrix $T \in \mathbb{R}^{c_{out}n^2 \times c_{in}n^2}$ with the blocks $T_{k,l} \in \mathbb{R}^{c_{out} \times c_{in}}$, we consider a matrix-valued generating function $F\colon [-\pi, \pi]^2 \to \mathbb{C}^{c_{out} \times c_{in}}$ such that

$$T_{k,l} = \frac{1}{(2\pi)^2} \int_{-\pi}^{\pi} \int_{-\pi}^{\pi} F(\tau_1, \tau_2) e^{-i(k\tau_1 + l\tau_2)} d\tau_1 d\tau_2.$$

The generating function can be explicitly written as [29]:

$$F(\tau_1, \tau_2) = \sum_{k=-h_1}^{h_2} \sum_{l=-w_1}^{w_2} T_{k,l} e^{i(k\tau_1 + l\tau_2)},$$

which is a generalization of [23, 24]. The function $F(\tau_1, \tau_2)$ is also called the spectral density matrix. The spectral norm of $F(\tau_1, \tau_2)$ is defined [23, 29] as

$$\|F\|_2 = \sup_{\tau_1, \tau_2 \in [-\pi, \pi]} \|F(\tau_1, \tau_2)\|_2.$$

Lemma 2 (Lemma 4, [29]). $\|T\|_2 \leq \|F\|_2$.

The set of singular values of circular convolution is obtained as $\sigma_j(F(\tau_1, \tau_2))$ with $(\tau_1, \tau_2) = (-\pi + \frac{2\pi j_1}{n}, -\pi + \frac{2\pi j_2}{n}), \forall j_1, j_2 \in [n] - 1$ [29]. However, by contrast to the circular case, for the zero-padded convolution the values of τ_1, τ_2 that lead to the singular values are not known a priori.

4 Main Results

The Lipschitz constant of a neural network f with respect to the Euclidean norm for an input space $\mathcal{X}$ is defined as

$$\mathrm{Lip}(f) = \sup_{x, x' \in \mathcal{X}, x \neq x'} \frac{\|f(x) - f(x')\|_2}{\|x - x'\|_2}.$$

It determines, e.g., how much the output value of the network changes relative to input perturbations. Computation of the Lipschitz constant of the whole network f is a challenging task [25]. However, assuming that f is a composition of maps ϕ_i:

$$f = \phi_\ell \circ \cdots \circ \phi_1,$$

it can be bounded from above with the product of the Lipschitz constants of individual layers:

$$\mathrm{Lip}(f) \leq \prod_{i=1}^{\ell} \mathrm{Lip}(\phi_i). \qquad (6)$$

Lipschitz constant of a linear layer $\phi(x) = Wx + b$ has the explicit representation:
$$\text{Lip}(\phi) = \|W\|_2 = \sigma_1(W).$$

In the case of a convolutional layer, we have $W = T$ from (1). Naive computation of its largest singular value by computing SVD appears to be intractable, as such a matrix does not even fit into GPU memory for practical sizes of layers and the computational complexity grows cubically with its size. Alternative iterative approaches, e.g., the power method that take the structure of T into account require multiple applications of convolution, so their complexity depends on input image shape n.

In this work, we take benefit of the matrix structure and obtain an upper bound that is independent of n, valid for any padding and can be efficiently computed in practice. This bound is formulated in the next theorem.

Theorem 1. *Let $T \in \mathbb{R}^{c_{out} n^2 \times c_{in} n^2}$ be the Jacobian of a convolutional layer with stride $s = 1$ and zero padding with the parameter $p \geq 0$ or circular padding. Let $K \in \mathbb{R}^{c_{out} \times c_{in} \times h \times w}$ be the convolution kernel. Then*
$$\|K\|_\sigma \leq \|T\|_2 \leq \sqrt{hw}\|K\|_\sigma.$$

Proof. Firstly, let us prove the upper bound. We can rewrite $F(\tau_1, \tau_2)$ as a convolution of the kernel with vectors z_1, z_2:

$$F(\tau_1, \tau_2) = \sum_{k=-h_1}^{h_2} \sum_{l=-w_1}^{w_2} K_{:,:,h_1+k,w_1+l} e^{i(k\tau_1 + l\tau_2)} = [\![K; I_{c_{out}}, I_{c_{in}}, z_1, z_2]\!],$$

$$z_1 = [e^{-h_1(i\tau_1)}, e^{(-h_1+1)(i\tau_1)} \ldots e^{h_2(i\tau_1)}], \quad z_2 = [e^{-w_1(i\tau_2)}, e^{(-w_1+1)(i\tau_2)} \ldots e^{w_2(i\tau_2)}]$$

and for the matrices in square brackets the summation is done along their last indices. For complex vectors u_i, we have,

$$\|F\|_2 = \sup_{\substack{\|u_i\|_2 = 1 \\ i=1,2}} |u_1^T F u_2| = \sup_{\substack{\|u_i\|_2 = 1 \\ i=1,2}} |[\![F; u_1, u_2]\!]| = \sup_{\substack{\|u_i\|_2 = 1 \\ i=1,2}} |[\![K; u_1, u_2, z_1, z_2]\!]|$$

$$= \sup_{\substack{\|u_i\|_2 = 1 \\ i=1,2}} \left| \sqrt{hw} \left[\!\!\left[K; u_1, u_2, \frac{z_1}{\sqrt{h}}, \frac{z_2}{\sqrt{w}} \right]\!\!\right] \right| \leq \sup_{\substack{\|u_i\|_2 = 1 \\ i=1,\ldots,4}} \left| \sqrt{hw} [\![K; u_1, u_2, u_3, u_4]\!] \right|$$

$$= \sqrt{hw}\,\|K\|_\sigma. \tag{7}$$

Thus, $\|T\|_2 \leq \sqrt{hw}\|K\|_\sigma$.

Let us consider an unfolding $R = K_{(1,234)}$. Note that it is a submatrix of a doubly Toeplitz matrix T up to a permutation of columns. In particular, we can obtain R by choosing a block row in T which starts with $T_{-h_1, -w_1}$ and excluding zero entries from it, see Sect. 3.1. Since the spectral norm of a matrix upper bounds spectral norm of any of its submatrices, we have

$$\|K\|_\sigma \leq \|R\|_2 = \|K_{(1,234)}\|_2 \leq \|T\|_2,$$

which completes the proof.

Remark 1. Note that the matrix F and the vectors z_1, z_2 from (7) are complex. Therefore, all suprema in this context are taken over complex vectors u_i. As a result, we need a complex rank-1 approximation of the real kernel K to find $\|K\|_\sigma$ when applying HOPM Algorithm 1. This is because complex and real spectral norms of a real tensor do not always coincide [9,10]. In Appendix B we provide an example of a tensor for which $\sqrt{hw}\|K\|_\sigma$ with supremum taken over real vectors does not bound from above the spectral norm of a convolution.

Remark 2. As a direct corollary of Theorem 1, the bounds are exact for $h = w = 1$, meaning that $\|T\|_2 = \|K\|_\sigma$.

Note that Theorem 1 also suggests that in the worst case, our TN bound does not overestimate the exact singular value of the convolution Jacobian by more than $\sqrt{hw}$ times, i.e., $\sqrt{hw}\|K\|_\sigma \leq \sqrt{hw}\|T\|_2$.

Let us now discuss how our bound compares to [20,29]. The work [20] proposed to bound the singular value of circular convolution using the minimum of spectral norms of four unfoldings of the kernel:

$$\|T\|_2 \leq \sqrt{hw}\min\left(\|K_{(13;24)}\|_2, \|K_{(14;23)}\|_2, \|K_{(1;234)}\|_2, \|K_{(2;134)}\|_2\right). \quad (8)$$

Lemma 1 shows that the tensor spectral norm never exceeds the norm of any of its unfoldings, including the ones that are not present in (8). Therefore, our bound is provably more accurate than the one in [20].

Theorem 1 naturally extends to convolutions with any number of dimensions.

Theorem 2. *Let us consider convolution with a kernel $K \in \mathbb{R}^{c_{out} \times c_{in} \times h_1 \times \ldots \times h_d}$, $d \in \mathbb{N}_+$, with arbitrary padding and stride $s = 1$. Then*

$$\|K\|_\sigma \leq \|T\|_2 \leq \sqrt{h_1 \ldots h_d}\|K\|_\sigma.$$

Proof. The proof is provided in Appendix D.

The following theorem extends Theorem 1 to the case of strided convolutions.

Theorem 3. *Let $T_s \in \mathbb{R}^{c_{out}\frac{n^2}{s^2} \times c_{in}n^2}$ be the Jacobian of a convolution with the kernel $K \in \mathbb{R}^{c_{out} \times c_{in} \times h \times w}$ and stride s. Then*

$$\|Q\|_\sigma \leq \|T_s\|_2 \leq \sqrt{\left\lceil\frac{h}{s}\right\rceil\left\lceil\frac{w}{s}\right\rceil}\|Q\|_\sigma,$$

where $Q \in \mathbb{R}^{c_{out} \times c_{in}s^2 \times \lceil\frac{h}{s}\rceil \times \lceil\frac{w}{s}\rceil}$ is padded with zeros and reshaped kernel K

$$K_{c,d,a,b} = Q_{c,ds^2+s(a \ (\text{mod } s))+b \ (\text{mod } s),\lfloor\frac{a}{s}\rfloor,\lfloor\frac{b}{s}\rfloor}.$$

Proof. The proof is provided in Appendix A.1.

5 Computation of the Spectral Norm

The power method is a standard algorithm for the computation of the largest singular value and the respective singular vector of a matrix. Works [3,4] extended the power method from matrices to tensors of higher dimension. The algorithm starts with either randomly initialized vectors $u_1, u_2, \ldots, u_d$, or it can start with a good approximation to the singular vectors. For example, in our case, the kernel weights change slowly during training, so we may utilize vectors from previous iterations. The i-th substep, $i = 1, \ldots, d$ of an iteration is equivalent to the minimization of the functional from (5) with respect to a single u_i. This leads to simple formulas with contractions, see Algorithm 1. Note that according to Remark 1, we need to utilize complex vectors u_i, which is why we have the complex conjugate operation $\text{conj}(u_i)$. The operations in HOPM can be reorganized to speed up computations for the certain hardware. In our repository with code, you can find several implementations of HOPM that have proven to be faster in practice when running on either CPU or GPU. It is also advantageous to rerun the HOPM algorithm from the beginning several times if no good initial guess is available. This will help ensure convergence to the global optimum. Other initialization strategies also exist, see [4].

Algorithm 1. Higher-Order Power Method (HOPM)

Input kernel: $K \in \mathbb{R}^{c_{out} \times c_{in} \times h \times w}$; number of iterations: n_iters; initial unit vectors: $u_1 \in \mathbb{C}^{c_{out}}$, $u_2 \in \mathbb{C}^{c_{in}}$, $u_3 \in \mathbb{C}^h$, $u_4 \in \mathbb{C}^w$.
Return $\|K\|_\sigma$
for $1 \ldots$ n_iters **do**
    $u_1 = [\![K; I, u_2, u_3, u_4]\!]$, $u_1 = \text{conj}(u_1)/\|u_1\|$ ▷ $\mathcal{O}(c_{out}c_{in}hw)$
    $u_2 = [\![K; u_1, I, u_3, u_4]\!]$, $u_2 = \text{conj}(u_2)/\|u_2\|$ ▷ $\mathcal{O}(c_{out}c_{in}hw)$
    $u_3 = [\![K; u_1, u_2, I, u_4]\!]$, $u_3 = \text{conj}(u_3)/\|u_3\|$ ▷ $\mathcal{O}(c_{out}c_{in}hw)$
    $u_4 = [\![K; u_1, u_2, u_3, I]\!]$, $u_4 = \text{conj}(u_4)/\|u_4\|$ ▷ $\mathcal{O}(c_{out}c_{in}hw)$
end for
return $|[\![K; u_1, u_2, u_3, u_4]\!]|$ ▷ $\mathcal{O}(c_{out}c_{in}hw)$

Note that, although the algorithm involves complex vectors, the norm $\|K\|_\sigma$ itself is a real number. The following proposition provides explicit formulas for the norm and its gradient, which avoid complex arithmetic calculations.

Proposition 1. *Let u_1, u_2, u_3, u_4 be the complex singular vectors corresponding to $\|K\|_\sigma$. Let $u_j = a_j + ib_j$, where a_j and b_j are the real and imaginary parts. The gradient of the bound $\sqrt{hw}\|K\|_\sigma$ with respect to the kernel is computed as*

$$\nabla_K \sqrt{hw}\|K\|_\sigma = \sqrt{hw}\nabla_K \sqrt{real^2 + im^2} = \frac{\sqrt{hw}}{\|K\|_\sigma}\left(real\nabla_K real + im\nabla_K im\right),$$

where real and im are the real and imaginary parts of $[\![K; u_1, u_2, u_3, u_4]\!]$. For part $\in \{real, im\}$, we can write it and its gradient in terms of predefined tensors

$P_{real}, P_{im} \in \{-1, 0, 1\}^{2 \times 2 \times 2 \times 2}$ (see Appendix C for their explicit representation):

$$part = \langle P_{part}, [\![K; [a_1, b_1], [a_2, b_2], [a_3, b_3], [a_4, b_4]]\!] \rangle_F,$$
$$\nabla_K part = [\![P_{part}; [a_1, b_1]^T, [a_2, b_2]^T, [a_3, b_3]^T, [a_4, b_4]^T]\!]. \quad (9)$$

6 Experiments

The source code is available at https://github.com/GrishKate/conv_norm.

6.1 Spectral Norm Computation

Table 2 compares our TN bound with the closest competitor, "fantastic four" bound [20] ($F4$ for short), in terms of precision. Values in each of the kernels are sampled from $\mathcal{N}(0, 1)$. We observe that TN bound produces close to exact values, whereas $F4$ exceeds exact spectral norm 1.7–2.6 times. Results for strided convolutions are presented in Appendix A.2.

Table 3 shows that the TN bound provides an accurate approximation for the spectral norm of convolutional layers for the pre-trained ResNet18. Our bound performs systematically better for most layers compared to the $F4$ bound.

Figure F.4 in Appendix demonstrates comparison with other existing methods in terms of precision, memory consumption and time performance. Although methods [6,18] obtain a tight upper bound, they require significantly more memory than the other approaches. Algorithms based on power iteration [7,17] are highly accurate, but their time and memory consumption grows with the input size n. While the bounds [1,20] are independent of n, they are noticeably less accurate. Our TN bound does not depend on n and therefore is both memory efficient and fast to compute, while being more precise than [1,20].

6.2 Spectral Norm Regularization

Following [20], we apply the TN bound for regularization and study its effect on image classification accuracy. We train ResNet18 on CIFAR100 and ResNet34 on ImageNet for 90 epochs. We use weight decay 1e−4, SGD with momentum 0.9 and initial learning rate of 0.1, which is reduced 10 times every 30 epochs. We apply the sum of estimates of the spectral norms of all convolutional layers of the network as a regularizer. The objective loss function becomes:

$$\mathcal{L} = \mathcal{L}_{train} + \beta \sum_i \sigma_i,$$

where β denotes the regularization coefficient, $\mathcal{L}_{train}$ is cross-entropy loss, σ_i denotes the bound on the largest singular value of the ith convolutional layer computed with different methods. During the epoch, one iteration step is sufficient to update σ_i and the singular vectors. However, for reliability, we recompute singular vectors every epoch from random initialization. We obtain the highest

Table 2. Comparison of the TN bound and the $F4$ bound for Gaussian kernels. Reference values for spectral norm of zero-padded convolution $\|T\|_2$ were computed with power method [17] for the image size 32×32. We used 100 iterations for all methods, stride equals to 1.

Kernel size	$\|T\|_2$ (Reference)	$F4$	TN(Ours)	$\dfrac{F4}{\|T\|_2}$	$\dfrac{TN}{\|T\|_2}$
64, 64, 3, 3	49.35	81.26	51.51	1.646	**1.044**
128, 128, 3, 3	67.14	114.68	69.99	1.708	**1.042**
256, 256, 3, 3	95.71	164.44	96.47	1.718	**1.008**
512, 512, 3, 3	135.62	234.55	136.99	1.729	**1.01**
64, 64, 5, 5	81.55	175.74	88.25	2.155	**1.082**
128, 128, 5, 5	113.5	250.91	119.29	2.211	**1.051**
256, 256, 5, 5	160.07	354.46	165.37	2.214	**1.033**
512, 512, 5, 5	227.1	501.8	229.66	2.21	**1.011**
64, 64, 7, 7	112.94	293.34	127.69	2.597	**1.131**
128, 128, 7, 7	158.36	415.81	170.99	2.626	**1.08**
256, 256, 7, 7	222.45	587.98	235.28	2.643	**1.058**
512, 512, 7, 7	315.25	834.35	326.2	2.647	**1.035**

accuracy with $\beta = 0.0022$ for CIFAR100 and $\beta = 1e{-}4$ for ImageNet. The results of hyperparameter tuning are presented in Table E.2 in Appendix. We compare the effect of regularization both with and without weight decay for CIFAR100. Table 4 suggests that TN reduces generalization error and gives systematically better results than $F4$.

Training time comparison is presented in Table 5. The wall clock time difference between training with and without the TN regularization is almost negligible. The time performance of TN is comparable with $F4$ [20]. Methods of [6,18] require much more memory and cause "Out of Memory" error for large input sizes. The singular value clipping method [7] is computationally expensive and is applied only once per 50 steps. The advantage of TN over the Gram iteration-based approach [6] is that TN bound can be updated with one iteration of HOPM during training, while the Gram iteration needs to start from scratch every time. Hence, Gram iteration increases time of training multiple times, e.g., see Sect. 5.5 in [6], where time per epoch increased by a factor of 3. By contrast, with one iteration per training step, TN bound can be used as a computationally efficient regularization with a negligible increase in time and memory.

Figure 1 (left) presents a comparison of our TN bound with $F4$. Throughout training, the TN bound appears to be more precise. Regularization with the TN bound significantly decreases the singular value of convolutions and makes the bound more precise (middle). One may wonder how regularization affects the Lipschitz constant of the entire network. The right figure suggests that reg-

Table 3. Comparison of the TN bound and the $F4$ bound for kernels from ResNet18 pretrained on ImageNet. We use 100 iterations for all methods. For layers with $h = w = 1$ both bounds give an exact value, as proved in Remark 2. The $F4$ bound does not take the stride into account, which results in even looser bound for strided convolution (see the first row, where the ratio $F4/\|T\|_2 = 3.5$).

Kernel size	Stride	$\|T\|_2$ (Exact)	$F4$	TN(Ours)	$\frac{F4}{\|T\|_2}$	$\frac{TN}{\|T\|_2}$
64, 3, 7, 7	2	8.2	28.89	14.91	3.526	**1.819**
64, 64, 3, 3	1	5.99	9.33	8.56	1.558	**1.43**
64, 64, 3, 3	1	5.3	6.3	5.36	1.187	**1.01**
64, 64, 3, 3	1	6.95	8.71	8.63	1.253	**1.242**
64, 64, 3, 3	1	3.8	5.4	4.04	1.42	**1.062**
128, 64, 3, 3	2	2.8	6.01	3.16	2.149	**1.129**
128, 128, 3, 3	1	5.69	7.21	6.67	1.267	**1.173**
128, 64, 1, 1	2	1.91	1.91	1.91	1.0	**1.0**
128, 128, 3, 3	1	4.38	6.78	5.46	1.549	**1.247**
128, 128, 3, 3	1	4.86	7.57	6.07	1.558	**1.248**
256, 128, 3, 3	2	3.95	8.45	4.32	2.14	**1.093**
256, 256, 3, 3	1	6.45	8.04	7.07	1.246	**1.096**
256, 128, 1, 1	2	1.22	1.22	1.22	1.0	**1.0**
256, 256, 3, 3	1	6.23	7.57	6.43	1.216	**1.033**
256, 256, 3, 3	1	7.53	9.17	8.25	1.218	**1.095**
512, 256, 3, 3	2	5.53	11.0	5.99	1.989	**1.082**
512, 512, 3, 3	1	8.38	10.45	9.54	1.247	**1.139**
512, 256, 1, 1	2	2.05	2.05	2.05	1.0	**1.0**
512, 512, 3, 3	1	16.26	18.37	17.95	1.13	**1.104**
512, 512, 3, 3	1	6.8	7.6	7.52	1.117	**1.106**

ularization reduces the norm of composition of convolution with the subsequent BatchNorm, meaning that the upper bound (6) on the Lipschitz constant of the whole network declines. The same effect is observed for all layers, see Figures F.1, F.2, F.3.

6.3 Orthogonal Regularization

The work [26] proposes to use orthogonal regularization for training CNNs. The method utilizes convolution of the kernel with itself to estimate the divergence of the layer's Jacobian from orthogonal in the Frobenius norm, which leads to the following loss function:

$$\mathcal{L}_{OCNN} = \|Conv(K, K) - I\|_F$$

Table 4. Test accuracy with different regularizers.

Method	Acc. w/o wd	Acc. w/ wd
Baseline	73.10	73.84
$F4$	**73.96**	74.91
TN (Ours)	**73.96**	**74.99**

(a) ResNet18 trained on CIFAR100.

Method	Acc@1	Acc@5
Baseline	73.368	**91.438**
$F4$	73.388	91.300
TN (Ours)	**73.510**	91.420

(b) ResNet34 trained on ImageNet.

Table 5. Average time per epoch for training ResNet18 on CIFAR100 with different regularizers. For computing LipBound [1] we use 10 samples. We present time for clipping once per 50 iterations with PowerQR [7].

Method	Baseline	TN (Ours)	$F4$	LipBound	PowerQR	Gram iter.	Sedghi
Time (s)	141.0	143.6	143.7	242.5	149.6	OOM	OOM

However, using the spectral norm of the Jacobian may seem more natural in this case, as it is an operator norm. Let T be the Jacobian of the circular convolution with a kernel K. It is known that the product of two circulant matrices is also a circulant, thus $T^T T$ is also a Jacobian of some convolution. Theorem 1 suggests that $\|T^T T - I\|_2 \leq \sqrt{h'w'}\|Conv(K, K, \texttt{padding=``circular''}) - I\|_\sigma$. Thus, the following regularizer penalizes the largest difference between the singular values of the convolution and 1:

$$\mathcal{L}_{2norm} = \|Conv(K, K, \texttt{padding=``circular''}) - I\|_\sigma$$

We introduce one more regularizer based on the spectral norm of the kernel. As is known, the squared Frobenius norm of a matrix is equal to the sum of squares of its singular values and the ratio $\frac{\|A\|_2}{\|A\|_F}$ is minimized when all the singular values of matrix A are equal. The Frobenius norm of the kernel can be used as approximation for the Frobenius norm of convolution Jacobian up to a constant

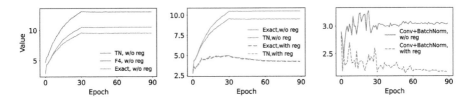

Fig. 1. The behaviour of spectral norm bounds for the third layer of ResNet-18 during training on CIFAR100. The left figure compares tightness of TN and $F4$ bounds when training without regularization. The middle figure shows the effect of training with and without TN regularization. The right one demonstrates the influence of regularization on the spectral norm of composition of convolution and the subsequent BatchNorm layer. Similar plots for all layers are presented in Figs. F.1, F.2, F.3 in Appendix.

factor, e.g., in the case of circular padding $n\|K\|_F = \|T\|_F$. Hence, we propose

$$\mathcal{L}_{Ratio} = \frac{\sqrt{hw}\|K\|_\sigma}{\|K\|_F}.$$

We train Resnet18 on CIFAR100 dataset with the same hyperparameters as in the previous section and weight decay 1e-4. The regularized objective loss function is given as follows:

$$\mathcal{L} = \mathcal{L}_{train} + \beta \sum_i \mathcal{L}^i_{reg},$$

where $\mathcal{L}^i_{reg}$ denotes one of the regularization losses for the ith layer described above. The results are reported in Table 6. Our regularizers improve generalization performance. $\mathcal{L}_{2norm}$ yields the best top-1 accuracy gain of more than 2%.

Table 6. ResNet18 CIFAR100 accuracy and average training time per epoch with different orthogonal regularizers.

Method	β	Acc@1	Acc@5	Time (s)
Baseline	0	73.84	93.16	141.0
OCNN	0.1	75.32	93.45	150.1
OCNN	0.01	75.32	93.30	150.1
Ratio (Ours)	1.0	74.98	93.50	147.0
Ratio (Ours)	0.1	74.71	93.52	147.0
Ratio (Ours)	0.01	74.71	93.45	147.0
2norm (Ours)	1e−2	74.43	93.47	171.8
2norm (Ours)	5e−3	**75.99**	**94.08**	171.8
2norm (Ours)	1e−3	75.83	93.92	171.8

7 Conclusion

We propose to use the spectral norm of the kernel tensor to bound the spectral norm of convolutional layers. Our bound noticeably improves the accuracy of existing upper bounds that are independent of the resolution of input images. It also provides a trade-off between accuracy and computation efficiency compared with the other methods. We demonstrate that this bound can be used as a regularizer for improving generalization of NNs. Furthermore, we propose two new regularizers based on it that enforce orthogonality of convolutions.

Acknowledgments. The publication was supported by the grant for research centers in the field of AI provided by the Analytical Center for the Government of the Russian Federation (ACRF) in accordance with the agreement on the provision of subsidies (identifier of the agreement 000000D730321P5Q0002) and the agreement with HSE University №70-2021-00139. The calculations were performed in part through the computational resources of HPC facilities at HSE University [12].

References

1. Araujo, A., Negrevergne, B., Chevaleyre, Y., Atif, J.: On Lipschitz regularization of convolutional layers using Toeplitz matrix theory. In: Proceedings of the AAAI Conference on Artificial Intelligence, vol. 35, pp. 6661–6669 (2021)
2. Cisse, M., Bojanowski, P., Grave, E., Dauphin, Y., Usunier, N.: Parseval networks: improving robustness to adversarial examples. In: International Conference on Machine Learning, pp. 854–863. PMLR (2017)
3. De Lathauwer, L., Comon, P., De Moor, B., Vandewalle, J.: Higher-order power method. In: Nonlinear Theory and its Applications, NOLTA 1995, vol. 1, p. 4 (1995)
4. De Lathauwer, L., De Moor, B., Vandewalle, J.: On the best rank-1 and rank-$(r_1, r_2, \ldots, r_n)$ approximation of higher-order tensors. SIAM J. Matrix Anal. Appl. **21**(4), 1324–1342 (2000)
5. Delattre, B., Barthélemy, Q., Allauzen, A.: Spectral norm of convolutional layers with circular and zero paddings. arXiv preprint arXiv:2402.00240 (2024)
6. Delattre, B., Barthélemy, Q., Araujo, A., Allauzen, A.: Efficient bound of Lipschitz constant for convolutional layers by gram iteration. In: International Conference on Machine Learning, pp. 7513–7532. PMLR (2023)
7. Ebrahimpour-Boroojeny, A., Telgarsky, M., Sundaram, H.: Spectrum extraction and clipping for implicitly linear layers. In: NeurIPS 2023 Workshop on Mathematics of Modern Machine Learning (2023)
8. Farnia, F., Zhang, J., Tse, D.: Generalizable adversarial training via spectral normalization. In: International Conference on Learning Representations (2019)
9. Friedland, S., Lim, L.H.: Nuclear norm of higher-order tensors. Math. Comput. **87**(311), 1255–1281 (2018)
10. Friedland, S., Wang, L.: Spectral norm of a symmetric tensor and its computation. Math. Comput. **89**(325), 2175–2215 (2020)
11. Golub, G.H., Van Loan, C.F.: Matrix Computations. JHU Press (2013)
12. Kostenetskiy, P., Chulkevich, R., Kozyrev, V.: HPC resources of the higher school of economics. In: Journal of Physics: Conference Series, vol. 1740, p. 012050 (2021)
13. Lim, L.H.: Singular values and eigenvalues of tensors: a variational approach. In: 2005 1st IEEE International Workshop on Computational Advances in Multi-Sensor Adaptive Processing, pp. 129–132. IEEE (2005)
14. Meunier, L., Delattre, B.J., Araujo, A., Allauzen, A.: A dynamical system perspective for lipschitz neural networks. In: International Conference on Machine Learning, pp. 15484–15500. PMLR (2022)
15. Miyato, T., Kataoka, T., Koyama, M., Yoshida, Y.: Spectral normalization for generative adversarial networks. In: International Conference on Learning Representations (2018)
16. Neyshabur, B., Bhojanapalli, S., Srebro, N.: A PAC-Bayesian approach to spectrally-normalized margin bounds for neural networks. In: International Conference on Learning Representations (2018)

17. Ryu, E., Liu, J., Wang, S., Chen, X., Wang, Z., Yin, W.: Plug-and-play methods provably converge with properly trained denoisers. In: International Conference on Machine Learning, pp. 5546–5557. PMLR (2019)
18. Sedghi, H., Gupta, V., Long, P.M.: The singular values of convolutional layers. In: International Conference on Learning Representations (2018)
19. Senderovich, A., Bulatova, E., Obukhov, A., Rakhuba, M.: Towards practical control of singular values of convolutional layers. Adv. Neural. Inf. Process. Syst. **35**, 10918–10930 (2022)
20. Singla, S., Feizi, S.: Fantastic four: differentiable and efficient bounds on singular values of convolution layers. In: International Conference on Learning Representations (2020)
21. Singla, S., Feizi, S.: Skew orthogonal convolutions. In: International Conference on Machine Learning, pp. 9756–9766. PMLR (2021)
22. Singla, S., Singla, S., Feizi, S.: Improved deterministic l2 robustness on CIFAR-10 and CIFAR-100. In: International Conference on Learning Representations (2021)
23. Tilli, P.: Singular values and eigenvalues of non-Hermitian block Toeplitz matrices. Linear Algebra Appl. **272**(1–3), 59–89 (1998)
24. Tyrtyshnikov, E.E.: A unifying approach to some old and new theorems on distribution and clustering. Linear Algebra Appl. **232**, 1–43 (1996)
25. Virmaux, A., Scaman, K.: Lipschitz regularity of deep neural networks: analysis and efficient estimation. Adv. Neural. Inf. Process. Syst. **31** (2018)
26. Wang, J., Chen, Y., Chakraborty, R., Yu, S.X.: Orthogonal convolutional neural networks. In: Proceedings of the IEEE/CVF Conference on Computer Vision and Pattern Recognition, pp. 11505–11515 (2020)
27. Wang, M., Duc, K.D., Fischer, J., Song, Y.S.: Operator norm inequalities between tensor unfoldings on the partition lattice. Linear Algebra Appl. **520**, 44–66 (2017)
28. Xiao, L., Bahri, Y., Sohl-Dickstein, J., Schoenholz, S., Pennington, J.: Dynamical isometry and a mean field theory of CNNs: how to train 10,000-layer vanilla convolutional neural networks. In: International Conference on Machine Learning, pp. 5393–5402. PMLR (2018)
29. Yi, X.: Asymptotic spectral representation of linear convolutional layers. IEEE Trans. Signal Process. **70**, 566–581 (2022)
30. Yoshida, Y., Miyato, T.: Spectral norm regularization for improving the generalizability of deep learning. Stat **1050**, 31 (2017)

Deciphering the Role of Representation Disentanglement: Investigating Compositional Generalization in CLIP Models

Reza Abbasi, Mohammad Hossein Rohban, and Mahdieh Soleymani Baghshah(✉)

Sharif University of Technology, Tehran, Iran
{reza.abbasi,rohban,soleymani}@sharif.edu

Abstract. CLIP models have recently shown to exhibit Out of Distribution (OoD) generalization capabilities. However, Compositional Out of Distribution (C-OoD) generalization, which is a crucial aspect of a model's ability to understand unseen compositions of known concepts, is relatively unexplored for the CLIP models. Our goal is to address this problem and identify the factors that contribute to the C-OoD in CLIPs. We noted that previous studies regarding compositional understanding of CLIPs frequently fail to ensure that test samples are genuinely novel relative to the CLIP training data. To this end, we carefully synthesized a large and diverse dataset in the single object setting, comprising attributes for objects that are highly unlikely to be encountered in the combined training datasets of various CLIP models. This dataset enables an authentic evaluation of C-OoD generalization. Our observations reveal varying levels of C-OoD generalization across different CLIP models. We propose that the disentanglement of CLIP representations serves as a critical indicator in this context. By utilizing our synthesized datasets and other existing datasets, we assess various disentanglement metrics of text and image representations. Our study reveals that the disentanglement of image and text representations, particularly with respect to their compositional elements, plays a crucial role in improving the generalization of CLIP models in out-of-distribution settings. This finding suggests promising opportunities for advancing out-of-distribution generalization in CLIPs. For more details and access to our dataset, please visit https://github.com/abbasiReza/CLIP-COoD.

Keywords: Compositional Out-of-Distribution (C-OoD) Generalization · CLIP · Disentanglement

---

Supplementary Information The online version contains supplementary material available at https://doi.org/10.1007/978-3-031-73024-5_3.

© The Author(s), under exclusive license to Springer Nature Switzerland AG 2025
A. Leonardis et al. (Eds.): ECCV 2024, LNCS 15147, pp. 35–50, 2025.
https://doi.org/10.1007/978-3-031-73024-5_3

1 Introduction

Out-of-Distribution (OoD) generalization which is the ability of a model to generalize to the data distributions differing from the training distribution is very important for most learning models [1]. In recent years, several studies suggested that some Vision-Language Models (VLMs) such as the CLIPs [2], exhibit OoD generalization [2,3]. Specifically, several studies reported that CLIP models demonstrate enhanced zero- and few-shot accuracies on parallel versions of ImageNet, comprising images with various style shifts with respect to the original ImageNet [3,4].

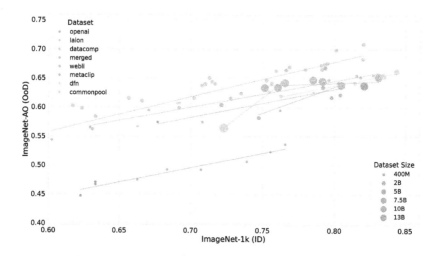

Fig. 1. Comparing zero-shot compositional out-of-distribution (C-OoD) generalization across diverse CLIP models and training sets. In-distribution (ID) performance is evaluated on the ImageNet validation set with object name labels, while the C-OoD generalization is assessed on our designed compositional dataset using attribute-object pair labels. Noticeably, CLIP models trained on the Common Pool dataset exhibit a steeper accuracy slope when transitioning from the ID to the OoD compositional setting compared to models trained on other datasets like WebLI. CLIPs trained on the LAION and DataComp datasets also show significantly higher C-OoD across ID accuracy. Despite improved in-distribution accuracy, models pretrained on WebLI do not demonstrate substantial gains in generalizing to the novel compositional out-of-distribution test cases.

In particular, Compositional OoD (C-OoD) generalization is a main branch of the OoD generalization, focusing specifically on the ability of models to generalize to unseen combinations of known concepts or entities. Essentially, compositional generalization relates to human-like inductive biases that leads to more efficient learning via composing seen concepts [5]. Recently, some studies have worked on evaluating or improving compositional generalization in the NLP tasks [5–7]. However, C-OoD generalization for vision tasks is less explored since the unseen compositions of concepts can not be easily created visually for investigation.

In the recent years, evaluating the ability of VLMs in encoding objects, attributes, and their relations has recently received attention [8,9]. Some benchmarks such as VL-Checklist [10], Winoground [11], and Attribute-Relation-Order (ARO) [8] have been introduced to assess the image-text matching ability of VLMs in compositional setups more exactly. VL-Checklist provides a benchmark to evaluate VLMs capabilities in three categories of objects, attributes, and relations. ARO showcases that the reordering of words in the text does not highly impact on the similarity of the text with the corresponding image. Some of these studies [8,11] discussed shortcommings of VLMs in encoding the compositional relationships between objects and attributes and [9] showed that VLMs can compose concepts in a single-object setting including single attribute-object compositions. Nonetheless, most of the work around compositional reasoning [12–15] were more concerned about compositional understanding of the inputs, and less attention has been paid to the OoD generalization in which the generalization ability are evaluated against truly novel compositions with respect to the training set. In a nutshell, the literature suggests that compositional understanding in VLMs might be more feasible in the single-object setups. However, until now the C-OoD capability of CLIPs is unexplored. This makes us ask the question:

Do CLIPs really have nontrivial C-OoD generalization in the single-object setting? and where does this ability stem from in such models?

We propose a new benchmark to evaluate the C-OoD performance of CLIP models. Our approach involves generating a dataset, called ImageNet-AO (Attribute Object), distinct from the CLIPs training data. We gather comprehensive lists of objects and attributes, then generate images by combining these objects and attributes using a text-to-image model. The generated images undergo several filtering processes to ensure they are aligned with their intended and specified object-attribute description, and are novel compared to the combined CLIP training datasets both in the text and image domains. We then evaluate different CLIP models on our OoD dataset to classify an input image into its composition constituents. Figure 1 gives an overview of this result, in which certain CLIPs, such as the ones trained on the LAION and DataComp, yielded strong C-OoD performance.

Finally, we analyze the factors that contribute to better performance in our benchmark. We found that the CLIPs that show higher C-OoD generalization typically exhibit strong disentangled text representations with respect to the composition constituents. We backed this observation by assessing numerous disentanglement metrics, and the intrinsic dimensionality of the composition text embeddings. We found that CLIPs with strong C-OoD accuracy also enjoy a more disentangled image representation, albeit at a lower level compared to that of the text embedding. Based on these results, we hypothesize that the inherent disentanglement of the text is induced from the text representation space to that of the images through contrastive learning. We elaborate on this

Fig. 2. Examples of images from our generated dataset. This dataset is created by combining attributes and objects that do not appear in the CLIP training sets, specifically designed for benchmarking compositional OoD generalization purposes.

hypothesis in Sect. 4. Consistently, various disentanglement metrics of the text and image representations are observed to be highly correlated in CLIPs. We also repeat all these experiments in datasets that were previously designed for evaluating disentanglement, and contain factors at a more fine-grained level, and note that all these observations hold.

Our contributions are summarized as follows:

- Designing an image test dataset of attribute-object pairs that are unseen in common CLIP training datasets.
- Benchmarking the compositional generalization of various CLIPs in the carefully designed and controlled setting.
- Discovering that the CLIP representation space is decomposable into embedding of concepts (e.g., objects and attributes) especially for the embeddings obtained by the text encoder, and suggesting that it is the source of compositional generalization.
- Demonstrating a strong connection between CLIPs text/image disentanglements and better C-OoD generalization through different disentanglement metrics, on both our ImageNet-AO datasets and exisiting datasets designed previously for disentanglement evaluation.

2 Methodology

In this section, we explain how we conducted our study step-by-step. We first describe how we created our challenging benchmark dataset, ImageNet-AO, which involves finding new combinations and making images with text-to-image models (Sect. 2.1). Examples of images in ImageNet-AO are shown in Fig. 2. Then, we dive into how we test CLIP models in the zero-shot setting, and the chosen criteria to evaluate the models (Sect. 2.2).

2.1 ImageNet-AO Dataset Design

To rigorously evaluate the compositional generalization capabilities of vision-language models, we devised an innovative dataset featuring compositions that are out-of-distribution with respect to the training datasets of these models. Our dataset is crafted to include rare and unique compositions, thus ensuring it presents novel challenges to the VLMs under study. The dataset construction process is meticulously designed and involves several key steps, as depicted in Fig. 3 and detailed below:

Selection of Objects (Nouns). Our initial step involved curating objects by extracting class names from the ImageNet dataset. This choice facilitates a direct comparison between the performance of models on our dataset and their performance on the well-established ImageNet validation set. By selecting a diverse array of class names, we aim to increase the complexity and richness of the generated compositional images.

Selection of Attributes (Adjectives). We then selected 140 adjectives from the Visual Attributes Words (VAW) dataset [16]. These adjectives span various categories, including color, material, and texture, allowing us to create a wide range of descriptive combinations for image generation. A complete list of the 140 adjectives used from the Visual Attributes Words (VAW) dataset is provided in Appendix .

Image Generation with Attribute-Object Prompts. Utilizing the SD-XL Turbo, one of the most advanced and efficient text-to-image models available, we generated images based on combinations of the selected attributes and objects. By pairing 140 adjectives with 1,000 nouns, we created 140,000 unique prompts, which were then used to produce corresponding images, enriching our dataset with a vast array of compositional variety.

Filtering Process. To guarantee the integrity and the intended OoD characteristics of our dataset, we implemented a meticulous three-step filtering process. This approach ensures that our dataset not only accurately represents the specified attribute-object combinations but also stands apart from existing datasets in terms of composition and novelty. The steps are as follows:

Step 1 - Initial Validation: Each generated image was subjected to an initial evaluation to verify its accuracy in depicting the intended attribute-object

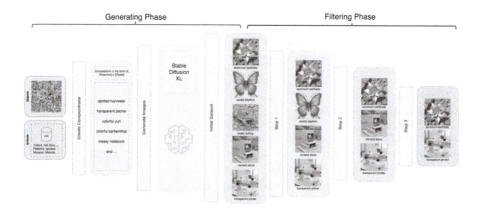

Fig. 3. Dataset Design Stages: The data design process involves a generation phase that makes the initial dataset from the whole set of the object and attribute compositions, and three distinct filtration steps. In the first filtration step, images where the target attribute or object lacks clear visibility are eliminated. In the second filtration step, the process removes images whose captions are already present in public datasets specifically curated for CLIP training. In the third filtration step, the faiss k-nearest neighbors algorithm is employed to identify and filter out images exhibiting similarities.

pair, exclusively through human assessment. During this process, evaluators were tasked with answering two critical questions: "Is this an image of [object]?" and "Does it exhibit [attribute]?" If at least one of these questions was answered with a "no", the image was removed from consideration. This step ensured that only images accurately representing the specified characteristics were retained for further processing.

Step 2 - Exclusion of Known Combinations: To ensure the exclusivity of our dataset, we conducted a comprehensive search across several datasets (LAION, CommonPool, YFCC, and CC) to identify and eliminate any attribute-object combinations already present. This was achieved through a relaxed matching criterion, where combinations were removed if both the object and attribute appeared in a caption of an image, even if not in direct association.

Step 3 - Verification of OoD Status: The final step in our filtering process was to ensure the OoD nature of our dataset. We used the Faiss library [17] for a K-nearest neighbors search to compare our generated images against those in the LAION, CommonPool, YFCC, and CC datasets. Images were considered unique and retained in our dataset if no closely matching analogs were found, based on human evaluation. This rigorous approach ensured the novelty and uniqueness of our dataset by excluding combinations that had similar matches in the referenced datasets.

The dataset design process culminates in around 21,000 novel combinations of attributes and objects. The final generated dataset, after passing through the filtering process, comprises approximately 60,000 images representing 21,000 unique attribute-object combinations. Detailed properties and statistics about the dataset, including the list of attributes and objects used, can be found in the appendix. Additionally, another filtered version of the dataset is also available in the appendix.

2.2 Model/Data Zoo and Evaluation Criteria

In our experiments, we evaluate CLIP models trained on a diverse selection of datasets, including OpenAI's private dataset, LAION, YFCC15m, CC12m, DataComp, DFN-5B, WebLI, and CommonCrawl. These models leverage a variety of backbone image encoders such as ResNet50, ResNet101, ViT-B-32, ViT-B-16, ViT-L-14, ViT-H-14, ViT-g-14, and ViT-BigG-14. Our evaluation also extends to new CLIP variations, including EVA CLIP, SigLIP, and CLIPA, allowing for a comprehensive assessment of their performance and generalization capabilities across different tasks and datasets.

3 Comparison of CLIP Models on ImageNet-AO

To evaluate the CLIP model performance in the classification tasks, we adopted the evaluation method developed by [18], similar to the zero-shot evaluation approach described in [2]. Our evaluation involves providing the model with the actual images and various captions, obtaining embeddings for both the images and texts, and calculating their cosine similarities. This allows us to estimate the relevance of the captions to the image content, similar to a classification task. Given that our dataset only provided class labels (attribute-object pairs) for images, we expanded on this by creating 80 captions per class using various templates. This approach, inspired by the methodology described in [2], allows for a more comprehensive representation of each class. We generated embeddings for these captions and averaged them to produce a final embedding for each class, which was then used in our zero-shot evaluation. For the test sets, all 1000 classes of ImageNet were used as the in-distribution set and expanded the number of classes to approximately 21000 for the OoD set. The CLIP evaluations are shown in Fig. 1.

While our results generally showed that models trained on larger datasets exhibited improved accuracy in both in-distribution and out-of-distribution settings, supporting the notion that larger training datasets can enhance compositional out-of-distribution generalization performance, it is crucial to note that dataset size alone does not directly predict model strength. The performance of models varied significantly with not only the dataset size but also the quality and curation of the data. For instance, CLIP trained on the unfiltered CommonPool-XL dataset performed weaker than CLIP trained on the CommonPool-XL dataset filtered using ClipScore, despite the unfiltered dataset containing an additional 7 billion images. This further reinforces that simply increasing dataset size does not necessarily lead to improved model performance, and carefully curating and filtering the data can be more effective than merely accumulating vast amounts of unfiltered data.

Additionally, as evident from Fig. 1, models with different configurations trained on various datasets exhibited different training slope trajectories. The models trained on CommonPool-XL with different data filtering techniques demonstrated particularly steep performance trends, suggesting that the combination of a large dataset and effective data curation can yield significant performance gains.

Interestingly, the SigLip (denoted as WebLI) models presented a unique case with a somewhat negative slope, indicating that while enhancements to the backbone architecture improve in-distribution data performance, they may adversely affect out-of-distribution data performance. This highlights the nuanced relationship between architectural improvements and model generalization capabilities.

This extensive analysis, which encompasses the performance of diverse CLIP models across a broad spectrum of datasets, underscores the complexity of factors influencing model behavior and the pivotal role of dataset characteristics in achieving optimal performance in both in- and out-of-distribution settings. Further details on the performance evaluation of various CLIP models can be found in Sect. 7.4 of the Appendix.

4 Why CLIP Has Compositional Generalization?

Having established the superior C-OoD performance of certain CLIPs, we next try to investigate the reasons behind these observations. It has been widely known that disentangled representations make meaningful construction of known concept mixtures in the embedding space feasible, hence resulting in better C-OoD generalization [19–21]. Here, disentanglement means assignment of separate and independent embedding dimensions to different factors of variations, which in this case are the objects and attributes.

We hypothesize that the discrete nature of the language, and large and diverse training datasets promote a more decomposable text representation. On the other hand, alignment of the text and image embeddings through contrastive learning in CLIPs induces this decomposability in the image domain. Based on these insights, we posit that representation decomposability is the key to the CLIP unseen compositional generalization. This claim is supported by two main arguments:

- Decomposability of the CLIP text embedding, measured through a comprehensive set of metrics, is correlated to the CLIP C-OoD generalization (Fig. 4, bottom row).
- Text representation disentanglement is induced in the image encoding, due to implicit maximization of the mutual information of text and image representations through contrastive learning. We elaborate on this claim empirically (Fig. 4, top row), and theoretically in what follows.

Why Disentanglement is Induced from One View to Another in the Contrastive Learning? We next try to give some theoretical insight on why and how the disentanglement emerges in the CLIP vision encoder. Several studies have shown the relation between minimizing the contrastive loss and maximizing the mutual information [22]. Therefore, the CLIP training implicitly maximizes the mutual information between text and image embeddings. We claim that disentanglement in the text representation, which was evidenced previously, may encourage disentanglement in the image encoding. To see this, let y_1 and y_2 be the text embeddings for the objects and attributes, respectively. Let x_1 and x_2 be the corresponding image embeddings. Assuming a decomposable text embedding means $y_1 \perp y_2$, i.e. $p(y_1, y_2) = p(y_1)p(y_2)$. Now by minimizing the contrastive loss, the mutual information $I(x_1, x_2; y_1, y_2)$ is maximized. By letting $x = (x_1, x_2)$, and $y = (y_1, y_2)$, we have:

$$I(x_1, x_2; y_1, y_2) = D_{KL}(p(x,y) \parallel p(x)p(y))$$
$$= D_{KL}(p(x_1|x_2,y)p(x_2|y)p(y) \parallel p(x_1|x_2)p(x_2)p(y))$$
$$= \mathbb{E}_{x_1,x_2,y}[\log(p(x_1|x_2,y)/p(x_1|x_2))] + \mathbb{E}_{x_2,y}[\log(p(x_2|y)/p(x_2))]$$
$$= \mathbb{E}_{x_2,y}[D_{KL}(p(x_1|x_2,y) \parallel p(x_1|x_2))] + \mathbb{E}_y[D_{KL}(p(x_2|y) \parallel p(x_2))]$$

Maximization of the latter term makes x_2 and y dependent random variables, otherwise if $x_2 \perp y$, the expected KL divergence would be minimum (or zero), which is against maximizing the mutual information. Note that however, x_2 does not ideally depend on both y_1 and y_2, otherwise the two distributions in the KL divergence in the first term become similar, which is also against maximizing the mutual information. Putting these together, x_2 mostly depends on y_2 if the mutual information is maximized. Using a symmetric argument, x_1 mostly depends on y_1. Finally, because $y_1 \perp y_2$, we conclude that x_1 and x_2 tend to become independent. Therefore, maximizing $I(x_1, x_2; y_1, y_2)$ decomposes x if y is already decomposed.

5 Decomposable Representation of CLIP Models

In this section, our primary objective is to leverage the generated dataset and other synthtic datasets to analyze our hypotheses, focusing on the decomposable CLIP representation space and its impact on the compositional OoD performance.

5.1 Attribute-Object Decomposition of Representation Space

In this section, we show that the representation space of the CLIP models on the proposed dataset can be decomposable into the representations of the objects and the attributes.

Disentanglement of Attributes and Objects. Here, we aim to assess the level of embeddings disentanglement in various CLIPs on ImageNet-AO. We utilize some common disentanglement metrics, namely the Z-Diff Score [23], DCI [24] and Explicitness score [25] to quantitatively evaluate the embeddings. These metrics are typically employed for supervised disentanglement assessment and require access to the latent factors of data. Since we have a compositional text specifying the attribute and the object for each image, we can consider two super latent factors corresponding to attributes and objects respectively. More details about these disentanglement metrics and their formulas can be found in Appendix 7.5.

We calculate these metrics for each CLIP model on our ImageNet-AO dataset. Subsequently, in Fig. 4 (bottom), we visualize the relationship between the C-OoD accuracy and the disentanglement metrics. Each point in the plot represents a CLIP model, with the x-axis denoting the C-OoD accuracy and the y-axis representing the disentanglement metric. As observed in bottom row of the plot, there is a discernible pattern where models with higher C-OoD accuracy tend to exhibit more disentangled text and image representations. This empirical observation aligns with our initial hypothesis. Notably, the disentanglement in the text embedding (blue points), is more pronounced compared to the image embeddings (green points). Additionally, in Fig. 4 (top), we show the correlation between the image encoder and the text encoder for different disentanglement metrics. This figure demonstrates that by increasing the disentanglement in the text encoder, the disentanglement in the image encoder also increases, indicating a correlation between them.

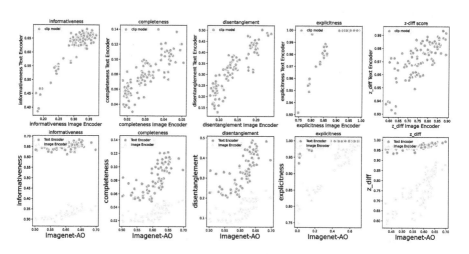

Fig. 4. Top: Representation disentanglments are correlated in text and image embeddings of CLIPs. Bottom: Disentanglment metrics vs. C-OoD Accuracy.

Intrinsic Dimensionality of the Composition Representations. The previously reported metrics of disentanglement focus on the correspondence between embedding dimensions and latent factors, and hence often require training an auxiliary classifier, in which a given representation is classified into levels of any latent factor. One could alternatively take a training-free approach through measuring relative intrinsic dimensionality of the composition patterns. This could be achieved by measuring the soft rank of the embeddings of attribute-object pairs. The soft rank is defined by the number of singular values of a given matrix that are greater than a pre-specified positive threshold. The soft rank is then normalized and made comparable across CLIPs by being divided to the number of embedding dimensions. This way the soft rank measures the relative intrinsic dimensionality of the embedding space. If the representation is entirely disentangled, huge combinations of attribute-objects would only result in a small intrinsic dimensionality, i.e. sum of the intrinsic dimensionalities of object and attribute spaces. Otherwise, each attribute-object embedding would appear to be *novel* with respect to other composition embeddings, resulting in a near full-rank space.

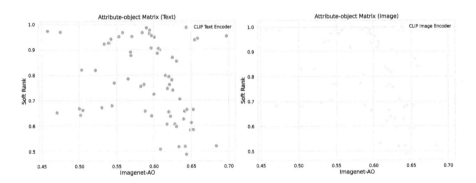

Fig. 5. The decrease in the soft rank of attribute-object representations relative to the embedding size correlates with improved C-OoD accuracy. This indicates that decomposing representations of attributes and objects results in a low dimensional representation of CLIPs that exhibits robust C-OoD performance. This highlights the representation disentanglement in CLIPs with strong C-OoD generalization.

For this experiment, we use ImageNet-AO, which provides around 21,000 unique combinations of attributes and objects. We utilize their image embeddings, obtained from the CLIP image encoder, and caption embeddings, obtained from the CLIP text encoder, to calculate the soft rank with a threshold of 0.1. Figure 5 shows that the intrinsic dimensionality is decreasing as the C-OoD accuracy increases, in both text and image domains.

Image Retrieval with Image±text Queries. Inspired by the work of [26], we designed an experiment to evaluate the compositional nature of embeddings

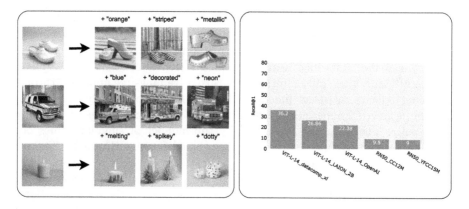

Fig. 6. The performance of various CLIP models in the task of image±text retrieval. A model's superior performance in this task indicates that its representation is more decomposable.

learned by the CLIP models. Our primary objective is to assess the representation disentanglement of the CLIP models trained on diverse datasets. To accomplish this goal, we devised a test in which we input an image from our dataset into the image encoder of the model, and obtain its corresponding embedding. Next, we employed the text encoder of the model to compute the embedding of an adjective, ensuring that the adjective differed from those associated with the current image. These two embeddings were then combined through summation and used as a query in a process similar to the image retrieval. We then show the image closest to the generated query embedding. A total of 200 random images were used to conduct this test for each model.

In order to evaluate the accuracy of the models predictions, we consider the image that is most similar to the query as the correct prediction if it possess both the intended object and adjective. A higher level of accuracy in the image retrieval task indicates that the model embeddings are more disentangled. Model evaluations are demonstrated in Fig. 6. The Recall@1 performance of various models aligns with our expectations. Specifically, we anticipated that models excelling in C-OoD tasks would also exhibit more disentangled representations. We previously observed in Fig. 1 that CLIPs associated with LAION and DataComp datasets stand out as having highest C-OoD accuracies. These two CLIPs also performed best in this experiment.

5.2 Disentanglement of Fine-Grained Factors

In the field of Disentanglement Representation Learning, the concept of disentanglement is explored from two distinct perspectives: fine-grained factors at the dimension level and coarse-grained factors at the vector level [27]. Our initial investigation into CLIP models, utilizing our curated dataset, provided insights into coarse-grained disentanglement (e.g. separating attributes and objects as two factors) and revealed multifaceted evaluation metrics. Moving forward, we aim to delve into the realm of fine-grained disentanglement at the dimension level. However, our current dataset poses inherent limitations in segregating factors at such a granular level. Consequently, to facilitate a comprehensive evaluation of fine-grained disentanglement, it becomes necessary to adopt specialized datasets designed explicitly for disentanglement studies within this domain.

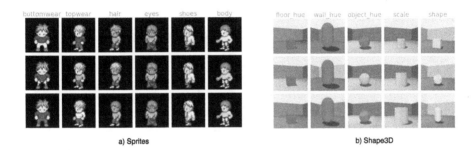

Fig. 7. Disentanglement datasets. a: Sprites dataset, consist of 6 factor and 54,000 images b: Shape3D, consist of 5 factor and 32,000 images

For our in-depth analysis of the fine-grained disentanglement, we selected two distinguished datasets: Sprites [28], Shapes3D [29] as they are specifically designed for disentanglement studies in image-centric models. Examples from these datasets can be seen in Fig. 7.

Since our focus extends beyond image-centric models to evaluate disentanglement in both the text encoder and image encoder components of CLIP models, we generated captions for each image based on the vector of factors associated with that image. This approach enables us to assess the disentanglement capabilities of CLIP models in both the visual and textual domains.

Figure 8 shows the text encoder exhibits higher disentanglement than the image encoder. As models improve on the C-OoD task, disentanglement tends to increase for both encoders.

More Analysis on Decomposability of the Representation Space. Using the Shapes 3D, we conducted two experiments to investigate the representation of factors more accurately.

In first experiment, we employ the 480,000 images of Shapes 3D dataset, each with specific latent factors such as floor hue, wall hue, object hue, scale, shape and orientation. We train a classifier to calculate the Z-Diff Score and utilize it to determine which dimensions are most critical for each latent factor. In the process of calculating the Z-Diff score, we train a classifier that can determine, for a group of data points that have a fixed specific value for one of the latent factors, what that factor is. By using this classifier, we can identify which dimensions are more important for determining each factor. Subsequently, we extract the

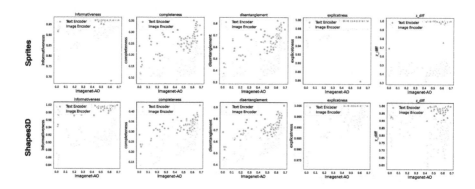

Fig. 8. Disentanglment metrics vs. C-OoD Accuracy on Sprites and Shapes3D dataset.

top 100 important dimensions for each factor and calculate how many dimensions are common across factors. Our results, presented in Table 1, demonstrate that models with higher C-OoD accuracy tend to exhibit fewer common dimensions across factors. This finding suggests that improved C-OoD generalization is associated with more disentangled representations.

In the second experiment, we looked at the impact of disentanglement on zero-shot object color manipulation using two identical images except for the object color. We calculated the embeddings using the CLIP and used the classifier of the first experiment to identify the most important dimensions for detecting object color. By switching the top k dimensions between the two image embeddings, we tested the models' ability to detect captions matching the switched new color. The results are summarized in Table 1 showing that models with higher C-OoD accuracy require fewer dimension switches to achieve the color change, indicating that disentangled representations enable more effective zero-shot modifications.

Table 1. Number of common dimensions across factors and switching dimensions for color manipulation in the Shapes 3D dataset

Dataset	Architecture	C-OoD Acc.	# Com. Dims	# Sw. Dims
LAION	ViT-L/14	64.61%	2	40
LAION	ViT-B/16	61.55%	5	60
LAION	ViT-B/32	61.05%	7	90
OpenAI	ViT-L/14	52.28%	3	5
OpenAI	ViT-B/16	49.22%	4	10
OpenAI	ViT-B/32	47.07%	6	30
CC	RN50	26.64%	15	200
YFCC	RN50	12.23%	21	250

6 Conclusion

This study examines how well CLIPs can generalize to new compositions of objects and attributes. We created an authentic benchmark of compositional images that are truly novel with respect to the CLIP training sets, and found that CLIPs ability to decompose the text/images representation space (into the embedding of concepts) is crucial for the compositional generalization. We have assessed the decomposability through the lens of several well-known metrics, as well the composition representation intrinsic dimensionality. These experiments were conducted on a wide range of datasets, from our attribute-object dataset to the ones previously designed specifically to evaluate disentanglement. We also covered a wide variety of problem setups in this direction, ranging from factor classification, and image±text retrieval, to factor manipulation. All mentioned assessments consistently demonstrate a strong connection between text and image representation disentanglement and C-OoD generalization.

References

1. Liu, J., et al.: Towards out-of-distribution generalization: a survey. arXiv preprint arXiv:2108.13624 (2021)
2. Radford, A., et al.: Learning transferable visual models from natural language supervision. In: International Conference on Machine Learning, pp. 8748–8763. PMLR (2021)
3. Fang, A., et al.: Data determines distributional robustness in contrastive language image pre-training (CLIP). In: International Conference on Machine Learning, pp. 6216–6234. PMLR (2022)
4. Nguyen, T., Ilharco, G., Wortsman, M., Oh, S., Schmidt, L.: Quality not quantity: on the interaction between dataset design and robustness of CLIP. arXiv preprint arXiv:2208.05516 (2022)
5. Wang, Z., Hershcovich, D.: On evaluating multilingual compositional generalization with translated datasets. In: Rogers, A., Boyd-Graber, J., Okazaki, N. (eds.) Proceedings of the 61st Annual Meeting of the Association for Computational Linguistics (Volume 1: Long Papers), , Toronto, Canada, pp. 1669–1687. Association for Computational Linguistics (2023)
6. Shaw, P., Chang, M.-W., Pasupat, P., Toutanova, K.: Compositional generalization and natural language variation: can a semantic parsing approach handle both? In: Zong, C., Xia, F., Li, W., Navigli, R., (eds.) Proceedings of the 59th Annual Meeting of the Association for Computational Linguistics and the 11th International Joint Conference on Natural Language Processing (Volume 1: Long Papers), pp. 922–938. Association for Computational Linguistics (2021)
7. Mehta, S.V., Rao, J., Tay, Y., Kale, M., Parikh, A.P., Strubell, E.: Improving compositional generalization with self-training for data-to-text generation. arXiv preprint arXiv:2110.08467 (2021)
8. Yuksekgonul, M., Bianchi, F., Kalluri, P., Jurafsky, D., Zou, J.: When and why vision-language models behave like bags-of-words, and what to do about it? In: The Eleventh International Conference on Learning Representations (2023)
9. Lewis, M., et al.: Does clip bind concepts? Probing compositionality in large image models (2023)

10. Zhao, T., et al.: VL-CheckList: evaluating pre-trained vision-language models with objects, attributes and relations (2023)
11. Thrush, T., et al.: WinoGround: probing vision and language models for visio-linguistic compositionality (2022)
12. Ossowski, T., Jiang, M., Hu, J.: Prompting large vision-language models for compositional reasoning (2024)
13. Zhang, J., Cai, M., Xie, T., Lee, Y.J.: CounterCurate: enhancing physical and semantic visio-linguistic compositional reasoning via counterfactual examples (2024)
14. Wang, H., Si, H., Shao, H., Zhao, H.: Enhancing compositional generalization via compositional feature alignment (2024)
15. Doveh, S., et al.: Dense and aligned captions (DAC) promote compositional reasoning in VL models (2023)
16. Pham, K., et al.: Learning to predict visual attributes in the wild (2021)
17. Douze, M., et al.: The faiss library (2024)
18. Ilharco, G., et al.: OpenCLIP (2021)
19. Yang, T., Wang, Y., Lan, C., Yan, L., Zheng, N.: Vector-based representation is the key: a study on disentanglement and compositional generalization (2023)
20. Montero, M.L., Ludwig, C.J.H., Costa, R.P., Malhotra, G., Bowers, J.: The role of disentanglement in generalisation. In: International Conference on Learning Representations (2021)
21. Xu, Z., Niethammer, M., Raffel, C.A.: Compositional generalization in unsupervised compositional representation learning: A study on disentanglement and emergent language. In: Advances in Neural Information Processing Systems, vol. 35, pp. 25074–25087 (2022)
22. Chen, T., Kornblith, S., Norouzi, M., Hinton, G.: A simple framework for contrastive learning of visual representations. In: International Conference on Machine Learning, pp. 1597–1607. PMLR (2020)
23. Higgins, I., et al.: beta-VAE: learning basic visual concepts with a constrained variational framework. In: International Conference on Learning Representations (2016)
24. Eastwood, C., Williams, C.K.I.: A framework for the quantitative evaluation of disentangled representations. In: International Conference on Learning Representations (2018)
25. Ridgeway, K., Mozer, M.C.: Learning deep disentangled embeddings with the F-statistic loss. In: Advances in Neural Information Processing Systems, vol. 31 (2018)
26. Jia, C., et al.: Scaling up visual and vision-language representation learning with noisy text supervision. In: International Conference on Machine Learning, pp. 4904–4916. PMLR (2021)
27. Wang, X., Chen, H., Tang, S., Wu, Z., Zhu, W.: Disentangled representation learning (2023)
28. Li, Y., Mandt, S.: Disentangled sequential autoencoder (2018)
29. Burgess, C., Kim, H.: 3D shapes dataset (2018). https://github.com/deepmind/3dshapes-dataset/

Modality Translation for Object Detection Adaptation Without Forgetting Prior Knowledge

Heitor Rapela Medeiros[(✉)], Masih Aminbeidokhti,
Fidel Alejandro Guerrero Peña, David Latortue, Eric Granger,
and Marco Pedersoli

Laboratoire d'imagerie, de vision et d'intelligence artificielle (LIVIA), Department of Systems Engineering, ETS Montreal, Montreal, Canada
{heitor.rapela-medeiros.1,masih.aminbeidokhti.1,
fidel-alejandro.guerrero-pena.1,david.latortue.1}@ens.etsmtl.ca,
{eric.granger,marco.pedersoli}@etsmtl.ca

Abstract. A common practice in deep learning involves training large neural networks on massive datasets to achieve high accuracy across various domains and tasks. While this approach works well in many application areas, it often fails drastically when processing data from a new modality with a significant distribution shift from the data used to pre-train the model. This paper focuses on adapting a large object detection model trained on RGB images to new data extracted from IR images with a substantial modality shift. We propose Modality Translator (ModTr) as an alternative to the common approach of fine-tuning a large model to the new modality. ModTr adapts the IR input image with a small transformation network trained to directly minimize the detection loss. The original RGB model can then work on the translated inputs without any further changes or fine-tuning to its parameters. Experimental results on translating from IR to RGB images on two well-known datasets show that our simple approach provides detectors that perform comparably or better than standard fine-tuning, without forgetting the knowledge of the original model. This opens the door to a more flexible and efficient service-based detection pipeline, where a unique and unaltered server, such as an RGB detector, runs constantly while being queried by different modalities, such as IR with the corresponding translations model. Our code is available at: https://github.com/heitorrapela/ModTr.

Keywords: Object Detection · Modality Translation · Infrared · Visual

1 Introduction

Powerful pre-trained models have become essential in the field of computer vision, particularly in object detection (OD) tasks [31, 32]. These OD models are typically

Supplementary Information The online version contains supplementary material available at https://doi.org/10.1007/978-3-031-73024-5_4.

Fig. 1. Bounding box predictions over different adaptations of the RGB detector (Faster R-CNN) for IR images on two benchmarks: LLVIP and FLIR. Yellow and red boxes show the ground truth and predicted detections, respectively. In a) we see the RGB data. In b) FastCUT is an unsupervised image translation approach that takes as input infrared images (IR) and produces pseudo-RGB images. It does not focus on detection and requires both modalities for training. In c) we have fine-tuning, which is the standard approach to adapting the detector to the new modality. It requires only IR data but forgets the original knowledge of the original RGB detector. Finally, in d) is the ModTr, which focuses the translation on detection, requires only IR data and does not forget the original knowledge so that it can be reused for other tasks. Bounding box predictions for other detectors are provided in the supplementary material.

pre-trained on extensive natural-image RGB datasets, such as COCO [28]. Moreover, the knowledge encoded by these models can be leveraged for various tasks in a zero-shot way or with additional fine-tuning for downstream tasks [43]. However, adding new modalities to these models, such as infrared (IR), without losing the intrinsic knowledge of the detector remains a challenge [29].

These additional modalities, though not as common as RGB images, are still important in various tasks, like surveillance [5,8], autonomous driving [33,41], and robotics [23,38], which strive to achieve robust performance in real-world environments, where capture conditions change, such as different illumination conditions [2]. The dominant way to adapt pre-trained detectors to these novel conditions is by fine-tuning the model [29]. However, fine-tuning often results in catastrophic forgetting and can destroy the intrinsic knowledge of the detector [25]. Ideally, we would like to adapt the detector to new modalities without changing the original model. This is most useful for server-side applications, where a single model runs uninterrupted and can be queried by different inputs, ideally on different modalities. The main challenge lies in the significant distribution shift introduced by the new modality. This shift occurs because the pre-trained knowledge, such as the visual information in RGB images, differs markedly from the thermal data in IR images. This shift can degrade model performance when applied directly as input to the model, since the features learned from one modality may not be relevant or present in another. This can ultimately impact the resulting OD performance [44].

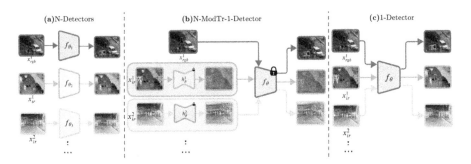

Fig. 2. Different approaches to deal with multiple modalities and/or domains. (a) The simplest approach is to use a different detector adapted to each modality. This can lead to a high level of accuracy but requires storing several models in memory. (b) Our proposed solution uses a single pre-trained model normally trained on the more abundant data (RGB) and then adapts the input through our ModTr model. (c) A single detector is jointly trained on all modalities. This allows using of a single model but requires access to all modalities jointly, which is often impossible, especially when dealing with large pre-trained models.

Image translation methods [35,36] have emerged as powerful tools to overcome the downsides of fine-tuning and narrowing the gap between source and target modalities [17]. These methods do not directly work on the weight space of the original detector but rather adapt the input values to reduce the discrepancy between the source and target modalities. However, such methods often require access to source data or some statistics about it during training. Furthermore, their primary focus is on image reconstruction quality rather than the final OD task, which can cause a significant drop in performance. For instance, Fig. 1 shows different ways to adapt the RGB detector (see the caption for more details).

Our work aims to improve the image translation paradigm while addressing its limitations. Our proposed approach, Modality Translation for OD (ModTr), incorporates the detector's knowledge into the translation module by training directly for the final detection task. Unlike traditional image translation methods, ModTr does not require any source data. It is a conceptually simple approach that can be easily integrated with any detector, be it a one-stage or two-stage detector. A notable application of ModTr is using a pre-trained RGB detector as a server that incorporates different ModTr blocks as input translators for new modalities such as IR. This new detector generates the desired output with performance comparable to full fine-tuning without losing the original knowledge of the pre-trained model. In Fig. 2, we present several options for integrating IR modalities into an RGB system. Figure 2a illustrates the N-Detectors approach, where each detector is trained for a specific case. This method effectively demands more memory and forgets previously learned information. Figure 2c shows a single detector trained on combined modalities. This method does not incur additional memory, yet it requires simultaneous access to all modalities, which may not always be feasible. Figure 2b illustrates our proposed approach, which involves training a specialized translator for each condition without altering the parameters of the original detector. The N-ModTr-1-Detector strikes a balance between the previous methods, addressing their shortcomings by requiring only a single detector. Importantly, it retains the original pre-training knowledge, as it leaves the detector unchanged. In this work, we focus on the effectiveness of our approach for the IR modality, commonly used in surveillance and robotics, and the incremental modality detector server-based application, which is crucial for many settings that require uninterrupted detection predictions.

Our main contributions can be summarized as follows.

(1) We present ModTr, a method for adapting pre-trained ODs from large RGB datasets to new scarce modalities like IR, without requiring access to any source dataset, by translating the input signal.

(2) In contrast to standard fine-tuning, our approach does not modify the original detector weights. This allows the detector to retain the knowledge of the source data while adapting to a new modality. As a result, a single model can be used to handle multiple modalities across various translators. For instance, the same model can be used to process RGB during the daytime and IR at nighttime.

(3) An extensive empirical evaluation of ModTr in several scenarios, showcasing its advantages and flexibility. In particular, with our different proposed fusion strategies, ModTr achieves OD accuracy that is competitive when compared with image translation methods on two challenging RGB/IR datasets (LLVIP and FLIR).

2 Related Work

(a) Object Detection. OD is a computer vision task that aims to provide labels and localization for the objects in the image [47]. Two-stage detectors, exemplified by Faster R-CNN [39], generate regions of interest and then use a second classifier to confirm object presence within those regions. On the other hand, one-stage detectors streamline the detection process by eliminating the proposal generation stage, aiming for end-to-end training and real-time inference speeds. RetinaNet [27] is a one-stage OD model that utilizes a focal loss function to address class imbalance during training. Also, models like FCOS [42] have emerged in this category, eliminating predefined anchor boxes to potentially enhance inference efficiency. The proposed work investigates these three traditional and powerful detectors: Faster R-CNN, RetinaNet, and FCOS. The choice of such detectors was due to the simplicity in implementation and integration among other methods, as well as a different range of pre-trained backbone weights, such as ResNet [13] and MobileNet [15].

(b) Image Translation. Image translation is a pivotal task in computer vision, aiming to map images from a source domain to a target domain while preserving inherent content [35]. The goal is to discover a transformation function such that the distribution of images in the translated domain is aligned with the distribution of images in the target domain. The commonly used approaches for image translation are based on variational autoencoders (VAEs) [24] and generative adversarial network (GANs) [11,35]. Isola et al. developed the Pix2Pix [21], a method that consists of a generator (based on U-Net) and a discriminator (based on GANs architecture) that work together to generate images based on input data and labels. Then, Zhu et al. proposed a method called CycleGAN [50], which is based on GANs, with the objective of unsupervised domain translation. Even though CycleGAN can produce quite visual results, it's hard to optimize due to the adversarial mechanism and memory footprint needed. In contrast, VAEs are easier to train than GANs but require more constraints in the optimization to produce images of good quality than GAN-based approaches. Recent advancements include diffusion models known for their high-quality image generation, although they may not inherently suit domain translation tasks. To enhance models such as CycleGAN, novel methods like Contrastive Unpaired Translation (CUT) [36] and FastCUT [36] have been introduced. CUT, in particular, accelerates the image translation process by maximizing mutual information between image patches, achieving competitive results quickly. In the context of RGB/IR modality, InfraGAN presents an image-level adaptation for RGB to IR conversion, prioritizing image quality [34]. This approach is distinct in its focus on optimizing image quality losses. Moreover, Herrmann et al. have explored OD in RGB/IR modality by adapting IR images to RGB using traditional image preprocessing techniques, allowing the use of RGB object detectors without parameter modification [14]. Despite significant advances in image translation, these techniques do not specifically address OD tasks. In our previous work, we introduced HalluciDet [29], which employs an image translation mechanism for OD. However, this approach requires prior access to the source RGB data from the same domain as the target for pre-training the detector.

(c) Adapting Without Forgetting. Catastrophic forgetting (CF) is the idea that a neural network tends to forget knowledge when sequentially trained on a different task and replaces it with knowledge tailored to the new objective [45]. CF can be harmful or beneficial. Researchers identified harmful learning as situations where retaining the original knowledge while adapting to a different task is necessary. In that case, it is imperative to mitigate the risk of CF. However, some CF can also be beneficial, for instance, to prevent privacy leakage from large pretrained models, to enhance the generalization, or to remove noisy information from the originally, acquired knowledge that is negatively affecting the new tasks. In our case, knowledge-forgetting is harmful. There are different ways to address this issue including simple techniques like decreasing the learning rate [16], use weight decay [4,49] or mixout regularization [26] during fine-tuning or more complex approaches like Recall and learn [6], Robust Information Fine-tuning [46] or CoSDA [10]. Some adaptation methods use techniques based on replay of the source data or even using the weights of the initial model to keep some prior information [30]. Some of these works focus on adding continually different tasks in an incremental learning setting. However, these methods may still produce a loss of knowledge since the original parameters are not frozen. Furthermore, in adapting without forgetting, an adapter, which adopts a frozen pre-trained backbone to generate a representation followed by a different classifier for each downstream task [45], can be seen as a powerful method to preserve knowledge. Even though our ModTr shares some similarities, we work in the input space to adapt to the new modalities, and address this incremental modality adaptation, optimizing the translation directly for the final OD task.

3 Proposed Method

(a) Preliminary Definitions. The training set for OD is denoted as $\mathcal{D} = \{(x, Y)\}$, where $x \in \mathbb{R}^{W \times H \times C}$ represents an image in the dataset, with dimensions $W \times H$ and C channels. Subsequently, the OD model aims to identify N regions of interest within these images, denoted as $Y = \{(b_i, c_i)\}_{i=1}^{N}$. The top-left corner coordinates and the width and height of the object define each region of interest b_i. Additionally, a classification label c_i is assigned to each detected object, indicating its corresponding class within the dataset. In this study, the number of input channels for the detector is fixed at three, corresponding to RGB-like inputs. In terms of optimization, the primary goal of this task is to maximize detection accuracy, often measured using the average precision (AP) metric across all classes. An OD is formally represented as the mapping $f_\theta : \mathbb{R}^{W \times H \times C} \to \hat{Y}$, where θ denotes the parameter vector. To effectively train a detector, a differentiable surrogate for the AP metric, referred to as the detection cost function, $\mathcal{C}_{det}(\theta)$, is employed. The typical structure of such a cost function involves computing the average detection loss over dataset $\mathcal{D}$, denoted as $\mathcal{L}_{det}$, described as:

$$\mathcal{C}_{det}(\theta) = \frac{1}{|\mathcal{D}|} \sum_{(x,Y) \in \mathcal{D}} \mathcal{L}_{det}(f_\theta(x), Y). \quad (1)$$

(b) Modality Translation Module. Our approach primarily consists of an image-to-image translation network responsible for converting the input modality into an RGB-like space intelligible to the detector. These networks typically adopt an encoder-decoder structure to synthesize and reconstruct knowledge in a pixel-wise manner. While we employ U-Net [40] as the translation network, with parameters ϑ, in this work, our framework is general and not limited by the translation architecture. In general terms, this mapping is denoted as $h_\vartheta^d : \mathbb{R}^{W \times H \times C} \to \mathbb{R}^{W \times H \times 3}$, with a translation network assigned to each available input modality d. Unlike the detection network, the number of input channels varies depending on the modality, for instance, $C = 1$ for IR and depth images. It's important to note that, being a pixel-level architecture, the output of such a network retains the spatial resolution of the input. However, the number of output channels is consistently fixed at three, corresponding to RGB-like images ($C = 3$).

Unlike other image-to-image translation approaches, we drive the process using the aforementioned detection cost (Equation (1)). Thus, the underlying optimization problem is formulated as $\vartheta^* = \arg\min \mathcal{L}_{det}(\vartheta)$, incorporating the output of the composition $(f_\theta \circ h_\vartheta^d)(x)$ at the loss function level. To streamline the learning process, we utilize a residual learning strategy in which the function h_ϑ^d focuses on capturing the small variations in the input that are necessary to solve the task. This approach is similar to the one employed on diffusion models, which inspired our work. For the sake of simplicity, we separate the fusion step from the translation mapping in our notation, as various types of fusion are investigated. Consequently, the proposed image-to-image translation loss function is defined as:

$$\mathcal{L}_{\text{ModTr}}(x, Y; \vartheta) = \mathcal{L}_{det}(f_\theta \left(\Phi(h_\vartheta^d(x), x)\right), Y), \tag{2}$$

where $\Phi(.,.)$ is a non-parametric fusion function. Note that the output of $h_\vartheta^d(x)$ is an RGB-like image, whereas x may only consist of a single channel, depending on the input modality. We have chosen this definition to simplify the notation, but appropriate reshaping should be performed during implementation to ensure compatibility.

In addition, note that, while a detection loss is employed to update the translation network, the weight vector θ remains constant. This constraint is consistent with the premise of this study, where a pre-trained detector is solely available on the server side and remains unaltered. An overview of the proposed approach can be seen in Fig. 2b).

(c) Fusion Strategy. As previously mentioned, we utilize a non-parametric fusion of the intermediate representation $h_\vartheta^d(x)$ and the original input x to simplify the learning process of the translation network. In this context, we employ an element-wise product, also known as the Hadamard product, which is particularly interesting for attention mechanisms and has been explored previously for re-calibrating feature maps based on their importance [19]. Although we investigated various fusion mechanisms, the element-wise product yielded the best results. For more details on different fusion strategies, please refer to supplementary materials.

ModTr$_\odot$: The Hadamard product-based fusion serves as a gating mechanism to filter or highlight information from the input image. In this approach, the output of the translation network acts as a weight map for the input, and they are fused using pixel-wise multiplication, $\odot$. Consequently, the translation network tends to highlight information from the input when the pixel value tends toward 1 or discard it when it approaches 0. Additionally, the output translation modality can be interpreted as an attention map, as described by the following Eq. (3):

$$\mathcal{L}_{\text{ModTr}_\odot}(x, Y; \vartheta) = \mathcal{L}_{det}(f_\theta \left(h_\vartheta^d(x) \odot x\right), Y). \tag{3}$$

In our design choices, we opt to utilize these straightforward non-parametric functions to assist in optimization while maintaining low inference costs.

4 Results and Discussion

4.1 Experimental Methodology

(a) **Datasets: LLVIP:** LLVIP is a surveillance dataset composed of 30,976 images, in which 24,050 (12,025 IR and 12,025 RGB paired images) are used for training and 6,926 for testing (3,463 IR and 3,463 RGB paired images) with only pedestrians annotated. **FLIR ALIGNED:** We used the sanitized and aligned paired sets provided by Zhang et al. [48]. It has 10,284 images, that is 8,258 for training (4,129 IRs and 4,129 RGBs) and 2,026 (1,013 IRs and 1,013 RGBs) for test. FLIR images are captured from the perspective of a camera in the front of a car, with a resolution of 640 by 512. It contains the bicycles, dogs, cars, and people classes. It has been found that with FLIR, the "dog" objects are inadequate for training [3], thus we decided to remove them.

(b) **Implementation Details:** In our experiments, we randomly selected 80% of the training set for training and the rest for validation. All results reported are on the test set. As starting pre-trained weights for the detectors, we used Torchvision models with COCO [28] weights and for the U-Net translation network, we used PyTorch Segmentation Models [20] and we changed the last layer for 3-channel (RGB-like) with a Sigmoid function, to be closer to an image with values between 0 and 1, to perform translation instead of traditional segmentation. For the translation network backbones, we explored our default ResNet$_{34}$, and for subsequent studies on reducing parameters, we dive into ResNet and MobileNet-family. All the code is available on GitHub for reproducibility in the experiments. To ensure fairness, we trained the detectors under the library version and the same experimental design, i.e., data order, augmentations, etc. Furthermore, we trained with PyTorch Lightning [9] training framework, evaluated the APs with TorchMetrics [7], and logged all experiments with WandB [1] logging tool. The different performance measures (e.g., APs) can be found in suppl. materials.

4.2 Comparison with Translation Approaches

In this section, ModTr is compared with different image-to-image translation methods employing different learning strategies. These include basic image processing strategy [14], reconstruction strategies such as CycleGAN [51], CUT [37], and FastCUT [37], which employs a contrastive learning approach, as well as HalluciDet [29], which utilizes a detection-based loss. As outlined in Table 1, we evaluated the methods based on their final detection performance across three commonly used detectors: FCOS, RetinaNet, and Faster R-CNN. The reported results are derived from the IR test set and are averaged over three different seeds, which helps mitigate the impact of randomness across runs and splits of the training and validation datasets.

For each method, we also consider its dependency on the prior knowledge data (RGB) and ground truth bounding boxes (bboxes) on the IR images. Methods that rely on reconstruction techniques do not require bbox annotations on IR images but cannot provide accurate translations for detection purposes. However, HalluciDet and ModTr require bbox annotations to adjust the input image in a discriminative manner. The main difference between HalluciDet and ModTr is the use of source images. HalluciDet requires RGB images for an initial fine-tuning of the model, while our approach can work without that fine-tuning by reusing the detector's zero-shot knowledge.

The proposed ModTr displays robustness across the three detectors and consistently exhibits improvement on two different datasets: LLVIP [22] and FLIR aligned [12]. Note that each algorithm described in Table 1 employs different training supervisions. For instance, CycleGAN employs an adversarial mechanism with both RGB and infrared modalities in an unpaired setting. Similarly, CUT and FastCUT operate with positive and negative patches in an unpaired setting. In contrast, HalluciDet doesn't require the presence of both modalities during training but employs a detection mechanism during training similar to ours. In our approach, we solely require examples from the target modality. In this section, we present the performance of our best approach ModTr$_\odot$. For additional results, refer to suppl. materials.

As reported in Table 1, the detection performance of ModTr over the LLVIP dataset exhibited significant improvements. Specifically, it surpassed HalluciDet, the second best, by more than 29.0 AP with both FCOS and RetinaNet architectures, while obtaining comparable results with Faster R-CNN. Such disparity with the previous technique can be attributed to the loss of previous knowledge inherent in HalluciDet, which necessitates a pre-fine-tuning strategy on the source modality. Although the performance of the FLIR dataset also improved, the dataset's inherent challenges, such as changing the background from a moving car setup, make detection more difficult. Nonetheless, our proposal consistently enhances results, with improvements of more than 11 AP for FCOS and RetinaNet, and over 7 AP for Faster R-CNN. We also observed improvements on the AP_{50} and AP_{75}. Because of the space constraint, we include these in supplementary materials. These promising results indicate that our proposal can effectively translate images from the original IR modality to an RGB-like representation, sufficiently close to the source data to be usable by the detector.

4.3 Translation vs. Fine-Tuning

In this section, we further show that the proposed approach can be trained jointly with both translation and detector, which preserves the detector's knowledge. Here, ModTr is compared to three baselines fully fine-tuning (FT), FT of the head and LoRA [18], and our best ModTr fusion strategy, as shown in Table 2.

We conduct LoRA fine-tuning using two settings. In the first, we apply LoRA across all layers; in the second, only to the last layer of detectors. The latter results in superior performance, so we have adopted it as our default LoRA setting. The Table 2 shows AP for the LLVIP and FLIR datasets, with a consistent trend across all detectors (FCOS, RetinaNet, and Faster R-CNN). Furthermore, in the case of the FLIR dataset, we observed enhancements of ModTr over the

Table 1. Detection performance (AP) of ModTr versus baseline image-to-image methods to translate the IR to RGB-like images, using three different detectors (FCOS, RetinaNet, and Faster R-CNN). The methods were evaluated on IR test set of LLVIP and FLIR datasets. The RGB column indicates if the method required access to RGB images during training, and Box refers to the use of ground truth boxes during training.

Image translation	RGB	Box	Test Set IR (Dataset: LLVIP)		
			FCOS	RetinaNet	Faster R-CNN
Histogram Equal. [14]			31.69 ± 0.00	33.16 ± 0.00	38.33 ± 0.02
CycleGAN [51]	✓		23.85 ± 0.76	23.34 ± 0.53	26.54 ± 1.20
CUT [37]	✓		14.30 ± 2.25	13.12 ± 2.07	14.78 ± 1.82
FastCUT [37]	✓		19.39 ± 1.52	18.11 ± 0.79	22.91 ± 1.68
HalluciDet [29]	✓	✓	28.00 ± 0.92	19.95 ± 2.01	57.78 ± 0.97
ModTr$_\odot$ (ours)		✓	**57.63 ± 0.66**	**54.83 ± 0.61**	**57.97 ± 0.85**

Image translation	RGB	Box	Test Set IR (Dataset: FLIR)		
			FCOS	RetinaNet	Faster R-CNN
Histogram Equal. [14]			22.76 ± 0.00	23.06 ± 0.00	24.61 ± 0.01
CycleGAN [51]	✓		23.92 ± 0.97	23.71 ± 0.70	26.85 ± 1.23
CUT [37]	✓		18.16 ± 0.75	17.84 ± 0.75	20.29 ± 0.48
FastCUT [37]	✓		24.02 ± 2.37	22.00 ± 2.73	26.68 ± 2.59
HalluciDet [29]	✓	✓	23.74 ± 2.09	22.29 ± 0.45	29.91 ± 1.18
ModTr$_\odot$ (ours)		✓	**35.49 ± 0.94**	**34.27 ± 0.27**	**37.21 ± 0.46**

standard detector FT. As demonstrated, our approach surpasses standard fine-tuning while maintaining the detector's performance in the original modality. It is worth noting that our method also improves performance in terms of localization metrics such as AP_{50} and AP_{75} compared to fine-tuning alone, and we provide detailed results in the supplementary materials.

4.4 Different Backbones for ModTr

In this context, we evaluate ModTr and examine the trade-off between performance and parameter cost. It is widely recognized that increasing the number of parameters can enhance performance, but this relationship is not strictly linear. We demonstrated that models with fewer parameters can still achieve good performance; for example, MobileNet$_{v2}$, with fewer parameters than ResNet$_{18}$, sometimes outperformed it. This trade-off highlights the versatility of the model, which can be deployed with MobileNet-based architectures and utilized in low-cost devices. In Table 3, the default number of parameters is successfully reduced

Table 2. Detection performance (AP) of ModTr versus baseline fine-tuning (FT) of the detector, FT of the head and LoRA [18], using three different detectors (FCOS, RetinaNet, and Faster R-CNN. The methods were evaluated on IR test set of LLVIP and FLIR datasets. Results with "-" diverged from the optimization.

Method	Test Set IR (Dataset: LLVIP)		
	FCOS	RetinaNet	Faster R-CNN
Fine-Tuning (FT)	57.37 ± 2.19	53.79 ± 1.79	**59.62 ± 1.23**
FT Head	49.11 ± 0.70	44.00 ± 0.28	59.33 ± 2.17
LoRA [18]	47.72 ± 0.58	-	54.83 ± 1.30
ModTr$_\odot$ (ours)	**57.63 ± 0.66**	**54.83 ± 0.61**	57.97 ± 0.85

Method	Test Set IR (Dataset: FLIR)		
	FCOS	RetinaNet	Faster R-CNN
Fine-Tuning (FT)	27.97 ± 0.59	28.46 ± 0.50	30.93 ± 0.46
FT Head	27.40 ± 0.12	26.78 ± 0.70	33.53 ± 0.36
LoRA [18]	-	-	29.44 ± 0.61
ModTr$_\odot$ (ours)	**35.49 ± 0.94**	**34.27 ± 0.27**	**37.21 ± 0.46**

from 24.4M (ResNet$_{34}$) to 6.6M using MobileNet$_{v2}$ while maintaining similar performance. For instance, on LLVIP, MobileNet$_{v2}$ achieved a mean AP of 56.15, comparable to 56.35 AP$_{50}$ from ResNet$_{34}$ (others APs and detectors are reported in the supplementary material).

This approach opens up new possibilities, particularly in scenarios where using one translation network and one detector (e.g., one ModTr and one detector for RGB/IR) proves advantageous. This setup requires a total of 44.9M parameters, compared to 83.6M parameters, when employing two detectors—one for each modality (for example, for Faster R-CNN). Similar reductions in parameter costs were observed for FCOS (from 66.4M to 36.3M) and RetinaNet (from 68M to 37.1M) when using one detector for both modalities while preserving the knowledge of the previous modality and incorporating a new one. These numbers are based on MobileNet$_{v3s}$, which strikes a balance between performance and the number of parameters, making it suitable for memory-restricted systems. The complete evaluations for FCOS and RetinaNet are included in the supplementary material.

4.5 Knowledge Preservation Through Input Modality Translation

ModTr is designed to prevent catastrophic forgetting by keeping the weights of the pre-trained detector fixed. In this section, we demonstrate how various adaptation paradigms, shown in Fig. 2, effectively solve the final task while preserv-

Table 3. Detection performance (AP) of ModTr with different backbones for the translation networks with different numbers of parameters, using three different detectors (FCOS, RetinaNet, and Faster R-CNN). The methods were evaluated on IR test set of LLVIP and FLIR datasets.

Method	Parameters	AP ↑
Test Set IR (Dataset: LLVIP)		
Faster R-CNN	41.8 M	
MobileNet$_{v3s}$	+ 3.1 M	54.51 ± 0.28
MobileNet$_{v2}$	+ 6.6 M	56.15 ± 0.51
ResNet$_{18}$	+ 14.3 M	55.53 ± 1.14
ResNet$_{34}$	+ 24.4 M	56.35 ± 0.65
Test Set IR (Dataset: FLIR)		
Faster R-CNN	41.8 M	
MobileNet$_{v3s}$	+ 3.1 M	32.06 ± 0.75
MobileNet$_{v2}$	+ 6.6 M	36.77 ± 0.67
ResNet$_{18}$	+ 14.3 M	36.68 ± 0.22
ResNet$_{34}$	+ 24.4 M	37.21 ± 0.46

ing intrinsic knowledge. We compare our proposed method, ModTr, with two fine-tuning baseline methods. The first baseline method involves N-detectors, each fine-tuning the target modality individually. The second baseline method employs a single detector trained on the joint modality using balanced sampling. Note that while a copy of the original detector can be used in the N-detectors paradigm, it is unavailable in the 1-detector paradigm because the original modality is assumed to be inaccessible during training.

In all scenarios, we use COCO as the pre-training dataset and LLVIP and FLIR as target domains. Specifically, in the N-detectors scenario (Fig. 2a), we fine-tune one detector on each dataset and use a copy of the original detector for the RGB modality. In the 1-detector scenario (Fig. 2c), we fine-tune one detector on the combined FLIR and LLVIP datasets. In the N-ModTr-1-Detector scenario (Fig. 2b), two translators are trained, one per dataset. To assess catastrophic forgetting, we re-evaluate each scenario on COCO-val.

Table 4 shows the final performance. While all adaptation paradigms achieve relatively similar performance, the 1-detector method completely fails in the zero-shot scenario. The N-detectors method mitigates this by duplicating the detector three times. In contrast, ModTr preserves knowledge using a single detector and three efficient translators, demonstrating its practicality for embedded devices, as it requires less memory. Based on the average performance on all datasets, ModTr obtains the best results.

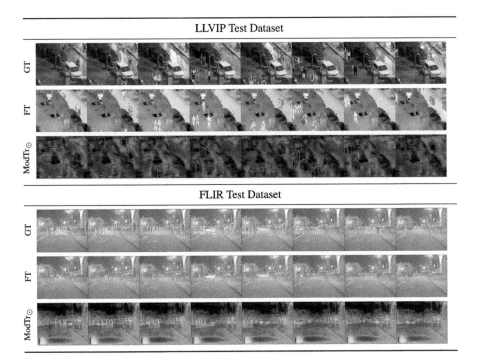

Fig. 3. Illustration of a sequence of 8 images of LLVIP and FLIR dataset for Faster R-CNN. For each dataset, the first row is the RGB modality, followed by the IR modality and different representations created by ModTr. For visualizations of other detectors and variants of ModTr, please refer to the supplementary materials.

4.6 Visualization of ModTr Translated Images

In Fig. 3, we present qualitative results for LLVIP and FLIR, alongside a comparison with fine-tuning. Each dataset section includes three rows: the first row displays the ground-truth RGB images, the second row showcases the results of fine-tuning using IR, and the last row features ModTr with a Hadamard product-based fusion over the Faster R-CNN detector. Due to space constraints, additional visualizations for other detectors and fusion strategies are provided in the supplementary materials. Notably, the IR results exhibit some false positives, particularly when detected objects overlap. Our method mitigates this issue effectively. Further insights, provided in the supplementary materials, reveal how our method effectively blurs or removes objects that do not belong to the target classes, thereby enhancing detection accuracy. Although the obtained intermediate representations are not visually pleasant, they prove more efficient for incorporating the knowledge necessary for the OD. Additionally, we conducted experiments with loss function terms aimed at enhancing the visual effects of the image, but they were not conclusive in terms of helping the detection performance.

Table 4. Detection performance (AP) of knowledge preserving techniques N-Detectors, 1-Detector, and N-ModTr-1-Detector, using three different detectors (FCOS, RetinaNet, and Faster R-CNN). The methods were evaluated on COCO and IR test sets of LLVIP and FLIR datasets.

Detector	Dataset	N-Detectors	1-Detector	N-ModTr-1-Det.
FCOS	LLVIP	57.37 ± 2.19	**58.55 ± 0.89**	57.63 ± 0.66
	FLIR	27.97 ± 0.59	26.70 ± 0.48	**35.49 ± 0.94**
	COCO	**38.41 ± 0.00**	00.33 ± 0.04	**38.41 ± 0.00**
	AVG.	41.25 ± 0.92	28.52 ± 0.47	**43.84 ± 0.53**
RetinaNet	LLVIP	53.79 ± 1.79	53.26 ± 3.02	**54.83 ± 0.61**
	FLIR	28.46 ± 0.50	25.19 ± 0.72	**34.27 ± 0.27**
	COCO	**35.48 ± 0.00**	00.29 ± 0.01	**35.48 ± 0.00**
	AVG.	39.24 ± 0.76	26.24 ± 1.28	**41.52 ± 0.29**
Faster R-CNN	LLVIP	59.62 ± 1.23	**62.50 ± 1.29**	57.97 ± 0.85
	FLIR	30.93 ± 0.46	28.90 ± 0.33	**37.21 ± 0.46**
	COCO	**39.78 ± 0.00**	00.40 ± 0.00	**39.78 ± 0.00**
	AVG.	43.44 ± 0.56	30.60 ± 0.54	**44.98 ± 0.43**

4.7 Fine-Tuning of ModTr and the Detector

The main reason to use ModTr is to avoid fine-tuning the detector for a specific task so that it can preserve its knowledge and be used for multiple modalities. However, in this section, we consider what would happen if we learn jointly ModTr and the detector weights. Results are reported in Fig. 4. We see that fine-tuning the detector can further boost performance. Thus, another application

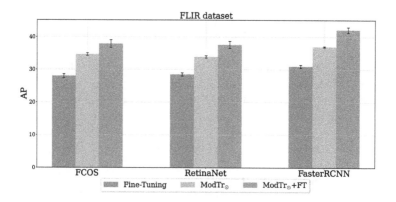

Fig. 4. Comparison of the performance of fine-tuning the ModTr and normal fine-tuning on the FLIR dataset for the three different detectors (FCOS, RetinaNet, and Faster R-CNN). In blue, the Fine-tuning; in orange, the ModTr$_\odot$, and in green, ModTr$_\odot$ + FT. (Color figure online)

of ModTr could be used to improve the fine-tuning of a detector with a reduced additional computational cost.

5 Conclusion

In this paper, a novel method called ModTr is proposed for adapting RGB object detectors (ODs) for IR modality without changing their parameters. A key advantage of our approach is that it preserves the full knowledge of the detector, allowing the translation network to act as a node that changes the modality for an unaltered detector. This is much more flexible and computationally efficient than having a specialized OD for each modality. Our approach performs well in various settings, outperforming powerful image-to-image models and previous competitors. We evaluated ModTr for different tasks, including detection based on image translation, comparison with traditional fine-tuning, and incremental IR modality application. Experimental results show the high performance and versatility of our method in all these settings.

Additionally, to explore integrating modalities beyond IR, we applied ModTr to Canny edges extracted from IR images as detailed in the supplementary material. While ModTr significantly enhances the performance of zero-shot RGB OD on edges, it still does not match the effectiveness of full fine-tuning on this other modality. We believe this shortfall arises from the limited information provided by edges compared to the richer data in the IR modality, leading to lower initial zero-shot OD performance. A potential solution is to replace the deterministic translator module within ModTr with a generative model. This substitution could enrich modality information by generating the missing data, potentially improving the zero-shot detector's performance. This promising direction will be explored in future research.

Acknowledgments. This work was supported in part by Distech Controls Inc., the Natural Sciences and Engineering Research Council of Canada, the Digital Research Alliance of Canada, and MITACS.

References

1. Biewald, L.: Experiment tracking with weights and biases (2020). wandb.com
2. Bustos, N., Mashhadi, M., Lai-Yuen, S.K., Sarkar, S., Das, T.K.: A systematic literature review on object detection using near infrared and thermal images. Neurocomputing 126804 (2023)
3. Cao, Y., Bin, J., Hamari, J., Blasch, E., Liu, Z.: Multimodal object detection by channel switching and spatial attention. In: Proceedings of the IEEE/CVF Conference on Computer Vision and Pattern Recognition, pp. 403–411 (2023)
4. Chelba, C., Acero, A.: Adaptation of maximum entropy capitalizer: little data can help a lot. Comput. Speech Lang. **20**(4), 382–399 (2006). https://doi.org/10.1016/j.csl.2005.05.005, https://www.sciencedirect.com/science/article/pii/S0885230805000276

5. Chen, J., Li, K., Deng, Q., Li, K., Philip, S.Y.: Distributed deep learning model for intelligent video surveillance systems with edge computing. IEEE Trans. Ind. Inform. (2019)
6. Chen, S., Hou, Y., Cui, Y., Che, W., Liu, T., Yu, X.: Recall and learn: fine-tuning deep pretrained language models with less forgetting. CoRR abs/2004.12651 (2020). https://arxiv.org/abs/2004.12651
7. Detlefsen, N.S., et al.: Torchmetrics-measuring reproducibility in pytorch. J. Open Sour. Softw. **7**(70), 4101 (2022)
8. Dubail, T., Guerrero Peña, F.A., Medeiros, H.R., Aminbeidokhti, M., Granger, E., Pedersoli, M.: Privacy-preserving person detection using low-resolution infrared cameras. In: Karlinsky, L., Michaeli, T., Nishino, K. (eds.) ECCV 2022. LNCS, vol. 13805, pp. 689–702. Springer, Cham (2022). https://doi.org/10.1007/978-3-031-25072-9_46
9. Falcon, W.: The PyTorch lightning team: PyTorch lightning (2019). https://doi.org/10.5281/zenodo.3828935
10. Feng, H., et al.: CoSDA: continual source-free domain adaptation (2023)
11. Goodfellow, I., et al.: Generative adversarial nets. Adv. Neural Inf. Processing Syst. **27** (2014)
12. Group, F., et al.: FLIR thermal dataset for algorithm training (2018)
13. He, K., Zhang, X., Ren, S., Sun, J.: Deep residual learning for image recognition. In: Proceedings of the IEEE Conference on Computer Vision and Pattern Recognition, pp. 770–778 (2016)
14. Herrmann, C., Ruf, M., Beyerer, J.: CNN-based thermal infrared person detection by domain adaptation. In: Autonomous Systems: Sensors, Vehicles, Security, and the Internet of Everything, vol. 10643, p. 1064308. International Society for Optics and Photonics (2018)
15. Howard, A.G., et al.: MobileNets: efficient convolutional neural networks for mobile vision applications. arXiv preprint arXiv:1704.04861 (2017)
16. Howard, J., Ruder, S.: Fine-tuned language models for text classification. CoRR abs/1801.06146 (2018). http://arxiv.org/abs/1801.06146
17. Hsu, H.K., et al.: Progressive domain adaptation for object detection. In: Proceedings of the IEEE/CVF Winter Conference on Applications of Computer Vision, pp. 749–757 (2020)
18. Hu, E.J., et al.: LoRA: low-rank adaptation of large language models. In: International Conference on Learning Representations (2022). https://openreview.net/forum?id=nZeVKeeFYf9
19. Hu, J., Shen, L., Sun, G.: Squeeze-and-excitation networks. In: Proceedings of the IEEE Conference on Computer Vision and Pattern Recognition, pp. 7132–7141 (2018)
20. Iakubovskii, P.: Segmentation models pytorch (2019)
21. Isola, P., Zhu, J.Y., Zhou, T., Efros, A.A.: Image-to-image translation with conditional adversarial networks. In: 2017 IEEE Conference on Computer Vision and Pattern Recognition (CVPR) (2017)
22. Jia, X., Zhu, C., Li, M., Tang, W., Zhou, W.: LLVIP: a visible-infrared paired dataset for low-light vision. In: Proceedings of the IEEE/CVF International Conference on Computer Vision, pp. 3496–3504 (2021)
23. Jing, C., Potgieter, J., Noble, F., Wang, R.: A comparison and analysis of RGB-D cameras' depth performance for robotics application. In: 2017 24th International Conference on Mechatronics and Machine Vision in Practice (M2VIP), pp. 1–6. IEEE (2017)

24. Kingma, D.P., Welling, M.: Auto-encoding variational bayes (2022)
25. Kirkpatrick, J., et al.: Overcoming catastrophic forgetting in neural networks. Proc. Natl. Acad. Sci. **114**(13), 3521–3526 (2017)
26. Lee, C., Cho, K., Kang, W.: Mixout: effective regularization to finetune large-scale pretrained language models (2020)
27. Lin, T.Y., Goyal, P., Girshick, R., He, K., Dollár, P.: Focal loss for dense object detection. In: Proceedings of the IEEE International Conference on Computer Vision, pp. 2980–2988 (2017)
28. Lin, T.-Y., et al.: Microsoft COCO: common objects in context. In: Fleet, D., Pajdla, T., Schiele, B., Tuytelaars, T. (eds.) ECCV 2014. LNCS, vol. 8693, pp. 740–755. Springer, Cham (2014). https://doi.org/10.1007/978-3-319-10602-1_48
29. Medeiros, H.R., Pena, F.A.G., Aminbeidokhti, M., Dubail, T., Granger, E., Pedersoli, M.: HalluciDet: hallucinating RGB modality for person detection through privileged information. In: Proceedings of the IEEE/CVF Winter Conference on Applications of Computer Vision, pp. 1444–1453 (2024)
30. Menezes, A.G., de Moura, G., Alves, C., de Carvalho, A.C.: Continual object detection: a review of definitions, strategies, and challenges. Neural Netw. (2023)
31. Minderer, M., Gritsenko, A., Houlsby, N.: Scaling open-vocabulary object detection. Adv. Neural Inf. Process. Syst. **36** (2024)
32. Minderer, M., et al.: Simple open-vocabulary object detection. In: Avidan, S., Brostow, G., Cissé, M., Farinella, G.M., Hassner, T. (eds.) ECCV 2022. LNCS, vol. 13670, pp. 728–755. Springer, Cham (2022). https://doi.org/10.1007/978-3-031-20080-9_42
33. Natan, O., Miura, J.: End-to-end autonomous driving with semantic depth cloud mapping and multi-agent. IEEE Trans. Intell. Veh. **8**(1), 557–571 (2022)
34. Özkanoğlu, M.A., Ozer, S.: InfraGAN: a GAN architecture to transfer visible images to infrared domain. Pattern Recogn. Lett. **155**, 69–76 (2022)
35. Pang, Y., Lin, J., Qin, T., Chen, Z.: Image-to-image translation: methods and applications (2021)
36. Park, T., Efros, A.A., Zhang, R., Zhu, J.-Y.: Contrastive learning for unpaired image-to-image translation. In: Vedaldi, A., Bischof, H., Brox, T., Frahm, J.-M. (eds.) ECCV 2020. LNCS, vol. 12354, pp. 319–345. Springer, Cham (2020). https://doi.org/10.1007/978-3-030-58545-7_19
37. Park, T., Efros, A.A., Zhang, R., Zhu, J.-Y.: Contrastive learning for unpaired image-to-image translation. In: Vedaldi, A., Bischof, H., Brox, T., Frahm, J.-M. (eds.) ECCV 2020. LNCS, vol. 12354, pp. 319–345. Springer, Cham (2020). https://doi.org/10.1007/978-3-030-58545-7_19
38. Pierson, H.A., Gashler, M.S.: Deep learning in robotics: a review of recent research. Adv. Robot. **31**(16), 821–835 (2017)
39. Ren, S., He, K., Girshick, R., Sun, J.: Faster R-CNN: towards real-time object detection with region proposal networks. Adv. Neural. Inf. Process. Syst. **28**, 91–99 (2015)
40. Ronneberger, O., Fischer, P., Brox, T.: U-net: convolutional networks for biomedical image segmentation. In: Navab, N., Hornegger, J., Wells, W.M., Frangi, A.F. (eds.) MICCAI 2015. LNCS, vol. 9351, pp. 234–241. Springer, Cham (2015). https://doi.org/10.1007/978-3-319-24574-4_28
41. Stilgoe, J.: Machine learning, social learning and the governance of self-driving cars. Soc. Stud. Sci. **48**(1), 25–56 (2018)
42. Tian, Z., Shen, C., Chen, H., He, T.: FCOS: fully convolutional one-stage object detection. In: Proceedings of the IEEE/CVF International Conference on Computer Vision, pp. 9627–9636 (2019)

43. Vasconcelos, C., Birodkar, V., Dumoulin, V.: Proper reuse of image classification features improves object detection. In: Proceedings of the IEEE/CVF Conference on Computer Vision and Pattern Recognition, pp. 13628–13637 (2022)
44. Wang, Q., Chi, Y., Shen, T., Song, J., Zhang, Z., Zhu, Y.: Improving RGB-infrared object detection by reducing cross-modality redundancy. Remote Sens. **14**(9), 2020 (2022)
45. Wang, Z., Yang, E., Shen, L., Huang, H.: A comprehensive survey of forgetting in deep learning beyond continual learning (2023)
46. Wortsman, M., et al.: Robust fine-tuning of zero-shot models. In: Proceedings of the IEEE/CVF Conference on Computer Vision and Pattern Recognition, pp. 7959–7971 (2022)
47. Zhang, A., Lipton, Z.C., Li, M., Smola, A.J.: Dive into deep learning. arXiv preprint arXiv:2106.11342 (2021)
48. Zhang, H., Fromont, E., Lefèvre, S., Avignon, B.: Multispectral fusion for object detection with cyclic fuse-and-refine blocks. In: 2020 IEEE International Conference on Image Processing (ICIP), pp. 276–280. IEEE (2020)
49. Zhang, T., Wu, F., Katiyar, A., Weinberger, K.Q., Artzi, Y.: Revisiting few-sample BERT fine-tuning (2021)
50. Zhu, J.Y., Park, T., Isola, P., Efros, A.A.: Unpaired image-to-image translation using cycle-consistent adversarial networks. In: 2017 IEEE International Conference on Computer Vision (ICCV) (2017)
51. Zhu, J.Y., Park, T., Isola, P., Efros, A.A.: Unpaired image-to-image translation using cycle-consistent adversarial networks. In: Proceedings of the IEEE International Conference on Computer Vision, pp. 2223–2232 (2017)

FroSSL: Frobenius Norm Minimization for Efficient Multiview Self-supervised Learning

Oscar Skean[1](✉)[iD], Aayush Dhakal[2][iD], Nathan Jacobs[2][iD], and Luis Gonzalo Sanchez Giraldo[1][iD]

[1] University of Kentucky, Lexington, USA
oscar.skean@uky.edu
[2] Washington University in St. Louis, St. Louis, USA

Abstract. Self-supervised learning (SSL) is a popular paradigm for representation learning. Recent multiview methods can be classified as sample-contrastive, dimension-contrastive, or asymmetric network-based, with each family having its own approach to avoiding informational collapse. While these families converge to solutions of similar quality, it can be empirically shown that some methods are epoch-inefficient and require longer training to reach a target performance. Two main approaches to improving efficiency are covariance eigenvalue regularization and using more views. However, these two approaches are difficult to combine due to the computational complexity of computing eigenvalues. We present the objective function FroSSL which reconciles both approaches while avoiding eigendecomposition entirely. FroSSL works by minimizing covariance Frobenius norms to avoid collapse and minimizing mean-squared error to achieve augmentation invariance. We show that FroSSL reaches competitive accuracies more quickly than any other SSL method and provide theoretical and empirical support that this faster convergence is due to how FroSSL affects the eigenvalues of the embedding covariance matrices. We also show that FroSSL learns competitive representations on linear probe evaluation when used to train a ResNet18 on several datasets, including STL-10, Tiny Imagenet, and Imagenet-100. Github.

1 Introduction

The problem of learning representations without human supervision is fundamental in machine learning. Unsupervised representation learning is particularly useful when label information is difficult to obtain or noisy. It requires the identification of structure in data with limited knowledge about what the structure is. One common way of learning structure without labels is joint embedding self-supervised learning (SSL) [4,5,14,16,17,24,33,36]. The basic goal of SSL is to

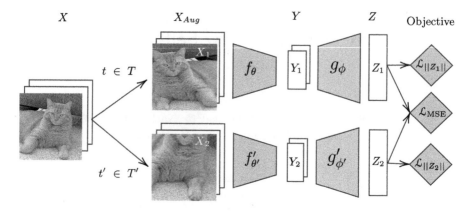

Fig. 1. The SSL pipeline used in this work. In general, the encoder and projector may be asymmetric. We use symmetric encoders with shared weights and the same augmentation set for each view. We refer to X_1 as view 1 of X, and X_2 as view 2. Only two views are shown here, though more may be used in practice.

train neural networks to capture *semantic* input features that are *augmentation-invariant*. This goal is appealing for representation learning because the inference set often has similar semantic content to the training set.

A trivial solution to learning augmentation-invariant features is to encode all images to the same point, often called trivial or informational collapse. The resulting networks are essentially useless for downstream tasks. Different mechanisms have been proposed to handle collapse in SSL. These can be grouped into three families: sample-contrastive, dimension-contrastive, and asymmetric network methods.

A less studied problem in all current SSL methods is their speed of convergence. When compared to traditional supervised learning, SSL methods must be trained for large numbers of iterations to reach competitive performance on downstream tasks. For example, a typical experiment in the literature is to train for 1000 epochs on ImageNet which can take several weeks even with many GPUs. An imperative direction of research is to investigate how to reduce SSL training time. An observation that is often hidden by only reporting the final epoch accuracy is that, empirically, certain SSL methods require more training time to reach competitive accuracies. This phenomenon has been observed for many dimension-contrastive methods by Simon *et al.* [27] but not discussed in detail. We provide additional support for this claim in Sect. 4.1. Our work attempts to answer the following research question: Does there exist an SSL method with dimension-contrastive advantages, namely simplicity via avoidance of both negative sampling and architectural restrictions, while achieving competitive accuracies more quickly than other existing SSL methods?

We propose an SSL objective which we call FroSSL. Similar to many dimension-contrastive methods, FroSSL consists of a variance and invariance term. The invariance term is simply a mean-squared error between the views and is identical to VICReg's invariance term [1]. The variance term is the logarithm of the squared Frobenius norm of the normalized covariance embedding matrices. Using the Frobenius norm of covariance matrices for improving learned representations has not been explored in SSL.

Our contribution can be summarized as:

- We introduce the FroSSL objective function and show that it is *both* dimension-contrastive and sample-contrastive up to a normalization of the embeddings.
- We introduce a theoretical framework that unifies dimension-contrastive methods that scale linearly in the number of views.
- We show that FroSSL combines two techniques to reduce training time: using more views and improving eigenvalue dynamics. We examine covariance eigenvalue trajectories during training on STL-10 to show that FroSSL learns useful, high-rank representations more quickly than other dimension-contrastive methods.
- We evaluate FroSSL on the standard setup of SSL pretraining and linear probe evaluation on CIFAR-10, CIFAR-100, STL-10, Tiny Imagenet, and Imagenet-100. We find that FroSSL achieves strong performance, especially when models are trained for fewer epochs.

2 Background and Notation

Consider a matrix $A \in \mathbb{R}^{m \times n}$. Let $A_{ij} \in \mathbb{R}$ denote the element at the ith row and jth column of A, and $A_{i,:} \in \mathbb{R}^m$ denote the ith column vector representing the ith row of A, and $A_{:,j}$ the jth column of A. Let $\sigma_k(A)$ denote the kth singular value of A ordered non-increasingly. The entry-wise product (also known as Hadamard product) is denoted as $A \odot B$. The Ky Fan p norm of A is defined as [18]:

$$\|A\|_p = \left(\sum_k^{\min(m,n)} \sigma_k^p(A) \right)^{1/p}, \qquad (1)$$

which is a unitarily invariant norm. For $p = 2$, we have the Frobenius norm $\|A\|_2 = \|A\|_F = \sqrt{\sum_i \sum_j A_{ij}^2}$.

2.1 The Joint Embedding Self-supervised Learning Problem

The goal of self-supervised learning is to learn useful representations without external supervision. Many visual joint embedding SSL methods follow a similar procedure which was first introduced in [4]. An example of this procedure is depicted in Fig. 1. Let $\mathbf{X} = \{x_i\}_{i=1}^N$ be a mini-batch with N samples, V the number of augmented views, $T(\cdot)$ a function that applies randomly selected augmentations to an image, f a visual encoder network, and g a projector network.

First, each image $x_i \in \mathbf{X}$ is paired with augmented versions of itself, making the augmented dataset $\mathbf{X}_{\text{aug}} = \{T_1(x_i), \cdots, T_V(x_i)\}_{i=1}^N = \{X_1, \cdots, X_V\}$ With ideal augmentations, $(X_1)_{i,:}$ and $(X_2)_{i,:}$ have identical *semantic* content and different *style* content. Note that typically $V = 2$, but we make no such assumptions. For each augmentation, the embedding set is given by $Y_v = \{f((X_v)_{i,:})\}_{i=1}^N$ and projection set $Z_v = \{g((Y_v)_{i,:})\}_{i=1}^N$. Finally, an SSL objective is computed on the projections and backpropagated through both networks. The goal of the objective is to ensure that encoded augmentations for the same image are mapped close together by the projector, i.e. $(Z_a)_{i,:}$ and $(Z_b)_{i,:}$ are close in some sense of distance for all $a, b = 1, 2, \ldots, V$. At the same time, projections should capture the variability among images. Thus the goal of SSL is to train the networks f and g to extract *semantic* features that are invariant to any augmentations induced by $T(\cdot)$. In the following, we take a closer look at choices for the SSL objective.

2.2 The Three Families of Joint Embedding SSL Objectives

Objective functions for joint embedding self-supervised learning can be divided into three families. The first family consists of *sample-contrastive* methods [2, 4,16,17,33] which use a contrastive loss to learn a representation that maps positive samples (augmentation of the same image) close together while pushing negative samples (different images) apart. These methods avoid collapse at the expense of making comparisons between positive and negative samples.

The second family consists of *asymmetric network* methods [3,5,14] which place restrictions on the architecture of the mapping network used, including asymmetrical encoders [5,14], momentum encoders [17], and stop gradients [5,15]. While these methods can achieve great results, they are rooted in implementation details and there is no clear theoretical understanding of how they avoid collapse [1].

The third, and most recent, family are the *dimension-contrastive* methods, which are sometimes called negative-free contrastive [32] or feature decorrelation methods [29]. These methods operate by reducing the redundancy in feature dimensions. Instead of examining *where* samples live in feature space, these methods examine *how* feature dimensions are being used. Methods in this family can avoid the use of negative samples while also not requiring restrictions in the network architecture to prevent collapse. Barlow Twins objective pushes the normalized cross-covariance between views towards the identity matrix [36]. VICReg consists of three terms: the invariance term enforces similarity in embeddings across views, while the variance/covariance terms regularize the covariance matrices of each view to prevent collapse [1]. W-MSE whitens and projects embeddings to the unit sphere before maximizing cosine similarity between positive samples [11]. I-VNE maximizes the von Neumann entropy of the embedding covariance matrices [20]. Finally, CorInfoMax maximizes the log det entropy of both views while minimizing mean-squared error [25].

2.3 A Framework for Dimension-Contrastive Methods

Many recent works in dimension-contrastive SSL, whether explicitly or implicitly, consist of a combination of two competing objectives: an augmentation

Table 1. Taxonomy of dimension-contrastive SSL methods describing how they avoid informational collapse and achieve augmentation invariance in the D_{inv} and D_{var} framework of Sect. 2.3.

Method	Variance $D_{\text{var}}(\Sigma_v\|\mathbf{I})$	Invariance $D_{\text{inv}}(Z_v, Z_r)$
VICReg	(Variance) Hinge loss on auto-covariance diagonal	MSE $\frac{1}{N}\|Z_v - Z_r\|_F^2$
	(Covariance) covariance off-diagonals per view	
	$\sum_k^D \max\left(0, 1 - \sqrt{(\Sigma_v)_{k,k} + \epsilon}\right) + \nu \|\Sigma_v - \Sigma_v \odot \mathbf{I}\|_F^2$	
W-MSE	Implicit through whitening that	
	$D_{\text{var}}(\Sigma_v\|\mathbf{I}) = 0$ since $\Sigma_v = \mathbf{I}$ for all v	
CorInfoMax	Log Det Divergence: $D_{\log \det}(A\|B) = \text{trace}(AB^{-1}) - D - \log \det(AB^{-1})$	
	$D_{\log \det}(\Sigma_v + \epsilon\mathbf{I} \| \mathbf{I}) = \text{trace}(\Sigma_v + \epsilon\mathbf{I}) - D - \log\det(\Sigma_v + \epsilon\mathbf{I})$	
I-VNE	von Neumann Relative Entropy: $S_1(A\|B) = \text{trace}(A(\ln A - \ln B))$	Cosine Similarity
	$S_1(\Sigma_v\|\mathbf{I}) = \text{trace}(\Sigma_v \ln \Sigma_v)$	
FroSSL (ours)	2-Order Petz-Rényi Relative Entropy: $S_2(A\|B) = \log \text{trace}(A^2 B^{-1})$	MSE $\frac{1}{N}\|Z_v - Z_r\|_F^2$
	$S_2(\Sigma_v\|\mathbf{I}) = \ln\left(\sum_{i=1}^N \sigma_i^2\right) = \ln \|\Sigma_v\|_F^2$	

invariance term that pulls different augmentations from the same image close together, and a **variance** term that avoids collapse of the mapping by regulating variance. Below, we unify dimension-contrastive methods into a general framework that is parameterized by choices of two distances. By carefully selecting these distances, specific dimension-contrastive methods can be recovered.

Let $Z_v \in \mathbb{R}^{N \times D}$ be a batch of projections and $\Sigma_v = \frac{1}{N}\hat{Z}_v^T \hat{Z}_v$ the corresponding covariance, where $\hat{Z}_v$ are the centered projections. A dimension-contrastive objective can be written as follows:

$$\min \sum_{v=1}^{V-1} \sum_{r=v+1}^{V} D_{\text{inv}}(Z_v, Z_r) + \gamma \sum_{v=1}^{V} D_{\text{var}}(\Sigma_v \| \mathbf{I}). \quad (2)$$

The first term of (2) is the **invariance** term which minimizes the distance $D_{\text{inv}} : \mathbb{R}^{N \times D} \times \mathbb{R}^{N \times D} \mapsto \mathbb{R}_{\geq 0}$ between all pairs of augmentations. The second term of (2) is a **variance** factor that forces the covariance of each augmentation to be close to identity according to a dissimilarity $D_{\text{var}} : \mathbb{R}^{D \times D} \times \mathbb{R}^{D \times D} \mapsto \mathbb{R}_{\geq 0}$. For instance, in VICReg [1], $D_{\text{inv}}(Z_v, Z_r) = \|Z_v - Z_r\|_F^2$ and $D_{\text{var}}(\Sigma_v\|\mathbf{I}) = \sum_k^D \max\left(0, 1 - \sqrt{(\Sigma_v)_{k,k} + \epsilon}\right) + \nu \|\Sigma_v - \Sigma_v \odot \mathbf{I}\|_F^2$. Similarly, in CorInfoMax [25] $D_{\text{inv}}(Z_v, Z_r)$ is the same as VICReg, but $D_{\text{var}}(\Sigma_v\|\mathbf{I})$ can be related to the log det divergence $D_{\log \det}(A\|B) = \text{trace}(AB^{-1}) - D - \log \det(AB^{-1})$ setting $A = \Sigma_v + \epsilon\mathbf{I}$ and B to a scaling of identity due to the normalization step in their projector. In Table 1, we show the dimension-contrastive methods which fit into this framework. We provide derivations in the Supp. Material.

Multiview Invariance Term. In (2) the invariance term requires $V(V-1)/2$ comparisons which scales quadratically with the number of views. However, if $D_{\text{inv}}(Z_v, Z_r) = \|Z_v - Z_r\|_F^2$, then the invariance term may be simplified to

$$\sum_{v=1}^{V-1} \sum_{r=v+1}^{V} D_{\text{inv}}(Z_v, Z_r) = V \sum_{v=1}^{V} D_{\text{inv}}\left(Z_v, \overline{Z}\right) \quad (3)$$

Table 2. Taxonomy of dimension-contrastive SSL methods showing which desirable criteria they fulfill.

	Invariant to Projection Rotations	Manipulates Eigenvalues Explicitly	Quadratic in Batch Size and Dimension	Linear in Views
Barlow Twins	✗	✗	✓	✗
VICReg	✗	✗	✓	✓
W-MSE	✓	✗	✗	✓
CorInfoMax	✓	✓	✗	✓
I-VNE	✓	✓	✗	✓
MMCR	✓	✓	✗	✓
FroSSL (ours)	✓	✓	✓	✓

where $\overline{Z} = \frac{1}{V}\sum_{i=1}^{V} Z_i$ is the average projection across all views. If a method has a D_{inv} that can be rewritten this way, we say the method scales linearly with views. All methods displayed in Table 1 have this property.

3 The FroSSL Objective

To motivate FroSSL, we begin by posing four desirable criteria of dimension-contrastive methods.

1. **Invariant to Projection Rotations** We argue that dimension-contrastive methods should be invariant to rotations in the projections because the orientation of the covariance does not affect the relationships between principal components. In other words, redundancy in the embedding dimensions is invariant to the rotation of the embeddings. Thus the choices of D_{var} and D_{inv} should be rotationally invariant as well.
2. **Manipulates Eigenvalues Explicitly** Several works have shown that regularizing projection covariance eigenvalues in SSL can lead to reduced training time and improved downstream performance [15,20,34]. We provide empirical support for this in Sect. 4.1.
3. **Scales Quadratically in Batch Size and Dimension** The time complexity of the objective function scale *at most* quadratically with respect to N and D. This is often in opposition to the prior criteria which typically requires cubic eigendecomposition.
4. **Scales Linearly in Views** The time complexity of the objective function should be linear in the number of views V. This is advantageous because recent work has shown that using more views can reduce training time and improve downstream performance [2,34]. We provide empirical support for this in Sects. 4.2 and 5. Any dimension-contrastive method with D_{inv} that satisfies Eq. (3) fulfills this criterion.

As shown in Table 2, no prior method meets all four criteria. We provide proofs in the Supp. Material. To construct a method that fulfills all criteria, we modify the I-VNE objective function:

$$\max \mathcal{L}_{\text{I-VNE}} = \sum_{v=1}^{V} \text{Tr} \Sigma_v \ln \Sigma_v + \sum_{v=1}^{V-1} \sum_{r=v+1}^{V} \frac{Z_v^T Z_r}{\|Z_v\|_2 \|Z_r\|_2} \quad (4)$$

The invariance term maximizes the cosine similarity between views. The variance term maximizes the von Neumann entropy of each view covariance matrix. The only criteria that I-VNE does not fulfill is being subcubic in batch size and dimension. This is due to the eigendecomposition needed to compute the matrix logarithm for the entropy. To begin addressing this, we first notice that the von Neumann entropy is a limit case of matrix-based α-order entropy [19,26,28]:

$$S_\alpha(\Sigma_v) = \frac{1}{1-\alpha} \log\left[\text{Tr}(\Sigma_v^\alpha)\right] = \frac{1}{1-\alpha} \log\left(\sum_{i}^{\min(N,D)} \lambda_i^\alpha(\Sigma_v)\right) \quad (5)$$

Here, we do not require $\text{trace}(\Sigma_v) = 1$ as is typically required by α-order entropy. The von Neumann entropy is equivalent to $S_1(\Sigma_v)$ in the limit. Another special case is collision entropy, given by $S_2(\Sigma_v)$ below:

$$S_2(\Sigma_v) = -\log\left(\sum_{i=1}^{\min(N,D)} \lambda_i^2(\Sigma_v)\right) = -\log(\|\Sigma_v\|_F^2) = -\log \sum_i \sum_j (\Sigma_v)_{ij}^2 \quad (6)$$

Notice how the left-hand side in the above equation explicitly uses the eigenvalues, while the right-hand side only uses matrix elements. This is made possible by the Frobenius norm, which offers an equivalency between a sum over eigenvalues and a sum over matrix elements. This has immediate impacts on the loss time complexity by relaxing the $O(\min(D,N)^3)$ eigendecomposition to the $O(\min(D,N)^2)$ Frobenius norm computation. The case of 2-order entropy is the only matrix-based α-entropy which does not require eigendecomposition. One potential downside is $S_2(\Sigma_v)$ does not penalize outlying eigenvalues as heavily as in $S_1(\Sigma_v)$. This is akin to the difference between mean-absolute error and mean-squared error. However, this has no significant impact in our experiments.

The variance term D_{var} for our objective will minimize the log Frobenius norm of normalized embeddings, causing the embeddings to spread out equally in all directions. Normalizing the embeddings is crucial because otherwise, minimizing the Frobenius norm will lead to trivial collapse. For the invariance term D_{inv}, we opt to use the mean-squared error between views. Our objective function FroSSL is given below:

$$\text{minimize } \mathcal{L}_{\text{FroSSL}} = \sum_{v=1}^{V} \log(\|\Sigma_v\|_F^2) + \gamma \|Z_v - \overline{Z}\|_F^2 \quad (7)$$

Note we simplify the pairwise mean-squared error via Eq. (3). Because the Frobenius norm is invariant to transposition, we can choose to compute either

$\left\|Z_v^T Z_v\right\|_F^2$ or $\left\|Z_v Z_v^T\right\|_F^2$ depending on if $D > N$. The former has time complexity $O(ND^2)$ while the latter has complexity $O(N^2D)$. For consistency, we always use the former in our experiments. We provide pseudocode in the Supp. Material.

3.1 The Role of the Logarithm

The log in Eq. (7) ensures that the contributions of the variance term to the gradient of the objective function become self-regulated ($\frac{d\log f(x)}{dx} = \frac{1}{f(x)}\frac{df(x)}{dx}$) with respect to the invariance term. We later compare the experimental performance of Equation (7) with and without the logarithms, showing that using logarithms leads to a gain in performance. Prior work has shown that Eq. (7) with no logarithms causes dead neurons in the final encoder layer [20].

3.2 FroSSL is both Sample-Contrastive and Dimension-Contrastive

It can be shown, up to an embedding normalization, that FroSSL is both dimension-contrastive and sample-contrastive. First, we provide formal definitions of dimension-contrastive and sample-contrastive SSL, following Garrido et al. [13].

Definition 1 (Dimension-Contrastive Method). *An SSL method is said to be dimension-contrastive if it minimizes the non-contrastive criterion $\mathcal{L}_{nc}(Z) = \left\|Z^T Z - diag(Z^T Z)\right\|_F^2$, where $Z \in \mathbb{R}^{N \times D}$ is a matrix of embeddings as defined above. This may be interpreted as penalizing the off-diagonal terms of the embedding covariance.*

Definition 2 (Sample-Contrastive Method). *An SSL method is said to be sample-contrastive if it minimizes the contrastive criterion $\mathcal{L}_c(Z) = \left\|ZZ^T - diag(ZZ^T)\right\|_F^2$. This may be interpreted as penalizing the similarity between pairs of different images.*

Next, we use the duality of the Frobenius norm, given by $\left\|Z^T Z\right\|_F = \left\|ZZ^T\right\|_F$, to show that FroSSL satisfies the qualifying criteria of both dimension-contrastive and sample-contrastive methods.

Proposition 1. *If every embedding dimension is normalized to have equal variance, then FroSSL is a dimension-contrastive method. See Supp. Material. for the proof.*

Proposition 2. *If every embedding is normalized to have equal norm, then FroSSL is a sample-contrastive method. See Supp. Material. for the proof.*

Proposition 3. *If the embedding matrices are doubly stochastic, then FroSSL is simultaneously dimension-contrastive and sample-contrastive.*

Proposition 3 allows for interpreting FroSSL as either a sample-contrastive or dimension-contrastive method, up to a normalization of the data embeddings. The choice of normalization strategy is not important to the performance of an SSL method [13]. Unless otherwise specified, we only normalize the variance and not the embeddings. These same proof techniques can be used to show that TiCo, MMCR, I-VNE, and CorInfoMax also belong to both families [20, 25, 34, 37]. Additionally, variants of the dimension-contrastive VICReg have been proposed [13] that allow it to be rewritten as the sample-contrastive SimCLR. However, VICReg cannot be rewritten in such a way due to the hinge loss.

While Proposition 3 is interesting theoretically, it also offers empirical benefits to FroSSL. We examine overall wall-training time to reach competitive accuracies (4.3), robustness to augmentations (5.1), and performance on little pretraining data (5.2). In all of these experiments, sample-contrastive methods outperform dimension-contrastive methods. FroSSL shares the advantages observed empirically in sample-contrastive methods.

4 On Efficiency in Self-Supervised Learning

It is well-known that traditional SSL algorithms need hundreds or thousands of epochs to reach competitive accuracies. To compare the efficiency of different SSL algorithms, we can borrow theoretical and practical tools from the broader field of algorithmic complexity. In the context of machine learning, there are two measurements of particular interest to practitioners: *wall-time* needed to reach a given accuracy and VRAM *space* used. The former can be decomposed into the atomic quantities of average *wall-time per minibatch* and the *number of epochs* needed to reach a given accuracy. To emphasize why this decomposition matters, consider a scenario where wall-time per minibatch differs between two methods but overall wall-time does not. In such a scenario, using the method with the slower minibatch wall-time is advantageous for using fewer disk reads and less distributed network traffic. This is not obvious without observing the atomic quantities. Note we are careful to specify "epochs to reach a given accuracy" rather than "epochs to convergence". One reason for this is that classical experiments in SSL train for a fixed number of epochs rather than until convergence. Another reason is algorithms that more quickly reach a target performance, such as FroSSL or I-VNE, do not necessarily converge in fewer epochs.

The design of an SSL algorithm is a balancing act between minibatch time, space, and number of epochs. While conversations involving minibatch time and space have been rarely discussed in the SSL literature, discussion about the number of epochs has seen renewed interest [15, 27, 31]. However, if SSL algorithm design is indeed a balancing act of the three quantities above, then space and minibatch time deserve discussion too. Methods that boast reductions to one quantity may come with significant penalties to a different quantity. For example, dimension-contrastive methods use less space in practice than sample-contrastive methods, which prefer large minibatch sizes, or asymmetric methods like BYOL, which need an additional prediction network. However, the improved space efficiency comes at the cost of requiring a higher number of epochs [27]. In Sects. 4.1 and 4.2, we discuss the advantages and drawbacks of two approaches to reducing the number of epochs. In Sect. 4.3, we compare a variety of SSL algorithms and visualize their time, space, and epoch tradeoffs.

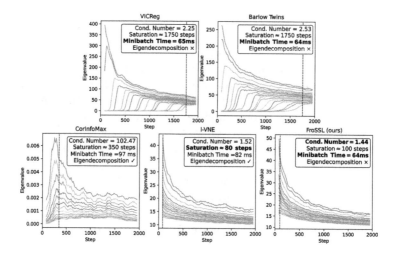

Fig. 2. The choice of variance term, $D_{\text{var}}(\Sigma_v \| \mathbf{I})$, has a significant impact on training dynamics. Each subplot visualizes the trajectories of the top 20 eigenvalues of the embedding covariance matrix Σ_1 when trained with dimension-contrastive methods. These trajectories show how quickly Σ_v converges to γI, which has eigenvalues equal to $\frac{\gamma}{D}$. VICReg, Barlow Twins, and CorInfoMax converge slowly. FroSSL and I-VNE have similar training dynamics, but FroSSL has significantly lower computational complexity because it avoids explicitly computing the eigendecomposition.

4.1 Reducing the Number of Epochs with Eigenvalue Dynamics

Recent work has examined the training dynamics of SSL models [27]. In particular, they find that the eigenvalues of the covariance exhibit *stepwise* behavior, meaning that one eigendirection is learned at a time. This is readily seen in Fig. 2 for VICReg and Barlow Twins. This phenomenon contributes to slowness in SSL optimization with the smallest eigendirections taking the longest to be learned. Other work shows that high-rank representations lead to better downstream performances [12]. It directly follows that if an SSL method requires a high number of epochs to learn high-rank representations, then it also needs a high number of epochs to learn useful representations.

We hypothesize that by carefully choosing the variance term $D_{\text{var}}(\Sigma_v \| \mathbf{I})$ to reduce stepwise eigenvalue dynamics, useful representations can be learned more quickly. Indeed, this behavior has already been observed in several SSL objectives already. CorInfoMax optimizes the log-determinant of each covariance Σ_v, which is defined as the log of the product of the Σ_v eigenvalues [25]. IsoLoss uses Σ_v eigenvalues as learning rate multipliers to equalize the convergence rate of different eigenmodes [15]. MMCR optimizes the nuclear norm of the average view embedding, which is defined as the sum of the singular value magnitudes [34]. I-VNE optimizes the von Neumann entropy of Σ_v, which is equal to the Shannon entropy of the Σ_v eigenvalues [20].

It is straightforward to show that FroSSL also directly influences the covariance eigenvalue dynamics. However, FroSSL is unique from prior methods because it does so while avoiding explicit eigendecomposition. This can be seen from Eq. 6. Additionally, using the Frobenius norm eliminates numerical instabilities typically associated with eigendecomposition [9].

To highlight the existence and remedy of stepwise phenomena in practical scenarios, we create an experiment similar to the one used by Simon et al. [27]. In Fig. 2, we plot the trajectories of the top 20 eigenvalues of Σ_1 when trained with different dimension-contrastive objectives. For all SSL objectives, a ResNet18 was trained for 5 epochs on STL-10 using SGD with $lr = 0.01$ and a batch size of 256. Further details are given in the Supp. Material.

The eigenvalues trajectories show how quickly Σ_v is approaching γI, which has eigenvalues equal to $\frac{\gamma}{D}$. We say that an objective is saturated once the stepwise learning phase is ended. This is marked as the step where λ_{20} has increased from zero and started decreasing. It is clear to see that SSL objectives that directly influence eigenvalues, namely CorInfoMax, I-VNE, and FroSSL, saturate much quicker than the others. Interestingly, the condition number for CorInfoMax, computed as $\frac{\lambda_1}{\lambda_{20}}$, is much larger than any other tested method. We hypothesize this is due to the choice of the ϵ hyperparameter for the regularization term when computing the determinant as $\det(\Sigma_1 + \epsilon I)$.

4.2 Reducing the Number of Epochs by Using More Views

Multiview with 3 or More Views. In contrastive learning, using more views has been shown to have significant impacts on representation quality and downstream performance [30]. In SSL, using more augmentations for each image has the effect of averaging out noise from the mean embedding, which acts as a target for many SSL objectives as shown in Eq. (3). This differs from increasing the batch size, which would instead average out noise across samples and not across targets. While using more views is promising, it has not seen widespread adoption in self-supervised learning. This is in part due to many sample-contrastive methods being quadratic in the number of views. However, this problem is circumvented for the dimension-contrastive methods shown in Table 1, which are instead linear in the number of views. One such method, W-MSE, has shown performance improvements when the number of views is increased from 2 to 4 [11]. Interestingly, MMCR is constant in the number of views because it operates only on the mean embedding [34].

Multi-patch and Multi-crop Methods. An approach in a different vein is to extract and augment image patches to serve as views, rather than using full images. EMP-SSL has shown that this drastically reduces the number of epochs and overall wall-time needed to reach competitive accuracies by utilizing a bag-of-features model that embeds hundreds of small augmented patches per image [6,31]. However, EMP-SSL comes at the cost of major penalties to time-per-minibatch and space in both training time and inference time.

Table 3. Comparison of the time/space/epoch tradeoffs for SSL algorithms trained on STL-10. FroSSL with 8 views achieves 80% top-1 accuracy in the least wall-time.

	SimCLR	MocoV2	BYOL	VICReg	Barlow	CorInfoMax	MMCR			FroSSL (ours)		
Num. Views	2	2	2	2	2	2	2	4	8	2	4	8
Loss Time Complexity	$O(V^2N^2)$			$O(VD^2)$	$O(V^2D^2)$	$O(VD^3)$	$O(\min(D,N)^3)$			$O(V\min(D,N)^2)$		
VRAM Space (GB)	1.6	2.8	2.0	1.6	1.6	1.7	1.7	2.9	5.3	1.7	2.9	5.3
Minibatch Wall-time (ms)	60	79	76	65	64	97	71	108	187	64	105	187
Number of Epochs to 80% Acc	347	180	187	360	370	405	380	211	63	290	144	**55**
Wall-time to 80% Acc (hours)	2.4	1.6	1.6	2.7	2.7	4.5	3.1	2.6	1.3	2.1	1.7	**1.2**

Table 4. Top-1 accuracies on STL-10 using an online linear classifier during training for specific numbers of epochs (left) and specific elapsed times (right).

	Epochs				
Method	3	10	30	50	100
SimCLR	40.7	44.8	61.5	66.2	70.1
MoCo v2	24.6	45.0	63.8	69.4	75.2
BYOL	28.8	32.7	59.6	64.7	70.6
VICReg	43.6	51.1	61.2	67.5	71.1
Barlow Twins	32.1	46.6	62.0	62.6	69.0
CorInfoMax	39.0	49.1	58.0	62.5	66.2
MMCR (2 views)	39.6	53.3	62.8	63.3	67.0
MMCR (4 views)	46.0	61.5	70.2	71.5	75.7
MMCR (8 views)	51.1	64.7	72.9	77.2	79.4
FroSSL (2 Views)	44.8	56.9	64.8	67.1	72.0
FroSSL (4 Views)	49.3	60.7	70.3	67.1	76.9
FroSSL (8 Views)	47.6	**65.5**	**74.5**	**78.4**	**81.8**

	Training Wall-Time (min.)		
Method	10	30	60
SimCLR	61.5	68.8	73.9
MoCo v2	57.4	70.0	76.2
BYOL	50.2	65.3	75.0
VICReg	63.4	70.3	72.9
Barlow Twins	55.5	66.1	68.0
CorInfoMax	56.1	64.9	65.6
MMCR (2 views)	54.7	68.4	70.9
MMCR (4 views)	**69.8**	**74.7**	76.6
MMCR (8 views)	64.7	73.5	78.0
FroSSL (2 Views)	63.4	69.7	74.9
FroSSL (4 Views)	**68.5**	73.7	76.3
FroSSL (8 Views)	58.6	74.1	**79.0**

As an alternative to using full-sized images or tiny patches for each view, multi-crop methods strike a balance [2]. A certain number of views are full-sized images while the remaining views are smaller crops. A typical setup for ImageNet is using two 224 × 224 views and six 96 × 96 views. These approaches differ from our experiments which use all full-sized views with FroSSL. However, we expect that FroSSL should work well as an objective function for these paradigms too.

4.3 Exploring Time, Space, and Epoch Tradeoffs

We now compare the efficiency of different SSL algorithms. We train a ResNet-18 for 500 epochs on STL-10 and measure the number of epochs needed to reach a top-1 accuracy of 80%. This threshold of 80% was chosen because all methods achieve that accuracy within 500 epochs. We used $N = 256$ and $D = 1024$ for all methods. Because these models were trained on a distributed cluster, it is important to compensate for different compute when measuring minibatch wall-time. In particular, we measure minibatch time by averaging over 2000 iterations of training on one NVIDIA A5000 GPU. We measure VRAM space as the maximum space requested by the training script. We calculate wall-time to 80% accuracy by multiplying minibatch time, epochs, and iterations per epoch.

In Table 3 we show the resources needed for each SSL objective. There are several observations to glean from this table. First, increasing the number of views reduces epochs and overall wall-time, even though space and minibatch time become larger. FroSSL with 8 views reaches 80% top-1 accuracy faster than

Table 5. Comparison of SSL methods on small datasets. All CorInfoMax and MMCR results are from our implementation. All Tiny ImageNet and STL-10 results are from our implementation. CIFAR-10 and CIFAR-100 results are reported from [8,11]. IN-100 baseline results are from [8]. We observed negligible improvements from using more views for FroSSL on the CIFAR datasets; we used the 2-view CIFAR accuracies to compute 4/8 view averages. An asterisk (*) denotes Tiny ImageNet results where weak augmentations outperformed strong ones. Results within 0.5% of best are **bolded**.

Method	CIFAR-10	CIFAR-100	STL-10	Tiny-IN	IN-100	Average
Sample-Contrastive						
SimCLR	91.8	65.8	85.9	41.9	77.6	72.6
SwAV	89.2	64.9	82.6	41.2	74.3	70.5
MoCo v2	**92.9**	69.9	83.2	41.9	79.3	73.4
Asymmetric Network						
SimSiam	90.5	66.0	88.5	**45.6***	78.7	73.9
BYOL	**92.6**	70.5	88.7	40.1	**80.3**	74.4
DINO	89.5	66.8	78.9	34.9	78.9	69.8
Dimension-Contrastive						
VICReg	92.1	68.5	85.9	37.5	79.4	65.8
Barlow Twins	92.1	**70.9**	85.0	**45.3**	**80.2**	74.7
W-MSE 2	91.6	66.1	72.4	28.8*	69.0	65.6
CorInfoMax	**92.6**	69.7	83.1	43.9	74.7	72.8
I-VNE	89.7	65.7	87.4	**45.2**	77.6	73.1
MMCR (2 views)	88.6	65.8	84.3	41.2	76.7	71.3
MMCR (4 views)	89.6	67.3	88.2	42.8	78.8	73.3
MMCR (8 views)	89.3	68.3	90.3	43.2	**80.3**	74.2
FroSSL (2 views) [no log]	88.9	62.3	82.4	36.4	78.3	69.7
FroSSL (2 views)	**92.8**	70.6	87.3	44.2	78.2	74.6
FroSSL (4 views)	–	–	90.0	**45.3**	79.4	75.6
FroSSL (8 views)	–	–	**90.9**	**45.3**	79.8	**75.9**

any other tested method. Second, asymmetric methods require the least overall wall-time for any method using 2 views, at the cost of space. We hypothesize this is due to enhanced training stability from momentum encoders. Third, doubling the number of views does not necessarily double the minibatch wall-time. This is because some parts of the training script, such as data loading and logging, do not get slower as the number of views increases. In Table 4, we show top-1 accuracies over epochs and over time. In both scenarios, FroSSL with 8 views has the highest accuracy after training is finished.

5 Experimental Results

In this section, we use a linear probe to evaluate learned representations on CIFAR-10 [22], CIFAR-100, STL-10 [7], Tiny ImageNet [23], and ImageNet-100 [30]. Our implementation is based on the solo-learn SSL framework [8].

Table 6. The top-1% accuracies after training on Tiny-Imagenet using weak or strong augmentations.

Method	Weak	Strong	Δ%
SimCLR	39.5	41.9	2.4
SwAV	39.9	41.2	1.5
MoCo v2	40.9	41.9	1.0
SimSiam	45.6	39.7	−5.9
BYOL	39.4	40.1	0.7
DINO	32.2	34.9	2.7
VICReg	18.1	37.5	19.6
Barlow Twins	36.8	45.3	8.5
CorInfoMax	33.1	43.9	10.8
MMCR (2 views)	24.2	41.2	17.0
FroSSL (2 views)	39.4	44.2	4.8

Table 7. Accuracies after pretraining on ImageNet-1k for 100 epochs with only 10% of data.

	Top-1	Top-5
SimCLR	31.1	56.6
BYOL	12.7	29.1
SimSiam	22.7	46.3
Barlow Twins	23.6	46.9
FroSSL (2 Views)	33.4	59.1
FroSSL (8 Views)	38.2	64.1

In Table 5, we show linear probe evaluation results on these datasets. It is readily seen that FroSSL learns competitive representations in comparison to other SSL methods. The implementation details can be summarized as:

- **Optimizer** The backbone uses LARS optimizer [35] with an initial learning rate of 0.3, weight decay of 1e-6, and a warmup cosine learning rate scheduler. The linear probe uses the SGD optimizer [21] with an initial learning rate of 0.3, no weight decay, and a step learning rate scheduler with decreases at 60 and 80 epochs.
- **Epochs** For CIFAR-10 and CIFAR-100, we pretrain the backbone for 1000 epochs. For STL-10, we pretrain for 500 epochs. For Tiny Imagenet, we pretrain for 800 epochs. For Imagenet-100, we pretrain for 800 epochs. All linear probes were trained for 100 epochs.
- **Hyperparameters** A batch size of $N = 256$ is used for all datasets, except for Tiny ImageNet which used $N = 512$. For FroSSL, we used $\gamma = 1.4$ for 2 views and $\gamma = 2$ for 4 and 8 views. We used an MLP with output dimension $D = 1024$ for FroSSL. Details about augmentations and method-specific hyperparameters are given in the Supp. Material.

5.1 Robustness to Augmentations

We trained models on Tiny ImageNet using both weak and strong augmentations. Weak augmentations had Gaussian blur probabilities (0.5, 0.5) and solarization probabilities (0, 0) for each view. Strong augmentations had Gaussian blur probabilities (1.0, 0.1) and solarization probabilities (0.2, 0). As shown in Table 6, FroSSL is more robust to changes in augmentations than any other dimension-contrastive method.

5.2 Performance In Low Data Regime

We trained models on ImageNet-1K [10] using only 10% of the data and evaluated them using the standard linear probe. Note that limited data was used in both pretaining and evaluation. As shown in Table 7, FroSSL achieves a better downstream performance on limited data than any other tested method.

6 Conclusion

We introduced FroSSL, a self-supervised learning method that can be seen as both sample- and dimension-contrastive. We showed that FroSSL enjoys the simplicity of dimension-contrastive methods while achieving the empirical advantages of sample-contrastive methods. In particular, we discovered that FroSSL can achieve substantially stronger performance than alternative SSL methods when trained with less overall wall-time. To better understand why this is happening, we presented empirical results based on eigenvalue trajectories. We demonstrated the effectiveness of FroSSL through extensive experiments on standard datasets.

Acknowledgements. This research is based upon work supported in part by the Office of the Director of National Intelligence (ODNI), Intelligence Advanced Research Projects Activity (IARPA), via 2021-2011000005 and the Office of the Under Secretary of Defense for Research and Engineering under award number FA9550-21-1-0227. The views and conclusions contained herein are those of the authors and should not be interpreted as necessarily representing the official policies, either expressed or implied, of ODNI, IARPA, the U.S. Department of Defense, or the U.S. Government. The U.S. Government is authorized to reproduce and distribute reprints for governmental purposes notwithstanding any copyright annotation therein.

References

1. Bardes, A., Ponce, J., LeCun, Y.: VICReg: variance-invariance-covariance regularization for self-supervised learning. In: International Conference on Learning Representations (2022)
2. Caron, M., Misra, I., Mairal, J., Goyal, P., Bojanowski, P., Joulin, A.: Unsupervised learning of visual features by contrasting cluster assignments. Adv. Neural Inf. Process. Syst. **33** (2020)
3. Caron, M., et al.: Emerging properties in self-supervised vision transformers. In: IEEE/CVF International Conference on Computer Vision, pp. 9650–9660 (2021)
4. Chen, T., Kornblith, S., Norouzi, M., Hinton, G.: A simple framework for contrastive learning of visual representations. In: International Conference on Machine Learning, pp. 1597–1607. PMLR (2020)
5. Chen, X., He, K.: Exploring simple Siamese representation learning. In: IEEE/CVF Conference on Computer Vision and Pattern Recognition, pp. 15750–15758 (2021)
6. Chen, Y., Bardes, A., Li, Z., LeCun, Y.: Bag of image patch embedding behind the success of self-supervised learning. Trans. Mach. Learn. Res. (2023)

7. Coates, A., Ng, A., Lee, H.: An analysis of single-layer networks in unsupervised feature learning. In: International Conference on Artificial Intelligence and Statistics, pp. 215–223 (2011)
8. da Costa, V.G.T., Fini, E., Nabi, M., Sebe, N., Ricci, E.: Solo-learn: a library of self-supervised methods for visual representation learning. J. Mach. Learn. Res. **23**(56), 1–6 (2022)
9. Dang, Z., Yi, K.M., Hu, Y., Wang, F., Fua, P., Salzmann, M.: EigendeComposition-free training of deep networks with zero eigenvalue-based losses. In: European Conference on Computer Vision, pp. 768–783 (2018)
10. Deng, J., Dong, W., Socher, R., Li, L.J., Li, K., Fei-Fei, L.: ImageNet: a large-scale hierarchical image database. In: IEEE Conference on Computer Vision and Pattern Recognition, pp. 248–255. IEEE (2009)
11. Ermolov, A., Siarohin, A., Sangineto, E., Sebe, N.: Whitening for self-supervised representation learning. In: International Conference on Machine Learning, pp. 3015–3024 (2021)
12. Garrido, Q., Balestriero, R., Najman, L., Lecun, Y.: RankMe: assessing the downstream performance of pretrained self-supervised representations by their rank. In: International Conference on Machine Learning, pp. 10929–10974 (2023)
13. Garrido, Q., Chen, Y., Bardes, A., Najman, L., LeCun, Y.: On the duality between contrastive and non-contrastive self-supervised learning. In: International Conference on Learning Representations (2023)
14. Grill, J.B., et al.: Bootstrap your own latent-a new approach to self-supervised learning. Adv. Neural Inf. Process. Syst. **33** (2020)
15. Halvagal, M.S., Laborieux, A., Zenke, F.: Implicit variance regularization in non-contrastive SSL. Adv. Neural Inf. Process. Syst. **36** (2023)
16. HaoChen, J.Z., Wei, C., Gaidon, A., Ma, T.: Provable guarantees for self-supervised deep learning with spectral contrastive loss. Adv. Neural Inf. Process. Syst. **34** (2021)
17. He, K., Fan, H., Wu, Y., Xie, S., Girshick, R.: Momentum contrast for unsupervised visual representation learning. In: IEEE/CVF Conference on Computer Vision and Pattern Recognition, pp. 9729–9738 (2020)
18. Horn, R.A., Johnson, C.R.: Matrix Analysis, 2nd edn. Cambridge University Press, Cambridge (2013)
19. Hoyos-Osorio, J.K., Sanchez-Giraldo, L.G.: The representation Jensen-Shannon divergence. arXiv preprint arXiv:2305.16446 (2023)
20. Kim, J., Kang, S., Hwang, D., Shin, J., Rhee, W.: VNE: an effective method for improving deep representation by manipulating eigenvalue distribution. In: Proceedings of the IEEE/CVF Conference on Computer Vision and Pattern Recognition, pp. 3799–3810 (2023)
21. Kingma, D.P., Ba, J.: Adam: a method for stochastic optimization. arXiv preprint arXiv:1412.6980 (2014)
22. Krizhevsky, A., Hinton, G.: Learning multiple layers of features from tiny images. Technical report, University of Toronto (2009)
23. Le, Y., Yang, X.: Tiny ImageNet visual recognition challenge. CS 231N **7**(7), 3 (2015)
24. Li, Y., Pogodin, R., Sutherland, D.J., Gretton, A.: Self-supervised learning with kernel dependence maximization. Adv. Neural Inf. Process. Syst. **34** (2021)
25. Ozsoy, S., Hamdan, S., Arik, S., Yuret, D., Erdogan, A.: Self-supervised learning with an information maximization criterion. Adv. Neural Inf. Process. Syst. **35** (2022)

26. Sanchez Giraldo, L.G., Rao, M., Principe, J.C.: Measures of entropy from data using infinitely divisible kernels. IEEE Trans. Inf. Theory **61**(1), 535–548 (2015)
27. Simon, J.B., Knutins, M., Liu, Z., Geisz, D., Fetterman, A.J., Albrecht, J.: On the stepwise nature of self-supervised learning. In: International Conference on Machine Learning (2023)
28. Skean, O., Osorio, J.K.H., Brockmeier, A.J., Giraldo, L.G.S.: DiME: maximizing mutual information by a difference of matrix-based entropies. arXiv preprint arXiv:2301.08164 (2023)
29. Tao, C., et al.: Exploring the equivalence of Siamese self-supervised learning via a unified gradient framework. In: IEEE/CVF Conference on Computer Vision and Pattern Recognition, pp. 14431–14440 (2022)
30. Tian, Y., Krishnan, D., Isola, P.: Contrastive multiview coding. In: Vedaldi, A., Bischof, H., Brox, T., Frahm, J.-M. (eds.) ECCV 2020. LNCS, vol. 12356, pp. 776–794. Springer, Cham (2020). https://doi.org/10.1007/978-3-030-58621-8_45
31. Tong, S., Chen, Y., Ma, Y., Lecun, Y.: EMP-SSL: towards self-supervised learning in one training epoch. arXiv preprint arXiv:2304.03977 (2023)
32. Tsai, Y.H.H., Bai, S., Morency, L.P., Salakhutdinov, R.: A note on connecting Barlow twins with negative-sample-free contrastive learning. arXiv preprint arXiv:2104.13712 (2021)
33. Tsai, Y.H.H., Wu, Y., Salakhutdinov, R., Morency, L.P.: Self-supervised learning from a multi-view perspective. In: International Conference on Learning Representations (2021)
34. Yerxa, T., Kuang, Y., Simoncelli, E., Chung, S.: Learning efficient coding of natural images with maximum manifold capacity representations. Adv. Neural Inf. Process. Syst. **36** (2024)
35. You, Y., Gitman, I., Ginsburg, B.: Large batch training of convolutional networks. arXiv preprint arXiv:1708.03888 (2017)
36. Zbontar, J., Jing, L., Misra, I., LeCun, Y., Deny, S.: Barlow twins: self-supervised learning via redundancy reduction. In: International Conference on Machine Learning, pp. 12310–12320 (2021)
37. Zhu, J., Moraes, R.M., Karakulak, S., Sobol, V., Canziani, A., LeCun, Y.: TiCo: transformation invariance and covariance contrast for self-supervised visual representation learning. arXiv preprint arXiv:2206.10698 (2022)

Learning Multimodal Latent Generative Models with Energy-Based Prior

Shiyu Yuan[1](✉), Jiali Cui[2], Hanao Li[2], and Tian Han[2]

[1] School of Systems and Enterprises, Stevens Institute of Technology, Hoboken, NJ 07310, USA
syuan14@stevens.edu
[2] Department of Computer Science, Stevens Institute of Technology, Hoboken, NJ 07310, USA
{jcui7,hli136,than6}@stevens.edu

Abstract. Multimodal generative models have recently gained significant attention for their ability to learn representations across various modalities, enhancing joint and cross-generation coherence. However, most existing works use standard Gaussian or Laplacian distributions as priors, which may struggle to capture the diverse information inherent in multiple data types due to their unimodal and less informative nature. Energy-based models (EBMs), known for their expressiveness and flexibility across various tasks, have yet to be thoroughly explored in the context of multimodal generative models. In this paper, we propose a novel framework that integrates the multimodal latent generative model with the EBM. Both models can be trained jointly through a variational scheme. This approach results in a more expressive and informative prior, better-capturing of information across multiple modalities. Our experiments validate the proposed model, demonstrating its superior generation coherence.

Keywords: EBM · Multimodal latent generative model

1 Introduction

Generative model (GM) has made remarkable progress in generating high-quality image [12,30], text [8,18], and video [9], and recently, the multimodal GM [1,7,16,19,21,27,38] has garnered significant interest for providing a powerful framework that intrigues popular applications of *cross generation*, such as the image-to-text [17], text-to-image [21,26]. However, these prior advances often focus on cross-modality learning by modeling the conditional dependency from one to the other while ignoring learning meaningful semantic representations

---

Supplementary Information The online version contains supplementary material available at https://doi.org/10.1007/978-3-031-73024-5_6.

© The Author(s), under exclusive license to Springer Nature Switzerland AG 2025
A. Leonardis et al. (Eds.): ECCV 2024, LNCS 15147, pp. 86–100, 2025.
https://doi.org/10.1007/978-3-031-73024-5_6

shared across multimodalities. Learning a shared representation can play a critical role in enabling the downstream tasks [14], such as *joint generation*, thus representing an active ongoing research area.

To tackle the challenge of learning data representations, various methods of latent generative models [15,29,33] have been explored. In particular, the latent variable generative model consists of low-dimensional latent variables and a generation network, where the latent variables are learned to capture the data representation, and thus, the generation network can construct the high-dimensional data by these learned latent variables. For multimodality learning, the multimodal VAEs [28,35], have recently been developed. Given a set of modalities, these models primarily focus on encoding (inferring) latent variables from different modalities and fusing them into a single latent space. Specifically, MVAE [35] and MMVAE [28] factorize *joint posterior* that infers independent latent variables from different modalities and fuses them into one latent space. Such a single latent space thus needs to represent the data representation of multiple modalities and aims to capture their shared representations. However, these multimodal VAEs only consider less informative Gaussian or Laplacian prior to modeling the latent distribution, which can be limited in expressivity for complex data representations [3,4,25], resulting in an ineffectively learned latent generative model.

The energy-based model (EBM), on the other hand, is shown to be expressive and known to be powerful in modelling complex data distribution [5,6]. In high-dimensional data space, it can be difficult for EBM learning as it typically involves Markov Chain Monte Carlo (MCMC) sampling for the EBM density [2,11], while for latent space, the EBM can be formulated as the EBM prior [25,36], reducing the burden of EBM prior sampling. Specifically, the EBM prior can be represented as an exponential tilting of the less informative reference distribution, where the energy function serves as a correction of reference distribution (e.g., Gaussian or Laplacian), rendering a more expressive prior distribution. However, most existing EBM works only consider modelling a single modality, and for multimodality, the latent generative model with the EBM prior still remains under-developed.

In this paper, we intend to explore the EBM prior to the more challenging multimodal learning task. In particular, we present a joint framework that is capable of leveraging the expressivity of EBM prior for modelling the shared semantic latent space from different multimodalities. With a set of modalities, we factorize a latent generative model with the EBM prior, in which the latent representation extracted from different modalities can be well-captured by the EBM prior. Compared to the uni-modal Gaussian or Laplacian prior, the EBM prior can be multi-modal and render more modelling capacity for complex representation learning, which in turn improves the generative power of the whole model and thus maintains semantic coherence for generated samples. Learning such EBM prior typically requires samples from the EBM prior and generator posterior, which is usually achieved by performing MCMC sampling for the EBM prior and generator posterior distribution. However, for the multimodal learning task, MCMC posterior sampling can be time-consuming as it may involve an additional inner-loop for computing the gradient through multiple generation networks from different modalities.

To ensure efficient posterior sampling and facilitate EBM prior learning, we employ the variational learning scheme and introduce an inference model [15, 28, 35] to approximate the generator posterior. With the inference model matching with the generator posterior, we can directly sample from the inference model and thus circumvent the burden of conducting MCMC posterior sampling. The EBM prior sampling can be achieved by MCMC sampling because of the low-dimensional latent space and, more importantly, the lightweight energy function used. We demonstrate the proposed method can render superior performance in various benchmarks.

Our contributions can be summarized as follows:

- We propose the energy-based prior model for multimodal latent generative models to capture complex shared information within multiple modalities.
- We develop the variational training scheme where the generation model, inference model, and energy-based prior can be jointly and effectively learned.
- We conduct various experiments and ablation studies and demonstrate superior performance compared to Laplacian prior baselines.

2 Related Works

In this section, we present the background of multimodal variational autoencoders and energy-based models in detail.

Multimodal Variational Autoencoders: Multimodal Variational Autoencoders (VAEs) [28,35] are based upon the standard VAE and have become the building blocks of all later works. Specifically, in these works, multiple pairs of encoder-decoder for the multi-modalities are learned, with the joint posterior obtained through the particular design of *product-of-experts* or *mixture-of-experts*. Inspired by such idea, mmJSD [31], MoPoE [32], MVTCAE [13], and MMVAE+ [24] have explored VAE-based methods further to tackle the multimodal learning problem. mmJSD [31] adopted dynamic prior and modality-specific latent subsets, MVTCAE [13] used multi-view correlation shared representation built upon product of expert [35], MoPoE [32] includes subsets based on both product and mixture of expert [28,35], and recent baseline MMVAE+ [24] incorporates modality-specific priors that are based on the mixture of expert [28]. However, these previous works used less informative uni-modal prior (e.g., Gaussian or Laplacian), which can be ineffective in capturing the complex latent representation shared across different modalities. To tackle this challenge, we propose learning latent space energy-based prior model, which can be more expressive in capturing the shared latent representation and rendering better synthesis across multi-modalities.

Energy-Based Models: The energy-based model (EBM) [6,37] offers a flexible framework for approximating complex data distributions and is shown to be expressive in capturing the data regularities. In addition to data space EBMs, [25] proposes to apply EBM on latent space, which is capable of improving the generative power of the whole model. Learning such latent space EBM requires MCMC posterior and prior sampling, for which the MCMC posterior sampling can be computationally expensive as it requires inner loops of computing the backward gradient of the generation model. Compared to LEBM [25] that only studies one (single-) modality, our work targets the challenging multimodal learning problem by learning latent space EBM to capture the shared latent representations. For multimodal learning problems, MCMC posterior sampling can be more difficult as it involves multiple generation networks during the learning. To alleviate the burden of MCMC posterior sampling, we further develop a variational learning scheme to facilitate efficient EBM sampling and learning. We show that the expressive EBM prior can be useful in the challenging multi-modal learning task.

3 Methodology

In this paper, we present a novel framework for effectively and efficiently modelling the multimodalities. In particular, we study learning the expressive EBM prior to capturing the shared and complex information across different modalities, for which the less informative Gaussian or Laplacian prior model can be limited in expressivity to effectively model. To facilitate efficiency for our EBM learning and sampling, we further develop a variational learning scheme that incorporates both the generator and inference model for jointly learning with the proposed EBM prior.

3.1 Energy-Based Prior for Multimodalities

Let $\mathbf{z}$ be the latent variable and $\mathbf{X}$ be the observation example that contains m modalities, i.e., $\mathbf{X} = \{\mathbf{x}^{(1)}, \ldots, \mathbf{x}^{(m)}\}$. A joint distribution can be specified as

$$p_{\beta,\alpha}(\mathbf{X}, \mathbf{z}) = p_\beta(\mathbf{X}|\mathbf{z})p_\alpha(\mathbf{z}) \quad \text{where}$$
$$p_\beta(\mathbf{X}|\mathbf{z}) = p_{\beta_{(1)}}(\mathbf{x}^{(1)}|\mathbf{z})p_{\beta_{(2)}}(\mathbf{x}^{(2)}|\mathbf{z}) \cdots p_{\beta_{(m)}}(\mathbf{x}^{(m)}|\mathbf{z}) \quad (1)$$

in which $p_\beta(\mathbf{X}|\mathbf{z})$ (β collect $\{\beta_{(1)}, \ldots, \beta_{(m)}\}$) and $p_\alpha(\mathbf{z})$ are the generation model and prior model parameterized by β and α, respectively. This factorization considers m modalities to be conditionally independently distributed while sharing the same latent space. The modality-common information is modelled by shared latent space $p_\alpha(\mathbf{z})$, and the modality-specific information is modelled by each $p_{\beta_{(i)}}(\mathbf{x}^{(i)}|\mathbf{z})$.

Generation Model. The $p_\beta(\mathbf{X}|\mathbf{z}) (= \prod_{i=1}^m p_{\beta_{(i)}}(\mathbf{x}^{(i)}|\mathbf{z}))$ consists of multiple generation models that seek to explain the high-dimensional $\mathbf{x}^{(i)} \in \mathbb{R}^{D^{(i)}}$ by a shared low-dimensional latent vector $\mathbf{z} \in \mathbb{R}^d$ ($d < D^{(i)}$), i.e.,

$$\mathbf{x}^{(i)} = G_{\beta_{(i)}}(\mathbf{z}) + \epsilon \quad \text{where} \quad \epsilon \sim \mathcal{N}(0, I_{D^{(i)}}) \tag{2}$$

which implies $p_{\beta_{(i)}}(\mathbf{x}^{(i)}|\mathbf{z}) \sim \mathcal{N}(G_{\beta_{(i)}}(\mathbf{z}), I_{D^{(i)}})$ with $G_{\beta_{(i)}}$ being a top-down neural network that maps from $\mathbf{z}$ to $\mathbf{x}^{(i)}$. We adopt such a generation model for its simplicity (also adopted in [15,25,36]), but it can also be other choices [29].

With a set of m multimodalities $\mathbf{X} = \{\mathbf{x}^{(1)}, \dots, \mathbf{x}^{(m)}\}$, each generation model is designed to be modal-specific, forming the joint distribution $p_{\beta_{(i)},\alpha}(\mathbf{x}^{(i)}, \mathbf{z}) = p_{\beta_{(i)}}(\mathbf{x}^{(i)}, \mathbf{z})p_\alpha(\mathbf{z})$. If $\mathbf{z} \sim p_\alpha(\mathbf{z})$ can successfully capture the shared information from different modalities, $\mathbf{x}^{(i)} \sim p_{\beta_{(i)},\alpha}(\mathbf{x}^{(i)}, \mathbf{z})$ and $\mathbf{x}^{(j)} \sim p_{\beta_{(j)},\alpha}(\mathbf{x}^{(j)}, \mathbf{z})$ can be drawn with semantic coherence.

Energy-Based Prior. To effectively capture the semantic information shared across the multimodalities, we intend to learn an expressive prior model. In particular, we study learning the energy-based prior model defined as

$$p_\alpha(\mathbf{z}) = \frac{1}{\mathbb{Z}(\alpha)} \exp[f_\alpha(\mathbf{z})] p_0(\mathbf{z}) \tag{3}$$

where $\mathbb{Z}(\alpha) \ (= \int_\mathbf{z} \exp[f_\alpha(\mathbf{z})] p_0(\mathbf{z}) d\mathbf{z})$ is the normalizing constant or partition function, $f_\alpha(\cdot)$ is the energy function parameterized with α, and $p_0(\mathbf{z})$ is the referenced distribution usually assumed to be standard Laplacian [24,28,31]. Such EBM prior generative model has seen success in modelling the data distribution of *single* modality [25,36] while in this paper, we intend to explore its effectiveness in modelling the multimodalities.

For multimodalities, capturing the shared content across modalities may serve as the key ingredient toward generating semantic coherent samples. [28,31,32,35] adopt different frameworks for multimodal learning but only consider the less informative, uni-modal Laplacian prior, which in turn limits the model expressivity, leading to an ineffectively learned model. The proposed latent generative model, on the other hand, is learned with the EBM prior, which is known to be powerful in capturing the data regularity and complex distribution. With well-captured shared information for multimodalities, generated samples of each modality thus can maintain strong semantic coherence across different modalities.

3.2 Learning and Sampling

Maximum Likelihood Estimation. Given n observed examples $\{\mathbf{X}_1, \dots, \mathbf{X}_n\}$ with each $\mathbf{X}_i$ containing m modalities $\mathbf{X}_i = \{\mathbf{x}_i^{(1)}, \dots, \mathbf{x}_i^{(m)}\}$, the generator model (Eq. 1) can be learned by maximum likelihood estimation (MLE) as

$$\begin{aligned} L(\theta) &= \sum_{i=1}^n \log p_\theta(\mathbf{X}_i) = \sum_{i=1}^n \log \int_\mathbf{z} p_\beta(\mathbf{X}_i|\mathbf{z}) p_\alpha(\mathbf{z}) d\mathbf{z} \\ &= \sum_{i=1}^n \log \int_\mathbf{z} p_{\beta_{(1)}}(\mathbf{x}_i^{(1)}|\mathbf{z}) \cdots p_{\beta_{(m)}}(\mathbf{x}_i^{(m)}|\mathbf{z}) p_\alpha(\mathbf{z}) d\mathbf{z} \end{aligned} \tag{4}$$

where θ collect learning parameters (β, α). With a large number of n, maximizing Eq. 4 is equivalent to minimizing the KL-divergence, i.e., $\mathrm{KL}(p_{\mathrm{data}}(\mathbf{X})\|p_\theta(\mathbf{X}))$, and learning θ can be done by computing the gradient as

$$\frac{\partial}{\partial \theta} L(\theta) = \mathbb{E}_{p_\theta(\mathbf{z}|\mathbf{X})}[\frac{\partial}{\partial \theta} \log p_\theta(\mathbf{X}, \mathbf{z})] \quad (5)$$

which requires samples from the generator posterior [10]. To obtain the posterior samples, it can be typically achieved by performing MCMC sampling for $p_\theta(\mathbf{z}|\mathbf{X}) \propto p_\beta(\mathbf{X}|\mathbf{z})p_\alpha(\mathbf{z})$. However, for the proposed model on the multimodalities task, $p_\beta(\mathbf{X}|\mathbf{z})$ consists of m of generation models of each modality, which makes the MCMC posterior sampling inefficient.

Variational Learning Scheme. To ensure efficient learning and posterior sampling, we introduce the inference model $q_\phi(\mathbf{z}|\mathbf{X})$ as an approximation model for the generator posterior. To facilitate learning with the generation model for multimodalities, our inference model is defined to be

$$q_\phi(\mathbf{z}|\mathbf{X}) = \frac{1}{m} \sum_{i=1}^{m} q_{\phi_{(i)}}(\mathbf{z}|\mathbf{x}^{(i)}) \quad (6)$$

where $q_{\phi_{(i)}}(\mathbf{z}|\mathbf{x}^{(i)}) \sim \mathcal{N}(u_{\phi_{(i)}}(\mathbf{x}^{(i)}), V_{\phi_{(i)}}(\mathbf{x}^{(i)}))$. Such an inference model serves as the *mixture of experts* and is also adopted in [28]. Specifically, [28] parameterize pairs of generation and inference model to be modal-specific, i.e., $\beta = (\beta_1, \ldots, \beta_m)$ and $\phi = (\phi_1, \ldots, \phi_m)$, such that each $p_{\beta_i}(\mathbf{x}^{(i)}|\mathbf{z})$ and $q_{\phi_i}(\mathbf{z}|\mathbf{x}^{(i)})$ only focus on one modality as a pair of *decoder* and *encoder*. We adopt such parameterization for its effectiveness.

With such introduced inference model $q_\phi(\mathbf{z}|\mathbf{X})$, a joint KL-divergence can be minimized,

$$-L(\beta, \phi, \alpha) = \mathrm{KL}(p_{\mathrm{data}}(\mathbf{X})q_\phi(\mathbf{z}|\mathbf{X}) \| p_\beta(\mathbf{X}|\mathbf{z})p_\alpha(\mathbf{z})) \quad (7)$$

which is equivalent to maximizing

$$L(\beta, \phi, \alpha) = \frac{1}{m} \sum_{i=1}^{m} \mathbb{E}_{q_{\phi_{(i)}}(\mathbf{z}|\mathbf{x}^{(i)})}[\log \frac{p_{\beta,\alpha}(\mathbf{X}, \mathbf{z})}{q_\phi(\mathbf{z}|\mathbf{X})}] \quad (8)$$

Therefore, we compute the gradient $\frac{\partial}{\partial \beta, \phi} L(\beta, \phi, \alpha)$ as,

$$\frac{\partial}{\partial \beta, \phi} \frac{1}{m} \sum_{i=1}^{m} [\mathbb{E}_{q_{\phi_{(i)}}(\mathbf{z}|\mathbf{x}^{(i)})}[\log p_{\beta_{(i)}}(\mathbf{x}^{(i)}|\mathbf{z})] + \sum_{j=1, j \neq i}^{m} \mathbb{E}_{q_{\phi_{(i)}}(\mathbf{z}|\mathbf{x}^{(i)})}[\log p_{\beta_{(j)}}(\mathbf{x}^{(j)}|\mathbf{z})]$$
$$+ \mathbb{E}_{q_{\phi_{(i)}}(\mathbf{z}|\mathbf{x}^{(i)})}[\log \frac{p_\alpha(\mathbf{z})}{q_\phi(\mathbf{z}|\mathbf{X})}]] \quad (9)$$

the gradient $\frac{\partial}{\partial \alpha} L(\beta, \phi, \alpha)$ is computed as

$$\frac{\partial}{\partial \alpha} L(\beta, \phi, \alpha) = \mathbb{E}_{q_\phi(\mathbf{z}|\mathbf{X})}\left[\frac{\partial}{\partial \alpha} f_\alpha(\mathbf{z})\right] - \mathbb{E}_{p_\alpha(\mathbf{z})}\left[\frac{\partial}{\partial \alpha} f_\alpha(\mathbf{z})\right] \quad (10)$$

where $\mathbf{z} \sim q_\phi(\mathbf{z}|\mathbf{X})$ is inferred from the inference model (Eq. 6), which is a fused joint posterior (i.e., $\mathbf{z} \sim \frac{1}{m} \sum_{i=1}^{m} q_{\phi_{(i)}}(\mathbf{z}|\mathbf{x}^{(i)})$ as average weighted over inferred latent vectors from all modalities).

Sampling from EBM Prior. Learning EBM (Eq. 10) requires samples from the EBM prior, which can be accomplished by conducting MCMC sampling, such as Langevin dynamics (LD) [20]. It iterates as

$$\mathbf{z}_{\tau+1} = \mathbf{z}_\tau - \frac{s^2}{2} \frac{\partial}{\partial \mathbf{z}} [\log p_\alpha(\mathbf{z}_\tau)] + s \cdot \epsilon_\tau \quad \text{where} \quad \epsilon_\tau \sim \mathcal{N}(0, I_d) \quad (11)$$

where s is the step size, ϵ is the Gaussian noise, and τ is the time step of Langevin dynamics. As $s \to 0$, and $\tau \to \infty$, the marginal distribution of $\mathbf{z}$ can asymptotically converge to the target $p_\alpha(\mathbf{z})$ as the stationary distribution. In this work, we employ Laplacian as the initial distribution (i.e. $\mathbf{z}_0 \sim \mathcal{N}(0, I_d)$) and conduct short-run Langevin dynamics, which can also provide meaningful learning signals [22,23].

Connection to ELBO. The VAEs compute the evidence lower bound (ELBO) as the learning objective, which is

$$\mathbb{E}_{q_\phi(\mathbf{z}|\mathbf{X})}[\log p_\beta(\mathbf{X}|\mathbf{z})] - \mathrm{KL}(q_\phi(\mathbf{z}|\mathbf{X}) \| p_0(\mathbf{z})) \quad (12)$$

where $p_0(\mathbf{z})$ is usually assumed to be Gaussian. Whereas, our objective $L(\beta, \phi, \alpha)$ (Eq. 7) can be decomposed into the form

$$L(\beta, \phi, \alpha) = \mathbb{E}_{q_\phi(\mathbf{z}|\mathbf{X})}[\log p_\beta(\mathbf{X}|\mathbf{z})] - \mathrm{KL}(q_\phi(\mathbf{z}|\mathbf{X}) \| p_\alpha(\mathbf{z})) \quad (13)$$

which is closely related to the ELBO of VAEs.

Different from Gaussian or Laplacian prior $p_0(\mathbf{z})$ in Eq. 12, we consider learning the EBM prior $p_\alpha(\mathbf{z})$. With a set of m modalities, each inference model $q_{\phi_{(i)}}(\mathbf{z}|\mathbf{x}^{(i)})$ together forms a *mixture of experts*, and thus the Gaussian or Laplacian prior can be limited in expressivity to capture and match with the posterior distribution, while the EBM prior can be more expressive and multi-modal, leading to a well-learned shared latent space.

Algorithm 1. Learning Energy-Based Multimodal Model

Input: observation examples $\{\mathbf{x}_i^{(1)}, \ldots, \mathbf{x}_i^{(m)}\}_{i=1}^n$, iteration number T, Langevin steps K and step size s. iteration step $t = 0$
repeat
    Posterior Sampling: Given $\{\mathbf{x}_i^{(1)}, \ldots, \mathbf{x}_i^{(m)}\}$, obtain $\{\mathbf{z}_i^{(1)}, \ldots, \mathbf{z}_i^{(m)}\}$ from inference model (Eq. 6).
    Prior Sampling: Obtain EBM prior samples $\mathbf{z}^-$ by performing Langevin dynamics (Eq. 11) with K and s.
    Learn ϕ and β: Update learning parameter ϕ, β with $\{\mathbf{x}_i^{(1)}, \ldots, \mathbf{x}_i^{(m)}\}$ and $\{\mathbf{z}_i^{(1)}, \ldots, \mathbf{z}_i^{(m)}\}$ using Eq. 9.
    Learn α: Update learning parameter α with inferred samples $\{\mathbf{z}_i^{(1)}, \ldots, \mathbf{z}_i^{(m)}\}$ using Eq. 6 and $\mathbf{z}^-$ using Eq. 10.
    Let $t = t + 1$;
until $t = T$

4 Experiments

In this section, we conduct various experiments to demonstrate the expressiveness of the EBM prior in capturing shared latent information across multimodalities more effectively than less-informative unimodal priors (e.g., Gaussian or Laplacian priors). In our experiments, we follow the prior arts [25] and train our model on standard multimodal datasets, such as the PolyMNIST [32] and MNIST-SVHN [28]. We evaluate the performance of our model in terms of joint coherence (Sect. 4.1) and cross coherence (Sect. 4.2). Additionally, we demonstrate the applicability and flexibility of the proposed method, showing its capacity for generalization to various factorizations (Sect. 4.3), including the incorporation of model-specific factors [24]. And we visualize the Markov transition on image synthesis and corresponding generated text applied to Caltech UCSD Birds (CUB) dataset [34]. Furthermore, we conduct ablation studies (Sect. 4.4) to gain a deeper understanding of our approach. We provided our code in https://github.com/syyuan2021/Learning-Multimodal-Latent-Generative-Models-with-EBM.

Baseline Method. For comparisons, our direct baselines include MVAE [35] and MMVAE [28], and we also compare with recent mmJSD [31], MoPoE [32], MVTACE [13] and MMVAE+ [24] that are developed based on the foundation methods of *product-of-expert* [35] and *mixture-of-expert* [28]. For a fair comparison, we adopt the same generation and inference network structures, as well as the pre-trained classifiers from [32] for PolyMNIST, and follow [28]'s implementation to train the classifier for MNIST-SVHN.

4.1 Joint Coherence

With a set of m modalities, we asses our model in generating *unconditional* image synthesis that maintains strong coherence across different modalities. If the EBM prior is well-learned, it should be capable of sampling the latent variables that capture semantic representations and thus generating the image synthesis semantically consistent between multimodalities. We measure such coherence by computing the classification accuracy of generated images of each modality with corresponding classifiers. Higher accuracy indicates better consistency between predicted categories of multimodal synthesis.

We report the quantitative results in Table 1, where our EBM prior shows superior performance compared to baseline models. Both numerically and visually, the EBM prior successfully captures complex latent representations, whereas the unimodal prior (Laplacian) is less expressive and limited in modeling meaningful semantic representations. Synthesis of joint generation can be found in Fig. 1.

Table 1. Accuracy of Joint Coherence and Cross Coherence for PolyMNIST and MNIST-SVHN datasets. The PolyMNIST results for MVAE are referenced from [24], and MNIST-SVHN results for MVAE are referenced from [28]. MMVAE* refers to *mixture of expert* without important sampling.

Model	Joint Coherence		Cross Coherence		
	PolyMNIST	MNIST-SVHN	PolyMNIST	MNIST-SVHN	
				M->S	S->M
Ours	**0.735** ↑	**0.412** ↑	**0.857** ↑	**0.195** ↑	**0.657** ↑
MVAE	0.080	0.127	0.298	0.095	0.093
MMVAE*	0.232	0.215	0.844	0.169	0.523

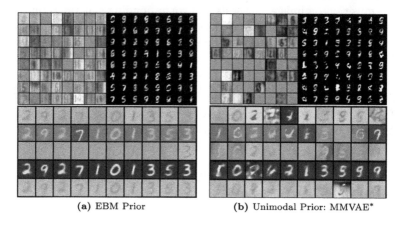

(a) EBM Prior (b) Unimodal Prior: MMVAE*

Fig. 1. Qualitative Results of Joint Generation on MNIST-SVHN (top) and PolyMNIST (bottom)

4.2 Cross Coherence

Next, we evaluate our model in cross-modal generation. Cross-modal generation is to generate image synthesis from input modality $\mathbf{x}^{(i)}$ to target modality $\mathbf{x}^{(j)}$ ($i \neq j$). Specifically, given input $\mathbf{x}^{(i)}$, we first obtain inferred latent vector $\mathbf{z} \sim q_\phi(\mathbf{z}|\mathbf{x}^{(i)})$ and then generate synthesis via generation model $\mathbf{x}^{(j)} \sim p_{\beta_j}(\mathbf{x}^{(j)}|\mathbf{z})$. With our EBM prior, the inference and generation model can be learned to integrate both the modality bias and the energy-based refinement and thus the latent vectors inferred from $\mathbf{x}^{(i)}$ carry the semantic representation shared with $\mathbf{x}^{(j)}$, leading to coherent input $\mathbf{x}^{(i)}$ and output $\mathbf{x}^{(j)}$. To measure the coherence, we compute the classification accuracy (same classifiers used in Sect. 4.1) of the predicted category of $\mathbf{x}^{(j)}$ and true category of $\mathbf{x}^{(i)}$. Higher accuracy means both the $\mathbf{x}^{(i)}$ and $\mathbf{x}^{(j)}$ share higher similarity in semantic features (e.g., digit classes in PolyMNIST). The results can be found in Table 1, in which the proposed method can render competitive performance compared to the baseline methods, suggesting the effectiveness of our EBM prior in multimodalities learning.

4.3 Model Generalization

This paper studies learning a novel framework for foundation multimodal latent generative models, such as MVAE and MMVAE, which serve as the building blocks of various recent works. In this section, we highlight the applicable capability of the proposed method toward other prior advances [24]. These advances usually factorize additional *modal-specific prior* or *modal-subset* to benefit the model complexity in learning the complex multimodalities. In particular, MMVAE+ develop their latent generative model as

$$p_\beta(\mathbf{X}, \mathbf{z}, \mathbf{W}) = p_\beta(\mathbf{X}|\mathbf{z}, \mathbf{W})p_0(\mathbf{z})p_0(\mathbf{W}) \quad (14)$$

where the prior model $p_0(\mathbf{W})$ is modal-specific (i.e., $\mathbf{W} = \{\mathbf{w}^{(1)}, \ldots, \mathbf{w}^{(m)}\}$) and is introduced to improve the expressivity of the whole prior model (compared to Eq. 1. However, the shared latent representation is still modelled by $p_0(\mathbf{z})$, which is assumed to be Laplacian and can be less informative.

Table 2. Accuracy of Joint Coherence and Cross Coherence for PolyMNIST. Results of MMVAE+, MoPoE, MVTCAE, and mmJSD are reported in [24]. Ours+ refers to generalizing our model with modality-specific priors.

Model	Joint Coherence	Cross Coherence
Ours+	**0.848** ↑	**0.889** ↑
MMVAE+	0.344	0.869
MoPoE	0.141	0.720
MVTCAE	0.003	0.591
mmJSD	0.060	0.778

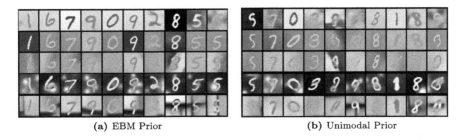

(a) EBM Prior   (b) Unimodal Prior

Fig. 2. Qualitative results of Joint Generation on PolyMNIST

Generalization. Our model is flexible and can be adapted to incorporate such a modal-specific factor, i.e., $p_{\beta,\alpha}(\mathbf{X}, \mathbf{z}, \mathbf{W}) = p_\beta(\mathbf{X}|\mathbf{z}, \mathbf{W})p_\alpha(\mathbf{z})p_0(\mathbf{W})$, such that

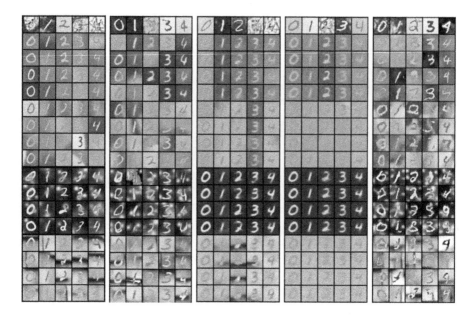

Fig. 3. Qualitative results of Cross Generation on PolyMNIST. From left to right, cross generation results are from (a) EBM prior, (b) MMVAE+, (c) MoPoE, (d) mmJSD, and (e) MVTCAE.

the shared latent representation is modelled by the proposed EBM prior $p_\alpha(\mathbf{z})$, which can be multi-modal and more expressive than the unimodal prior model $p_0(\mathbf{z})$. To examine the effectiveness, we follow MMVAE+ and train our model on PolyMNIST with the same network structures. We report the results of joint coherence and cross coherence in Table 2 and the qualitative results in Fig. 2 and Fig. 3.

Caltech UCSD Birds Dataset (CUB). To examine the scalability, we further train our model on the challenging multimodal dataset Caltech UCSD Birds (CUB) [34]. The proposed EBM can be viewed as an exponential tilting of the reference distribution and thus can correct the less informative prior model toward being more expressive. In this section, we intend to demonstrate the expressivity of our EBM prior in correcting the unimodal prior model. In practice, we utilize the same network structure and train MMVAE+ on CUB, and then we learn our EBM prior with the pre-trained generation and inference model. If our EBM prior can be learned well, the quality of synthesis generated by our EBM prior samples should be better than samples of the unimodal prior.

For illustration, we visualize the Markov chain transition during Langevin dynamics on generated images and text in Fig. 4. Specifically, the Markov chain transition starts from the latent drawn from the unimodal prior $p_0(\mathbf{z})$ and then progresses as an iterative sampler for our EBM prior $p_\alpha(\mathbf{z})$. It can be seen that the quality of image and text synthesis becomes better as the Langevin dynamics progresses and, more importantly, at the final step, the generated images and text render better semantic coherence, which further indicates the effectiveness of the proposed EBM prior.

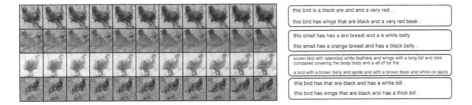

Fig. 4. Visualization of Markov chain transition on image synthesis and corresponding generated text. **Left:** visualization of Markov chain transition where left column images are generated by *unimodal* prior sample, and right column images are generated by *EBM* prior sample. **Right:** jointly generated text corresponding to each row in left figure. The sentence at the top is generated by *unimodal* prior sample, and the sentence at the bottom is generated by *EBM* prior sample.

4.4 Ablation Studies

In the previous sections, we demonstrated the effectiveness of using an EBM prior in multimodal contexts, illustrating that our model can better capture complex data representations across different modalities. To further investigate our model's capabilities, we examine the impact of our EBM prior settings on performance through a series of experiments conducted on the PolyMNIST dataset.

Energy-Based Model MCMC Steps: We first examine the influence of the number of MCMC steps, denoted as S, when sampling the latent variables from EBM prior. We fixed the network architecture to consist of 4 layers with 400 hidden units each. As observed in the middle two rows of Table 3, a smaller number of MCMC steps results in poorer performance. However, increasing the number of MCMC steps is computationally expensive, leading to a trade-off between computation time and model performance.

Energy-Based Model Layers: Then we examine whether more information transformation and interaction will increase coherence scores. So we fixed hidden dimension and Langevin steps while increase EBM layer L from 4 to 6. As shown in the first two rows of Table 3, increasing the number of layers L leads to better coherence results.

Energy-Based Model Complexity: Increasing the model dimension allows the model to learn more complex and representative features. To examine scalability, we increase the dimension of hidden units D to investigate the performance impact of capacity in capturing and processing information. We maintain the number of layers at 4 and use 50 Langevin steps. As observed in the bottom two rows of Table 3, increasing the number of units results in better coherence performance.

Table 3. Ablation studies for energy-based models with different numbers of hidden units D, MCMC steps S, and network layers L.

Model	Joint Coherence	Cross Coherence
Ours (D = 200, L = 4, S = 50)	0.580	0.854
Ours (D = 200, L = 6, S = 50)	0.682	0.840
Ours (D = 400, L = 4, S = 30)	0.641	0.825
Ours (D = 400, L = 4, S = 50)	**0.735**	**0.857**
Ours (D = 200, L = 4, S = 50)	0.580	0.854
Ours (D = 400, L = 4, S = 50)	**0.735**	**0.857**

5 Conclusions and Future Work

In this paper, we present a novel framework for multimodal latent generative models with an EBM prior. This expressive and flexible prior can better represent multimodal data complexity and capture shared information among modalities. Our experiments demonstrate significantly improved coherence of synthesized samples across different modalities compared to baseline models. Our proposed model also facilitates cross-generation between modalities, as validated by experimental results.

However, the proposed method is based on a simple *mixture of experts* scheme with an EBM prior to optimize the multimodal ELBO. Approaches that provide a tighter bound on ELBO, such as importance sampling and stratified sampling with an EBM prior, have not been fully explored. Additionally, to gain a better understanding of shared information learning schemes under multimodal contexts, other expressive priors such as normalizing flow or hierarchical priors will be considered in our future research. For EBM learning, we use latent variables from the variationally inferred posterior, which is less accurate in approximating the true posterior compared to methods such as MCMC sampling. While the latter can be time-consuming, this trade-off either sacrifices generative performance or computational efficiency. This dilemma is non-trivial in multimodal generative problems. Lastly, we will explore the scalability of our EBM prior in future work by investigating its performance on realistic multimodal datasets. This will allow us to evaluate the model's effectiveness in more complex and varied real-world scenarios, further validating its applicability and robustness.

Acknowledgements. This work is supported in part by NSF IIS-2339604. Any opinions, findings, and conclusions or recommendations expressed in this material are those of the authors and do not necessarily reflect the views of the NSF.

References

1. Baltrušaitis, T., Ahuja, C., Morency, L.P.: Multimodal machine learning: a survey and taxonomy. IEEE Trans. Pattern Anal. Mach. Intell. **41**(2), 423–443 (2018)
2. Cui, J., Han, T.: Learning energy-based model via dual-MCMC teaching. In: 37th Conference on Neural Information Processing Systems (2023)
3. Cui, J., Wu, Y.N., Han, T.: Learning hierarchical features with joint latent space energy-based prior. In: Proceedings of the IEEE/CVF International Conference on Computer Vision (ICCV), pp. 2218–2227 (2023)
4. Cui, J., Wu, Y.N., Han, T.: Learning joint latent space EBM prior model for multi-layer generator. In: Proceedings of the IEEE/CVF Conference on Computer Vision and Pattern Recognition (CVPR), pp. 3603–3612 (2023)
5. Du, Y., Li, S., Tenenbaum, J., Mordatch, I.: Improved contrastive divergence training of energy based models. arXiv preprint arXiv:2012.01316 (2020)
6. Gao, R., Song, Y., Poole, B., Wu, Y.N., Kingma, D.P.: Learning energy-based models by diffusion recovery likelihood. arXiv preprint arXiv:2012.08125 (2020)
7. Gu, S., et al.: Vector quantized diffusion model for text-to-image synthesis. In: Proceedings of the IEEE/CVF Conference on Computer Vision and Pattern Recognition, pp. 10696–10706 (2022)
8. Guo, J., Lu, S., Cai, H., Zhang, W., Yu, Y., Wang, J.: Long text generation via adversarial training with leaked information. In: Proceedings of the AAAI Conference on Artificial Intelligence, vol. 32 (2018)
9. Gupta, A., et al.: Photorealistic video generation with diffusion models. arXiv preprint arXiv:2312.06662 (2023)
10. Han, T., Lu, Y., Zhu, S.C., Wu, Y.N.: Alternating back-propagation for generator network. In: Proceedings of the AAAI Conference on Artificial Intelligence, vol. 31 (2017)
11. Han, T., Nijkamp, E., Zhou, L., Pang, B., Zhu, S.C., Wu, Y.N.: Joint training of variational auto-encoder and latent energy-based model. In: Proceedings of the IEEE/CVF Conference on Computer Vision and Pattern Recognition, pp. 7978–7987 (2020)
12. Ho, J., Jain, A., Abbeel, P.: Denoising diffusion probabilistic models. Adv. Neural. Inf. Process. Syst. **33**, 6840–6851 (2020)
13. Hwang, H., Kim, G.H., Hong, S., Kim, K.E.: Multi-view representation learning via total correlation objective. Adv. Neural. Inf. Process. Syst. **34**, 12194–12207 (2021)
14. Jiang, Q., et al.: Understanding and constructing latent modality structures in multi-modal representation learning. In: Proceedings of the IEEE/CVF Conference on Computer Vision and Pattern Recognition, pp. 7661–7671 (2023)
15. Kingma, D.P., Welling, M.: Auto-encoding variational bayes. arXiv preprint arXiv:1312.6114 (2013)
16. Li, B., Qi, X., Lukasiewicz, T., Torr, P.: Controllable text-to-image generation. Adv. Neural. Inf. Process. Syst. **32** (2019)
17. Li, J., Selvaraju, R., Gotmare, A., Joty, S., Xiong, C., Hoi, S.C.H.: Align before fuse: Vision and language representation learning with momentum distillation. Adv. Neural. Inf. Process. Syst. **34**, 9694–9705 (2021)
18. Lin, K., Li, D., He, X., Zhang, Z., Sun, M.T.: Adversarial ranking for language generation. Adv. Neural. Inf. Process. Syst. **30** (2017)
19. Lin, X., Bertasius, G., Wang, J., Chang, S.F., Parikh, D., Torresani, L.: VX2Text: end-to-end learning of video-based text generation from multimodal inputs. In:

Proceedings of the IEEE/CVF Conference on Computer Vision and Pattern Recognition, pp. 7005–7015 (2021)
20. Neal, R.M., et al.: MCMC using Hamiltonian dynamics. Handb. Markov Chain Monte Carlo **2**(11), 2 (2011)
21. Nichol, A., et al.: GLIDE: towards photorealistic image generation and editing with text-guided diffusion models. arXiv preprint arXiv:2112.10741 (2021)
22. Nijkamp, E., Hill, M., Han, T., Zhu, S.C., Wu, Y.N.: On the anatomy of MCMC-based maximum likelihood learning of energy-based models. In: Proceedings of the AAAI Conference on Artificial Intelligence, vol. 34, pp. 5272–5280 (2020)
23. Nijkamp, E., Hill, M., Zhu, S.C., Wu, Y.N.: Learning non-convergent non-persistent short-run MCMC toward energy-based model. Adv. Neural. Inf. Process. Syst. **32** (2019)
24. Palumbo, E., Daunhawer, I., Vogt, J.E.: MMVAE+: enhancing the generative quality of multimodal VAEs without compromises. In: The Eleventh International Conference on Learning Representations. OpenReview (2023)
25. Pang, B., Han, T., Nijkamp, E., Zhu, S.C., Wu, Y.N.: Learning latent space energy-based prior model. Adv. Neural. Inf. Process. Syst. **33**, 21994–22008 (2020)
26. Radford, A., et al.: Learning transferable visual models from natural language supervision. In: International Conference on Machine Learning, pp. 8748–8763. PMLR (2021)
27. Saharia, C., et al.: Photorealistic text-to-image diffusion models with deep language understanding. Adv. Neural. Inf. Process. Syst. **35**, 36479–36494 (2022)
28. Shi, Y., Paige, B., Torr, P., et al.: Variational mixture-of-experts autoencoders for multi-modal deep generative models. Adv. Neural. Inf. Process. Syst. **32** (2019)
29. Sø nderby, C.K., Raiko, T., Maalø e, L., Sø nderby, S.R.K., Winther, O.: Ladder variational autoencoders. In: Lee, D., Sugiyama, M., Luxburg, U., Guyon, I., Garnett, R. (eds.) Advances in Neural Information Processing Systems, vol. 29. Curran Associates, Inc. (2016). https://proceedings.neurips.cc/paper/2016/file/6ae07dcb33ec3b7c814df797cbda0f87-Paper.pdf
30. Song, Y., Sohl-Dickstein, J., Kingma, D.P., Kumar, A., Ermon, S., Poole, B.: Score-based generative modeling through stochastic differential equations. In: International Conference on Learning Representations (2020)
31. Sutter, T., Daunhawer, I., Vogt, J.: Multimodal generative learning utilizing Jensen-Shannon-divergence. Adv. Neural. Inf. Process. Syst. **33**, 6100–6110 (2020)
32. Sutter, T.M., Daunhawer, I., Vogt, J.E.: Generalized multimodal ELBO. In: International Conference on Learning Representations (2021)
33. Vahdat, A., Kautz, J.: NVAE: a deep hierarchical variational autoencoder. Adv. Neural. Inf. Process. Syst. **33**, 19667–19679 (2020)
34. Wah, C., Branson, S., Welinder, P., Perona, P., Belongie, S.: The caltech-UCSD birds-200-2011 dataset (2011). authors.library.caltech.edu
35. Wu, M., Goodman, N.: Multimodal generative models for scalable weakly-supervised learning. Adv. Neural. Inf. Process. Syst. **31** (2018)
36. Xiao, Z., Han, T.: Adaptive multi-stage density ratio estimation for learning latent space energy-based model. In: NeurIPS (2022). http://papers.nips.cc/paper_files/paper/2022/hash/874a4d89f2d04b4bcf9a2c19545cf040-Abstract-Conference.html
37. Xiao, Z., Kreis, K., Kautz, J., Vahdat, A.: VAEBM: a symbiosis between variational autoencoders and energy-based models. arXiv preprint arXiv:2010.00654 (2020)
38. Zhang, H., Koh, J.Y., Baldridge, J., Lee, H., Yang, Y.: Cross-modal contrastive learning for text-to-image generation. In: Proceedings of the IEEE/CVF Conference on Computer Vision and Pattern Recognition, pp. 833–842 (2021)

On Learning Discriminative Features from Synthesized Data for Self-supervised Fine-Grained Visual Recognition

Zihu Wang[1]([✉]), Lingqiao Liu[2], Scott Ricardo Figueroa Weston[1], Samuel Tian[3], and Peng Li[1]

[1] University of California, Santa Barbara, CA 93106, USA
{zihu_wang,scottricardo,lip}@ucsb.edu
[2] The University of Adelaide, Adelaide, South Australia 5001, Australia
lingqiao.liu@adelaide.edu.au
[3] Carnegie Mellon University, Pittsburgh, PA 15213, USA

Abstract. Self-Supervised Learning (SSL) has become a prominent approach for acquiring visual representations across various tasks, yet its application in fine-grained visual recognition (FGVR) is challenged by the intricate task of distinguishing subtle differences between categories. To overcome this, we introduce an novel strategy that boosts SSL's ability to extract critical discriminative features vital for FGVR. This approach creates synthesized data pairs to guide the model to focus on discriminative features critical for FGVR during SSL. We start by identifying non-discriminative features using two main criteria: features with low variance that fail to effectively separate data and those deemed less important by Grad-CAM induced from the SSL loss. We then introduce perturbations to these non-discriminative features while preserving discriminative ones. A decoder is employed to reconstruct images from both perturbed and original feature vectors to create data pairs. An encoder is trained on such generated data pairs to become invariant to variations in non-discriminative dimensions while focusing on discriminative features, thereby improving the model's performance in FGVR tasks. We demonstrate the promising FGVR performance of the proposed approach through extensive evaluation on a wide variety of datasets.

Keywords: Self-Supervised representation learning · fine-grained visual recognition · learning from generated data

1 Introduction

In computer vision, Fine-grained Visual Recognition (FGVR) focuses on identifying subcategories of visual data, such as bird species [1,38], aircraft variants

Supplementary Information The online version contains supplementary material available at https://doi.org/10.1007/978-3-031-73024-5_7.

[29], and vehicle models [25]. Different from studies on large-scaled general image datasets [26,32,37], FGVR tasks highlight the challenge of distinguishing subtle visual patterns.

Self-Supervised Learning (SSL) methods have recently largely advanced the domain of visual representation learning, circumventing the necessity for human-provided annotations. In SSL, many contrastive learning approaches achieve state-of-the-art performance by learning the similarities between data pairs derived from augmentations of identical source images [3,4,6,15,16,27,40]. These methods facilitate the transferability of the learned representations across a wide range of visual recognition problems [4,13,14,16]. Despite these advancements, it has been suggested that SSL may prioritize general visual similarities, instead of critical subtle features in FGVR tasks, which leads to SSL's 'coarse-label bias' [9]. Furthermore, recent studies [24,35,36] have highlighted a tendency among existing SSL methods to be distracted by task-irrelevant features, consequently failing to capture FGVR-relevant patterns.

To this end, we propose an innovative self-supervised learning strategy focusing on selectively extracting highly discriminative features while disregarding less informative, noisy ones. Our approach involves the generation of new contrastive data pairs from the latent feature space of the encoder, training the encoder to prioritize critical objects within these pairs. To facilitate this, a decoder is employed to generate data based on the latent feature space, reconstructing an image's feature vector and its perturbed counterpart to form each data pair. As the data pairs are generated to guide the encoder to learn key features and to be invariant to variations in non-discriminative features, in each feature vector, only those dimensions associated with non-discriminative patterns are perturbed.

Therefore, at the core of our methods are two non-discriminative feature identification techniques. Firstly, Grad-CAM [33] induced from the SSL loss to the latent feature space highlights dimensions relevant to FGVR [35,36]. Thus, we introduce greater perturbation to those less highlighted dimensions. Despite the conventional view that dimensional collapse in SSL—characterized by some encoder dimensions producing constant outputs—is undesirable [6,23,28], recent literature [10,45] suggests that inducing such collapse in task-irrelevant feature dimensions yields beneficial outcomes. As shown in Fig. 2, our empirical studies indicate that in encoders pre-trained by SSL methods, there are always dimensions with low variance across the dataset which cannot effectively separate data from different categories. We thus treat these low-variance dimensions as task-irrelevant and introduce perturbations to them. The two aforementioned perturbation components are then combined and applied to the latent feature vector of each image. Images are then reconstructed from the perturbed and original feature vectors to form contrastive pairs for a contrastive loss [4,30]. Such a framework encourages the encoder to learn the key features highlighted by Grad-CAM and to reduce variance and induce collapse in those non-discriminative dimensions with low variance across the dataset in the latent space.

Our proposed fine-grained feature learning method can be incorporated into various existing SSL methods. We use SimSiam [6] and MoCo v2 [16] as baseline methods and incorporate our proposed technique into these methods. Experiments across various fine-grained visual datasets show the effectiveness of our method. The proposed method provides a great improvement over baseline methods. Our methods built on MoCo v2 outperforms existing state-of-the-art Self-Supervised fine-grained visual recognition methods in numerous downstream tasks.

2 Related Works

2.1 Self-supervised Contrastive Learning

Self-Supervised Learning (SSL) facilitates the learning of visual representations without the need for labeled data. Among various SSL methodologies, contrastive learning has emerged as a promising technique. With the InfoNCE loss [30] and its variants [3,4,6,7,16] being introduced as the objectives for optimization, contrastive approaches treat different views of the same image as positive data pairs, while views from different images are considered negative data pairs. The goal for the encoder is to minimize the distance between positive pairs and maximize it between negative pairs within its representation space [2,4,16,39]. Methods such as BYOL [15] and SimSiam [6] rely exclusively on positive pairs. Additionally, the issue of dimensional collapse, where some encoder dimensions output constant values, is discussed in [6,23,28], along with proposed solutions to mitigate this phenomenon. Nonetheless, recent studies [10,45] have shown that the collapse of dimensions associated with task-irrelevant features can enhance the performance in downstream visual recognition tasks.

2.2 Fine-Grained Visual Recognition in Self-supervised Learning

While encoders pre-trained by Self-Supervised Learning (SSL) methods demonstrate transferability and generalizability in many tasks [4,6,16,21,41], studies [9,24,35] reveal SSL's limitations in capturing essential features for Fine-Grained Visual Recognition (FGVR). To enhance SSL's capability in identifying critical features, several works concentrate on refining data augmentations. Approaches such as SAGA [43], CAST [34], and ContrastiveCrop [31] adopt attention-guided heatmaps to locate and better crop key objects in images. DiLo [44] introduces a novel augmentation by merging key image objects with different backgrounds to generate additional views. Contrary to methods that modify images directly, our approach involves perturbing feature vectors and generating realistic images from the latent feature space to enhance the encoder's discriminative capacity. Another line of research employs auxiliary neural networks connected to the encoder's convolutional layers for improving encoder's attention on salient regions. LEWEL [20] trains an additional head to adaptively aggregate features. Techniques such as CVSA [12,42] train a network to fit segmentation annotations or outputs of pre-trained saliency detectors. Similarly, LCR [35] and SAM [36] train the network to align with Grad-CAM, treating Grad-CAM as the ground truth for the encoder's attention maps. Our method proposes training the encoder on generated data pairs to learn critical features. In addition to Grad-CAM, we use dimension variance as a criterion for identifying non-discriminative features. Low-variance dimensions where data points across the dataset are not well separated are treated less crucial. Besides, SimCore [24] pre-trains an encoder on a target dataset, then using it to select more relevant data from a large-scaled dataset to expand the training set, upon which a new encoder is retrained for downstream tasks.

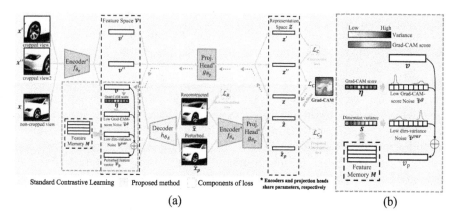

Fig. 1. The overview of the proposed method. (a) Our method can be incorporated into various existing SSL methods. A decoder is utilized to generate images from both the original feature vector and its perturbed counterpart to form data pairs. The overall loss consists three terms: a conventional contrastive loss, a reconstruction loss (ensuring the decoder evolves with the encoder), and a proposed contrastive loss on the generated pairs. (b) We propose two techniques to identify and perturb non-discriminative features in a feature vector, i.e., features with low variance that fail to effectively separate data and those deemed less important by Grad-CAM induced from the SSL loss.

3 Method

3.1 Background

Self-supervised Contrastive Learning. Without need for labels, self-supervi- sed contrastive learning learns to represent data $\mathbf{x} \in \mathbb{R}^m$ in a lower dimensional space $\mathbb{R}^n$ by learning the similarities among data samples. Typically, in contrastive learning, the model consists two components, an encoder $f_{\theta_e} : \mathbb{R}^m \to \mathbb{R}^n$ that maps data to a latent feature space $\mathcal{V} \subseteq \mathbb{R}^n$, and a projection head $g_{\theta_p} : \mathbb{R}^n \to \mathbb{R}^k$ which projects the latent feature vectors in $\mathcal{V} \subseteq \mathbb{R}^n$ to a lower dimensional representation space $\mathcal{Z} \subseteq \mathbb{R}^k$ where the contrastive loss is applied. Formally, given a batch $\mathcal{B}$ of unlabeled data, every image $\mathbf{x}$ in it is augmented by two random augmentation $\mathcal{T}_1$ and $\mathcal{T}_2$ to acquire two views, i.e., $\mathbf{x}' = \mathcal{T}_1(\mathbf{x})$, $\mathbf{x}'' = \mathcal{T}_2(\mathbf{x})$. Views augmented from the same image are considered a positive pair, while those acquired from different images form negative pairs. Two augmented views of all images form a new batch $\mathcal{B}_a$ which doubles the size of $\mathcal{B}$. The encoder and projection head are then used to represent the views in $\mathcal{Z}$, i.e., $\mathbf{z}' = g_{\theta_p}(f_{\theta_e}(\mathbf{x}'))$, $\mathbf{z}'' = g_{\theta_p}(f_{\theta_e}(\mathbf{x}''))$. A contrastive loss $\mathcal{L}_C$ is then defined in $\mathcal{Z}$ space:

$$\mathcal{L}_C(\mathbf{z}', \mathbf{z}'') = -\log \frac{\exp(\mathbf{z}' \cdot \mathbf{z}''/\tau)}{\sum_{\mathbf{z}_i \in \mathcal{B}_a, \mathbf{z}_i \neq \mathbf{z}', \mathbf{z}_i \neq \mathbf{z}''} \exp(\mathbf{z}' \cdot \mathbf{z}_i/\tau)} \quad (1)$$

where τ is the temperature hyperparameter.

Although there are subtle differences between different contrastive methods, the contrastive loss are defined similarly. MoCo [5,16] introduces a large memory of negative representations. SimSiam [6] and BYOL [15] discard negative pairs and learn solely from positive pairs.

3.2 Overview

The overview of our method is illustrated in Fig. 1. The essence of the proposed method is learning discriminative fine-grained visual features from synthesized data pairs which are reconstructed from latent feature vectors by a decoder h_{θ_d}. During training, to ensure that the decoder follows the evolution of the encoder, we use Mean Square Error (MSE) as the loss function to optimize the decoder. Our empirical studies show that the decoder produces images of better quality when it is trained on non-augmented images. The following reconstruction loss $\mathcal{L}_R$ is calculated for each image.

$$\mathcal{L}_R = \frac{1}{m} \|\mathbf{x} - \hat{\mathbf{x}}\|_2^2 \quad (2)$$

where $\mathbf{x} \in \mathbb{R}^m$ is non-augmented image, and $\hat{\mathbf{x}} = h_{\theta_d}(f_{\theta_e}(\mathbf{x}))$ is the reconstruction of it.

In addition to producing $\hat{\mathbf{x}}$, we generate $\hat{\mathbf{x}}_p$ from $\mathbf{v}_p \in \mathcal{V}$, which is a perturbed version of $\mathbf{v}$ where non-discriminative features are perturbed. A positive data pair is formed between $\hat{\mathbf{x}}$ and $\hat{\mathbf{x}}_p$ from which the encoder learns discriminative features while disregarding task-irrelevant ones. In Sects. 3.3 and 3.4, we introduce two methods of identifying crucial discriminative dimensions in the latent feature space $\mathcal{V}$.

3.3 Identifying Key Dimensions via Grad-CAM

Grad-CAM [33], a widely used saliency detection technique, uses the gradient of the target loss with respect to intermediate features of the network to produce an attention map highlighting regions in the features that contribute to minimizing the loss. As in our proposed self-supervised method, labels are not available during training, we thus choose the contrastive loss as the target. To identify important features within an image's feature vector $\mathbf{v} = f_{\theta_e}(\mathbf{x}) \in \mathbb{R}^n$, we form positive pair between $\mathbf{x}$ and an augmented view $\mathbf{x}''$ to calculate a contrastive loss $\mathcal{L}_C(\mathbf{z}, \mathbf{z}'')$, where $\mathbf{z} = g_{\theta_p}(f_{\theta_e}(\mathbf{x}))$, $\mathbf{z}'' = g_{\theta_p}(f_{\theta_e}(\mathbf{x}''))$. As it is shown in Fig. 1, the Grad-CAM score vector $\boldsymbol{\eta} = \{\eta_i\}_{i=1}^n \in \mathbb{R}^n$ is calculated by gradient of the contrastive loss with respect to the feature vector $\mathbf{v}$.

$$\eta_i = \text{ReLU}(\frac{\partial \mathcal{L}_C(g_{\theta_p}(\mathbf{v}), g_{\theta_p}(\mathbf{v}''))}{\partial v_i} \cdot v_i) \qquad (3)$$

Here, $\mathbf{v} = f_{\theta_e}(\mathbf{x})$, $\mathbf{v}'' = f_{\theta_e}(\mathbf{x}'')$. v_i denotes the i_{th} element of $\mathbf{v}$. The application of ReLU($\cdot$) makes $\eta_i > 0$ for all $i \in \{1, 2, \ldots, n\}$. Note that the original Grad-CAM calculates gradient with respect to feature maps of the last convolutional layer. To measure the saliency of feature dimensions, we calculate gradient with respect to feature vectors. Higher Grad-CAM scores in $\boldsymbol{\eta}$ represent corresponding dimension's higher contribution to data discrimination in contrastive learning.

With the Grad-CAM scores, random Gaussian noise is then introduced as perturbation to $\mathbf{v}$. We first scale all elements in $\boldsymbol{\eta}$ to $[0, 1]$ by min-max normalization.

$$\bar{\eta}_i = \frac{\eta_i - \min\{\eta_j, j = 1, 2, \ldots, n\}}{\max\{\eta_j, j = 1, 2, \ldots, n\} - \min\{\eta_j, j = 1, 2, \ldots, n\}} \qquad (4)$$

where $\bar{\eta}_i$ is the i_{th} element of the normalized Grad-CAM score vector $\bar{\boldsymbol{\eta}}$. After the normalization, a random Gaussian noise perturbation vector $\tilde{\mathbf{v}}^g \in \mathbb{R}^n$ is calculated as follows.

$$\tilde{\mathbf{v}}^g = \{\tilde{v}_i^g : \tilde{v}_i^g \sim \mathcal{N}(0, \epsilon_g \cdot (1 - \bar{\eta}_i))\}_{i=1}^n \qquad (5)$$

where ϵ_g is a hyperparameter that controls the standard deviation of Gaussian noise. In such a perturbation, all elements are sampled from i.i.d. zero-mean Gaussian distributions. Importantly, dimensions with lower normalized Grad-CAM scores $\bar{\eta}_i$ receive noise from a Gaussian distribution with a greater standard deviation, implying a higher likelihood of more significant noise affecting dimensions with lower Grad-CAM scores. And on average, key dimensions with higher Grad-CAM scores are affected less which helps preserve crucial features in the original images.

3.4 Determining Feature's Task-Relevance via Dimension Variance

In addition to the feature perturbation technique elaborated in Sect. 3.3, we propose another technique to determine and perturb task-irrelevant dimensions.

In SSL, dimensional collapse is a phenomenon where some encoder dimensions output constant values [6,19,23,28]. As data points are not separated along such dimensions, these dimensions can not be used to perform downstream visual recognition. However, recent works [10,45] suggest that collapse of dimensions which are related to downstream task-irrelevant features can be beneficial. In

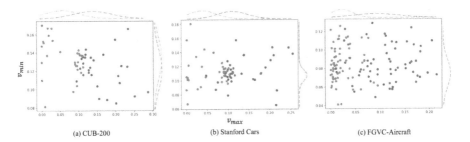

Fig. 2. An illustration of data distribution in the feature space of encoders pre-trained by MoCo v2 [5]. Blue and red dots represent feature vectors of two categories' data from 3 fine-grained datasets, CUB-200 [38], Stanford Cars [25], and FGVC-Aircraft [29]. v_{min} and v_{max} are the dimensions in the feature space where data has the minimal and maximal variance across the dataset. Probability density curve fitting of each category along each dimension is attached to the corresponding axis. Different classes are separated much better along v_{max} than v_{min}.

spite of the potential benefit, how to induce beneficial dimensional collapse is not illustrated by existing SSL studies.

As it is illustrated in Fig. 2, our empirical studies show that, in the latent feature space of encoders pre-trained by SSL methods, typically, data points are not well separated along dimensions with low variance. Variance along these dimensions thus introduces noise to downstream classification. Therefore, we treat such dimensions as task-irrelevant and propose a technique to induce collapse in these dimensions to guide the encoder's to be invariant to variations of such features. This technique start with estimating the dataset's variance along each feature dimension in a feature vector memory bank $\mathbf{M} \in \mathbb{R}^{D \times n}$ of size D. During training, whenever the encoder is provided with a batch of data, its feature vectors will be stored in the memory to replace the oldest batch of feature vectors in it. Variance of each dimension across the dataset can be approximated in $\mathbf{M}$ to get a variance vector $\mathbf{s} = \{s_i : \sigma^2(\bar{\mathbf{w}}_i), i = 1, 2, \ldots, n\} \in \mathbb{R}^n$. Here $\bar{\mathbf{w}}_i \in \mathbb{R}^D$ is the ℓ_2-normalized i_{th} column vector of $\mathbf{M}$. A feature represented by dimension i is considered less discriminative if its corresponding variance $s_i < \kappa$ where κ is a threshold hyperparameter. To introduce random noise to those less discriminative dimensions, similar to Eq. (4), we first apply min-max normalization to $\mathbf{s}$ to acquire $\bar{\mathbf{s}} = \{\bar{s}_i\}_{i=1}^n$. We then calculate the random noise vector $\tilde{\mathbf{v}}^{var} = \{\tilde{v}_i^{var}\}_{i=1}^n$ to be applied.

$$\tilde{v}_i^{var} = \begin{cases} u_i \sim \mathcal{N}(0, \epsilon_{var} \cdot (1 - \bar{s}_i)) & \text{if } s_i < \kappa \\ 0 & \text{otherwise} \end{cases} \quad (6)$$

Here, ϵ_{var} defines the standard deviations of the i.i.d. Gaussian distributions. Similar to $\tilde{\mathbf{v}}^g$, $\tilde{\mathbf{v}}^{var}$ introduces greater noise to lower-variance dimensions.

3.5 Learning from Reconstructed Data Pairs

With the two feature perturbation techniques proposed in Sects. 3.3 and 3.4, we can finally perturb the feature vector $\mathbf{v}$ by adding the two random Gaussian noise to it to obtain its perturbed version $\mathbf{v}_p$, i.e., $\mathbf{v}_p = \mathbf{v} + \tilde{\mathbf{v}}^g + \tilde{\mathbf{v}}^{var}$. As it is shown in Fig. 1, $\mathbf{v}$ and $\mathbf{v}_p$ are reconstructed by the decoder to produce $\hat{\mathbf{x}}$ and $\hat{\mathbf{x}}_p$, respectively. In $\hat{\mathbf{x}}_p$, key patterns from $\hat{\mathbf{x}}$ are preserved, while those contribute less to Grad-CAM or deemed less discriminative across the dataset by the low-variance criterion are perturbed.

We then form a positive pair between $\hat{\mathbf{x}}$ and $\hat{\mathbf{x}}_p$, addressing extracting key features and disregarding non-discriminative ones. To this end, we propose to pass the representations $\hat{\mathbf{z}} = g_{\theta_p}(f_{\theta_e}(\hat{\mathbf{x}}))$ and $\hat{\mathbf{z}}_p = g_{\theta_p}(f_{\theta_e}(\hat{\mathbf{x}}_p))$ to a contrastive loss $\mathcal{L}_{C_p}$.

$$\mathcal{L}_{C_p}(\hat{\mathbf{z}}, \hat{\mathbf{z}}_p) = -\log \frac{\exp(\hat{\mathbf{z}} \cdot \hat{\mathbf{z}}_p / \tau)}{\sum_{\hat{\mathbf{z}}_i \in \mathcal{B}_r, \hat{\mathbf{z}}_i \neq \hat{\mathbf{z}}, \hat{\mathbf{z}}_i \neq \hat{\mathbf{z}}_p} \exp(\hat{\mathbf{z}} \cdot \hat{\mathbf{z}}_i / \tau)} \tag{7}$$

where $\mathcal{B}_r$ is the set of all reconstructed and perturbed images from the current batch $\mathcal{B}$. By forming a positive pair between $\hat{\mathbf{x}}$ and $\hat{\mathbf{x}}_p$, Eq. (7) requires the encoder to be invariant to those perturbed less discriminative features.

Finally, we write the training loss $\mathcal{L}$ of our method as follows.

$$\mathcal{L} = \mathcal{L}_C + \alpha \cdot \mathcal{L}_R + \nu \cdot \mathcal{L}_{C_p} \tag{8}$$

α and ν are hyperparameters that control the weight of $\mathcal{L}_R$ and $\mathcal{L}_{C_p}$ in training. The pseudocode of our method is provided in Appendices.

Table 1. Performance comparison on three datasets. Our method is compared with MoCo v2 [5] and ResNet50 supervised pre-trained on ImageNet-1k [11]. Top-1 classification accuracy (in%) is reported when model is evaluated on 100%, 50%, and 20% of all labels. Rank-1, rank-5, and mAP (in%) in image retrieval are reported.

Dataset	Methods	Classification			Image Retrieval		
		100%	50%	20%	rank-1	rank-5	mAP
CUB-200	ResNet50	63.06	55.71	42.24	40.39	68.94	15.88
	MoCo V2	63.98	56.35	42.63	39.72	67.14	15.91
	Ours	**66.17**	**60.84**	**49.69**	**42.06**	**69.59**	**19.70**
Stanford Cars	ResNet50	61.41	49.85	33.57	29.28	54.66	7.01
	MoCo V2	62.02	51.08	35.44	30.51	56.15	7.13
	Ours	**65.60**	**54.36**	**40.24**	**35.81**	**61.94**	**10.02**
FGVC-Aircraft	ResNet50	49.79	41.02	33.61	29.48	51.39	10.17
	MoCo V2	51.13	44.34	36.42	30.02	52.87	11.24
	Ours	**55.28**	**49.37**	**41.10**	**33.27**	**56.80**	**12.69**

4 Experiments

4.1 Experiment Settings

Datasets. Experiments are conducted across five fine-grained visual datasets. We adopt on three widely used fine-grained datasets. `Caltech-UCSD Birds 200 -2011 (CUB-200)` dataset [38] contains 5994 training data and 5794 testing data of 200 categories of birds. `Stanford Cars (Cars)` [25] has 196 classes of car models where 8144 data and 8041 data are in its training and testing split respectively. `FGVC-Aircraft (Aircraft)` [29] has 100 classes where 6667 images are for training and 3333 images are for testing. Additionally, we consider `German Traffic Sign Recognition Benchmark (GTSRB)` [18] which contains 43 classes of traffic signs. This dataset is usually used in autonomous driving and smart cities development. We take 4800 images for training and 3750 images for testing from GTSRB. We also evaluate the effectiveness of our method on `ISIC2017` [8], a medical image dataset with 3 categories of skin lesion analysis where 2000 images are for training and 600 images are for testing.

Training Settings. All methods adopt ResNet-50 [17] as the encoder backbone where weights are initialized by loading ImageNet-1k [32] pre-trained model. Using two state-of-the-art SSL method, MoCo v2 [5] and SimSiam [6], as the baseline methods, we incorporate our proposed method in their framework. For the sake of fair comparison, encoders of all methods are pre-trained for 100 epochs. And the training batch size is set to 128 for all methods. More detailed encoder pre-training settings are provided in Appendices.

Additionally, in our method, we use a feature vector memory bank of size $D = 5632$. For the training loss in Eq. (8), we choose $\alpha = 1$ and $\nu = 0.5$. When introducing noise described in Eqs. (5) and (6), we choose $\epsilon_g = 0.1$, $\epsilon_{var} = 0.05$, and $\kappa = 0.02$. To ensure the quality of reconstructed images, before encoder training, we freeze the encoder parameters and pre-train the decoder on target datasets by a reconstruction loss. We provide decoder pre-training details in Appendices.

Performance Evaluation Protocols. Linear evaluation is a widely adopted protocol for assessing the performance of learned representations in visual recognition. This approach freezes the parameters of the pre-trained encoder and attaches a linear classifier to it. The classifier is then trained to perform classification. Our linear evaluation setup follows [16], detailed further in the Appendices.

The task of image retrieval [22,35,41] serves as another pivotal method for evaluating the performance of representation learning. Without adjusting any model parameters, it searches the nearest neighbors of a query image in the latent feature space for images that share the same label with the query image. The effectiveness of the evaluated model is quantified by recording the proportion of retrieved images that fall into the same categories as the query image. We present three commonly utilized metrics of retrieval performance: rank-1, rank-5, and mean Average Precision (mAP).

Furthermore, to illustrate the effect of the proposed method, attention maps generated by Grad-CAM on encoders trained by different methods are compared. Images from the training datasets, along with their corresponding reconstructed and perturbed versions, are also provided.

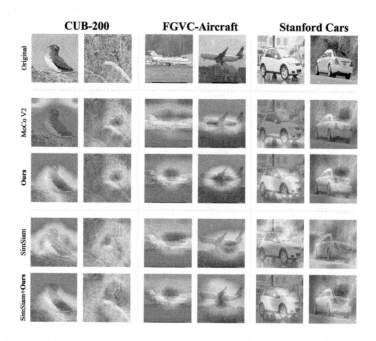

Fig. 3. Grad-CAM attention visualized on images. Our proposed method is incorporated into MoCo v2 and SimSiam and compared with them.

4.2 Main Results

Comparison Between the Proposed Method and Baseline Methods. As among all configurations of our method, the one based on MoCo v2 achieves the best overall performance, this configuration is thus called **'ours'** in the experiments. We first compare our method with MoCo v2 in Table 1. 'ResNet50' in Table 1 represents the encoder supervised pre-trained on ImageNet-1k [11].

In linear evaluation, we evaluate pre-trained encoders using different proportions of labels. Across all three datasets, our proposed method achieves an average top-1 accuracy improvement of 3.31%, 4.27%, and 5.51% over MoCo v2, with 100%, 50%, and 20% of all labels, respectively. This advancement highlights the efficacy of our data pairs generation technique, enabling great performance in classification tasks. Notably, the remarkable improvement in label-insufficient scenarios highlights our method's ability to learn more generalizable features from unlabeled data.

We further evaluate the effectiveness of our method by image retrieval tasks. Performance in these tasks serves as a measure of semantic consistency in the learned latent feature space. Our method outperforms MoCo v2 on all datasets in rank-1, rank-5, and mAP. Invariant to irrelevant patterns, our approach ensures that images from the same category, which exhibit discriminative patterns, are more closely clustered in the feature space. This distribution enhances performance in image retrieval.

Additionally, we visualize Grad-CAM attention in MoCo v2 and our method in Fig. 3. Unlike MoCo v2, which may concentrate on background regions irrelevant to visual recognition, our method exhibits enhanced precision in identifying and focusing on pivotal objects within the images, effectively minimizing the influence of background distractions.

Furthermore, we also integrate the proposed technique into a state-of-the-art negative pair-free method SimSiam [6], denoted as 'SimSiam+ours' in our experiments. Our framework significantly outperforms the original SimSiam in both linear evaluation and image retrieval tasks, as demonstrated in Table 3. Attention comparison is also visualized by Grad-CAM and compared with SimSiam in Fig. 3.

Table 2. Performance comparison on ISIC2017 [8] and GTSRB [18]. Top-1 classification accuracy (in %) and rank-1 (in %) of image retrieval is reported.

Methods	ISIC2017		GTSRB	
	Classification	Retrieval	Classification	Retrieval
MoCo V2 [5]	64.33	66.34	88.25	97.46
Ours	**66.92**	**67.90**	**91.67**	**98.59**
SimSiam [6]	58.83	64.46	81.10	92.81
SimSiam + Ours	**63.78**	**65.55**	**86.94**	**94.42**

To comprehensively assess the effectiveness of our proposed method, we utilize two more fine-grained datasets: the traffic sign visual dataset GTSRB and the medical image dataset ISIC2017. The results, as shown in Table 2, indicate that our approach enhances the performance of Self-Supervised Learning (SSL), demonstrating its potential in real-world applications for FGVR tasks.

Comparison with State-of-the-Art Self-supervised FGVR Methods. In this section, we compare our proposed method against existing self-supervised learning (SSL) techniques renowned for their enhanced fine-grained visual recognition capabilities [12,20,31,35,36,44], as detailed in Table 3. Methods like Dino [3], SimSiam [6], and MoCo v2 [16], which are not optimized for fine-grained feature extraction are also listed. Supervised training performance is included to provide a comprehensive comparison. Top-1 accuracy in linear evaluation and rank-1 in image retrieval tasks are reported.

Table 3. Comparison with state-of-the-art self-supervised FGVC methods. Supervised training is also included. Top-1 classification accuracy (in %) and rank-1 image retrieval (in %) are reported.

Method	Classification			Image Retrieval		
	CUB-200	Cars	Aircraft	CUB-200	Cars	Aircraft
supervised	77.46	88.60	85.93	–	–	–
Dino [3]	16.74	14.33	12.07	–	–	–
SimSiam [6]	46.75	45.72	38.52	16.24	12.45	18.49
MoCo V2 [5]	63.98	62.02	51.13	39.72	30.51	30.02
DiLo [44]	62.97	–	–	–	–	–
CVSA [12]	63.02	–	–	–	–	–
LEWEL [20]	64.59	62.91	51.90	39.91	32.36	31.09
ContrastiveCrop [31]	64.23	63.29	52.04	39.84	32.71	30.37
SAM-SSL-Bilinear [36]	64.94	62.85	52.83	40.08	33.19	30.52
LCR [35]	65.24	63.96	53.22	41.26	34.74	31.55
SimSiam+Ours	57.80	53.63	47.50	24.67	19.72	24.56
Ours	**66.17**	**65.60**	**55.28**	**42.06**	**35.81**	**33.27**

In Table 3, our proposed method achieves the best overall performance in both image retrieval and linear evaluation tasks. DiLo [44], CVSA [12], and ContrastiveCrop [31] innovate with novel data augmentation techniques which directly modify the original images. In contrary, our method generates more realistic images from the learned feature space, highlighting the learning of FGVR-related features. And unlike SAM [36] and LCR [35] which train an auxiliary network to directly fit the encoder's attention to Grad-CAM, our method learns from generated data to highlight dimensions with high Grad-CAM scores and introduce dimensional collapse to non-discriminative features.

Generated Data Pairs of the Proposed Method. To better understand why the proposed technique enhances the fine-grained visual feature extraction capability, we show the generated data pairs from different datasets in Fig. 4. The perturbed images, as illustrated in Fig. 4, show that they largely retain the original data's key objects, with modifications primarily appearing as subtle changes in background or less important regions, e.g., changes of the tree's branch behind a bird, alterations in vehicle light's textures and adjustments in an airplane's exterior finish. These modifications do not affect the defining features of the subjects. Remarkably, some perturbations lead to entirely new objects that maintain the original's identity. For instance, transforming a side view of a car into a front view, or depicting an aircraft in flight from a grounded position. These images, obtained by modifying latent semantics of original images, are difficult to obtain through traditional data augmentation techniques defined in the original data space. They efficiently guide the encoder in identifying which features to prioritize and which to ignore, enhancing its learning performance in FGVR.

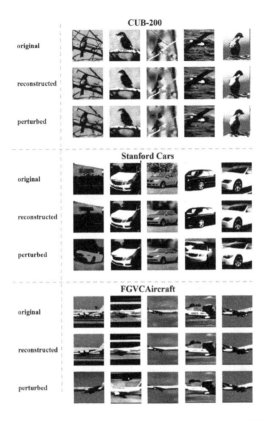

Fig. 4. Generated data pairs on CUB-200, Stanford Cars, and FGVC-Aircraft. The original images are also included.

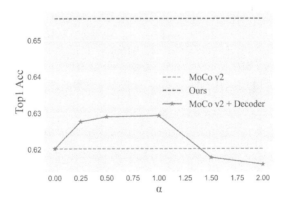

Fig. 5. Performance comparison of encoders trained by $\mathcal{L}_C + \alpha \cdot \mathcal{L}_R$ with respect to different α value (green solid line). Top-1 classification accuracy on Stanford Cars is reported. MoCo v2 (red dashed line) and our method (blue dashed line) are included for comparison.

4.3 Effectiveness of the Reconstruction Loss in Contrastive Learning

As described in Eq. (8), our model's overall training loss, $\mathcal{L}$, includes a reconstruction loss term, $\mathcal{L}_R$. To assess $\mathcal{L}_R$'s effect on self-supervised contrastive learning, we incorporate a decoder into MoCo v2 and train the encoder by a loss $\mathcal{L}_C + \alpha \cdot \mathcal{L}_R$ on the Stanford Cars dataset, varying α in the loss function. The results, shown in Fig. 5, indicate that α values of 0.5 and 1.0 enhance top-1 classification accuracy the most over MoCo v2. Generally, $\mathcal{L}_R$ provides a modest improvement (less than 1%) to Self-Supervised FGVC.

4.4 Effectiveness of the Two Feature Perturbation Techniques

As detailed in Sects. 3.3 and 3.4, two non-discriminative feature perturbation techniques are proposed to introduce noise $\tilde{\mathbf{v}}^g$ and $\tilde{\mathbf{v}}^{var}$, respectively. We conduct further experiments to assess the effectiveness of each technique in identifying and perturbing task-irrelevant features. In Table 4, we evaluate three different configurations of our method: (1) Ours: $\mathbf{v}_p$ is obtained by adding both noise components to $\mathbf{v}$, i.e., $\mathbf{v}_p = \mathbf{v} + \tilde{\mathbf{v}}^g + \tilde{\mathbf{v}}^{var}$; (2) Ours - Grad-CAM: $\mathbf{v}_p$ is obtained by adding only the noise generated by the low Grad-CAM scores criterion, i.e., $\mathbf{v}_p = \mathbf{v} + \tilde{\mathbf{v}}^g$; (3) Ours - low-var: $\mathbf{v}_p = \mathbf{v} + \tilde{\mathbf{v}}^{var}$.

As shown in Table 4, all configurations achieve competitive results comparing with existing state-of-the-art self-supervised FGVR methods. And when combining two noise components $\tilde{\mathbf{v}}^g$ and $\tilde{\mathbf{v}}^{var}$, our methods achieves the best overall performance.

Table 4. Comparison with state-of-the-art self-supervised FGVC methods. Supervised training is also included. Top-1 classification accuracy (in %) and rank-1 image retrieval (in %) are reported.

Method	Classification			Image Retrieval		
	CUB-200	Cars	Aircraft	CUB-200	Cars	Aircraft
MoCo V2 [5]	63.98	62.02	51.13	39.72	30.51	30.02
Ours - Grad-CAM	66.04	64.18	54.19	41.02	35.08	32.14
Ours - low-var	65.91	64.34	53.11	41.69	34.55	31.23
Ours	**66.17**	**65.60**	**55.28**	**42.06**	**35.81**	**33.27**

5 Conclusion

To enhance the performance of Self-Supervised Learning (SSL) in Fine-grained Visual Recognition (FGVR) tasks, this paper introduces a novel approach where an encoder learns discriminative features from generated images. By introducing noise to features deemed non-discriminative by two proposed criteria, we generate synthetic data from both the original and perturbed feature vectors by a decoder, thus forming data pairs that emphasize learning key features for FGVR. Our approach outperforms existing methods across various datasets in many downstream tasks. While the proposed approach also offers a modest boost to SSL performance in non-fine-grained visual recognition tasks—as detailed in the Appendices—the gains are notably more substantial in FGVR contexts. The refinement of our methodology for application to large-scaled general datasets remains an avenue for future research works.

Acknowledgment. This material is based upon work supported by the National Science Foundation under Grant No. 1956313.

References

1. Berg, T., Liu, J., Woo Lee, S., Alexander, M.L., Jacobs, D.W., Belhumeur, P.N.: Birdsnap: large-scale fine-grained visual categorization of birds. In: Proceedings of the IEEE Conference on Computer Vision and Pattern Recognition, pp. 2011–2018 (2014)
2. Caron, M., Misra, I., Mairal, J., Goyal, P., Bojanowski, P., Joulin, A.: Unsupervised learning of visual features by contrasting cluster assignments. In: Advances in Neural Information Processing Systems, vol. 33, pp. 9912–9924 (2020)
3. Caron, M., Touvron, H., Misra, I., Jégou, H., Mairal, J., Bojanowski, P., Joulin, A.: Emerging properties in self-supervised vision transformers. In: Proceedings of the IEEE/CVF International Conference on Computer Vision, pp. 9650–9660 (2021)
4. Chen, T., Kornblith, S., Norouzi, M., Hinton, G.: A simple framework for contrastive learning of visual representations. In: International Conference on Machine Learning, pp. 1597–1607. PMLR (2020)
5. Chen, X., Fan, H., Girshick, R., He, K.: Improved baselines with momentum contrastive learning. arXiv preprint arXiv:2003.04297 (2020)
6. Chen, X., He, K.: Exploring simple siamese representation learning. In: Proceedings of the IEEE/CVF Conference on Computer Vision and Pattern Recognition, pp. 15750–15758 (2021)
7. Chuang, C.Y., Robinson, J., Lin, Y.C., Torralba, A., Jegelka, S.: Debiased contrastive learning. In: Advances in Neural Information Processing Systems, vol. 33, pp. 8765–8775 (2020)
8. Codella, N.C., et al.: Skin lesion analysis toward melanoma detection: a challenge at the 2017 international symposium on biomedical imaging (ISBI), hosted by the international skin imaging collaboration (ISIC). In: 2018 IEEE 15th International Symposium on Biomedical Imaging (ISBI 2018), pp. 168–172. IEEE (2018)
9. Cole, E., Yang, X., Wilber, K., Mac Aodha, O., Belongie, S.: When does contrastive visual representation learning work? In: Proceedings of the IEEE/CVF Conference on Computer Vision and Pattern Recognition, pp. 14755–14764 (2022)
10. Cosentino, R., et al.: Toward a geometrical understanding of self-supervised contrastive learning. arXiv preprint arXiv:2205.06926 (2022)
11. Deng, J., Dong, W., Socher, R., Li, L.J., Li, K., Fei-Fei, L.: ImageNet: a large-scale hierarchical image database. In: CVPR09 (2009)
12. Di Wu, S.L., et al.: Align yourself: self-supervised pre-training for fine-grained recognition via saliency alignment. arXiv preprint arXiv:2106.15788 (2021)

13. Ericsson, L., Gouk, H., Hospedales, T.M.: How well do self-supervised models transfer? In: Proceedings of the IEEE/CVF Conference on Computer Vision and Pattern Recognition, pp. 5414–5423 (2021)
14. Gao, Y., et al.: Disco: remedying self-supervised learning on lightweight models with distilled contrastive learning. In: Avidan, S., Brostow, G., Cissé, M., Farinella, G.M., Hassner, T. (eds.) European Conference on Computer Vision, pp. 237–253. Springer, Cham (2022). https://doi.org/10.1007/978-3-031-19809-0_14
15. Grill, J.B., et al.: Bootstrap your own latent-a new approach to self-supervised learning. In: Advances in Neural Information Processing Systems, vol. 33, pp. 21271–21284 (2020)
16. He, K., Fan, H., Wu, Y., Xie, S., Girshick, R.: Momentum contrast for unsupervised visual representation learning. In: Proceedings of the IEEE/CVF Conference on Computer Vision and Pattern Recognition, pp. 9729–9738 (2020)
17. He, K., Zhang, X., Ren, S., Sun, J.: Deep residual learning for image recognition. In: Proceedings of the IEEE Conference on Computer Vision and Pattern Recognition, pp. 770–778 (2016)
18. Houben, S., Stallkamp, J., Salmen, J., Schlipsing, M., Igel, C.: Detection of traffic signs in real-world images: the German traffic sign detection benchmark. In: International Joint Conference on Neural Networks, No. 1288 (2013)
19. Hua, T., Wang, W., Xue, Z., Ren, S., Wang, Y., Zhao, H.: On feature decorrelation in self-supervised learning. In: Proceedings of the IEEE/CVF International Conference on Computer Vision, pp. 9598–9608 (2021)
20. Huang, L., You, S., Zheng, M., Wang, F., Qian, C., Yamasaki, T.: Learning where to learn in cross-view self-supervised learning. In: Proceedings of the IEEE/CVF Conference on Computer Vision and Pattern Recognition, pp. 14451–14460 (2022)
21. Islam, A., et al.: A broad study on the transferability of visual representations with contrastive learning. In: Proceedings of the IEEE/CVF International Conference on Computer Vision, pp. 8845–8855 (2021)
22. Jang, Y.K., Cho, N.I.: Self-supervised product quantization for deep unsupervised image retrieval. In: Proceedings of the IEEE/CVF International Conference on Computer Vision, pp. 12085–12094 (2021)
23. Jing, L., Vincent, P., LeCun, Y., Tian, Y.: Understanding dimensional collapse in contrastive self-supervised learning. arXiv preprint arXiv:2110.09348 (2021)
24. Kim, S., Bae, S., Yun, S.Y.: Coreset sampling from open-set for fine-grained self-supervised learning. In: Proceedings of the IEEE/CVF Conference on Computer Vision and Pattern Recognition, pp. 7537–7547 (2023)
25. Krause, J., Stark, M., Deng, J., Fei-Fei, L.: 3D object representations for fine-grained categorization. In: Proceedings of the IEEE International Conference on Computer Vision Workshops, pp. 554–561 (2013)
26. Krizhevsky, A., Hinton, G.: Learning multiple layers of features from tiny images (2009)
27. Lee, H., Lee, K., Lee, K., Lee, H., Shin, J.: Improving transferability of representations via augmentation-aware self-supervision. In: Advances in Neural Information Processing Systems, vol. 34, pp. 17710–17722 (2021)
28. Li, A.C., Efros, A.A., Pathak, D.: Understanding collapse in non-contrastive siamese representation learning. In: Avidan, S., Brostow, G., Cissé, M., Farinella, G.M., Hassner, T. (eds.) European Conference on Computer Vision, pp. 490–505. Springer, Cham (2022). https://doi.org/10.1007/978-3-031-19821-2_28
29. Maji, S., Rahtu, E., Kannala, J., Blaschko, M., Vedaldi, A.: Fine-grained visual classification of aircraft. arXiv preprint arXiv:1306.5151 (2013)

30. Oord, A.V.D., Li, Y., Vinyals, O.: Representation learning with contrastive predictive coding. arXiv preprint arXiv:1807.03748 (2018)
31. Peng, X., Wang, K., Zhu, Z., Wang, M., You, Y.: Crafting better contrastive views for siamese representation learning. In: Proceedings of the IEEE/CVF Conference on Computer Vision and Pattern Recognition, pp. 16031–16040 (2022)
32. Russakovsky, O., et al.: ImageNet large scale visual recognition challenge. Int. J. Comput. Vision **115**(3), 211–252 (2015). https://doi.org/10.1007/s11263-015-0816-y
33. Selvaraju, R.R., Cogswell, M., Das, A., Vedantam, R., Parikh, D., Batra, D.: Grad-CAM: visual explanations from deep networks via gradient-based localization. In: Proceedings of the IEEE International Conference on Computer Vision, pp. 618–626 (2017)
34. Selvaraju, R.R., Desai, K., Johnson, J., Naik, N.: Casting your model: learning to localize improves self-supervised representations. In: Proceedings of the IEEE/CVF Conference on Computer Vision and Pattern Recognition, pp. 11058–11067 (2021)
35. Shu, Y., van den Hengel, A., Liu, L.: Learning common rationale to improve self-supervised representation for fine-grained visual recognition problems. In: Proceedings of the IEEE/CVF Conference on Computer Vision and Pattern Recognition, pp. 11392–11401 (2023)
36. Shu, Y., Yu, B., Xu, H., Liu, L.: Improving fine-grained visual recognition in low data regimes via self-boosting attention mechanism. In: Avidan, S., Brostow, G., Cissé, M., Farinella, G.M., Hassner, T. (eds.) European Conference on Computer Vision, pp. 449–465. Springer, Cham (2022). https://doi.org/10.1007/978-3-031-19806-9_26
37. Thomee, B., et al.: YFCC100M: the new data in multimedia research. Commun. ACM **59**(2), 64–73 (2016)
38. Wah, C., Branson, S., Welinder, P., Perona, P., Belongie, S.: The caltech-UCSD birds-200-2011 dataset (2011)
39. Wang, T., Isola, P.: Understanding contrastive representation learning through alignment and uniformity on the hypersphere. In: International Conference on Machine Learning, pp. 9929–9939. PMLR (2020)
40. Wang, Z., Wang, Y., Hu, H., Li, P.: Contrastive learning with consistent representations. arXiv preprint arXiv:2302.01541 (2023)
41. Xiao, T., Wang, X., Efros, A.A., Darrell, T.: What should not be contrastive in contrastive learning. arXiv preprint arXiv:2008.05659 (2020)
42. Yao, Y., Ye, C., He, J., Elsayed, G.F.: Teacher-generated spatial-attention labels boost robustness and accuracy of contrastive models. In: Proceedings of the IEEE/CVF Conference on Computer Vision and Pattern Recognition, pp. 23282–23291 (2023)
43. Yeh, C.H., Hong, C.Y., Hsu, Y.C., Liu, T.L.: Saga: self-augmentation with guided attention for representation learning. In: ICASSP 2022-2022 IEEE International Conference on Acoustics, Speech and Signal Processing (ICASSP), pp. 3463–3467. IEEE (2022)
44. Zhao, N., Wu, Z., Lau, R.W., Lin, S.: Distilling localization for self-supervised representation learning. In: Proceedings of the AAAI Conference on Artificial Intelligence, vol. 35, pp. 10990–10998 (2021)
45. Ziyin, L., Lubana, E.S., Ueda, M., Tanaka, H.: What shapes the loss landscape of self-supervised learning? arXiv preprint arXiv:2210.00638 (2022)

LaWa: Using Latent Space for In-Generation Image Watermarking

Ahmad Rezaei[1](✉), Mohammad Akbari[2], Saeed Ranjbar Alvar[2], Arezou Fatemi[2], and Yong Zhang[2]

[1] University of British Columbia, Vancouver, BC, Canada
ahnr@mail.ubc.ca
[2] Huawei Technologies Canada Co. Ltd., Markham, ON, Canada
{mohammad.akbari,saeed.ranjbar.alvar1,yong.zhang3}@huawei.com

Abstract. With generative models producing high quality images that are indistinguishable from real ones, there is growing concern regarding the malicious usage of AI-generated images. Imperceptible image watermarking is one viable solution towards such concerns. Prior watermarking methods map the image to a latent space for adding the watermark. Moreover, Latent Diffusion Models (LDM) generate the image in the latent space of a pre-trained autoencoder. We argue that this latent space can be used to integrate watermarking into the generation process. To this end, we present LaWa, an in-generation image watermarking method designed for LDMs. By using coarse-to-fine watermark embedding modules, LaWa modifies the latent space of pre-trained autoencoders and achieves high robustness against a wide range of image transformations while preserving perceptual quality of the image. We show that LaWa can also be used as a general image watermarking method. Through extensive experiments, we demonstrate that LaWa outperforms previous works in perceptual quality, robustness against attacks, and computational complexity, while having very low false positive rate. Code is available here.

Keywords: Image Watermarking · Responsible AI · Image Generation

1 Introduction

With rapid advancements in generative models, AI-generated content (AIGC) in different modalities including images [35,38–40], video [21,45], text [4,9,48], and 3D [29,37] can be generated with high quality. Text-to-image diffusion models such as Stable Diffusion [39] and DALL·E 2 [38] are open to public and have shown stunning performance in generation of photo-realistic images that are indistinguishable from real ones. Such tools can be misused in different ways such as faking AI-generated images as human-created artworks, generating fake

---

A. Rezaei—Work done during an internship at Huawei Technologies Canada Co. Ltd.

Supplementary Information The online version contains supplementary material available at https://doi.org/10.1007/978-3-031-73024-5_8.

© The Author(s), under exclusive license to Springer Nature Switzerland AG 2025
A. Leonardis et al. (Eds.): ECCV 2024, LNCS 15147, pp. 118–136, 2025.
https://doi.org/10.1007/978-3-031-73024-5_8

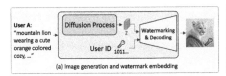 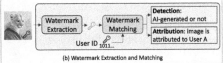

Fig. 1. An overview of in-generation image watermarking for LDMs. (a) generation and watermark (i.e., User ID) embedding. (b) Watermark extraction and matching for detection and attribution.

news, impersonation, and copyright infringement [5,50,67]. Such threats raise concerns about our confidence and trust in the authenticity of photo-realistic images. Thus, responsible implementation of generative AI services needs to be considered by service providers. Specifically, two problems including detection and attribution should be addressed. In detection, model developers detect if an image is generated by their model, i.e., if the image is AI-generated or not. In attribution, the developers attribute an image to the user who generated the image using their service.

Imperceptible image watermarking is a potential solution for the aforementioned problems. It involves embedding a message in the image without damaging its utility. A watermark extraction module is also required to accurately extract the embedded message from the watermarked image even if the it is modified. In the case of AI-generated images, the extracted message is used to detect such images and attribute them to the responsible user who created them (Fig. 1).

For AIGC images, existing watermarking methods can be applied after the images are generated (called post-generation methods in this paper) [14,60]. Such post-generation watermarking introduces an overhead on the generation service due to the extra watermark post-processing. The watermarking process generally includes embedding the watermark into a latent feature representation of an image [14,31,51,58]. The latent feature is either learned end-to-end in encoder-decoder-based watermarking methods [31,58] or it is extracted from the latent space of a pre-trained neural network [14,51]. In the case of Latent Diffusion Models (LDMs), generated images are in the latent space of a pre-trained autoencoder model. Thus, the latent features corresponding to the generation and watermarking procedures are de-coupled, which is a sub-optimal solution for combined image generation and watermarking problem.

In this paper, we propose LaWa, an in-generation watermarking method that effectively changes the latent feature of pre-trained LDMs to integrate watermarking into the generation process. Thus, generated images already conceal a watermark. Only a few prior works have addressed this problem [13,53]. Stable Signature [13] is a model watermarking method for the decoder of LDMs. Since it requires fine-tuning a new decoder model for each watermark message, its application for in-generation watermarking has high computational cost and limited scalability for an image generation service. Tree-Ring [53] watermarking modifies the initial noise vector used for image generation. It is only applicable to deterministic sampling [36] and requires image-to-noise inversion process for watermark extraction. Such process for text-guided diffusion does not always yield to the real noise vector [20,33].

To address these limitations, LaWa incorporates novel coarse-to-fine multi-scale embedding modules into the frozen intermediate layers of the LDM decoder to ensure robustness against severe geometrical attacks. LaWa achieves high-payload in-generation watermarking with only one decoder model. Thus, it can handle many users for an image generation service with low computational cost. The focus of LaWa is on black-box watermarking, where users do not have access to the image generation model. In comparison with post-generation watermarking, we show that using the same latent features for image generation and watermarking improves the quality vs. robustness trade-off. Moreover, we show that LaWa can convert a pre-trained autoencoder into a robust general image watermarking method. In summary, our contributions are as follows:

- A multi-scale latent modification mechanism for pre-trained generative autoencoders that is compatible with any autoencoder and is robust to a broad range of image modifications
- A simple yet effective spatial watermark coding that improves the trade-off between robustness and perceptual quality
- An in-generation image watermarking approach that can be used for any pre-trained LDM and any image generation task without further fine-tuning of the LDM
- Extending the proposed in-generation watermarking method to work as a general post-generation image watermarking technique
- Achieving state-of-the-art post- and in-generation results in perceptual quality, robustness, and computational complexity.

2 Related Work

Detection of AI-Generated Images. Considering the risks of AIGC, many works focus on the passive detection of generated/manipulated images. These methods are well-studied for deep-fakes using inconsistencies in generated images [7,17,28] as well as generator traces in the spatial [32,55] or frequency [15,64] domains. However, they have poor performance because they fall behind the rapid evolution of generative models. Similar approaches are proposed for diffusion models [10,42], but they are also shown to suffer from low detection accuracy and high false rate [13].

Image Watermarking. There are three main categories of image watermarking methods. Transform-based methods embed the watermark in a spatial [16,46] or frequency [22,23,34] domain, which can achieve great imperceptibility, but have low bit extraction robustness to even minor image modifications. Encoder-decoder-based methods use the encoder to concurrently create a latent feature and add the watermark before recreating the marked image [6,27,31,58,60,66]. Despite their good robustness, encoder-decoder networks may not generalize well

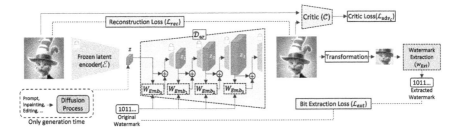

Fig. 2. Overall framework including the procedure of injecting the watermark embedding modules (W_{Emb}) to LDM's decoder ($\mathcal{D}$) as well as the end-to-end training process. During training, LDM's frozen encoder is used to generate latent z while at inference time, z is generated by the diffusion process.

to images out of the training data distribution and have strong trade-off between payload and utility. The third group uses the latent space of a fixed pre-trained network to add the watermark in several iterations [14,26,51,61], which substantially increases the watermarking time. RoSteALS proposes using latent space of autoencoders for steganography [6], but it does not consider image watermarking attack constraints. The proposed solution has no robustness to geometrical image modifications. Such limitation is critically important in the AIGC application scenario as users can easily evade the watermark.

In-Generation Image Watermarking. Previous works on integrating watermarking and image generation are mostly for model watermarking [49]. Some methods watermark the entire training dataset with one message [54,56,65] so that the output of the network carries the same message as well. Such methods are not extensible to new messages and require new training for every new message. Specifically for GANs, [57] proposes a solution to merge watermarking into the training procedure of the network. At the inference time, the generated images conceal a fingerprint specific to the user creating the image. This solution requires training the GAN model from scratch and is not applicable to pre-trained models.

In-generation watermarking for diffusion models is a recent research topic in the literature [13,53]. Tree-Ring watermarking [53] modifies the initial noise vector used in image generation. This method cannot conceal a specific bit string into the image and is only applicable to deterministic sampling [36] in the diffusion process. For watermark extraction, an inversion process should accurately map the watermarked image to the initial noise. [20,33] show that such inversion for text-guided diffusion does not always yield to the real noise vector, which can damage watermark extraction performance. Stable Signature [13] is a model watermarking method for the image decoder of a pre-trained LDM. If used for in-generation watermarking, a copy of the decoder model should be fine-tuned and stored for each watermark message, substantially increasing the computational cost and limiting the scalability of this solution.

3 Methodology

The overall framework of how LaWa embeds and extracts messages is shown in Fig. 1 (in-generation watermarking) and Fig. 4 (image watermarking). In the embedding phase, a latent is first generated by the diffusion process. Given the watermark message (i.e., user ID), LaWa enables the pre-trained LDM's latent decoder to simultaneously watermark and decode the latent, which results in generating the watermarked image. The watermark message can later be extracted using the extraction module to be used for detection and attribution by matching the message with one of existing watermarks.

Figure 2 illustrates the high-level procedure of modifying the LDM's decoder by injecting watermark embedding modules at intermediate layers of the decoder. The end-to-end training process of the added embedding and extraction networks is also shown, which will be described in details in the following subsections.

3.1 Modified Latent Decoder

In general, diffusion models are trained either in the image space [35,40] or in a compact latent space [18,38,39] for more computational efficiency. For LDMs, the output of the diffusion process is in this latent space. Besides the diffusion process, LDMs also include an image autoencoder. The image encoder, $\mathcal{E}$, is used to downsample the input image $x \in \mathbb{R}^{H \times W \times 3}$ by a factor of f to a latent feature, z, where $z = \mathcal{E}(x) \in \mathbb{R}^{H/f \times W/f \times C}$ [39]. The decoder model, $\mathcal{D}$, then upsamples z to create the reconstructed image $\hat{x} = \mathcal{D}(\mathcal{E}(x))$. At the generation time, the diffusion process generates the latent feature z, which is mapped to the pixel space using the decoder: $\hat{x} = \mathcal{D}(z)$. The upsampling in $\mathcal{D}$ is a multi-step process in which latent size is scaled up by a factor of 2 after each step.

To embed the watermark information, we feed the watermark message into the latent feature before each upsampling step of $\mathcal{D}$. This results in a multi-scale coarse-to-fine process of embedding the watermark into the latent space features. More specifically, assuming a k-bit watermark message $m \in \{0,1\}^k$, our watermark embedding module at ith upsampling step, W_{Emb_i}, accepts m and the latent z_i as input and generates a residual latent, δz_i, with the same size as $z_i : \delta z_i = W_{Emb_i}(z_i, m)$, where $i \in \left\{0, \ldots, \frac{f}{2} - 1\right\}$ shows the upsampling step of $\mathcal{D}$. The watermarked latent, z_{w_i} is then constructed as:

$$z_{w_i} = z_i + \delta z_i. \tag{1}$$

As illustrated in Fig. 3, W_{Emb_i} comprises of a linear layer that maps m to a noise block $b_i \in \mathbb{R}^{B \times B \times C_i}$. C_i is the latent channel size at the ith step and B is the height and width size of b_i that is kept constant for all upsampling steps. After b_i is obtained, we perform spatial watermark coding where b_i is repeated along the height and width dimensions to match the size of z_i. As a result of the applied spatial watermark coding, the watermark noise carries repeated watermark message with the idea to create sufficient redundancy in the noise to improve robustness to geometrical attacks while minimally damaging the

Fig. 3. Structure of the watermark embedding module W_{Emb_i}

Fig. 4. Diagram of general image watermarking process with LaWa

perceptual quality. The watermark noise is then passed through a convolution layer to generate δz_i. To ensure W_{Emb_i} does not initially change the latent feature, we follow [62] and initialize the weights and biases of the convolution layer to 0.

Unlike the original decoder $\mathcal{D}$ that reconstructs the image without watermark $\hat{x} = \mathcal{D}(z)$, the modified latent decoder $\mathcal{D}_w$ generates the watermarked image $\hat{x}_w = \mathcal{D}_w(z)$.

3.2 Training

In the training process, the LDM's pre-trained image encoder as well as the original layers of the decoder are kept frozen. This ensures that training the watermark embedding modules do not degrade the image generation quality.

To ensure the perceptual similarity of the watermarked image $\hat{x}_w$ and the original generated image $\hat{x}$, we use a combination of pixel-wise distortion and perceptual loss functions. For the distortion loss, MSE (l_2 distance) loss defined as $\mathcal{L}_I = \|\hat{x}_w - \hat{x}\|^2$ is used. For the perceptual loss function, we employ LPIPS loss [63] to minimize the perceptual distortion between $\hat{x}_w$ and $\hat{x}$.

To further improve the quality of the watermarked image, an adversarial critic network $\mathcal{C}$ using Wasserstein GAN [2] is also implemented. Thus, the corresponding adversarial training includes two loss functions as follows:

$$\mathcal{L}_{adv\mathcal{D}_w} = -\mathbb{E}_{\hat{x}_w \sim P_{\hat{x}_w}}[\mathcal{C}(\hat{x}_w)], \tag{2}$$

$$\mathcal{L}_{adv\mathcal{C}} = \mathbb{E}_{\hat{x}_w \sim P_{\hat{x}_w}}[\mathcal{C}(\hat{x}_w)] - \mathbb{E}_{\hat{x} \sim P_{\hat{x}}}[\mathcal{C}(\hat{x})], \tag{3}$$

where $\mathcal{L}_{adv\mathcal{C}}$ and $\mathcal{L}_{adv\mathcal{D}_w}$ are the adversarial losses for the critic $\mathcal{C}$ and the modified decoder $\mathcal{D}_w$, respectively. $P_{\hat{x}_w}$ and $P_{\hat{x}}$ are the distribution of the watermarked and original images. The overall reconstruction loss is then defined as:

$$\mathcal{L}_{rec} = \lambda_I \mathcal{L}_I + \lambda_{LPIPS} \mathcal{L}_{LPIPS}(\hat{x}_w, \hat{x}) + \lambda_{adv} \mathcal{L}_{adv\mathcal{D}_w}, \tag{4}$$

where λ_I, λ_{LPIPS}, and λ_{adv} are the relative loss weights.

Our watermark extractor, denoted by W_{Ext}, should be robust to different image modifications and attacks. As a result, after the watermarked image is generated, a transformation T is randomly sampled from a set of differentiable transformations, $\mathcal{T}$, including common image processing attacks (see supplementary material for more details). For the differentiability issue of JPEG compression, we used the forward simulation attack introduced in [59]. The selected transformation is then applied to the image and the transformed image is passed through the watermark extraction network, which is based on ResNet50 [19] architecture with the last linear layer changed to output a k-bit message. The output extracted message, $\hat{m} = W_{Ext}(\mathcal{T}(\hat{x}))$, should match the original embedded message m. We define the extraction loss as the binary cross entropy (BCE) loss between m and $\hat{m}$, $\mathcal{L}_{ext} = \mathcal{L}_{BCE}(m, \hat{m})$. The final watermark embedding and extraction networks are then optimized to minimize the following total loss:

$$\mathcal{L} = \mathcal{L}_{rec} + \lambda \mathcal{L}_{ext}, \tag{5}$$

where λ is the loss weight to control the trade-off between extraction accuracy and image reconstruction quality.

3.3 Image Watermarking

Figure 4 shows the general image watermarking process. Corresponding latent feature z of the cover image x is calculated using the frozen encoder, $z = \mathcal{E}(x)$. The feature z is separately processed by $\mathcal{D}$ and $\mathcal{D}_w$ to create a watermark mask that is computed as $\mathcal{D}(z) - \mathcal{D}_w(z)$. Using this watermark mask, the cover image is then watermarked as $\hat{x}_w = x + \mathcal{D}_w(z) - \mathcal{D}(z)$.

3.4 Watermark Matching

Given an image at the watermark extraction time, our watermark extractor W_{Ext} tries to decode the watermark message $\hat{m}$. To address the detection and attribution problems, $\hat{m}$ needs to be matched with one of existing original messages (i.e., m) stored in the database. Following [13,56] and to increase the matching robustness against image modifications, we use a soft matching method defined as:

$$M(\hat{m}, m), \geq n \quad n \in \{0, \ldots, k\}, \tag{6}$$

where $M(\hat{m}, m)$ counts the number of matching bits in $\hat{m}$ and m, and n is the threshold for flagging a match.

For a successful watermark matching, the rate of false detection and attribution should be at an acceptable level. Potential errors include detecting a vanilla image as watermarked (detection FPR), missing a watermarked image (detection false negative rate), and attributing an image to the wrong user (attribution FPR). Having low error rate is an important practical aspect of this problem.

Assuming that extracted bits from vanilla images are i.i.d Bernoulli random variables with parameter 0.5, the theoretical upper bound of FPR is as follows:

$$FPR_{det}(n) = Pr\left(M(\tilde{m}, m) > n \mid H_0\right) = \sum_{i=n+1}^{k} \binom{k}{i} 0.5^k, \tag{7}$$

$$FPR_{att}(N, n) = 1 - (1 - FPR_{det}(n))^N, \tag{8}$$

where FPR_{det} is the detection FPR, FPR_{att} is the attribution FPR, N is the total number of users, $\tilde{m}$ is any random message, and H_0 is the null hypothesis: "any random message, $\tilde{m}$, match with m in more than n bits". Clearly, the more the number of users, the higher the attribution FPR. By setting the FPR_{att} and total number of users to a desired level, we can obtain the minimum required n for watermark matching by using Eqs. (7) and (8) reversely. More information is provided in the supplementary material.

4 Experiments

In this section, the performance of the proposed method is quantitatively and qualitatively evaluated and compared with the previous works. We also study the generality of LaWa for different image generation tasks. A false positive analysis along with an extensive test for the trade-off between capacity, quality, and robustness is also presented. At the end, we will provide an ablation study to verify the effect of each loss term in our objective function.

4.1 Experimental Settings

To train LaWa's watermark embedding and extraction modules, we utilize 100K images from MIRFlickR dataset [24], where the images are randomly cropped to 256 × 256 resolution. The training is performed with an AdamW optimizer with the learning rate of $6e-5$, 40 epochs, and the batch size of 8. For all our experiments, we set the weights in Eqs. (4) and (5) to $\lambda_I = 0.1$, $\lambda_{LPIPS} = 1.0$, $\lambda_{adv} = 1.0$, and $\lambda = 2.0$, which are obtained experimentally. For the main experiments, we use the KL-f8 auto-encoder of Stable Diffusion [39] and 48-bit messages to train the LaWa method unless mentioned otherwise.

To numerically evaluate the distortion of the watermarked and original generated images, we use Peak Signal-to-Noise Ratio (PSNR) and Structural Similarity Index Measure (SSIM) [52]. Moreover, we evaluate the perceptual quality of images using LPIPS similarity score [63] and Single Image Frechet Inception Distance (SIFID) [43].

4.2 Comparison Results

We compare the performance of LaWa with previous in-generation and image watermarking methods using AI-generated and natural image datasets. For AI-generated dataset, we consider using LaWa as both in-generation and post-generation watermarking methods. For the baseline of in-generation methods, we used Stable Signature (SS) [13], which is a model watermarking method for in-generation image watermarking by creating different copies of the decoder. Alternatively, image watermarking methods can be used to watermark the generated images in a post-generation manner. We provide the comparison results with different post-generation techniques including encoder-decoder based (HiDDeN [66], RivaGAN [60], and RoSteALS [6]), optimization-based (SSL [14] and

Table 1. Comparison results of LaWa with existing post- and in-generation image watermarking methods in terms of quality, embedding time, and robustness against various attacks. *****: in-generation methods.

Method (bit#)	Image quality		Emb. time (ms)	Bit accuracy ↑										Ave.
	PSNR/SSIM ↑	LPIPS/SIFID ↓		None	C. Crop 0.1	R. Crop 0.1	Resize 0.7	Rot. 15	Blur	Contr. 2.0	Bright. 2.0	JPEG 70	Comb.	
AI-generated Images														
DCT-DWT (32) [11]	39.47/0.97	0.03/0.02	139	0.91	0.51	0.51	0.52	0.52	0.51	0.52	0.51	0.51	0.51	0.55
Hidden (30) [25]	32.59/0.95	0.03/0.05	17	0.91	0.91	0.91	0.82	0.79	0.76	0.75	0.74	0.53	0.59	0.77
SSL (32) [14]	33.23/0.89	0.15/0.28	870	**1.00**	0.74	0.72	0.99	0.99	**1.00**	0.96	0.95	0.99	0.85	0.92
RoSteALS (32) [6]	29.31/0.93	0.04/0.06	144	**1.00**	0.50	0.51	**1.00**	0.47	**1.00**	0.88	0.86	**1.00**	0.50	0.77
RivaGan (32) [60]	**40.53/0.98**	0.03/0.04	57	0.99	0.98	**0.98**	0.87	0.91	0.99	0.81	0.80	0.98	0.93	0.92
LaWa* (32)	34.25/0.89	**0.03/0.01**	1	**1.00**	**1.00**	**0.98**	**1.00**	**1.00**	**1.00**	**1.00**	**1.00**	**1.00**	0.97	**0.99**
LaWa-post-gen (32)	34.28/0.90	0.03/0.02	33	**1.00**	**1.00**	0.96	**1.00**	**1.00**	**1.00**	**1.00**	**1.00**	**1.00**	**0.98**	**0.99**
SSL (48) [14]	33.25/0.89	0.15/0.27	870	**1.00**	0.72	0.70	0.99	**0.99**	**1.00**	0.94	0.94	0.99	0.82	0.91
FNNS (48) [26]	**36.84/0.97**	0.04/0.05	645	**1.00**	**0.98**	**0.96**	0.95	0.75	0.54	0.86	0.85	0.94	0.91	0.87
RoSteALS (48) [6]	29.30/0.92	0.05/0.07	144	**1.00**	0.51	0.50	**1.00**	0.50	**1.00**	0.88	0.85	**1.00**	0.49	0.78
SS* (48) [13]	32.02/0.84	0.05/0.09	0	0.99	0.95	0.93	0.96	0.81	0.78	0.97	0.96	0.92	0.92	0.92
LaWa* (48)	33.52/0.86	**0.04/0.02**	1	**1.00**	0.95	0.91	0.99	0.96	0.99	**1.00**	**1.00**	**1.00**	0.94	**0.97**
LaWa-post-gen (48)	33.45/0.87	**0.04/0.02**	34	**1.00**	0.94	0.90	0.96	0.97	0.99	**1.00**	**1.00**	**1.00**	**0.95**	**0.97**
CLIC														
DCT-DWT (32) [11]	38.94/0.97	**0.02/0.02**	230	0.87	0.51	0.51	0.51	0.51	0.52	0.50	0.50	0.52	0.50	0.55
Hidden (30) [25]	33.33/0.96	0.04/0.04	28	0.89	0.89	0.88	0.86	0.76	0.64	0.76	0.74	0.54	0.58	0.75
SSL (32) [14]	35.03/0.92	0.16/0.19	1261	**1.00**	0.86	0.80	**1.00**	**1.00**	**1.00**	0.96	0.93	0.98	0.90	0.94
RoSteALS (32) [6]	28.22/0.91	0.08/0.04	240	**1.00**	0.50	0.50	**1.00**	0.51	**1.00**	0.90	0.87	**1.00**	0.50	0.78
RivaGan (32) [60]	**42.04/0.98**	0.04/0.06	84	**1.00**	0.98	0.96	0.99	0.95	0.99	0.89	0.87	0.99	0.96	0.96
LaWa (32)	36.19/0.94	0.05/0.03	32	**1.00**	**1.00**	**0.97**	0.98	**1.00**	**1.00**	**1.00**	0.99	**1.00**	**0.99**	**0.99**
SSL (48) [14]	35.03/0.92	0.16/0.19	1261	**1.00**	0.83	0.78	**1.00**	**1.00**	**1.00**	0.96	0.92	0.98	0.88	0.93
FNNS (48) [26]	35.54/0.96	0.07/0.20	1116	**1.00**	**0.99**	0.95	0.94	0.75	0.51	0.85	0.82	0.91	0.89	0.86
RoSteALS (48) [6]	27.83/0.91	0.08/**0.04**	240	**1.00**	0.50	0.50	**1.00**	0.48	**1.00**	0.82	0.80	**1.00**	0.50	0.76
LaWa (48)	**35.58**/0.92	0.05/0.05	34	**1.00**	**1.00**	**0.97**	0.98	**0.98**	0.98	**1.00**	0.99	0.98	**0.90**	**0.98**

FNNS [26]), and frequency-based (DCT-DWT [11]) methods. For HiDDeN and RivaGAN, 30-bit and 32-bit pre-trained weights in [25] and [44] are respectively used. For RoSteALS, we trained their model for 48- and 32-bit message lengths with the same dataset and attacks as in LaWa. For SSL and FNNS, we used their publicly available codes and their pre-trained modules to obtain the results.

The upper portion of Table 1 summarizes LaWa's results with 32- and 48-bit watermark messages compared to the baselines in terms of the watermarked image quality as well as the bit extraction accuracy after different attacks. The results are obtained by watermarking 1K generated images (with 1K text prompts obtained from MS-COCO [30] and MagicPrompt-SD [41] datasets). Resolution of generated images is set to 512 × 512. Each generated image is watermarked with 10 different messages and the average bit accuracy is reported. Combined attack includes a series of 40% area center cropping, brightness 2.0, and JPEG 80. LaWa* and LaWa-post-gen refer to the cases where the proposed method is used as in-generation and post-generation watermarking, respectively.

As shown in Table 1, LaWa achieves the best perceptual quality for AI-generated images with 32- and 48-bit versions in LPIPS and SIFID scores. For PSNR and SSIM, LaWa has comparable performance with other methods, while RivaGAN and FNNS outperform the others for 32- and 48-bit watermarks, respectively. For robustness to attacks, LaWa has the best bit accuracy for all attacks for 32-bit watermarks. For 48-bit watermarks, except for 10% random crop attack where FNNS works better (90% bit accuracy of LaWa compared to 96% of FNNS), LaWa has better or comparable bit accuracy. Overall, LaWa

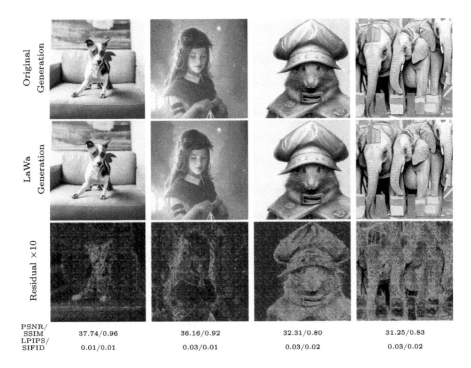

Fig. 5. Qualitative examples of generated images using LaWa, original generated images, and residual images (multiplied by 10). Extensive added noise by LaWa in high frequency textures reduces the PSNR/SSIM without damaging the perceptual quality (LPIPS/SIFID).

outperforms other methods in more number of attacks, specifically for the combined attack. Moreover, LaWa outperforms SS (as an in-generation method) in both image quality and robustness to a variety of attacks. More detailed set of results with different attack parameters is given in the supplementary material. When LaWa is used as a post-generation watermarking method (i.e., decoupled from the generation), the extraction accuracy is nearly unchanged compared to in-generation watermarking.

The computational complexity of all methods compared to ours in terms of averaged embedding time is also given in Table 1. Overall, it is seen that post-generation methods result in a significant overhead for watermarking the generated images, especially the iteration-based ones such as SSL and FNNS. On the other hand, the in-generation methods such as SS and ours provide near zero overhead as no post-processing is needed to add the watermark.

We also compare the performance of LaWa in general image watermarking against baseline methods using a natural (i.e., not AI-generated) dataset. Following earlier works in [6,14], CLIC dataset [47] is used, which has 428 test images with different resolutions. For all images, while maintaining the aspect ratio, larger dimension of each image is resized to 1024 to avoid memory limitation issue for some of the baselines. The results with CLIC dataset are given in the bottom part of Table 1. LaWa-post-gen achieves the highest bit accuracy for almost all the attacks and comparable or better visual quality compared to the baselines. Qualitative results for CLIC are in the supplementary material.

4.3 Qualitative Analysis

Figure 5 shows visual examples of 4 images generated by the original and modified decoders (i.e., original generated vs. watermarked images). The residual images (multiplied by 10) and the numerical quality metrics including PSNR, SSIM, LPIPS, and SIFID are also given for each pair.

Comparing the watermarked and original images, although the calculated PSNR is not very high, it is visually not possible to differentiate them as no visual artifact is evident. This is expected as we trained LaWa focusing on the perceptual quality of the watermarked images while giving lower weight to the MSE loss, which directly affects the PSNR. From the examples in Fig. 5, it is clear that our method has learned to add more noise to the high frequency textures, especially for the generated images with more texture (e.g., the elephants). In practice, when using in-generation watermarking, there is no original image available for comparison as all generated images conceal the watermark. This relaxes the requirement for watermarked images to have a high pixel-wise similarity to the original image. Therefore, we argue that if generated images have no visual artifact, watermark addition is successful even if PSNR is not very high. More qualitative results compared to the previous works are presented in the supplementary material.

4.4 Capacity, Quality, and Robustness Trade-Off

Effect of Number of Bits. We study the effect of the number of bits on generation quality and robustness in Table 2. LaWa can hide 64- and 96-bit watermark messages with minor drop in image quality, while achieving near 100% bit accuracy with no attack. With 128-bit messages, minor drop in the quality and bit accuracy against attack is observed compared to embedding shorter messages. Using higher number of bits can further improve the FPR of LaWa for its application in image generation services, which are designed to support a large number of users.

Table 2. Effect of the number of bits on generation quality and bit extraction accuracy.

Bit length	PSNR/SSIM	LPIPS/SIFID	None	C. Cro 0.1	Contr. 2.0	Rot. 15	JPEG 70	Comb.
32	34.25/0.89	0.03/0.01	1.00	1.00	1.00	1.00	1.00	0.97
48	33.52/0.86	0.04/0.02	1.00	0.95	1.00	0.96	1.00	0.94
64	33.11/0.83	0.04/0.02	1.00	0.93	0.99	0.92	0.96	0.95
96	32.82/0.83	0.04/0.02	0.99	0.87	0.98	0.77	0.92	0.95
128	31.87/0.82	0.08/0.09	0.95	0.93	0.95	0.84	0.92	0.87

Quality-Robustness Trade-Off. We can trade the generation quality for higher bit accuracy by changing λ in Eq. (5). Table 3 presents the performance

Table 3. Trade-off between image quality and bit accuracy. λ: trade-off weight in Eq. (5).

λ	1.0	1.7	2.0	4.0	7.0	10.0
PSNR ↑	34.76	33.84	33.52	33.06	30.56	28.47
Bit acc. on comb. ↑	0.65	0.75	0.94	0.96	0.97	0.99

Table 4. Ablation study on the effect of image quality loss weights.

λ_I	0.1	0.5	1.0
PSNR/SSIM ↑	33.52/0.86	33.60/0.88	**33.94/0.90**
LPIPS/SIFID ↓	**0.04/0.02**	0.05/0.04	0.06/0.05
Bit acc. on None/comb. ↑	**1.00/0.94**	1.00/0.88	0.99/0.84

of LaWa when trained with different λs. The corresponding PSNR and bit accuracy (after applying combined attack) on 1K images is reported. As observed, by increasing λ, the extraction robustness is improved, but the quality of the watermarked images drops.

4.5 LaWa's Performance for Different Tasks

In this sub-section, the generality of LaWa for different image generation tasks [39] including text-to-image, inpainting, super-resolution, and image editing is studied. The corresponding performance results are given in Table 5. The results are obtained for 1K images generated/watermarked with a separate random watermark message of 48 bits (see the supplementary material for more details). As summarized in the Table 5, the achieved bit accuracy averaged over all the tasks is higher than 97%, while the perceptual quality is preserved. This shows the applicability and generality of our proposed method to different generation tasks.

4.6 False Positive Analysis

The false positive analysis of LaWa's performance for both detection and attribution problems is given in this sub-section. As discussed in Sect. 3.4, in the detection problem, we only aim to detect if an image is watermarked while having a low FPR (detecting vanilla images as watermarked). On the other hand, in the attribution problem, we aim to find the user who created the image without making a false attribution to the wrong user (false attribution).

Detection. For detection, we use 1K watermarked images with a single 48-bit message to study the effect of the threshold n used for soft watermark matching on FPR_{det}. For each $n \in \{0, ..., 48\}$, we use Eq. (7) to calculate FPR_{det}, while TPR_{det} is experimentally obtained for the 1K images under various attacks. We experimentally validate the theoretical FPR_{det} in the supplementary material.

Figure 6a shows LaWa performance at different FPR_{det}. Under no attack, brightness 2.0, and JPEG 60, LaWa successfully detects 99% of the images while only flagging 1 out of 10^{15} vanilla images as watermarked. For more severe attacks such as crop 10% and combined attacks, there is a clear trade-off between detection accuracy and FPR_{det}. LaWa still detects more than 95% of the cropped and 83% of the combined-attacked images while only flagging 1 out

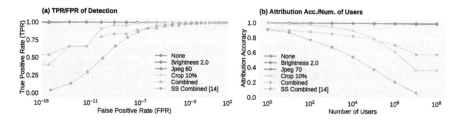

Fig. 6. Watermark matching performance in detection and attribution. (a) Detection TPR/FPR for different number of matching bits under different attacks. (b) Accuracy of attributing the watermarked image to the right user under different attacks and number of users assuming the detection FPR of 10^{-6}.

Table 5. Performance of LaWa in different image generation tasks in terms of quality and robustness.

Task	PSNR/SSIM	LPIPS/SIFID	C. Crop 0.1	Contr. 2.0	Rot. 15	JPEG 70
Text-to-Image	33.52/0.86	0.04/0.02	0.95	1.00	0.96	1.00
Inpainting	33.68/0.86	0.04/0.02	0.94	1.00	0.97	0.99
Super-Resolution	35.40/0.89	0.04/0.01	0.96	0.99	0.96	0.98
Image Editing	34.22/0.88	0.04/0.01	0.95	1.00	0.96	0.99

Table 6. Ablation study on the effect of each loss term on image quality and bit extraction robustness.

MSE	LPIPS	Critic	PSNR/SSIM	LPIPS/SIFID	Bit Acc. Comb.
✓			31.26/0.82	0.22/0.52	0.96
	✓		28.83/0.36	0.05/0.41	0.93
✓	✓		29.47/0.47	0.04/0.31	0.85
✓		✓	32.29/0.80	0.10/0.05	**0.99**
	✓	✓	33.05/0.85	0.05/0.02	0.73
✓	✓	✓	**33.52/0.86**	**0.04/0.02**	0.94

of 10^9 vanilla images. We also compare LaWa with Stable Signature (SS) [13] under combined attack and show that LaWa has a much higher TPR.

Attribution. For the attribution problem, we can use Eqs. (7) and (8) reversely to find the number of possible users for any FPR_{att}. For example, for $FPR_{att} = 10^{-6}$ in Eq. (8), we can calculate the required FPR_{det} for any number of users N. We further use the computed FPR_{det} and Eq. (7) to find the necessary watermark matching threshold n. For example, for $FPR_{att} = 10^{-6}$ and $N = 10^5$, we should use $n = 45$ as the threshold of soft watermark matching for 48-bit watermark messages.

We study LaWa's attribution accuracy for each N with 50K different 48-bit messages. We generate 2 images per message resulting in a total of 100K images. For $FPR_{att} = 10^{-6}$ and each N, the required threshold n is computed. Using this n, we perform watermark matching between the extracted message from each image and all the other messages. The match with the highest score above n is considered true positive (TP) if it corresponds to the correct user. Accuracy is calculated as $TP/100k$. For Ns larger than 50k, we assume some users have not generated any image.

Figure 6b shows the attribution accuracy of our method for different attacks. Under no attack, brightness 2.0, and JPEG 70, LaWa attributes 99% of the generated images to the correct user among 10^6 total users. Accuracy decreases to less than 70% and 60% for more severe combined and 10% crop attacks, respectively. Compared to Stable Signature (SS) [13], LaWa achieves a better performance with 20% accuracy for the combined attack. Moreover, LaWa has zero false attribution when $N = 10^6$, which means no image was attributed to the wrong user. This is expected as the watermark matching threshold is really high at $n = 47$. Overall, as we increase the number of users, the probability of false accusations also increases.

4.7 Learning-Based Attacks

We study the impact of learning-based watermark removal attacks, where the watermarked image is passed through an auto-encoder model [1, 3,8,12,39] as an attempt to remove the watermark. Figure 7 shows the robustness of LaWa compared to Stable Signature (SS) [13] against several auto-encoder models including Bmshj2018 [3], Cheng2020 [8], KL-VAE [39], and VQ-VAE [12].

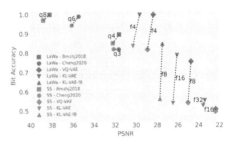

Fig. 7. Effect of the learning-based removal attacks. Results are reported for (SS) [13] and LaWa. q, f: quality/downsampling factors for each attack.

Each type of attack is applied using different quality/downsampling factors used in the auto-encoders. We observe that LaWa is quite robust against the attacks performed by Bmshj2018 with q8, Cheng2020 with q6, KL-VAE with f4, and VQ-VAE with f4. As the attack gets stronger (i.e., images are more compressed with larger quality or downsampling factors), both the bit accuracy and the PSNR reduce. Meaning if the attack is strong enough to remove the watermark, the quality of the watermarked image will inevitably experience a significant degradation (i.e., PSNR<24). It is also observed that even when the auto-encoder used in the attack is the same as the one we used to generate/watermark the images (i.e., KL-VAE f8), the attack will not erase the watermark and the bit accuracy remains as high as 85%. In contrast, the bit accuracy of SS [13] drops to 55%. It should be noted that unlike [13], we calculate the PSNR between the watermarked and the attacked images which is a more accurate representation of the attack's effect.

4.8 Ablation Study

Loss Terms. Table 6 shows the effect of each loss used in Eq. (4). We observe that the Critic network helps to improve the visual quality of the images. Using MSE and Critic losses achieves the best bit accuracy, but this setting leads to low perceptual quality and high visual artifacts in the image. By adding LPIPS loss, such visual artifacts are removed from the generated images.

Loss Weights. Table 4 shows the effect of changing loss weights in Eq. (4). We did this study by fixing both λ_{LPIPS} and λ_{adv} to 1.0 and changing λ_I. We observe that $\lambda_I > 0.1$ provides minor improvement in PSNR/SSIM, but damages LPIPS/SIFID and LaWa's robustness. More ablation studies over other components of LaWa along with qualitative results are given in the supplementary material.

5 Conclusion

In this paper, we proposed an in-generation image watermarking solution, which is applicable to LDMs and any image generation task. We showed that by injecting the watermark embedding modules into the intermediate layers of the frozen decoder, these modules can learn the decoder-specific mapping of latent space to RGB images. Thus, the watermark message is concealed in the generated images in a way that is imperceptible and robust to image modifications. Our solution is scalable to support a large number of users only with one single decoder model. With the fast evolution of generative AI models and its corresponding risks, model developers are expected to actively help with the safe adaptation of these models. Our work is a step towards responsible AI and we hope it can help model developers to implement their generative AI service while they have countermeasures in place to limit any improper use.

References

1. Akbari, M., Liang, J., Han, J., Tu, C.: Learned bi-resolution image coding using generalized octave convolutions. In: Proceedings of the AAAI Conference on Artificial Intelligence, vol. 35, pp. 6592–6599 (2021)
2. Arjovsky, M., Chintala, S., Bottou, L.: Wasserstein GAN. arXiv:1701.07875 (2017)
3. Ballé, J., Minnen, D., Singh, S., Hwang, S.J., Johnston, N.: Variational image compression with a scale hyperprior. arXiv:1802.01436 (2018)
4. Brown, T., et al.: Language models are few-shot learners. In: Advances in Neural Information Processing Systems, vol. 33, pp. 1877–1901 (2020)
5. Brundage, M., et al.: The malicious use of artificial intelligence: forecasting, prevention, and mitigation. arXiv preprint arXiv:1802.07228 (2018)
6. Bui, T., Agarwal, S., Yu, N., Collomosse, J.: Rosteals: robust steganography using autoencoder latent space. In: Proceedings of the IEEE/CVF Conference on Computer Vision and Pattern Recognition, pp. 933–942 (2023)
7. Chai, L., Bau, D., Lim, S.-N., Isola, P.: What makes fake images detectable? understanding properties that generalize. In: Vedaldi, A., Bischof, H., Brox, T., Frahm, J.-M. (eds.) ECCV 2020. LNCS, vol. 12371, pp. 103–120. Springer, Cham (2020). https://doi.org/10.1007/978-3-030-58574-7_7
8. Cheng, Z., Sun, H., Takeuchi, M., Katto, J.: Learned image compression with discretized gaussian mixture likelihoods and attention modules. In: Proceedings of the IEEE Conference on Computer Vision and Pattern Recognition (CVPR) (2020)
9. Chowdhery, A., et al.: Palm: scaling language modeling with pathways. arXiv preprint arXiv:2204.02311 (2022)

10. Corvi, R., Cozzolino, D., Zingarini, G., Poggi, G., Nagano, K., Verdoliva, L.: On the detection of synthetic images generated by diffusion models. In: ICASSP 2023-2023 IEEE International Conference on Acoustics, Speech and Signal Processing (ICASSP), pp. 1–5. IEEE (2023)
11. Cox, I., Miller, M., Bloom, J., Fridrich, J., Kalker, T.: Digital Watermarking and Steganography. Morgan Kaufmann (2007)
12. Esser, P., Rombach, R., Ommer, B.: Taming transformers for high-resolution image synthesis. arXiv:2012.09841 (2021)
13. Fernandez, P., Couairon, G., Jégou, H., Douze, M., Furon, T.: The stable signature: rooting watermarks in latent diffusion models. In: ICCV (2023)
14. Fernandez, P., Sablayrolles, A., Furon, T., Jégou, H., Douze, M.: Watermarking images in self-supervised latent spaces. In: ICASSP 2022-2022 IEEE International Conference on Acoustics, Speech and Signal Processing (ICASSP), pp. 3054–3058. IEEE (2022)
15. Frank, J., Eisenhofer, T., Schönherr, L., Fischer, A., Kolossa, D., Holz, T.: Leveraging frequency analysis for deep fake image recognition. In: International Conference on Machine Learning, pp. 3247–3258. PMLR (2020)
16. Ghazanfari, K., Ghaemmaghami, S., Khosravi, S.R.: LSB++: an improvement to LSB+ steganography. In: TENCON 2011-2011 IEEE Region 10 Conference, pp. 364–368. IEEE (2011)
17. Gragnaniello, D., Cozzolino, D., Marra, F., Poggi, G., Verdoliva, L.: Are GAN generated images easy to detect? a critical analysis of the state-of-the-art. In: 2021 IEEE international conference on multimedia and expo (ICME), pp. 1–6. IEEE (2021)
18. Gu, S., et al.: Vector quantized diffusion model for text-to-image synthesis. In: Proceedings of the IEEE/CVF Conference on Computer Vision and Pattern Recognition, pp. 10696–10706 (2022)
19. He, K., Zhang, X., Ren, S., Sun, J.: Deep residual learning for image recognition. In: Proceedings of the IEEE Conference on Computer Vision and Pattern Recognition, pp. 770–778 (2016)
20. Hertz, A., et al.: Prompt-to-prompt image editing with cross attention control. arXiv preprint arXiv:2208.01626 (2022)
21. Ho, J., et al.: Imagen video: high definition video generation with diffusion models. arXiv preprint arXiv:2210.02303 (2022)
22. Holub, V., Fridrich, J.: Designing steganographic distortion using directional filters. In: 2012 IEEE International Workshop on Information Forensics and Security (WIFS), pp. 234–239. IEEE (2012)
23. Holub, V., Fridrich, J., Denemark, T.: Universal distortion function for steganography in an arbitrary domain. EURASIP J. Inf. Secur. **2014**(1), 1–13 (2014). https://doi.org/10.1186/1687-417X-2014-1
24. Huiskes, M.J., Lew, M.S.: The MIR flickr retrieval evaluation. In: MIR '08: Proceedings of the 2008 ACM International Conference on Multimedia Information Retrieval. ACM, New York, NY, USA (2008)
25. Khachatryan, A.: Hidden (2019). https://github.com/ando-khachatryan/HiDDeN
26. Kishore, V., Chen, X., Wang, Y., Li, B., Weinberger, K.Q.: Fixed neural network steganography: train the images, not the network. In: International Conference on Learning Representations (2021)
27. Lee, J.E., Seo, Y.H., Kim, D.W.: Convolutional neural network-based digital image watermarking adaptive to the resolution of image and watermark. Appl. Sci. **10**(19), 6854 (2020)

28. Li, Y., Lyu, S.: Exposing deepfake videos by detecting face warping artifacts. arXiv preprint arXiv:1811.00656 (2018)
29. Lin, C.H., et al.: Magic3D: high-resolution text-to-3D content creation. In: Proceedings of the IEEE/CVF Conference on Computer Vision and Pattern Recognition, pp. 300–309 (2023)
30. Lin, T.-Y., et al.: Microsoft COCO: common objects in context. In: Fleet, D., Pajdla, T., Schiele, B., Tuytelaars, T. (eds.) ECCV 2014. LNCS, vol. 8693, pp. 740–755. Springer, Cham (2014). https://doi.org/10.1007/978-3-319-10602-1_48
31. Luo, X., Zhan, R., Chang, H., Yang, F., Milanfar, P.: Distortion agnostic deep watermarking. In: Proceedings of the IEEE/CVF Conference on Computer Vision and Pattern Recognition, pp. 13548–13557 (2020)
32. Marra, F., Gragnaniello, D., Verdoliva, L., Poggi, G.: Do GANs leave artificial fingerprints? In: 2019 IEEE Conference on Multimedia Information Processing and Retrieval (MIPR), pp. 506–511. IEEE (2019)
33. Mokady, R., Hertz, A., Aberman, K., Pritch, Y., Cohen-Or, D.: Null-text inversion for editing real images using guided diffusion models. In: Proceedings of the IEEE/CVF Conference on Computer Vision and Pattern Recognition, pp. 6038–6047 (2023)
34. Navas, K., Ajay, M.C., Lekshmi, M., Archana, T.S., Sasikumar, M.: DWT-DCT-SVD based watermarking. In: 2008 3rd International Conference on Communication Systems Software and Middleware and Workshops (COMSWARE'08), pp. 271–274. IEEE (2008)
35. Nichol, A., et al.: Glide: towards photorealistic image generation and editing with text-guided diffusion models. arXiv preprint arXiv:2112.10741 (2021)
36. Nichol, A.Q., Dhariwal, P.: Improved denoising diffusion probabilistic models. In: International Conference on Machine Learning, pp. 8162–8171. PMLR (2021)
37. Poole, B., Jain, A., Barron, J.T., Mildenhall, B.: DreamFusion: text-to-3D using 2D diffusion. arXiv arXiv:2209.14988 (2022)
38. Ramesh, A., Dhariwal, P., Nichol, A., Chu, C., Chen, M.: Hierarchical text-conditional image generation with clip latents. arXiv preprint arXiv:2204.06125 (2022)
39. Rombach, R., Blattmann, A., Lorenz, D., Esser, P., Ommer, B.: High-resolution image synthesis with latent diffusion models. In: Proceedings of the IEEE/CVF Conference on Computer Vision and Pattern Recognition, pp. 10684–10695 (2022)
40. Saharia, C., et al.: Photorealistic text-to-image diffusion models with deep language understanding. In: Advances in Neural Information Processing Systems, vol. 35, pp. 36479–36494 (2022)
41. Santana, G.: Magicprompt (2022). https://huggingface.co/Gustavosta/MagicPrompt-Stable-Diffusion
42. Sha, Z., Li, Z., Yu, N., Zhang, Y.: De-fake: detection and attribution of fake images generated by text-to-image diffusion models. arXiv preprint arXiv:2210.06998 (2022)
43. Shaham, T.R., Dekel, T., Michaeli, T.: SinGAN: learning a generative model from a single natural image. In: Proceedings of the IEEE/CVF International Conference on Computer Vision, pp. 4570–4580 (2019)
44. ShieldMnt: invisible-watermark (2021). https://github.com/ShieldMnt/invisible-watermark/tree/main
45. Singer, U., et al.: Make-a-video: text-to-video generation without text-video data. arXiv preprint arXiv:2209.14792 (2022)

46. Taha, M.S., Rahem, M.S.M., Hashim, M.M., Khalid, H.N.: High payload image steganography scheme with minimum distortion based on distinction grade value method. Multimedia Tools Appl. **81**(18), 25913–25946 (2022)
47. Toderici, G., et al.: Workshop and challenge on learned image compression (CLIC2020). In: CVPR (2020)
48. Touvron, H., et al.: LLaMA: open and efficient foundation language models. arXiv preprint arXiv:2302.13971 (2023)
49. Uchida, Y., Nagai, Y., Sakazawa, S., Satoh, S.: Embedding watermarks into deep neural networks. In: Proceedings of the 2017 ACM on International Conference on Multimedia Retrieval, pp. 269–277 (2017)
50. Vincent, J.: An online propaganda campaign used AI-generated headshots to create fake journalists. Verge. com (2020)
51. Vukotić, V., Chappelier, V., Furon, T.: Are deep neural networks good for blind image watermarking? In: 2018 IEEE International Workshop on Information Forensics and Security (WIFS), pp. 1–7. IEEE (2018)
52. Wang, Z., Bovik, A.C., Sheikh, H.R., Simoncelli, E.P.: Image quality assessment: from error visibility to structural similarity. IEEE Trans. Image Process. **13**(4), 600–612 (2004)
53. Wen, Y., Kirchenbauer, J., Geiping, J., Goldstein, T.: Tree-ring watermarks: fingerprints for diffusion images that are invisible and robust. arXiv preprint arXiv:2305.20030 (2023)
54. Wu, H., Liu, G., Yao, Y., Zhang, X.: Watermarking neural networks with watermarked images. IEEE Trans. Circuits Syst. Video Technol. **31**(7), 2591–2601 (2020)
55. Yu, N., Davis, L.S., Fritz, M.: Attributing fake images to GANs: learning and analyzing GAN fingerprints. In: Proceedings of the IEEE/CVF International Conference on Computer Vision, pp. 7556–7566 (2019)
56. Yu, N., Skripniuk, V., Abdelnabi, S., Fritz, M.: Artificial fingerprinting for generative models: rooting deepfake attribution in training data. In: Proceedings of the IEEE/CVF International Conference on Computer Vision, pp. 14448–14457 (2021)
57. Yu, N., Skripniuk, V., Chen, D., Davis, L., Fritz, M.: Responsible disclosure of generative models using scalable fingerprinting. In: International Conference on Learning Representations (ICLR) (2022)
58. Zhang, C., Benz, P., Karjauv, A., Sun, G., Kweon, I.S.: UDH: universal deep hiding for steganography, watermarking, and light field messaging. In: Advances in Neural Information Processing Systems, vol. 33, pp. 10223–10234 (2020)
59. Zhang, C., Karjauv, A., Benz, P., Kweon, I.S.: Towards robust deep hiding under non-differentiable distortions for practical blind watermarking. In: Proceedings of the 29th ACM International Conference on Multimedia, pp. 5158–5166 (2021)
60. Zhang, K.A., Xu, L., Cuesta-Infante, A., Veeramachaneni, K.: Robust invisible video watermarking with attention. arXiv preprint arXiv:1909.01285 (2019)
61. Zhang, L., Liu, X., Martin, A.V., Bearfield, C.X., Brun, Y., Guan, H.: Robust image watermarking using stable diffusion. arXiv preprint arXiv:2401.04247 (2024)
62. Zhang, L., Rao, A., Agrawala, M.: Adding conditional control to text-to-image diffusion models. In: Proceedings of the IEEE/CVF International Conference on Computer Vision, pp. 3836–3847 (2023)
63. Zhang, R., Isola, P., Efros, A.A., Shechtman, E., Wang, O.: The unreasonable effectiveness of deep features as a perceptual metric. In: Proceedings of the IEEE Conference on Computer Vision and Pattern Recognition, pp. 586–595 (2018)

64. Zhang, X., Karaman, S., Chang, S.F.: Detecting and simulating artifacts in GAN fake images. In: 2019 IEEE International Workshop on Information Forensics and Security (WIFS), pp. 1–6. IEEE (2019)
65. Zhao, Y., Pang, T., Du, C., Yang, X., Cheung, N.M., Lin, M.: A recipe for watermarking diffusion models. arXiv preprint arXiv:2303.10137 (2023)
66. Zhu, J., Kaplan, R., Johnson, J., Fei-Fei, L.: Hidden: hiding data with deep networks. In: Proceedings of the European Conference on Computer Vision (ECCV), pp. 657–672 (2018)
67. Zohny, H., McMillan, J., King, M.: Ethics of generative AI. J. Med. Ethics **49**, 79–80 (2023)

Hierarchical Conditioning of Diffusion Models Using Tree-of-Life for Studying Species Evolution

Mridul Khurana[1]($^{\boxtimes}$), Arka Daw[2], M. Maruf[1], Josef C. Uyeda[1], Wasila Dahdul[3], Caleb Charpentier[1], Yasin Bakış[4], Henry L. Bart Jr.[4], Paula M. Mabee[5], Hilmar Lapp[6], James P. Balhoff[7], Wei-Lun Chao[8], Charles Stewart[9], Tanya Berger-Wolf[8], and Anuj Karpatne[1]

[1] Virginia Tech, Blacksburg, VA, USA
{mridul,karpatne}@vt.edu
[2] Oak Ridge National Laboratory, Oak Ridge, TN, USA
[3] University of California, Irvine, CA, USA
[4] Tulane University, New Orleans, LA, USA
[5] Battelle, Columbus, OH, USA
[6] Duke University, Durham, NC, USA
[7] University of North Carolina at Chapel Hill, Chapel Hill, NC, USA
[8] The Ohio State University, Columbus, OH, USA
[9] Rensselaer Polytechnic Institute, Troy, NY, USA

Abstract. A central problem in biology is to understand how organisms evolve and adapt to their environment by acquiring variations in the observable characteristics or traits of species across the tree of life. With the growing availability of large-scale image repositories in biology and recent advances in generative modeling, there is an opportunity to accelerate the discovery of evolutionary traits automatically from images. Toward this goal, we introduce Phylo-Diffusion, a novel framework for conditioning diffusion models with phylogenetic knowledge represented in the form of HIERarchical Embeddings (HIER-Embeds). We also propose two new experiments for perturbing the embedding space of Phylo-Diffusion: trait masking and trait swapping, inspired by counterpart experiments of gene knockout and gene editing/swapping. Our work represents a novel methodological advance in generative modeling to structure the embedding space of diffusion models using tree-based knowledge. Our work also opens a new chapter of research in evolutionary biology by using generative models to visualize evolutionary changes directly from images. We empirically demonstrate the usefulness of Phylo-Diffusion in capturing meaningful trait variations for fishes and birds, revealing novel insights about the biological mechanisms of their evolution. (Model and code can be found at imageomics.github.io/phylo-diffusion)

Keywords: Evolution · Diffusion Models · Hierarchical Conditioning

---

Supplementary Information The online version contains supplementary material available at https://doi.org/10.1007/978-3-031-73024-5_9.

138   M. Khurana et al.

1 Introduction

Given the astonishing diversity of life forms on the planet, an important end goal in biology is to understand how organisms evolve and adapt to their environment by acquiring variations in their observable characteristics or *traits* (*e.g.*, beak color, stripe pattern, and fin curvature) over millions of years in the process of evolution. Our knowledge of species evolution is commonly represented in a graphical form as the "tree of life" (also referred to as the *phylogenetic tree* [14], see Fig. 1), illustrating the evolutionary history of species (leaf nodes) and their common ancestors (internal nodes). Discovering traits that are heritable across the tree of life, termed *evolutionary traits*, is important for a variety of biological tasks such as tracing the evolutionary timing of trait variations common to a group of species and analyzing their genetic underpinnings through gene-knockout or gene-editing/swapping (*e.g.*, CRISPR [22]) experiments. However, quantifying trait variations across large groups of species is labor-intensive and time-consuming, as it relies on expert visual attention and subjective definitions [28], hindering rapid scientific advancement [19].

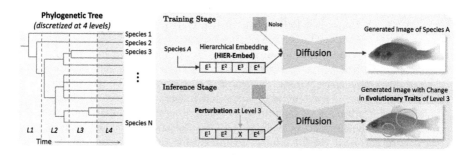

Fig. 1. Overview of Phylo-Diffusion framework. Every species in the tree of life (phylogenetic tree) is encoded to a HIERarchical Embedding (HIER-Embed) comprising of four vectors (one for each phylogenetic level), which is used to condition a latent diffusion model to generate synthetic images of the species. By structuring the embedding space with phylogenetic knowledge, Phylo-Diffusion enables visualization of changes in the evolutionary traits of a species (circled pink) upon perturbing its embedding. (Color figure online)

The growing deluge of large-scale image repositories in biology [9,30,31] presents a unique opportunity for machine learning (ML) methods to accelerate the discovery of evolutionary traits automatically from images. In particular, with recent developments in generative modeling such as latent diffusion models (LDMs) [26], we are witnessing rapid improvements in our ability to control the generation of high-quality images based on input conditioning of text or image prompts. This is facilitating breakthroughs in a variety of commercial use-cases of computer vision where we can analyze how changes in the input prompts affect variations in the generated images [7,27,34]. We ask the question: *can* we leverage LDMs to control the generation of biological images of organisms conditioned on the position of a species in the tree of life? In other words, can we encode the structure of evolutionary relationships among species and their ancestors as input conditions in LDMs? This can help us analyze trait variations in generated images across different branches in the phylogenetic tree, revealing novel insights into the biological mechanisms of species evolution.

Toward this goal, we introduce **Phylo-Diffusion**, a novel framework for discovering evolutionary traits of species from images by conditioning diffusion models with phylogenetic knowledge (see Fig. 1). One of the core innovations of Phylo-Diffusion is a novel HIERarchical Embedding (**HIER-Embed**) strategy that encodes evolutionary information of every species as a sequence of four vectors, one for each discretized level of ancestry in the tree of life (covering different evolutionary periods). We also propose two novel experiments for analyzing evolutionary traits by perturbing the embedding space of Phylo-Diffusion and observing changes in the features of generated images, akin to biological experiments involving genetic perturbations. First, we introduce **Trait Masking**, where one or more levels of information in HIER-Embed are masked out with noise to study the disappearance of traits inherited by species at those levels. This is inspired by *gene knockout* experiments [10], wherein one or more genes are deactivated or "knocked out" to investigate the gene's function, particularly its impact on the traits of the organism. Second, we introduce **Trait Swapping**, where a certain level of HIER-Embed in a reference species is swapped with the embedding of a sibling node at the same level, similar in spirit to *gene editing/swapping* experiments made possible by the CRISPR technology [22]. The goal of trait swapping is to visualize trait differences at every branching point in the tree of life that results in the diversification of species during evolution.

Here are the main contributions of this paper. Our work represents a novel methodological advance in the emerging field of knowledge-guided machine learning (KGML) [15–17] to structure the embedding space of generative models using tree-based knowledge. Our work also opens a new chapter of research in evolutionary biology by using generative models to visualize evolutionary changes directly from images, which can serve a variety of biological use-cases. For example, Phylo-Diffusion can help biologists automate the discovery of *synapomorphies*, which are distinctive traits that emerge on specific evolutionary branches and are crucial for systematics and classification [33]. Our proposed experiments of trait masking and swapping can also be viewed as novel image-based counterparts to genetic experiments, which traditionally take years. Our work thus enables biologists to rapidly analyze the impacts of genetic perturbations on particular branches of the phylogenetic tree–a grand challenge in developmental biology [5,20]. We empirically demonstrate the usefulness of Phylo-Diffusion in capturing meaningful trait changes upon perturbing its embedding for fishes and birds, generating novel hypotheses of their evolution.

2 Related Works and Background

Interpretable ML: Discovering evolutionary traits from images requires the identification and interpretation of fine-grained features in images that define and differentiate species. Several methodologies have recently been developed in the field of interpretable ML for localizing image regions that contain discriminatory information of classes, including ProtoPNet [3], PIP-Net [21] and INTR [23]. Despite their effectiveness and applicability across a wide range of applications, these methods are not directly suited for our target application of discovering evolutionary traits for two primary reasons. First, they are not designed to incorporate structured biological knowledge (*e.g.*, knowledge of the tree of life) in the learning of interpretable features, and thus are unable to provide biologically meaningful explanations of feature differences across groups of species in the phylogenetic tree, which is key to discovering evolutionary traits. Second, since most methods in interpretable ML are designed for the task of classification, it is non-trivial to integrate them in generative modeling frameworks to produce synthetic images with controlled perturbations in the embedding space similar to gene knockout and gene editing/swapping experiments, in contrast to our proposed experiments in Phylo-Diffusion.

Phylogeny-guided Neural Networks (Phylo-NN): A recent work closely aligned with our goal of discovering evolutionary traits directly from images is Phylo-NN [6]. Phylo-NN uses an encoder-decoder architecture to represent images of organisms as structured sequences of feature vectors termed "Imageomes", where different segments of Imageomes capture evolutionary information from varying levels of ancestry in the phylogenetic tree. While Phylo-NN shares several similarities with our proposed framework, Phylo-Diffusion, in terms of motivations and problem formulations, there are also prominent differences. The primary goal of Phylo-NN is specimen-level image reconstruction, whereas Phylo-Diffusion considers a different goal of controlling image generation at the species-level. As a result, Phylo-NN learns a unique Imageome sequence for every organism, enabling us to study the variability in individuals from the same species and the analysis of similarity in Imageome segments learned at shared ancestry levels. On the other hand, Phylo-Diffusion learns a unique embedding for every species and ancestor node in the tree of life, which serves as input conditions to generate distributions of synthetic images. Phylo-Diffusion thus uses hard constraints to ensure that all species with a common ancestor learn the exact same embeddings at their shared ancestry levels, making it easy to analyze trait commonalities and variations across groups of species, in contrast to Phylo-NN. Additionally, Phylo-Diffusion allows for perturbations in the embedding space of generative models in biologically meaningful ways inspired by gene knockout and gene editing/swapping experiments, going beyond the capabilities of Phylo-NN. We consider Phylo-NN as a baseline in our experiments to compare its performance with Phylo-Diffusion.

Background on Latent Diffusion Models (LDMs): One of the state-of-the-art approaches in generative modeling is the framework of Diffusion Models

[13], which learns a target distribution $p(x)$ by incrementally transforming a noisy sample x generated from a Gaussian distribution $\mathcal{N}(0, I)$ into one that is more likely to be generated from $p(x)$ over a series of timesteps T. While early frameworks of diffusion models (*e.g.*, DDPM [13], DDIM [29] and ADM [4]) suffered from high computational costs and long training/inference times, Latent Diffusion Models (LDMs) [26] are able to address these concerns to a large extent by operating in a compressed latent space, significantly accelerating their ability to generate high-resolution images. The basic idea of LDMs is to train a separate auto-encoder to map an input image x into its latent representation $z_0 = \mathcal{E}(x)$ using encoder $\mathcal{E}$, which when fed to decoder $\mathcal{D}$ produces a reconstruction of the original image, $\tilde{x} = \mathcal{D}(\tilde{z}_0)$. LDMs employ diffusion models in the compressed latent space z by modeling the conditional probability of the reverse diffusion process as $\tilde{z}_{t-1} \sim p_\theta(\tilde{z}_{t-1}|\tilde{z}_t, y, t)$, where y is the input condition. This is implemented using a conditional denoising U-Net backbone $\epsilon_\theta(z_t, y, t)$ with learnable parameters θ. LDMs also pre-process y using a domain-specific encoder $\mathbf{E} = \tau_\phi(y)$ trained alongside the U-Net backbone ϵ_θ to project y into the intermediate layers of ϵ_θ using cross-attention mechanisms. The learnable parameters of LDMs are trained by minimizing the following loss function:

$$\mathcal{L}(\theta, \phi) = \mathbb{E}_{z_t, y, t, \epsilon \sim \mathcal{N}(0, I)} \left[\|\hat{\epsilon}_\theta(z_t, \tau_\phi(y), t) - \epsilon\|^2 \right] \quad (1)$$

3 Proposed Framework of Phylo-Diffusion

3.1 Hierarchical Embedding (HIER-Embed)

Phylo-Diffusion uses a novel hierarchical embedding (HIER-Embed) strategy to structure the embedding E of every species node using phylogenetic knowledge. As a first step, we consider a discretized version of the phylogenetic tree involving four ancestral levels, level-1 to level-4, where every level corresponds to a different range of time in the process of evolution. (See Appendix for a detailed characterization of the four ancestry levels for fish species used in this study.) Given a set of n species, $\mathcal{S} = \{S_1, S_2, S_3, ..., S_n\}$, let us represent the position of species $S_i \in \mathcal{S}$ in the phylogenetic tree at the four ancestry levels as $\{S_i^1, S_i^2, S_i^3, S_i^4\}$, where S_i^l represents the ancestor node of S_i at level-l. Hence, if two species S_i and S_j share common ancestors till level-k, then $S_i^l = S_j^l$ for $l = 1$ to k. We define the level-l embedding of species S_i as:

$$\mathbf{E}_i^l = \mathtt{Embed}(S_i^l) \in \mathbb{R}^{d'}, \quad (2)$$

where $\mathtt{Embed}(.)$ is a learnable embedding layer that provides a simple way to store and look-up the trained embeddings of every node. The combined hierarchical embedding (HIER-Embed) of species S_i is obtained by concatenating its embeddings across all four levels as follows:

$$\mathbf{E}_i = \tau(S_i) = \mathtt{Concat}[\mathbf{E}_i^1, \mathbf{E}_i^2, \mathbf{E}_i^3, \mathbf{E}_i^4] \in \mathbb{R}^d, \quad (3)$$

where Concat[.] denotes the concatenation operation and $y = S_i$ is the input condition used in LDMs. Note that different segments of $\mathbf{E}_i$ capture information about the traits of S_i acquired at different time periods of evolution. In particular, we expect the embedding vectors learned at earlier ancestry levels of $\mathbf{E}_i$ to capture evolutionary traits of S_i common to a broader group of species. On the other hand, embeddings learned at later ancestry levels are expected to be more specific to S_i. In the following, we present two novel experiments for studying evolutionary traits by perturbing the embedding space learned by HIER-Embed.

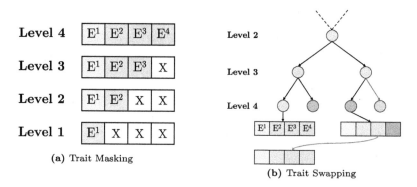

Fig. 2. Schematic diagrams of the two proposed experiments for discovering evolutionary traits using Phylo-Diffusion.

3.2 Proposed Experiment of Trait Masking

The goal of this experiment is to verify if HIER-Embed is indeed able to capture hierarchical information in its level-embeddings such that masking information at lower levels of the embedding only erases traits acquired at later stages of evolution while retaining trait variations learned at earlier levels. In other words, we want to verify that the embeddings learned by HIER-Embed at level-l capture information common to all descendant species that are part of the same subtree at level-l. Figure 2a represents a schematic diagram of the process followed for trait masking. We start with the combined embedding containing information at all four levels, $[\mathbf{E}^1, \mathbf{E}^2, \mathbf{E}^3, \mathbf{E}^4]$. To examine what is learned at the last level of this embedding, we mask it out by substituting it with Gaussian noise defined as $\mathbf{z_{noise}} \sim \mathcal{N}(0, I) \in \mathbb{R}^{d'}$. This results in the perturbed embedding $[\mathbf{E}^1, \mathbf{E}^2, \mathbf{E}^3, \mathbf{z_{noise}}]$, effectively eliminating the species-level (or $\mathbf{E}^4$) information. This masking should prompt the model to generate images that reflect only the information learned up to the third level while obscuring species-level details. We can extend this experiment by incrementally introducing *noise* at later levels, e.g., at both levels 3 & 4, and so on.

Expected Changes in Probability Distributions: Note that when all four level embeddings are used, *i.e.* $[\mathbf{E}^1, \mathbf{E}^2, \mathbf{E}^3, \mathbf{E}^4]$, the generated images are expected to be classified to a unique species S_i. In terms of probability distributions, the probability of predicting species S_i should be distinctly higher than the probability of predicting any other species. However, when we mask out certain level embeddings (*i.e.*, mask out information at level 4), we are intentionally removing information necessary to distinguish species S_i from its siblings species that are part of the same sub-tree (e.g., those that share a common ancestor at level 3). For this reason, we expect the generated images to show higher probabilities of being classified to any of the descendant species of the sub-tree, compared to the other species that are outside of the sub-tree. To quantify this behavior, we can measure the change in probability distributions for species within and outside the sub-tree after masking out an internal node. Since we expect probabilities to increase only for species within the sub-tree, the mean increase in probabilities for within-subtree species should be higher than that for out-tree species, as empirically demonstrated later in the Results Section.

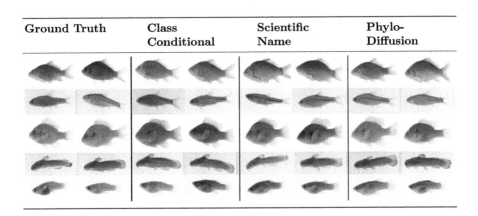

Fig. 3. Comparing the quality of synthetic images generated by different conditioning mechanisms in LDMs. Every row corresponds to a different species and we show two samples per species for every conditioning mechanism. The order of species from top to bottom is *Cyprinus carpio*, *Notropis hudsonius*, *Lepomis auritus*, *Noturus exilis*, and *Gambusia affinis*.

3.3 Proposed Experiment of Trait Swapping

In trait swapping, we substitute the level-l embedding of a source species with the level-l embedding of a sibling subtree at an equivalent level. Figure 2b shows a schematic representation of the trait swapping experiment where the level-3 embedding (green) of a source species is replaced with its sibling level-3 embedding (yellow). Images generated for this perturbed embedding are expected to retain all of the traits of the source species except the swapped embedding, which should borrow traits from the sub-tree rooted at the sibling node. Visualizing trait differences in the generated images before and after trait swapping can help us understand the evolutionary traits that branched at a certain level (e.g., those leading to the diversification of green and yellow sub-trees at level-3 in the example phylogeny of Fig. 2b). In terms of the probability distribution, similar to trait masking, we expect to see a drop in the probabilities of the source species (*pink*), and simultaneously, we expect an increase in probabilities for all the descendent species in subtree at node *yellow*, i.e. red and purple.

4 Evaluation Setup

Datasets: We use a collection of fish images as our primary dataset for evaluation. This dataset was procured from the Great Lakes Invasives Network (GLIN) [1] Project, comprising a total of 5434 images spanning 38 fish species. We obtained the phylogenetic tree of fish species from *opentree* [2] python package (see Appendix for details on the phylogenetic tree). The raw museum images were pre-processed and resized to 256 × 256 pixels and the dataset was partitioned into training and validation sets, following a 75-25 split. We provide additional results on the CUB-200-2011 dataset [32] of bird species in the Appendix.

Table 1. Quantitive comparison of generated images sampled using DDIM [29] (100 samples per class).

Model Type	Method	FID ↓	IS ↑	Prec. ↑	Recall ↑
GAN	Phylo-NN	28.08	2.35	0.625	0.084
Diffusion	Class Conditional	11.46	2.47	0.679	0.359
Diffusion	Scientific Name	11.76	2.43	0.683	0.332
Diffusion	Phylo-Diffusion (ours)	11.38	2.53	0.654	0.367

Baselines of Conditioning Mechanisms: (1) *Class Conditional:* One of the simplest ways of encoding information about a species class is to map class labels $y \in [1, N_c]$ to a fixed d-dimensional embedding vector $e \in \mathbb{R}^d$ using a trainable embedding layer. Note that the resulting embeddings are not designed to contain any hierarchical information in contrast to HIER-Embed. (2) *Scientific Name Encoding:* The scientific name of a species contains valuable biological information typically comprising of a combination of the *genus* name and *species* name. Since species that share their *genus* name are likely to contain common phylogenetic traits, we use them as a baseline for conditioning LDMs for discovering evolutionary traits. Specifically, we employ a pre-trained frozen CLIP model [25] to encode the scientific names of species into fixed d-dimensional embeddings.

Training details: We used $d' = 128$ as the embedding dimension for each level of HIER-Embed, which when concatenated across the four levels produces

Table 2. Classification F1-Score on the 100 samples generated per class. The base classifier has an accuracy of *85%* on the test set.

Method	F1-Score (in %) ↑
Phylo-NN	47.37
Class Conditional	81.99
Scientific Name	70.16
Phylo-Diffusion (ours)	82.21

the combined hierarchical embedding of $d = 512$ dimensions. Phylo-Diffusion uses this d-dimensional embedding to condition LDMs through cross-attention in denoising U-Net backbone and train LDMs without classifier-free guidance. We used VQGAN [8] as the backbone encoder-decoder to achieve the latent representations desired for LDMs with a downsampling factor of 4. All the models with different encoders are trained for 400k iterations, employing the best model checkpoint if convergence occurs early. Additional hyperparameters, such as learning rate, batch size, and U-Net architecture, are detailed in the Appendix.

5 Results

5.1 Quality of Generated Images

Table 1 compares the quality of generated images of baselines using the metrics of Fréchet Inception Distance (FID) score, Inception Score (IS), and Precision, Recall calculated in the feature space as proposed in [18]. Our results show that Phylo-Diffusion is at par with state-of-the-art generative models, achieving an FID of 11.38 compared to LDM's 11.46. We show a sample of generated images in Fig. 3, with additional images provided in the Appendix. We also show the robustness of Phylo-Diffusion's results with varying numbers of phylogenetic levels and embedding dimensions in the Appendix.

5.2 Classification Accuracy

We used a separate model for species classification, specifically a ResNet-18 model [12] trained using the same training/validation split as Phylo-Diffusion. The primary objective behind building this classifier is to verify if images generated by Phylo-Diffusion contain sufficient discriminatory information to be classified as their correct species classes. Table 2 compares the classification F1-scores over 100 samples generated by baseline conditioning schemes. We can see that the synthetic images generated by Phylo-Diffusion achieve the highest F1 score (82.21%), which is quite close to the F1 score of the base classifier on the original test images (85%). We present additional results showing the generalizability of Phylo-Diffusion in classifying generated images to unseen species in the Appendix.

5.3 Matching Embedding Distances with Phylogenetic Distances

We investigate the quality of embeddings produced by baseline methods by comparing distances in the embedding space with the ground-truth (GT) phylogenetic distances computed from the tree of life, as illustrated in Fig. 4. Ideally, we expect distances in the embedding space of species pairs to be reflective of their phylogenetic distances. For Class Conditional, we can see that the distance matrix does not show any alignment with the GT phylogenetic distance matrix. In the case of Scientific Name Encoding, the distance matrix exhibits notable similarities to the phylogenetic distances, thanks to the hierarchical nature of information contained in scientific names (i.e., *genus*-name & *species*-name). However, one limitation of this encoding is its inability to capture inter-genus similarities or differences. In contrast, HIER-Embed shows a distance matrix that closely aligns with the GT phylogenetic distance matrix, validating its ability to preserve evolutionary distances among species in its embedding space.

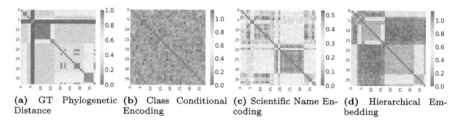

(a) GT Phylogenetic Distance (b) Class Conditional Encoding (c) Scientific Name Encoding (d) Hierarchical Embedding

Fig. 4. Comparing Cosine distances in the embedding space of species for varying conditioning mechanisms.

5.4 Trait Masking Results

To obtain classification probabilities or logits associated with generated images, we employ the classifier detailed in Sect. 5.2. For the masked embeddings of subtrees at level 3, defined as $[\mathbf{E}^1, \mathbf{E}^2, \mathbf{E}^3, \mathbf{z}_{\text{noise}}]$, we analyze logits of generated images and compare them with those generated without masking. Figure 5 demonstrates that for a specific subtree, in this case *Lepomis*, logits for species within the subtree are higher compared to those for species outside it. This outcome aligns with the expectation that Phylo-Diffusion, when provided with information up to Level 3, can capture overarching characteristics of all species within the given subtree. Additionally, Fig. 5 presents probability distributions for species within the *Lepomis* subtree when the full set of hierarchical encodings $[\mathbf{E}^1, \mathbf{E}^2, \mathbf{E}^3, \mathbf{E}^4]$ is provided. It demonstrates that the probabilities are significantly higher for the targeted class, as intended for image generation. After masking, we observe that the generated images are very similar and capture common features of the *Lepomis* genus. For all our calculations and plots, we

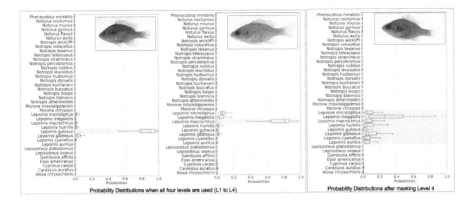

Fig. 5. Left: class probability distributions of images generated by using embeddings at all four levels for two species *Lepomis gulosus* and *Lepomis macrochirus* (shown in green) that are part of the same sub-tree till level 3. Right: class probability distributions of images generated by masking level 4 (descendant species that have common ancestry till level 3 are highlighted in green)

generate 100 images for each subtree and node. The appendix contains additional histograms that detail logit distributions across all different subtrees at each level, offering comprehensive insights into how the model discriminates and learns the hierarchical structure across different levels and nodes.

Quantitative Evaluation of Probability Distrubtions: To quantitatively evaluate the ability of Phylo-Diffusion to capture hierarchical information and show desired changes in probability distributions after masking, we compute the following metrics. Let us denote the set of all species in the data as $\mathcal{S}$ and for a given sub-tree at an internal node i of level l, let us denote the subset of descendant species as $\mathcal{S}_i^l = \{S_1, S_2, \ldots, S_n\}$. We first compute the reference probabilities P_{ref} of every species before masking (i.e., by using all four level embeddings). Let us denote the probability of predicting a generated image using all four embeddings of a descendant species $S_j \in \mathcal{S}_i^l$ into species class S_k as $P_{S_j}(S_k)$. The reference probability of a species S_k can then be given as:

$$P_{ref}(S_k) = \begin{cases} \frac{1}{|\mathcal{S}_i^l|-1} \sum_{S_j \in \mathcal{S}_i^l \setminus S_k} P_{S_j}(S_k), & \text{if } S_k \in \mathcal{S}_i^l, \\ \frac{1}{|\mathcal{S}_i^l|} \sum_{S_j \in \mathcal{S}_i^l} P_{S_j}(S_k), & \text{if } S_k \notin \mathcal{S}_i^l. \end{cases} \quad (4)$$

Note that when S_k is part of the sub-tree, i.e., $S_k \in \mathcal{S}_i^l$, $P_{ref}(S_k)$ is computed by averaging over $|\mathcal{S}_i^l| - 1$ probability values since we exclude the case when S_k is used to generate the images. On the other hand, when S_k is outside of the sub-tree, i.e., $S_k \notin \mathcal{S}_i^l$, we average over all $|\mathcal{S}_i^l|$ probability values. Given these reference probabilities values before masking, we can compute the change in probability of predicting species S_k after masking as $P_{diff}(S_k) = P_{mask}(S_k) - P_{ref}(S_k)$, where $P_{mask}(S_k)$ is the probability of predicting a generated image after masking to S_k. We expect P_{diff} to be larger for descendant species $S_k \in \mathcal{S}_i^l$

compared to species that are outside of the sub-tree because of the dispersion of probabilities in a sub-tree as a consequence of masking. We thus compute the average P_{diff} for species that belong to subtree $\mathcal{S}_i^l$ as $P_{diff}^{sub}(i, l)$ and species that are outside the subtree $\mathcal{S}_i^l$ as $P_{diff}^{out}(i, l)$.

Figure 6 shows the box plot of P_{diff}^{sub} and P_{diff}^{out} across internal nodes at levels 2 and 3. We can see that species within the subtree exhibit a more pronounced increase in probabilities compared to those outside the subtree, aligning with our expectations. Notably, this trend is consistently observed across both Levels 2 & 3. More details of the class-wise probability distribution shifts for each level are provided in the appendix. The outcomes of these experiments affirm that Phylo-Diffusion effectively identifies unique features at Levels 2 and 3, and captures shared features or traits of any chosen subtree at the internal nodes of the phylogeny.

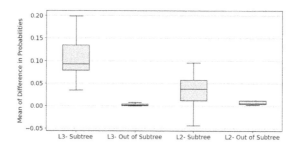

Fig. 6. Box plot for the mean of difference in probabilities for species within the subtree and out of subtree for level-3 and level-2.

5.5 Trait Swapping Results

Figure 7 shows examples of trait swapping results enabled by Phylo-Diffusion. For the first example (first row of Fig. 7a), we swap the level-2 encoding of source species *Noturus exilis* with level-2 encoding of its sibling group *Notropis/ Carassius*. The goal here is to discover traits of *Noturus exilis* inherited at level 2 that differentiates it from other groups of species that branched out at this point of time in evolution. We can see that the generated images of the perturbed embedding (center) exhibit the absence of barbels (whiskers) highlighted in purple, while the caudal (or tail) fin is beginning to fork (or split), a trait adopted from *Notropis* (right). In contrast, other fins such as the dorsal, pelvic, and anal fins highlighted in green remain similar to those of the source species, *Noturus exilis* (left). This suggests that at level-2, *Notropis* and *Noturus* species diverged by developing differences in two distinct traits, barbels and forked caudal fins while keeping other traits intact.

For the second example (Fig. 7a, row 2), we swap level-2 information of *Gambusia affinis* (left) with that of *Esox americanus* (right). The generated images

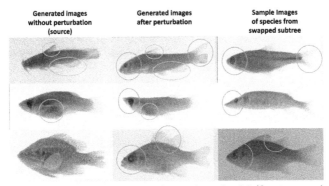

(a) Examples of traits swapping for species at level-2 (first two rows) and level-3 (last row). The order of species from top to bottom is *Noturus exilis* swapped with *Notropis* and *Gambusia affinis* swapped with *Esox americanus*. The third row shows trait swapping at level-3 for *Lepomis gulosus* swapped with *Morone*.

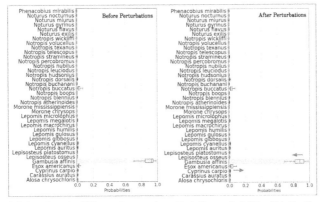

(b) For Row 2 of Figure 7a, we show that the probability distribution of *Gambusia affinis* decreases after the swapping traits at level-2, with an increase in the probability distribution of *Esox americanus*.

Fig. 7. Examples of trait swapping results.

of the perturbed embedding (center) exhibit a more pointed head highlighted in purple, and a slimmer body shape resembling *Esox americanus*. Notably, the perturbed species retains discoloration at the bottom from the source species highlighted in green. Figure 7b presents probability distributions (or logits) of *Gambusia affinis* before and after trait swapping using a separately trained classifier. We observe a slight decrease in logits for *Gambusia affinis* and an increase in logits for *Esox americanus*, consistent with our expectations. In the third row of Fig. 7a, we swap level-3 information of *Lepomis gulosus* (left) with that of *Morone* genus (right). The resulting images from the perturbed embedding (center) capture the horizontal line pattern characteristic of *Morone* genus, and the dorsal fin (purple circle) begins to split. All of these examples suggest novel scientific hypotheses about differences in evolutionary traits acquired by species at different ancestry levels, which can be validated by biologists in subsequent studies. Note that our experiments are most effective at levels near the species nodes, specifically at levels 2 & 3, since phylogenetic signal is known to diminish as we move toward the root of the tree [11,24]. Additional visualizations of trait swapping results are provided in the Appendix.

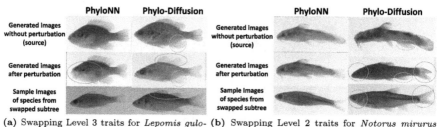

(a) Swapping Level 3 traits for *Lepomis gulosus* with *Morone*

(b) Swapping Level 2 traits for *Notorus mirurus* with *Notropis*

Fig. 8. Comparing Phylo-NN with Phylo-Diffusion for examples of trait swapping.

Comparisions with Phylo-NN: Figure 8 compares trait swapping results of Phylo-Diffusion and Phylo-NN for the same set of example species. Figure 8a shows trait swapping at level-3 for the source species of *Lepomis gulosus* (top) and target sub-tree of the *Morone* genus (bottom). In Phylo-NN, images generated by perturbing the Imageome sequences appear blurry (red circle), while Phylo-Diffusion effectively captures the splitting of dorsal fin (purple circle) and the horizontal stripe pattern of the *Morone* genus, while maintaining the fin structure of *Lepomis gulosus* (green circle). Similarly, Fig. 8b compares trait swapping for *Noturus miurus* (top) with the target sub-tree of *Notropis* genus (bottom) at level-2. For Phylo-NN, the perturbed images are almost identical to the source species. However, Phylo-Diffusion shows visible trait differences such as the absence of barbels and the caudal (or tail) fin beginning to fork or split (purple circle), which are traits picked from the target sub-tree of *Notropis* genus. Note that we had considered the same target sub-tree in Fig. 7a row 1 and observed similar trait differences in the generated images after perturbation, further validating the ability of Phylo-Diffusion to discover consistent evolutionary traits. We provide additional results comparing Phylo-Diffusion and Phylo-NN trait swapping results in the Appendix.

6 Conclusions and Future Work

In this work, we introduced Phylo-Diffusion, a novel framework for discovering evolutionary traits from images by structuring the embedding space of diffusion models using tree-based knowledge. In the future, our approach can be extended to work on other applications involving image data linked with phylogenies or pedigrees. Our work also has limitations that need to be addressed in future research. For example, while our current work is limited to discretized trees with a fixed number of levels, future works can focus on discovering evolutionary traits at every internal node of the phylogenetic tree with varying levels without performing any discretization. Future works can also attempt to capture convergent changes in evolution, i.e., changes that occur repeatedly in different branches of the tree, and perform ancestral state reconstruction with uncertainty estimates.

Acknowledgments. This research is supported by National Science Foundation (NSF) awards for the HDR Imageomics Institute (OAC-2118240). We are thankful for the support of computational resources provided by the Advanced Research Computing (ARC) Center at Virginia Tech.

This manuscript has been authored by UT-Battelle, LLC, under contract DE-AC05-00OR22725 with the US Department of Energy (DOE). The US government retains and the publisher, by accepting the article for publication, acknowledges that the US government retains a nonexclusive, paid-up, irrevocable, worldwide license to publish or reproduce the published form of this manuscript, or allow others to do so, for US government purposes. DOE will provide public access to these results of federally sponsored research in accordance with the DOE Public Access Plan (https://www.energy.gov/doe-public-access-plan).

References

1. Great lakes invasives network project.https://greatlakesinvasives.org/portal/index.php
2. Rees, J.A., Cranston, K.: Automated assembly of a reference taxonomy for phylogenetic data synthesis. Biodivers. Data J. **5**, e12581 (2017). https://doi.org/10.3897/BDJ.5.e12581
3. Chen, C., Li, O., Tao, D., Barnett, A., Rudin, C., Su, J.K.: This looks like that: deep learning for interpretable image recognition. In: Advances in Neural Information Processing Systems, vol. 32 (2019)
4. Dhariwal, P., Nichol, A.: Diffusion models beat GANs on image synthesis. In: Advances in Neural Information Processing Systems, vol. 34, pp. 8780–8794 (2021)
5. Edmunds, R., et al.: Phenoscape: identifying candidate genes for evolutionary phenotypes. Mol. Biol. Evol. **33**, 13–24 (2015). https://doi.org/10.1093/molbev/msv223
6. Elhamod, M., et al.: Discovering novel biological traits from images using phylogeny-guided neural networks. In: Proceedings of the 29th ACM SIGKDD Conference on Knowledge Discovery and Data Mining, pp. 3966–3978 (2023)
7. Esser, P., et al.: Scaling rectified flow transformers for high-resolution image synthesis. In: Forty-first International Conference on Machine Learning (2024)
8. Esser, P., Rombach, R., Ommer, B.: Taming transformers for high-resolution image synthesis. In: Proceedings of the IEEE/CVF Conference on Computer Vision and Pattern Recognition, pp. 12873–12883 (2021)
9. Gharaee, Z., et al.: A step towards worldwide biodiversity assessment: the BIOSCAN-1M insect dataset. In: Advances in Neural Information Processing Systems, vol. 36 (2024)
10. Griffiths, A.J.: An Introduction to Genetic Analysis. Macmillan (2005)

11. Harmon, L., et al.: Early bursts of body size and shape evolution are rare in comparative data. Evolution **64**, 2385–96 (2010). https://doi.org/10.1111/j.1558-5646.2010.01025.x
12. He, K., Zhang, X., Ren, S., Sun, J.: Deep residual learning for image recognition. In: Proceedings of the IEEE Conference on Computer Vision and Pattern Recognition, pp. 770–778 (2016)
13. Ho, J., Jain, A., Abbeel, P.: Denoising diffusion probabilistic models. In: Advances in Neural Information Processing Systems, vol. 33, pp. 6840–6851 (2020)
14. Kapli, P., Yang, Z., Telford, M.J.: Phylogenetic tree building in the genomic age. Nat. Rev. Genetics **21**(7), 428–444 (2020). https://doi.org/10.1038/s41576-020-0233-0
15. Karpatne, A., et al.: Theory-guided data science: a new paradigm for scientific discovery from data. IEEE Trans. Knowl. Data Eng. **29**(10), 2318–2331 (2017)
16. Karpatne, A., Jia, X., Kumar, V.: Knowledge-guided machine learning: current trends and future prospects. arXiv preprint arXiv:2403.15989 (2024)
17. Karpatne, A., Kannan, R., Kumar, V.: Knowledge Guided Machine Learning: Accelerating Discovery using Scientific Knowledge and Data. CRC Press (2022)
18. Kynkäänniemi, T., Karras, T., Laine, S., Lehtinen, J., Aila, T.: Improved precision and recall metric for assessing generative models. In: Advances in Neural Information Processing Systems, vol. 32 (2019)
19. Lürig, M.D., Donoughe, S., Svensson, E.I., Porto, A., Tsuboi, M.: Computer vision, machine learning, and the promise of phenomics in ecology and evolutionary biology. Front. Ecol. Evol. **9**, 642774 (2021)
20. Manda, P., Balhoff, J., Lapp, H., Mabee, P., Vision, T.: Using the phenoscape knowledgebase to relate genetic perturbations to phenotypic evolution. Genesis **53**, 561–571 (2015). https://doi.org/10.1002/dvg.22878
21. Nauta, M., Schlötterer, J., van Keulen, M., Seifert, C.: Pip-net: patch-based intuitive prototypes for interpretable image classification. In: Proceedings of the IEEE/CVF Conference on Computer Vision and Pattern Recognition (CVPR), pp. 2744–2753 (2023)
22. Nemudryi, A.A., Valetdinova, K.R., Medvedev, S.P., Zakian, S.M.: TALEN and CRISPR/Cas genome editing systems: tools of discovery. Acta Naturae **6**(3), 19–40 (2014)
23. Paul, D., et al.: A simple interpretable transformer for fine-grained image classification and analysis. arXiv preprint arXiv:2311.04157 (2023)
24. Pennell, M., FitzJohn, R., Cornwell, W., Harmon, L.: Model adequacy and the macroevolution of angiosperm functional traits. Am. Nat. **186**, E33–E50 (2015). https://doi.org/10.1086/682022
25. Radford, A., et al.: Learning transferable visual models from natural language supervision. In: International Conference on Machine Learning, pp. 8748–8763. PMLR (2021)
26. Rombach, R., Blattmann, A., Lorenz, D., Esser, P., Ommer, B.: High-resolution image synthesis with latent diffusion models. In: Proceedings of the IEEE/CVF Conference on Computer Vision and Pattern Recognition, pp. 10684–10695 (2022)
27. Ruiz, N., Li, Y., Jampani, V., Pritch, Y., Rubinstein, M., Aberman, K.: DreamBooth: fine tuning text-to-image diffusion models for subject-driven generation. In: Proceedings of the IEEE/CVF Conference on Computer Vision and Pattern Recognition, pp. 22500–22510 (2023)
28. Simões, T.R., Caldwell, M.W., Palci, A., Nydam, R.L.: Giant taxon-character matrices: quality of character constructions remains critical regardless of size. Cladistics **33**(2), 198–219 (2017)

29. Song, J., Meng, C., Ermon, S.: Denoising diffusion implicit models. arXiv preprint arXiv:2010.02502 (2020)
30. Stevens, S., et al.: Bioclip: a vision foundation model for the tree of life. In: Proceedings of the IEEE/CVF Conference on Computer Vision and Pattern Recognition, pp. 19412–19424 (2024)
31. Van Horn, G., Mac Aodha, O.: iNat Challenge 2021 - FGVC8 (2021). https://kaggle.com/competitions/inaturalist-2021
32. Wah, C., Branson, S., Welinder, P., Perona, P., Belongie, S.: Cub-200-2011. Tech. Rep. CNS-TR-2011-001, California Institute of Technology (2011)
33. Zelditch, M.L., Fink, W.L., Swiderski, D.L.: Morphometrics, homology, and phylogenetics: quantified characters as synapomorphies. Syst. Biol. **44**(2), 179–189 (1995)
34. Zhang, L., Rao, A., Agrawala, M.: Adding conditional control to text-to-image diffusion models. In: Proceedings of the IEEE/CVF International Conference on Computer Vision, pp. 3836–3847 (2023)

Open Access This chapter is licensed under the terms of the Creative Commons Attribution 4.0 International License (http://creativecommons.org/licenses/by/4.0/), which permits use, sharing, adaptation, distribution and reproduction in any medium or format, as long as you give appropriate credit to the original author(s) and the source, provide a link to the Creative Commons license and indicate if changes were made.

The images or other third party material in this chapter are included in the chapter's Creative Commons license, unless indicated otherwise in a credit line to the material. If material is not included in the chapter's Creative Commons license and your intended use is not permitted by statutory regulation or exceeds the permitted use, you will need to obtain permission directly from the copyright holder.

Markov Knowledge Distillation: Make Nasty Teachers Trained by Self-undermining Knowledge Distillation Fully Distillable

En-hui Yang and Linfeng Ye(✉)

University of Waterloo, Waterloo, ON N2L 3G1, Canada
{ehyang,l44ye}@uwaterloo.ca

Abstract. To protect intellectual property of a deep neural network (DNN), two knowledge distillation (KD) related concepts are proposed: distillable DNN and KD-resistant DNN. A DNN is said to be distillable if used as a black-box input-output teacher, it can be distilled by a KD method to train a student model so that the distilled student outperforms the student trained alone with label smoothing (LS student) in terms of accuracy. A DNN is said to be KD-resistant with respect to a specific KD method if used as a black-box input-output teacher, it cannot be distilled by that specific KD method to yield a distilled student outperforming LS student in terms of accuracy. A new KD method called Markov KD (MKD) is further presented. When applied to nasty teachers trained by self-undermining KD, MKD makes those nasty teachers fully distillable, although those nasty teachers are shown to be KD-resistant with respect to state-of-the-art KD methods existing in the literature before our work. When applied to normal teachers, MKD yields distilled students outperforming those trained by KD from the same normal teachers by a large margin. More interestingly, MKD is capable of transferring knowledge from teachers trained in one domain to students trained in another domain.

Keywords: Cross-domain knowledge transfer · Intellectual property protection · Knowledge Distillation · Markov matrix · Power transform

1 Introduction

Aiming originally for model compression, knowledge distillation (KD) was first introduced by Buciluă et al. [3], where a small model (student) was trained to match the logits of a large model (teacher). Later, a generalized version now known as KD was popularized by Hinton et al. [5], where temperature scaling

---

Supplementary Information The online version contains supplementary material available at https://doi.org/10.1007/978-3-031-73024-5_10.

is introduced to soften the logits of both the teacher and student, enabling the student to mimic the soft probabilities of the teacher. Since then, many KD variants have been proposed, including logit-based KD variants [7,8,11,15,32, 37,38] and feature-based KD variants [1,4,20–22,28,34]. They have been widely used in both industry and academia in recent years to train students, yielding distilled students outperforming the students trained alone with label smoothing in terms of accuracy [2,20,24].

In the above paradigm, the teacher is in a fully cooperative mode and willing to allow its knowledge in whatever form to be transferred to the student. In other cases, however, the teacher may not want to cooperate at all. For example, since it takes time, efforts, money, and resources including data and computing infrastructure to train a good teacher, it may be desirable in some scenarios to protect intellectual property (IP) of the teacher so that it is hard for the student to learn from and mimic the behavior of the teacher. Once the teacher is released to provide a "block-box" input-output service, regardless of whether or not the teacher is willing to cooperate, its input-output function as a form of knowledge is always available for the student to leverage. In this case, logit-based KD methods would pose a threat to the teacher since they may be applied to help the student to gain competitive advantage by either leaking proprietary information of the teacher (such as valuable training medical data and model parameters) or leveraging the teacher's input-output knowledge to improve the student performance [10,27,29,36,39].

To mitigate the threat posed by logit-based KD methods in the context of protecting IP of the teacher, the concept of nasty teacher was introduced in [16]. A teacher is said to be nasty if the distillation process of applying logit-based KD methods to the teacher degrades the accuracy of the student. In [16], a method called self-undermining knowledge distillation was further proposed to train and build nasty teachers; it was demonstrated that the distillation process of applying the standard KD method [5] to such a trained nasty teacher indeed results in significant loss in student accuracy. Subsequently, however, different variants of logit-based KD methods were proposed in [8,11,15,17] to recover fully or partially the loss in student accuracy.

To a large extent, the concept of nasty teacher is not well-defined for at least two reasons. First, logit-based KD methods are not fixed. There are many existing logit-based KD methods and many possible ways to design new logit-based KD methods. As shown in [8,11,15,17], it is possible that the distillation process of applying one logit-based KD method to a nasty teacher would degrade the accuracy of the student, while the distillation process of applying another logit-based method to the same nasty teacher would improve the accuracy of the student. Second, since the student trained alone with the cross entropy loss plus label smoothing penalty (LS student) [19] generally outperforms the student trained alone with the cross entropy (CE student) in accuracy [19,23,35], whether there is a benefit from applying a logit-based KD method to a teacher should not be determined by its comparison with the CE student, but rather with the LS student. If the distillation process yields a distilled student outperforming the LS student, then there is a benefit. Otherwise, there is no incentive to leverage the teacher through that particular logit-based KD method. Indeed, it was shown in [38] that by dropping the temperature from the student side, the distilled student from the standard KD method will never worse than the LS student if the temperature on the teacher side is approaching infinity, no matter what the teacher is.

In order to lay out a solid foundation for systematically studying how to protect IP of a deep neural network (DNN) when the DNN could be potentially used as a block-box input-output teacher, in this paper we use instead the LS student as the benchmark and introduce two KD related concepts: distillable DNN and KD-resistant DNN. Specifically, we have the following definitions:

Definition 1. *Used as a block-box input-output teacher, a DNN is said to be* distillable *with respect to a student model if there is a KD method which, when applied to the teacher and the student model, yields a distilled student outperforming the LS student in accuracy. Note that if such a KD method exists, it must be at least logit-based.*

Definition 2. *Used as a block-box input-output teacher, a DNN is said to be* KD-resistant *with respect to a specific KD method and a student model if the DNN cannot be distilled by that specific KD method into a distilled student outperforming the LS student in accuracy.*

Equipped with the above concepts, we then investigate the KD-resistance and distillability of nasty teachers trained by self-undermining KD. Our contributions can be summarized as follows:

- We introduce two new concepts of distillable DNN and KD-resistant DNN to study how to protect IP of a DNN when the DNN could be potentially used as a block-box input-output teacher.
- It is demonstrated that nasty teachers trained by self-undermining KD are KD-resistant with respect to state-of-the-art logit-based KD methods existing in the literature before our work in the paper.
- We propose a new generic KD method called Markov Knowledge Distillation (MKD) which is simple, powerful, and distribution-based, and can be applied in conjunction with any other distribution-based KD method[1].
- We show by experiments that nasty teachers trained by self-undermining KD are fully distillable by our proposed MKD.
- When applied to normal teachers, MKD outperforms the underlying distribution-based KD method by a good margin.
- More interestingly, MKD is able to transfer knowledge from one domain to another domain. Specifically, it is shown that a teacher trained in one domain (say a set of images with a label set) is fully distillable by MKD to a student trained in another domain (say a different set of images with a different non-overlapping label set), regardless of whether the teacher is normal or nasty.

---

[1] A distribution-based KD method uses only the softmax of logits, i.e., the output distribution of the teacher, and requires less knowledge from the teacher in comparison with a logit-base KD method.

2 Notation and Preliminaries

2.1 Notation

For a positive integer C, let $[C] \triangleq \{1,\ldots,C\}$. For a classification problem with C class labels, we will use $[C]$ to denote its set of class labels whenever there is no ambiguity. Let $\mathcal{P}([C])$ denote the set of all probability distributions over $[C]$. For any two probability distributions $p_1, p_2 \in \mathcal{P}([C])$, the cross entropy of p_1 and p_2 is defined as

$$H(p_1, p_2) = \sum_{i=1}^{C} -p_1(i) \ln p_2(i), \tag{1}$$

where ln denotes the logarithm with base e; the Kullback–Leibler (KL) divergence (or relative entropy) between p_1 and p_2 is defined as

$$D(p_1 \| p_2) = \sum_{i=1}^{C} p_1(i) \ln \frac{p_1(i)}{p_2(i)}. \tag{2}$$

For any $y \in [C]$ and $p \in \mathcal{P}([C])$, write the cross entropy of the one-hot probability distribution corresponding to y and p as

$$H(y, p) = -\ln p(y). \tag{3}$$

Following [31], we occasionally regard, mathematically, a classification student DNN as a mapping from raw data $x \in \mathbb{R}^d$ to a probability distribution $q_x \in \mathcal{P}([C])$. Given a student DNN: $x \in \mathbb{R}^d \to q_x$, let θ denote its weight vector consisting of all its connection weights; whenever there is no ambiguity, we also write q_x as $q_{x,\theta}$. For random variables X and Y, denote by P_X the probability distribution of X, by $P_{X|Y}$ the conditional probability distribution of X given Y, and by $\mathbb{E}_X[\cdot]$ the expectation computed w.r.t. X. We use $\mathfrak{M}^{[m,n]}$ to denote an $m \times n$ real matrix, and $\mathfrak{E}^n$ to denote the $n \times n$ identity matrix. Denote the ith row of the matrix $\mathfrak{M}^{\{m,n\}}$ by $\mathfrak{M}^{\{m,n\}}[i]$.

2.2 Power Transform of Probability Distribution

The power transform of probability distribution was introduced in [38]. Given $p \in \mathcal{P}([C])$ and $\gamma = 1/T > 0$, the power transform of p is another probability distribution $\hat{p} \in \mathcal{P}([C])$ defined as

$$\hat{p}(i) = \frac{p(i)^\gamma}{\sum_{j \in C} p(j)^\gamma}, \forall i \in [C].$$

If p is the softmax of the logit vector $(l_1, l_2, \cdots, l_C)$, then it is not hard to verify that the power transformed $\hat{p}$ is the softmax of the power scaled logit vector $(l_1/T, l_2/T, \cdots, l_C/T)$. Therefore, temperature scaling can be equivalently operated directly on the distribution itself.

2.3 Knowledge Distillation

Let (X, Y) be a pair of random variables, the distribution of which governs a training set, where $X \in \mathbb{R}^d$ represents the raw input data and Y is the ground truth label of X. Fix a pre-trained teacher: $x \in \mathbb{R}^d \to p_x$. Let $T > 0$ be the scaling temperature. In light of the equivalence between temperature scaling and power transform, the standard KD formulated in [5] trains the student DNN: $x \in \mathbb{R}^d \to q_x$ by solving the following minimization problem

$$\min_{\theta} \mathbb{E}_{(X,Y)}[H(Y, q_{X,\theta}) + \beta T^2 D(\hat{p}_X || \hat{q}_{X,\theta})] \tag{4}$$

where $\beta > 0$ is a hyperparameter, and $\hat{p}_x$ and $\hat{q}_{x,\theta}$ are the power transforms of the teacher's distribution p_x and student's distribution $q_{x,\theta}$ in response to the input x, respectively. Therefore, the standard KD is a distribution-based distillation method.

2.4 Self-undermining KD

To train and build nasty teachers, Ma et al. [16] proposed a method called self-undermining KD. Fix a pre-trained normal teacher: $x \in \mathbb{R}^d \to p_x$. Let $T > 0$ be the scaling temperature. In self-undermining KD, the student: $x \in \mathbb{R}^d \to q_x$ is a nasty teacher to be trained, normally has the same network architecture as the fixed pre-trained teacher, and is trained by solving the following minimization problem

$$\min_{\theta} \mathbb{E}_{(X,Y)}[H(Y, q_{X,\theta}) - \omega T^2 D(\hat{q}_{X,\theta} || \hat{p}_X)] \tag{5}$$

where ω is a hyperparameter. In comparison with (4), the training process in self-undermining KD tries to push $\hat{q}_{x,\theta}$ away from $\hat{p}_x$.

2.5 Markov Matrix and Conditional Mutual Information

A real-valued $C \times C$ matrix $\mathfrak{M}^{[C,C]}$ is said to be a Markov matrix if all of its elements are nonnegative, and the sum of each row is equal to 1. To facilitate our subsequent discussion, here we provide some context to understand the role of Markov matrix in the information theoretic paradigm of DNN [31].

Given a classification DNN: $x \in \mathbb{R}^d \to p_x$, let $\hat{Y}$ be the label predicted by the DNN with probability $p_X(\hat{Y})$ in response to the input X. It can be verified that $Y \to X \to \hat{Y}$ forms a Markov chain in the indicated order. For each label $y \in [C]$, consider all input samples x with its ground truth label being y. The DNN then maps this subset of input samples into a cluster of probability distributions p_x in the space $\mathcal{P}([C])$ (see Fig. 1 for visualization). It was demonstrated in [31] that the concentration of this cluster can be measured by the conditional mutual information $I(X, \hat{Y}|Y=y)$ between the input X and the predicted label $\hat{Y}$ given the ground truth label $Y = y$; the latter was further shown to be the conditional average KL divergence between p_x in the cluster and the centroid of the cluster given $Y = y$

$$I(X, \hat{Y}|Y = y) = \sum_x P_{X|Y}(x|y) D(p_x \| o_y) \tag{6}$$

where

$$o_y = \sum_x P_{X|Y}(x|y) p_x$$

is the conditional distribution of $\hat{Y}$ given $Y = y$, i.e., the centroid of the cluster. The conditional mutual information (CMI) $I(X; \hat{Y}|Y)$ is then the average concentration of all clusters corresponding to all labels.

To measure how far clusters corresponding to all labels are from each other, Yang et al. [31] further introduced the notion of separation distance Γ. Let (U, V) be a pair of random variables which are independent of (X, Y) and have the same joint distribution as (X, Y). The separation distance Γ of clusters corresponding to all labels is computed as

$$\Gamma = \mathbb{E}_{(X,Y,U,V)}[I_{Y \neq V} H(p_X, p_U)] \tag{7}$$

where $I_{Y \neq V}$ is the indicator function of the event $\{Y \neq V\}$.

How can we further make clusters each more concentrated and far apart from each other without modifying or re-training the DNN? One way is to multiply each p_x by a Markov matrix $\mathfrak{M}^{[C,C]}$, where p_x is regarded as a row vector. This will be further explored in Sect. 4, leading us to propose MKD.

3 KD-Resistance of Nasty Teachers

Using the concept of KD-resistant DNN, we now evaluate the KD-resistance of nasty teachers trained by self-undermining KD against popular logit-based KD methods existing in the literature.

As mentioned earlier, it was demonstrated in [16] that for student models tested therein, the distillation process of applying the standard KD method to nasty teachers trained by self-undermining KD results in significant loss in student accuracy. Therefore, these nasty teachers are KD-resistant with respect to the standard KD. Later on, more advanced logit-based KD methods were proposed in [8,11,15], and the resulting loss in student accuracy was fully or partially recovered. However, as shown in Table 1 (the last three columns), none of the distilled students from the KD methods in [8,11,15] outperforms the respective LS student. For more experiment results, please refer to Sect. 5. Therefore, nasty teachers trained by self-undermining KD are still KD-resistant with respect to these KD methods, even though they were designed specifically to distill nasty teachers.

In the literature, in addition to logit-based KD methods designed specifically to distill nasty teachers, there are also other state-of-the-art logit-based KD methods which deliver good distillation performance from normal teachers,

Table 1. Illustration of KD-resistance of nasty teachers against various state-of-the-art logit-based KD methods.

Cifar-10										
Tch.	Stu.	LS	KD [5]	DKD [37]	DIST [7]	NKD [32]	MLD [9]	Skep. [15]	HTC [8]	Avg. [11]
R18	CNN	87.40	82.11	86.89	85.28	85.64	85.59	86.71	87.33	83.02
	RC20	92.53	88.42	92.30	84.70	91.57	91.67	91.85	92.48	88.73
	RC32	93.36	89.61	92.49	85.85	92.11	92.56	92.98	93.26	90.33

notably KD methods from [7,9,32,37]. As such, it is instructive to evaluate KD-resistance of nasty teachers against these methods as well. Table 1 shows their respective results on Cifar-10 dataset for nasty teacher R18 and student models CNN, RC20, and RC32. As can be seen from Table 1, none of the resulting distilled students from these methods outperforms the respective LS student. Therefore, nasty teachers trained by self-undermining KD are KD-resistant with respect to these state-of-the-art KD methods as well.

4 Markov Knowledge Distillation

In this section, we present our proposed MKD. We begin with providing some insight into why nasty teachers are KD-resistant with respect to various state-of-the-art KD methods from the lens of information geometry of DNN [31].

4.1 Information Geometry

(a) **normal** teacher.   (b) **nasty** teacher.   (c) **Markov** transformed **nasty** teacher.

Fig. 1. Visualization of the three projected clusters of probability distributions corresponding to three randomly chosen labels (CIFAR-10 training dateset).

Consider a classification DNN: $x \in \mathbb{R}^d \to p_x$. It maps input samples x with different labels into clusters of probability distributions p_x in the space $\mathcal{P}([C])$, one cluster per label. To visualize these clusters, we follow the visualization approach advocated in [31]. Take the training set of CIFAR-10 dateset [13] as an example. Pick three labels randomly. For each probability distribution in the three clusters corresponding to these picked labels, consider only the probabilities of these three labels, normalize them so that they become a 3 dimensional probability vector, and further project the resulting probability vector into the 2 dimensional simplex. Then the three clusters corresponding to three picked labels are projected into and can be viewed in the two dimensional simplex [31].

Figure 1 shows the resulting three projected clusters for a normal teacher trained with the standard cross-entropy loss (Fig. 1a) and for a nasty teacher trained by self-undermining KD (Fig. 1b). From Fig. 1, it is quite clear that the three projected clusters of the nasty teacher look very different from those of the normal teacher in pattern. In particular, (i) the projected clusters of automobiles and trucks in the case of the nasty teacher have a strong linear relationship and are also interconnected and mixed together; (ii) probability vectors within each cluster of the nasty teacher also have a strong linear relationship, which aligns with the previous observation that the nasty teacher's output probability has multiple peak entries [16]. It is the first phenomenon that impacts negatively on the ability of existing state-of-the-art KD methods to distill nasty teachers.

Is there any way to correct the first phenomenon mentioned above without modifying the nasty teacher itself? If one can find a way to move distributions within each cluster around, then it could be possible to correct the first phenomenon. One such a way is to find a right $C \times C$ Markov matrix $\mathfrak{M}_c^{[C,C]}$ for each cluster corresponding to a label c and then multiply each p_x in that cluster by $\mathfrak{M}_c^{[C,C]}$. Such a transform is called Markov transform. Figure 1c shows the resulting projected three clusters after such a Markov transform with properly designed $\mathfrak{M}_c^{[C,C]}$, $\forall c \in [C]$. Comparing Figs. 1c with 1a) and Fig. 1b, it is clear that the three projected clusters in Fig. 1c are now well separated, look more similar to those in Fig. 1a), and are each more concentrated than those in Fig. 1b.

Markov transform can also be nicely interpreted in a probabilistic language. Recall that $Y \to X \to \hat{Y}$ forms a Markov chain, where $\hat{Y}$ is the label randomly predicted by the DNN with probability $p_X(\hat{Y})$ in response to the input X. Now we introduce the fourth random variable $\tilde{Y}$ such that

$$\Pr\{\tilde{Y} = j | Y = c, X = x, \hat{Y} = i\} = \mathfrak{M}_c^{[C,C]}(i,j), \forall c, x, i, j \qquad (8)$$

where $\mathfrak{M}_c^{[C,C]}(i,j)$ is the element at the ith row and j^{th} column of $\mathfrak{M}_c^{[C,C]}$. Note that the conditional distribution of $\tilde{Y}$ given $Y = c$ and $X = x$ is exactly $p_x \mathfrak{M}_c^{[C,C]}$. In other words, given $Y = c$ and $X = x$, $\tilde{Y}$ is the randomly predicted label after Markov transform $\mathfrak{M}_c^{[C,C]}$ is applied to p_x, whereas $\hat{Y}$ is the randomly predicted label before Markov transform.

As discussed in Subsect. 2.5, the concentration of the cluster of distributions p_x corresponding to the label c is the CMI $I(X; \hat{Y}|Y = c)$. After Markov transform, the concentration of the Markov transformed cluster corresponding to the label c is then equal to $I(X; \tilde{Y}|Y = c)$. Likewise, the separation distance Γ among Markov transformed clusters corresponding to all labels can be computed similarly by simply replacing $\hat{Y}$ with $\tilde{Y}$. As an example, Figs. 1b and 1c illustrate these values before and after Markov transform. From Figs. 1b and 1c, we see that after Markov transform, the CMI value of each cluster decreases while the separation distance Γ increases, which is consistent with the visual effect of those clusters shown in Figs. 1b and 1c.

While we are unable, at this point, to prove that the separation distance Γ always increases after Markov transform, which is left open as an information theoretic problem, the following theorem does guarantee that the CMI of each cluster indeed never increases after Markov transform.

Theorem 1. *For any Markov matrices $\mathfrak{M}_c^{[C,C]}$, $\forall c \in [C]$, we have*

$$I(X;\tilde{Y}|Y=c) \leq I(X;\hat{Y}|Y=c), \forall c \in [C]. \tag{9}$$

The proof Theorem 1 is presented in the Appendix. Now the question is how to design Markov matrices $\mathfrak{M}_c^{[C,C]}$, $\forall c \in [C]$ so that Markov transformed distributions can facilitate the distillation process to improve the student performance, regardless of whether the teacher is nasty or normal. This question is addressed in the next subsection.

4.2 Detailed Description of MKD

Based on the above discussion, we are now ready to describe MKD in details. From the outset, it should be emphasized that MKD is a generic method which can be applied in conjunction with any underlying distribution-based KD method. As shown in Fig. 3, MKD has several key components (in addition to the pre-trained teacher and student model): (1) power transform, (2) Markov transform, (3) student loss, and (4) co-learning of Markov transform and student. With reference to Fig. 3, these components are described next one by one.

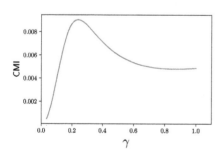

Fig. 2. CMI-γ curve of Resnet18 trained by self-undermining KD on Cifar-10 dataset, which has its maximum at $\gamma = 0.237$.

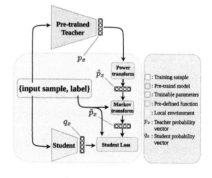

Fig. 3. The diagram of MKD, where the student loss function is identical to that of the underlying distribution-based KD method except it now takes $\tilde{p}_x$ instead of p_x as one of its inputs.

Power Transform. For each input sample x, the power transform converts the output probability distribution p_x of the teacher into the distribution $\hat{p}_x$. As discussed in Subsect. 2.2, the power transform is equivalent to temperature scaling proposed in [5]. The purpose here is to increase the CMI $I(X; \hat{Y}|Y)$ of the teacher[2]. Since the teacher is fixed, the power parameter $\gamma = 1/T$ can be determined in advance of student training. Figure 2 plots the CMI-γ curve of Resnet18 trained by self-undermining KD on Cifar-10 dataset. In this case, the optimal value for γ is $\gamma = 0.237$, at which the CMI value peaks and is 83.3% larger than the default value before the power transformation. In general, the optimal γ value can be easily found by ternary search.

Markov Transform. For each input sample x with its ground truth label c, Markov transform with Markov matrix $\mathfrak{M}_c^{[C,C]}$ further converts the power transformed distribution $\hat{p}_x$ into the distribution $\tilde{p}_x$:

$$\tilde{p}_x = \hat{p}_x \mathfrak{M}_c^{[C,C]}. \tag{10}$$

Note that different labels have their own different Markov matrices $\mathfrak{M}_c^{[C,C]}$, $\forall c \in [C]$.

How would each Markov matrix $\mathfrak{M}_c^{[C,C]}$ be represented? To make it to be easily trained, we limit each Markov matrix $\mathfrak{M}_c^{[C,C]}$ to the following form:

$$\mathfrak{M}_c^{[C,C]} = softmax\left(\mathfrak{Z}_c^{[C,C]}\right) \tag{11}$$

where $\mathfrak{Z}_c^{[C,C]}$ is any trainable $C \times C$ real-valued matrix, and $softmax$ is applied to each row of $\mathfrak{Z}_c^{[C,C]}$ so that $\mathfrak{M}_c^{[C,C]}$ is indeed a Markov matrix.

With the matrix form given in (11), the number of Markov transform parameters is C^3. Since Markov matrices have to be co-learned later, C^3 may be two large to be acceptable. To get the number of Markov transform parameters under control, one can further limit each matrix $\mathfrak{Z}_c^{[C,C]}$ to the following form:

$$\mathfrak{Z}_c^{[C,C]} = \alpha \times \mathfrak{E}^C + \mathfrak{A}^{[C,n]} \mathfrak{B}^{[n,C]} \tag{12}$$

where $\mathfrak{A}^{[C,n]}$ and $\mathfrak{B}^{[n,C]}$ are two real-valued matrices, α is a trainable scaling factor, and n is referred to as the intrinsic dimension of $\mathfrak{Z}_c^{[C,C]}$. In this case, the number of Markov transform parameters is reduced to $C(1 + 2nC)$, which is acceptable when n is small.

Student Loss. In MKD, the student loss function is identical to that of the underlying distribution-based KD method except that it now takes $\tilde{p}_x$ instead of the original p_x as one of the inputs. If temperature scaling is applied to p_x in the

---

[2] The recent works in [31,33] show that in the context of KD, it is beneficial to increase $I(X; \hat{Y}|Y)$ [33], whereas in the context of training a DNN for its own performance, it is beneficial to minimize $I(X; \hat{Y}|Y)$ [31].

underlying distribution-based KD method, then the same operation is applied to $\tilde{p}_x$ as well, but in its equivalent form of power transform. For example, if the underlying distribution-based KD method is the standard KD [5], in which case we refer to MKD specifically as MKD-KD, then the student loss function in MKD-KD is

$$\mathbb{E}_{(X,Y)}[H(Y, q_{X,\theta}) + \beta\tau^2 D(\hat{\tilde{p}}_X || \hat{q}_{X,\theta})] \tag{13}$$

where τ is the temperature used in KD. In the above, when the power transform is applied to both $\tilde{p}_x$ and $q_{x,\theta}$, its power parameter is $\gamma = 1/\tau$.

Co-Learning of Markov Transform and Student. MKD trains the student and Markov transform parameters simultaneously by minimizing the student loss. Specifically, if the student loss function in MKD is expressed as

$$\mathbb{E}_{(X,Y)}[L(Y, q_{X,\theta}, \tilde{p}_X)]$$

then the student and Markov transform are co-trained simultaneously by solving the following minimization problem

$$\min_{\theta} \min_{3_c^{[C,C]}, c\in[C]} \mathbb{E}_{(X,Y)}[L(Y, q_{X,\theta}, \tilde{p}_X)]. \tag{14}$$

For example, in MKD-KD, the student and Markov transform are co-trained simultaneously by solving the following minimization problem

$$\min_{\theta} \min_{3_c^{[C,C]}, c\in[C]} \mathbb{E}_{(X,Y)}[H(Y, q_{X,\theta}) + \beta\tau^2 D(\hat{\tilde{p}}_X || \hat{q}_{X,\theta})]. \tag{15}$$

Standard learning algorithms with backward propagation and stochastic gradient decent (SGD) can then be applied to solve (14). Note that Markov transform parameters $3_c^{[C,C]}, c \in [C]$ are not part of the student. Once the training process is done, the distilled student is completely decoupled from Markov transform and can be used in inference as usual.

In the next section, we will evaluate the distillation capability of MKD by applying it to both nasty teachers and normal teachers.

5 Experiments

Through extensive experiments, in this section, We demonstrate that (1) MKD makes nasty teachers trained by self-undermining KD fully distillable; (2) when applied to normal teachers, MKD outperforms the underlying distribution-based KD method; (3) the same remains valid in the case of few shots; and (4) more interestingly, MKD is able to transfer knowledge from one domain to another domain.

Experiment Setup: We largely followed the setup in [16]. For datasets, we used three common image classification datasets: Cifar-10, Cifar-100, and TinyImageNet [14,25]. For teacher-student pairs, we selected teachers from ResNet{18, 50} [6] (shortened as R18 and R50, resp.) and ResNext29 [30] (shortened as Rnt29), and students from CNN, ResNetC{20, 32} (shortened as RC20 and RC32, resp.), MobileNetV2 [26] (shortened as MV2), ShuffleNetV2 [18] (shorten as SV2), and R18, where CNN denotes a customized 5-layer convolutional neural network.

Training Settings: For all experiments, a batch size of 128 was used across all models. Unless specified otherwise, the intrinsic dimension n of Markov transform was set to 3 for Cifar-10,100 and 5 for TinyImageNet. The SGD optimizer with a learning rate of 0.1 and momentum of 0.99 was used for ResNet, ShuffleNetV2, and MobileNetV2. For the CNN model, we employed the Adam optimizer [12] with a learning rate of 10^{-3} and momentum of 0.99.

5.1 Distilling Nasty Teachers

Using the standard KD as the underlying distribution-based KD method in MKD, we applied MKD-KD to distill nasty teachers trained by self-undermining KD. Table 2 reports the resulting Top-1 validation accuracy results along with those delivered by CE student, LS student, KD, and those distribution-based KD methods designed specifically to distill nasty teachers, namely Skep. [15], HTC [8] and Avg. [11]}. All results are obtained by averaging 3 independent runs. From Table 2, it is clear that although Skep., HTC, and Avg. can recover fully or partially the accuracy loss from the standard KD process, the distilled students by them still perform worse than LS student in general. On the other hand, the distilled student by MKD-KD consistently outperforms LS student. In particular, for the student model ShuffleNetV2, the accuracy gain over LS student is non-trivial and can be as high as 3.57%. Thus, MKD-KD makes nasty teachers fully distillable!. Experimental results on ImageNet [25] are provided in the Appendix.

5.2 Distilling Normal Teachers

To evaluate the distillation capability of MKD as a generic distribution-based KD method, we also applied MKD to distill normal teachers. Specifically, we used the standard KD and DKD as the respective underlying distribution-based KD methods in MKD and applied MKD-KD and MKD-DKD to the same teacher-students pairs as in Subsect. 5.1 on Cifar-100 and TinyImageNet, except now all teachers are normal. Table 3 reports the resulting Top-1 validation accuracy results along with those delivered by KD and DKD themselves. As seen from Table 3, MKD in general outperforms its underlying distribution-based KD method. In particular, for TinyImageNet dataset and the student model ShuffleNetV2, MKD outperforms KD by a large margin (as high as over 3%).

Table 2. Top-1 validation accuracy (%) of CE student, LS student, and various distilled students from nasty teachers by different KD methods on Cifar-{10, 100} and TinyImageNet datasets (averaged over 3 runs). (↑) ((↓), resp.) indicates better (worse, resp.) performance than LS student. For each teacher-student pair, the best and the second best results are **bold** and underlined, respectively.

Cifar-10								
Tch.	Stu.	CE	LS	KD (NPIS 2015)	Skep. (NPIS 2021)	HTC (ECCV 2022)	Avg. (ICLR 2023)	MKD-KD
R18	CNN	86.64	87.40	82.11 (↓)	86.71 (↓)	87.33 (↓)	83.02 (↓)	**87.82** (↑)
	RC20	92.31	92.53	88.43 (↓)	91.85 (↓)	92.48 (↓)	88.73 (↓)	**92.91** (↑)
	RC32	93.01	93.36	89.61 (↓)	92.98 (↓)	93.26 (↓)	90.33 (↓)	**93.69** (↑)
Cifar-100								
R18	MV2	69.00	70.37	6.53 (↓)	66.47 (↓)	69.64 (↓)	**71.45** (↑)	71.41 (↑)
	SV2	71.52	71.56	64.43 (↓)	70.46 (↓)	71.62 (↑)	71.26 (↓)	**72.39** (↑)
	R18	78.11	78.96	74.21 (↓)	77.45 (↓)	78.64 (↓)	77.42 (↓)	**79.26** (↑)
R50	MV2	69.00	70.37	4.23 (↓)	66.94 (↓)	69.73 (↓)	67.42 (↓)	**71.68** (↑)
	SV2	71.52	71.56	64.69 (↓)	71.44 (↓)	71.58 (↑)	70.59 (↓)	**72.33** (↑)
	R18	78.11	78.96	71.71 (↓)	77.33 (↓)	78.47 (↓)	78.53 (↓)	**79.46** (↑)
Rnt29	MV2	69.00	70.37	2.15 (↓)	65.43 (↓)	69.23 (↓)	70.16 (↓)	**71.04** (↑)
	SV2	71.52	71.56	59.34 (↓)	70.85 (↓)	71.69 (↑)	71.35 (↓)	**72.57** (↑)
	R18	78.11	78.96	66.42 (↓)	75.48 (↓)	78.29 (↓)	78.93 (↓)	**79.27** (↑)
TinyImageNet								
R18	MV2	54.62	55.79	1.04 (↓)	54.87 (↓)	55.42 (↓)	55.63 (↓)	**55.74** (↑)
	SV2	56.76	57.91	25.42 (↓)	58.24 (↑)	56.78 (↓)	57.42 (↓)	**60.21** (↑)
R50	MV2	54.62	55.79	2.42 (↓)	55.20 (↓)	55.69 (↓)	**57.49** (↑)	57.35 (↑)
	SV2	56.76	57.91	36.22 (↓)	56.12 (↓)	57.42 (↓)	57.52 (↓)	**61.48** (↑)

Table 3. Top-1 validation accuracy (%) of various distilled students from normal teachers by KD, DKD, and MKD on Cifar-100 and TinyImageNet datasets.

Cifar-100					
Tch	Stu	KD	MKD-KD	DKD	MKD-DKD
R18	MV2	72.41	72.31	72.60	73.04
	SV2	74.53	74.61	75.38	75.49
	R18	79.35	79.45	79.62	79.74
R50	MV2	72.13	72.09	73.01	73.50
	SV2	73.61	74.02	75.42	75.46
	R18	79.46	79.92	80.09	80.23
Rnt29	MV2	72.43	72.65	72.68	72.97
	SV2	72.68	73.23	75.04	75.35
	R18	79.63	79.98	79.78	80.25
TinyImageNet					
R18	MV2	55.99	56.33	59.22	59.43
	SV2	58.09	60.96	61.60	62.09
R50	MV2	56.24	57.42	60.02	60.13
	SV2	58.45	61.51	61.60	62.17

5.3 Few-Shot Knowledge Distillation

The effectiveness of MKD is also evaluated in the case of few shot learning where only a small percentage of training samples are available in student learning. Detailed results are reported in the Appendix.

5.4 Cross-Domain Knowledge Distillation

Table 4. Top-1 validation accuracy (%) of various students in cross-domain distillation.

Cifar-50-50						
Stu.	CE	LS	Nasty R18		Normal R18	
			KD	MKD-KD	KD	MKD-KD
MV2	78.95	78.98	4.04	80.02	8.18	79.53
SV2	79.04	79.18	8.87	80.21	4.10	79.91

So far, in this paper and also in the literature, various KD methods have been mainly evaluated in the in-domain setting where both the teacher and student are trained on datasets with same class labels. Now we want to evaluate the distillation capability of MKD in the cross-domain setting where the pre-trained teacher was trained on one domain whereas the student will be trained on another domain and where the two domains do not overlap. To this end, we create two cross-domain datasets dubbed Cifar-50-50 and TinyImageNet-100-100 as explained below.

- **Cifar-50-50:** We randomly partition the label set of Cifar-100 into two equal-sized subsets, each containing 50 labels. All samples in Cifar-100 with labels from the first subset are grouped into Cifar-50-50 Domain-1 dataset which contains 25K training samples and 5K validation samples. Likewise, all samples in Cifar-100 with labels from the second subset are grouped into Cifar-50-50 Domain-2 dataset which also contains 25K training samples and 5K validation samples.
- **TinyImageNet-100-100:** Similar to **Cifar-50-50**, we randomly partition the label set of TinyImageNet into two equal-sized subsets, each containing 100 labels, and create TinyImageNet-100-100 Domain-1 dataset and TinyImageNet-100-100 Domain-2 dataset accordingly.

Note that in the above, Domain-1 and Domain-2 datasets do not share any single common label. In our experiments, the teacher, whether it is nasty or normal, was pre-trained on Domain-1 dataset. We then applied both KD and MKD-KD to distill such teachers and trained students on Domain-2 dataset. In other words, the input samples in Fig. 3 all come from Domain-2 dataset. In MKD, the scaling factor α is set to 0, and the intrinsic dimension n is set to 16 for Cifar-50-50 and 25 for TinyImageNet-100-100.

Table 4 reports the resulting Top-1 validation accuracy results of distilled students by KD and MKD-KD, CE student, and LS student on Cifar-50-50. Note that the distilled student by KD is completely confused due to cross-domain transfer. On the other hand, the distilled student by MKD-KD always outperforms LS student, which is a bit surprising given the fact that the teacher was trained in a completely different domain. More experiment results are reported in the Appendix. The cross-domain distillation capability of MKD is due to its Markov transform and co-learning feature.

5.5 Ablation Study

Table 5. Ablation effects of $\mathcal{PT}$ and $\mathcal{MT}$ on Top-1 accuracy (%) of distilled students from nasty teachers by MKD-KD.

Cifar-100

Tch.	$\mathcal{PT}$	x	x	✓	✓	Tch.	$\mathcal{PT}$	x	x	✓	✓	Tch.	$\mathcal{PT}$	x	x	✓	✓
	$\mathcal{MT}$	x	✓	x	✓		$\mathcal{MT}$	x	✓	x	✓		$\mathcal{MT}$	x	✓	x	✓
R18	MV2	6.53	71.37	42.76	71.41	R50	MV2	4.23	71.11	53.65	71.68	Rnt29	MV2	2.15	70.50	23.54	71.01
	SV2	64.43	71.76	65.23	72.39		SV2	66.93	72.15	66.93	73.33		SV2	59.34	72.57	69.42	72.57
	R18	74.21	79.26	76.3	79.26		R18	74.63	79.14	74.63	79.46		R18	66.42	79.27	78.41	79.27

Table 5 shows the ablation effects of power transform ($\mathcal{PT}$) and Markov transform ($\mathcal{MT}$) on Top-1 validation accuracy (%) of distilled students from nasty teachers by MKD-KD. Clearly, Markov transform is very instrumental. The effects of intrinsic dimension n on distilled students from nasty teachers by MKD-KD are reported in the Appendix.

6 Conclusion

Using LS student, the student trained alone with the cross entropy loss plus label smoothing penalty, as a benchmark, we have introduced two new concepts related to knowledge distillation (KD)—distillable DNN and KD-resistant DNN—to investigate the distillation capability of various KD methods, and KD-resistance of DNNs against block-box distillation. It has been shown that nasty teachers trained by self-undermining KD are KD-resistant with respect to state-of-the-art KD methods existing in the literature before our work. On the other hand, we have proposed a new KD method dubbed Markov KD (MKD), which makes nasty teachers fully distillable. MKD is generic and turns out to be very powerful. When applied to normal teachers, it outperforms its underlying distribution-based KD method by a large margin. More interestingly, due to its co-learning feature, MKD is even capable to transfer knowledge from teachers trained in one domain to students trained in another domain, opening a new research direction: cross-domain knowledge transfer.

Acknowledgments. This work was supported in part by the Natural Sciences and Engineering Research Council of Canada under Grant RGPIN203035-22, and in part by the Canada Research Chairs Program.

References

1. Ahn, S., Hu, S.X., Damianou, A., Lawrence, N.D., Dai, Z.: Variational information distillation for knowledge transfer. In: Proceedings of the IEEE/CVF Conference on Computer Vision and Pattern Recognition, pp. 9163–9171 (2019)
2. Anil, R., Pereyra, G., Passos, A., Ormandi, R., Dahl, G.E., Hinton, G.E.: Large scale distributed neural network training through online distillation. arXiv preprint arXiv:1804.03235 (2018)
3. Buciluǎ, C., Caruana, R., Niculescu-Mizil, A.: Model compression. In: Proceedings of the 12th ACM SIGKDD International Conference on Knowledge Discovery and Data Mining, pp. 535–541 (2006)
4. Chen, P., Liu, S., Zhao, H., Jia, J.: Distilling knowledge via knowledge review. In: Proceedings of the IEEE/CVF Conference on Computer Vision and Pattern Recognition, pp. 5008–5017 (2021)
5. Hinton, G., Vinyals, O., Dean, J.: Distilling the knowledge in a neural network. arXiv:1503.02531 (2015)
6. He, K., Zhang, X., Ren, S., Sun, J.: Deep residual learning for image recognition. In: Proceedings of the IEEE Conference on Computer Vision and Pattern Recognition, pp. 770–778 (2016)
7. Huang, T., You, S., Wang, F., Qian, C., Xu, C.: Knowledge distillation from a stronger teacher. arXiv preprint arXiv:2205.10536 (2022)
8. Jandial, S., Khasbage, Y., Pal, A., Balasubramanian, V.N., Krishnamurthy, B.: Distilling the undistillable: learning from a nasty teacher. In: Avidan, S., Brostow, G., Cissé, M., Farinella, G.M., Hassner, T. (eds.) Computer Vision - ECCV 2022, pp. 587–603. Springer Nature Switzerland, Cham (2022). https://doi.org/10.1007/978-3-031-19778-9_34
9. Jin, Y., Wang, J., Lin, D.: Multi-level logit distillation. In: Proceedings of the IEEE/CVF Conference on Computer Vision and Pattern Recognition, pp. 24276–24285 (2023)
10. Kaissis, G., et al.: End-to-end privacy preserving deep learning on multi-institutional medical imaging. Nat. Mach. Intell. **3**(6), 473–484 (2021)
11. Keser, R.K., Toreyin, B.U.: Averager student: distillation from undistillable teacher (2023). https://openreview.net/forum?id=4isz71_aZN
12. Kingma, D.P., Ba, J.: Adam: a method for stochastic optimization. arXiv preprint arXiv:1412.6980 (2014)
13. Krizhevsky, A., Hinton, G.: Learning multiple layers of features from tiny images. University of Toronto (2009)
14. Krizhevsky, A., Nair, V., Hinton, G.: Cifar-10 (Canadian institute for advanced research). University of Toronto (2012). http://www.cs.toronto.edu/~kriz/cifar.html
15. Kundu, S., Sun, Q., Fu, Y., Pedram, M., Beerel, P.: Analyzing the confidentiality of undistillable teachers in knowledge distillation. In: Ranzato, M., Beygelzimer, A., Dauphin, Y., Liang, P., Vaughan, J.W. (eds.) Advances in Neural Information Processing Systems, vol. 34, pp. 9181–9192. Curran Associates, Inc. (2021). https://proceedings.neurips.cc/paper_files/paper/2021/file/4ca82782c5372a547c104929f03fe7a9-Paper.pdf

16. Ma, H., Chen, T., Hu, T.K., You, C., Xie, X., Wang, Z.: Undistillable: making a nasty teacher that cannot teach students. In: International Conference on Learning Representations (2021). https://openreview.net/forum?id=0zvfm-nZqQs
17. Ma, H., et al.: Stingy teacher: sparse logits suffice to fail knowledge distillation (2022). https://openreview.net/forum?id=ae7BJIOxkxH
18. Ma, N., Zhang, X., Zheng, H.T., Sun, J.: ShuffleNet V2: practical guidelines for efficient CNN architecture design. In: Proceedings of the European Conference on Computer Vision (ECCV), pp. 116–131 (2018)
19. Müller, R., Kornblith, S., Hinton, G.E.: When does label smoothing help? In: Advances in Neural Information Processing Systems, vol. 32 (2019)
20. Park, W., Kim, D., Lu, Y., Cho, M.: Relational knowledge distillation. In: Proceedings of the IEEE/CVF Conference on Computer Vision and Pattern Recognition, pp. 3967–3976 (2019)
21. Passalis, N., Tefas, A.: Learning deep representations with probabilistic knowledge transfer. In: Proceedings of the European Conference on Computer Vision (ECCV), pp. 268–284 (2018)
22. Peng, B., et al.: Correlation congruence for knowledge distillation. In: Proceedings of the IEEE/CVF International Conference on Computer Vision, pp. 5007–5016 (2019)
23. Pereyra, G., Tucker, G., Chorowski, J., Kaiser, L., Hinton, G.: Regularizing neural networks by penalizing confident output distributions. arXiv preprint arXiv:1701.06548 (2017)
24. Radosavovic, I., Dollár, P., Girshick, R., Gkioxari, G., He, K.: Data distillation: towards omni-supervised learning. In: Proceedings of the IEEE Conference on Computer Vision and Pattern Recognition, pp. 4119–4128 (2018)
25. Russakovsky, O., et al.: ImageNet large scale visual recognition challenge. Int. J. Comput. Vision **115**, 211–252 (2015)
26. Sandler, M., Howard, A., Zhu, M., Zhmoginov, A., Chen, L.C.: MobileNetV2: inverted residuals and linear bottlenecks. In: Proceedings of the IEEE Conference on Computer Vision and Pattern Recognition, pp. 4510–4520 (2018)
27. Shokri, R., Shmatikov, V.: Privacy-preserving deep learning. In: Proceedings of the 22nd ACM SIGSAC Conference on Computer and Communications Security, pp. 1310–1321 (2015)
28. Tian, Y., Krishnan, D., Isola, P.: Contrastive representation distillation. In: International Conference on Learning Representations (2019)
29. Wu, B., et al.: P3SGD: patient privacy preserving SGD for regularizing deep CNNs in pathological image classification. In: Proceedings of the IEEE/CVF Conference on Computer Vision and Pattern Recognition, pp. 2099–2108 (2019)
30. Xie, S., Girshick, R., Dollár, P., Tu, Z., He, K.: Aggregated residual transformations for deep neural networks. In: Proceedings of the IEEE Conference on Computer Vision and Pattern Recognition, pp. 1492–1500 (2017)
31. Yang, E.H., Hamidi, S.M., Ye, L., Tan, R., Yang, B.: Conditional mutual information constrained deep learning for classification. arXiv:2309.09123 (2023)
32. Yang, Z., Zeng, A., Yuan, C., Li, Y.: From knowledge distillation to self-knowledge distillation: a unified approach with normalized loss and customized soft labels. In: Proceedings of the IEEE/CVF International Conference on Computer Vision, pp. 17185–17194 (2023)
33. Ye, L., Hamidi, S.M., Tan, R., Yang, E.H.: Bayes conditional distribution estimation for knowledge distillation based on conditional mutual information. In: The Twelfth International Conference on Learning Representations (2024). https://openreview.net/forum?id=yV6wwEbtkR

34. Zagoruyko, S., Komodakis, N.: Paying more attention to attention: improving the performance of convolutional neural networks via attention transfer. In: International Conference on Learning Representations (2016)
35. Zhang, C.B., et al.: Delving deep into label smoothing. IEEE Trans. Image Process. **30**, 5984–5996 (2021). https://doi.org/10.1109/TIP.2021.3089942
36. Zhang, J., et al.: Protecting intellectual property of deep neural networks with watermarking. In: Proceedings of the 2018 on Asia Conference on Computer and Communications Security, pp. 159–172 (2018)
37. Zhao, B., Cui, Q., Song, R., Qiu, Y., Liang, J.: Decoupled knowledge distillation. In: Proceedings of the IEEE/CVF Conference on Computer Vision and Pattern Recognition, pp. 11953–11962 (2022)
38. Zheng, K., Yang, E.H.: Knowledge distillation based on transformed teacher matching. In: The Twelfth International Conference on Learning Representations (2024). https://openreview.net/forum?id=MJ3K7uDGGl
39. Ziller, A., Usynin, D., Braren, R., Makowski, M., Rueckert, D., Kaissis, G.: Medical imaging deep learning with differential privacy. Sci. Rep. **11**(1), 13524 (2021)

Co-speech Gesture Video Generation with 3D Human Meshes

Aniruddha Mahapatra[1], Richa Mishra[1(✉)], Renda Li[2,3], Ziyi Chen[4], Boyang Ding[2,3], Shoulei Wang[2,3], Jun-Yan Zhu[1], Peng Chang[4], Mei Han[4], and Jing Xiao[3]

[1] Carnegie Mellon University, Pittsburgh, PA, USA
[2] University of Science and Technology of China, Hefei, China
[3] Ping An Technology, Shenzhen, China
[4] PAII Inc., Palo Alto, CA, USA
richamis@andrew.cmu.edu

Abstract. Co-speech gesture video generation is an enabling technique for many digital human applications. Substantial progress has been made in creating high-quality talking head videos. However, existing hand gesture video generation methods are primarily limited by the widely adopted 2D skeleton-based gesture representation and still struggle to generate realistic hands. We introduce an audio-driven co-speech video generation pipeline to synthesize human speech videos leveraging 3D human mesh-based representations. By adopting a 3D human mesh-based gesture representation, we present a mesh-grounded video generator that includes a mesh texture map optimization step followed by a conditional GAN network and outputs photorealistic gesture videos with realistic hands. Our experiments on the TalkSHOW dataset demonstrate the effectiveness of our method over 2D skeleton-based baselines.

1 Introduction

Co-speech video generation [10] has emerged as a challenging research topic in visual content creation, offering huge potential for a wide array of applications in human-AI interactions, especially in the new era of conversational agents [28]. Co-speech video synthesis aims to generate a photorealistic and synchronized human talking video that aligns with a given speech, particularly lip movements, facial expressions, and accompanying gestures. Recently, significant progress has been made in realistic talking head video generation, including approaches building character-specific models [12,23] and others focusing on general models applicable for any given character [11,31,37,43,56]. Nonetheless, those approaches are mostly designed around talking face generation and do not naturally extend to the upper body, especially the highly dynamic and flexible human hands.

Previous co-speech gesture video synthesis works heavily rely on 2D skeletons as the intermediate representation for image and video synthesis networks [10,

---

A. Mahapatra and R. Mishra—Equal contribution.

© The Author(s), under exclusive license to Springer Nature Switzerland AG 2025
A. Leonardis et al. (Eds.): ECCV 2024, LNCS 15147, pp. 172–189, 2025.
https://doi.org/10.1007/978-3-031-73024-5_11

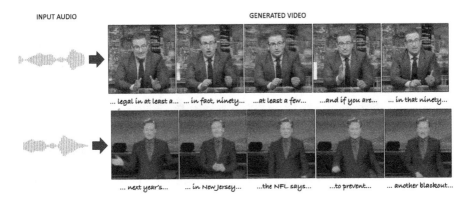

Fig. 1. Given an input audio sequence and a person's identity, our audio-driven co-speech video generation method can create videos with realistic appearances, natural facial expressions, and hand gestures. Here we present video sequences generated from input audio featuring two different speakers. Some areas may appear blurry, representing natural motion blur in the videos. For additional video results, please check our website https://cospeech-gesture-3d.github.io/website.

21,33]. One limitation of this 2D skeleton-based representation is its lack of 3D body shape modeling throughout the generation process, from motion prediction to video rendering steps. Consequently, those methods often struggle to produce natural or physically plausible gesture videos, resulting in blurred or sometimes even missing parts of arms and fingers.

Other works on speech-driven gesture generation [1,13,52,54,58] adopt 3D mesh-based representation for gestures. However, their main goal is to generate gestures in the chosen 3D parameterization space rather than generating the final videos, which is vital for downstream applications.

In this paper, we present a complete pipeline to generate a co-speech video for a target speaker with realistic facial animation and hand gesture movements. Starting with an audio input, we first predict human body shapes and movements for the face, body, and hands, using a 3D human mesh model [30]. Given the estimated 3D human body model (SMPL-X) and its gestures, we use an optimization-based approach to derive a UV texture map corresponding to this SMPL-X model. Finally, we introduce a GANs-based video generation network to transform the textured meshes into the final photorealistic video. We observe that these optimized 3D textured meshes are highly effective as an intermediate representation for co-speech video synthesis. Our code and additional results are available on our website.

2 Related Work

Co-speech Gesture Video Generation. remains under-explored compared to talking head generation research [11,23,24,31,37,43,48,53,56]. In early work,

Ginosar et al. [10] proposed a cross-modal translation network to generate 2D skeleton-based gestures from a given speech, and then the gestures are further synthesized into video with a standalone pose-to-image network [5]. Qian et al. [33] and Liu et al. [21] further improved the 2D skeleton-based gestures by introducing the learning of template vectors and codebooks containing both common gesture patterns and rhythmic movement details, respectively, followed by a pose-to-image generator. Although exciting results have been obtained, the 2D skeleton representation inherently lacks the 3D shape constraint and often leads to unnatural and physically inaccurate gesture configurations, degrading the quality of the generated videos. Concurrent with our work, Make-Your-Anchor [15] also explored the use of 3D representation but deployed a 2D diffusion model to perform the translation.

Co-speech Human Motion Generation. Co-speech motion generation has been widely explored over the years. While certain studies specialize in facial generation [7,8,18,35], other works specifically target body gestures [1,10,21, 33,52,54,58]. However, using a unified framework to generate holistic motions, including face, body, and hands, is still challenging. Habibie et al. [13] generates the 3D mesh of the face and the keypoints of the body and hands given an input voice, but the body mesh is not generated. The introduction of a series of human body parametric models [3,19,22,30] makes it possible to generate a coherent human body 3D mesh as a whole. Recently, Yi et al. [54] enables full-body 3D mesh generation by separately employing an encoder-decoder for facial motion generation and a cross-modal VQ-VAE [41] for synthesizing body and hand motion. We build upon this work but focus on co-speech video generation.

Conditonal Video Generation. Several works [5,38,46,57] proposed transferring motion from a source video to animate a target image. They identify the keypoint correspondence between the source frames and the target image and compute the dense optical flow using local deformations of the paired correspondences. In our setting, however, we do not have a source driving RGB video, but rather, the audio input. Other works focus on unconditional video generation from random latents [39,40,42,50], but do not support audio or motion conditions. Several recent and concurrent works explored the use of diffusion models for video generation and editing [4,6,9,32,49,51]. Despite their versatility across diverse objects, those methods have yet to show effectiveness and efficiency in generating human videos. The most closely related methods in video generation are Vid2Vid [44] and WC-Vid2Vid [25]. These approaches synthesize RGB videos from sequences of segmentation maps or edge maps. However, these methods require training on large datasets, which differs from our problem setup, where we have only a single, relatively short video of a character.

3 Method

Our goal is to create a co-speech gesture video given an input audio—a challenging task due to the limited data and camera viewpoints for each subject. To address this challenge, we decompose this task into three steps. First, given an audio input, we predict plausible motions of the face, hands, and body in Sect. 3.1. Second, we learn to recover the character's appearance given the estimated geometry and motion parameters. Given the estimated appearance, geometry, and motion, we train a conditional GAN model to synthesize the final video in Sect. 3.2.

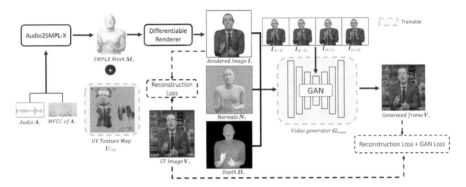

Fig. 2. Overview. Given an input audio A_t, we first train a network to predict the plausible human motion of face, hands, and body, as denoted as SMPL-X mesh M_t for each frame in Sect. 3.1. Here t is the frame index. Subsequently, we optimize the UV texture map U_{tex} for each character using differentiable rendering so that we can reconstruct all video frames given the texture map. Finally, in Sect. 3.2, we design a conditional GAN-based video generator G_{frame} to synthesize the final video given the current 2D rendered image I_t, the normal map N_t, the depth map D_t, as well as 2D rendered images from nearby frames. The video generator is trained with a combination of reconstruction loss and GAN loss.

3.1 Audio to Mesh

We first predict a plausible human motion sequence of face, hands, and body given the audio input based on the TALKSHOW model [54], which regresses the parameters of the SMPL-X model [30] from the audio. The parameters include shape, pose, and expression coefficients. Once we predict these parameters, we use the SMPL-X model to get the mesh corresponding to each frame. Below, we briefly describe the architecture of the TALKSHOW model [54].

Face Motion Generator. The face motion generator aims to predict facial movements and expressions, given input audio and the speaker's identity. It employs an encoder-decoder architecture, where the encoder uses the pre-trained WAV2VEC2 model [2], and the decoder consists of six layers of temporal convolutional networks (TCNs). The encoder first processes the audio sequence to extract audio embeddings. These embeddings are then concatenated with the speaker's identity and passed through the decoder to obtain the facial expressions and pose parameters. The network is trained with the mean squared error (MSE) loss between the predicted and ground truth parameters.

Body and Hand Motion Generator. The body and hand generator adopts two-stage training to generate realistic, diverse, and continuous body and hand poses. Specifically, in the first stage, two independent VQ-VAE models [41] are used to learn compositional quantized motion codebooks of the body and hand, which improves the diversity of the generation. To ensure the temporal consistency of poses generated from the separate body and hand codebooks, we adopt Gated PixelCNN [27] to model the interaction conditions between body and hand movements.

Consistent with the TALKSHOW model [54], we use standard VQ-VAE loss and reconstruction loss for both training stages. Once we have trained the model, given input audio A, for a particular speaker, we can obtain a sequence of SMPL-X meshes, $M_{1:T} = \{M_t\}_{t=1}^{T}$, featuring mouth movements and hand gestures, specific to this speaker, where $t = \{1...T\}$, and T denotes the total number of meshes in the sequence.

3.2 Mesh-Grounded Video Generation

Given the generated mesh sequence, our objective is to generate a photorealistic video following the speaker's audio input. We leverage conditional GAN [5,17,25,44] to generate the photorealistic video conditioned on the sequence of SMPL-X meshes $M_{1:T}$.

Challanges. However, directly using these untextured SMPL-X meshes as conditioning for the conditional GAN model can lead to artifacts in the generated video. Notably, inconsistencies arise in the hands of the speaker between the generated video and the input mesh. This discrepancy becomes more severe when the training video only consists of a limited number of frames. We hypothesize that generating realistic hands using conditional GAN from untextured meshes is a significant challenge, primarily due to the large motion blur in the ground-truth training frames. The artifacts are predominantly visible in the hands, as they exhibit more motion than other relatively stationary body parts. To address this issue, we propose a method that first optimizes the UV texture map from the training video for the SMPL-X meshes. Subsequently, we use these textured meshes as additional conditioning to generate the final video. Using these textured meshes helps us to ground the appearance of different body regions.

Mesh Texturing. Given a sequence of SMPL-X meshes, denoted as $M_{1:T}$, we aim to optimize a UV texture map, $U_{\text{tex}} \in \mathbb{R}^{N \times N \times 3}$, shared across all the meshes and initialized with standard normal distribution $\mathcal{N}(0, \mathbb{I}^{N \times N \times 3})$. More specifically, given a sequence of cameras, $P_{1:T}$, where the t-th camera corresponds to the t-th mesh, we project the meshes from 3D world space to a 2D image using a differentiable rendering function $\mathcal{F}(\cdot)$ [34]. We then measure the per-pixel difference, expressed as the **L2** norm between the 2D rendered image I_t of the mesh M_t, and the ground-truth frame V_t for the t-th frame. Formally, we optimize the UV texture map according to the following objective:

$$U_{\text{tex}} = \arg\min_{U_{\text{tex}}} \sum_{t=1}^{T} \|I_t - V_t\|^2, \qquad (1)$$

$$\text{where} \quad I_t = \mathcal{F}(M_t, P_t, U_{\text{tex}}).$$

Here, the differentiable rendering function $\mathcal{F}(\cdot)$ takes the camera pose P_t, the mesh M_t, and the shared texture map U_{tex}, and produces a 2D rendered image I_t. Once we learn a UV texture map U_{tex}, after many optimization iterations, we can render a sequence of textured meshes $I_{1:T}$, from their original untextured forms $M_{1:T}$.

As the meshes and the ground-truth exhibit small misalignment, we observe that the learned textures within the UV texture map are slightly blurry and tend to be an average representation of adjacent regions in the ground-truth images. However, this does not significantly impede our approach. We rely on the GAN model to learn the high-frequency details and adapt the mesh structure for generating photorealistic video. Notably, the textured meshes provide rough color and shape guidance for the GAN model.

Discussion. We have also experimented with more expressive neural 3D representations with view-dependent effects, such as dynamic NeRF representation (e.g., INST [59]). However, these models, while capable of producing highly detailed textures and fine-grained details for hair and face, tend to introduce severe artifacts for hands due to their complex deformations and self-occlusion. These artifacts can negatively impact the downstream video generation results. Therefore, we have chosen a less detailed but more robust 3D textured mesh representation and rely on a GANs-based video generator to add details, textures, and realistic lighting effects.

Additionally, we tried an alternate approach to learn per-vertex textures instead of a UV texture map using a similar optimization-based approach. We observed that both approaches yield very similar results.

Video Generation with Textured Meshes. As previously mentioned, we use an image-conditional GAN model [17,45] to generate a photorealistic video from a sequence of textured meshes. Specifically, we use a UNet [36] based generator G_{frame}. Instead of generating all the frames simultaneously, we generate each frame V_t sequentially conditioned on the rendered image I_t. We use SPADE [29]

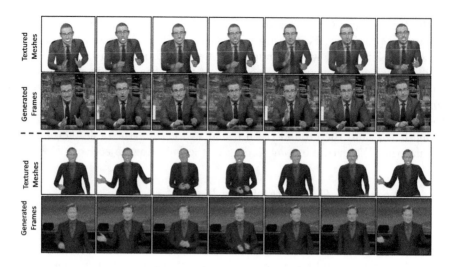

Fig. 3. Example sequences generated by our method. Here we show two video sequences generated by our method. For each sequence, we visualize both the textured meshes and the final image results. Our method is able to synthesize realistic video results with natural facial expressions and hand gesture movements. Please refer to our *website* for more video results.

normalization to incorporate conditioning in G_{frame}. To ensure temporal consistency across the generated frames, Vid2Vid [44] and WC-Vid2Vid [25] conditions the generation of frame $\widehat{V}_t$ on preceding k generated frames. However, we observe that this can lead to error propagation in the subsequent frames. Instead, we condition the generation of frame $\widehat{V}_t$ on a context window of $2k$ textured meshes: k prior and k subsequent frames.

$$\widehat{C}_t = \{I_{t-k}, I_{t-k+1}, ..., I_{t+k-1}, I_{t+k}\} \tag{2}$$

$$\widehat{V}_t = G_{\text{frame}}(I_t, \widehat{C}_t) \tag{3}$$

We find that this strategy ensures temporal consistency in the generated photorealistic video, V, without encountering the problem of error propagation. Similar to Mensha et al. [26], we also condition the generation of $\widehat{V}_t$ on the depth D_t and the rendered normal map N_t obtained from M_t.

$$\widehat{V}_t = G_{\text{frame}}(I_t, D_t.N_t, \widehat{C}_t). \tag{4}$$

Our training objective involves $L1$ reconstruction loss $\mathcal{L}_{\text{L1}}$, conditional GAN loss $\mathcal{L}_{\text{GAN}}$ [17], discriminator feature matching loss [45]$\mathcal{L}_{\text{FM}}$, and VGG-based perceptual loss [55] λ_{VGG}.

$$\mathcal{L} = \lambda_{\text{L1}}\mathcal{L}_{\text{L1}} + \lambda_{\text{VGG}}\mathcal{L}_{\text{VGG}} + \mathcal{L}_{\text{GAN}} + \lambda_{\text{FM}}\mathcal{L}_{\text{FM}}, \tag{5}$$

where $\lambda_{\text{L1}}, \lambda_{\text{FM}}$, and λ_{VGG} are scalar coefficients of $L1$ reconstruction loss, feature matching loss [45], and VGG-based perceptual loss [55], respectively. For more details, please refer to the Sect. 4.

4 Experiments

Datasets. For our work, we have used the dataset of speakers provided by TalkSHOW [54]. The dataset contains the identities of 4 different speakers, namely, "Oliver", "Chemistry", "Conan", and "Seth". We have taken the longest video out of a set of videos for each of the four speakers. This is because, except for Oliver, most of the videos for the other speakers contain less than 10K frames in total, which, in our observation, were insufficient to train the GAN model properly. Each video is at 30 fps and 720 x 1280 resolution. We take a square crop of the video to include the upper body of the speaker, including the hands, as we are mainly focused on generating a photorealistic talking video of a speaker with character-specific hand gestures. The cropped video frames are resized to 512×512 resolution. We also determine perspective cameras for each of the cropped videos for the four different speakers. This is essential as the meshes provided by the TalkSHOW [54] dataset contain full-body meshes. Hence, to use cropped videos, we need to align the projection of the meshes with our cropped videos. We split each video into training and testing splits such that the first 80% of the frames are used for training and the remaining for test. To train the frame generator model, G_{frame}, we also generate rendered textured meshes using Sect. 3.2 and their depth and normal maps.

Implementation Details. Following the settings in [54], the clip length and total epoch in the three training stages are 88 and 100, respectively. The batch size when training the face, VQ-VAEs, and body and hand generator are 1, 128, and 128, and the learning rates are set to constant 1×10^{-3}, 1×10^{-4} and 1×10^{-4} respectively. When training the body and hand generator, the weights of the two VQ-VAEs of the last epoch are loaded and frozen.

For mesh texturing, we optimize the parameters of a UV texture map as described in Sect. 3.2. The UV texture map is of shape NxNx3, where $N = 256$ is in our case. This is to make the optimization process faster. We also tried with $N = 512$ but did not notice visible improvements. We perform the optimization using **L2** loss of the mesh rendering w.r.t the corresponding ground-truth training video frames. Meshes are rendered at 256×256, and ground-truth video frames are resized to 256×256. We also tried rendering the meshes and optimizing at 512×512 but found the learning process very slow. Once the UV maps are learned, we render the textured meshes for the video frames at 512×512 resolution for training and evaluation.

For training the video frame generation model, G_{frame}, we use a learning rate of 2×10^{-3} and use the TTUR method to update the learning rate [14] and train for 50 epochs. We train the model with a batch size of 16. The training and inference are set at 512×512 resolution. We empirically set $\lambda_{\text{L1}} = 10.0$, $\lambda_{\text{FM}} = 10.0$, and $\lambda_{\text{VGG}} = 1.0$. We set the value of $k = 2$ in Sect. 3.2, i.e., we use two previous and two subsequent textured meshes as context to generate frame at time t. We train our model for 50K iterations on a single RTX6000 GPU (16GB VRAM) with a batch size of 8. Each training iteration takes 0.54 s. Inference requires 9GB VRAM and takes 0.013 s per frame.

Baselines. To verify the importance of using 3D human body representation for high-quality speaker video generation, we compare our method against a baseline that uses a 2D keypoint as the human body representation, a widely utilized method for human video generation. Our baseline utilization draws inspiration from prior works such as Ginosar et al. [10], which employs 2D keypoint as the representation of body gestures, and Lu et al. [23], which utilizes 2D facial keypoint to generate talking head video. In our baseline, we extract the joints from the corresponding SMPL-X mesh for the upper body, the face and the hands. This ensures that the temporal consistency for our baseline is the same as our method. In the context of the video frame generation model, we maintain the same training setup as our proposed approach. Although Ginosar et al. [10], Lui et al. [21], and Qian et al. [33] also generate co-speech gesture video, we could not compare them directly. The code is not publicly available for Lui et al. [21]. Additionally, Ginosar et al. [10] and Qian et al. [33], provide parts of their code for generating speaker-specific 2D skeleton keypoints from audio input but not for generating the RGB final video.

Evaluation. We evaluate the quality of video generated on the test sequence for each speaker in the dataset. Under this setting, we already have the ground-truth (untextured) SMPL-X meshes and corresponding ground-truth RGB video frames. We first texturize the (untextured) SMPL-X meshes using the learned UV texture map and then generate the photorealistic video with the GAN model. Since we already know the corresponding ground-truth RGB frames, we can use pixel-aligned metrics like PSNR, SSIM, and LPIPS [55] to compare the quality of the generated video. Note that in our work, we mainly focus on the photorealistic generation of the speaker and not the background. Thus we use modified PSNR and LPIPS that only take into account the speaker pixels, ignoring the background. In addition, not considering the background for evaluation is important because the ground-truth frames are clipped from a very long video, causing the background shifts over time due to subtle camera movement. This results in shifting and flickering of the background in the generated videos, which undermines the importance of evaluating the key component of our method. The mask of the foreground speaker is obtained using a Video Matting method [20] applied on the ground-truth sequence. The implementation of masked PSNR and LPIPS is taken from pix2latent [16].

User Study. In addition to PSNR, SSIM, and LPIPS [55], to further conduct a user study to assess the quality of our final method, where the video is generated directly from input audio against the baseline method, our method (w/o textured meshes), and our method (w/o depth and normal maps). Since we do not have the corresponding ground-truth video for end-to-end evaluation, we only rely on user study to verify the effectiveness of our full method. We perform a paired comparison, where the Amazon MTurkers are asked to select which video is more realistic for the same input audio.

Table 1. Baseline and ablation comparison. Here, we compare our full method (ours) with two ablated methods: ours (w/o textured mesh), ours (w/o depth and normal), and a 2D keypoint-based baseline. We conduct quantitative comparisons regarding video quality, regarding SSIM [47], PSNR, and LPIPS [55]. We observe that our method performs better than the baseline in all the metrics and that the majority of the performance gain in our method comes from the use of textured meshes.

Speaker	Method	SSIM (↑)	PSNR (↑)	LPIPS (↓)
Oliver	2D-keypoint Baseline	0.759	15.951	0.482
	Ours (w/o textured mesh)	0.783	16.233	0.411
	Ours (w/o depth and normal)	0.753	16.214	**0.407**
	Ours	**0.786**	**16.403**	0.408
Chemistry	2D-keypoint Baseline	0.759	14.862	0.411
	Ours (w/o textured mesh)	0.761	14.962	0.397
	Ours (w/o depth and normal)	0.776	14.969	**0.394**
	Ours	**0.777**	**15.021**	**0.394**
Conan	2D-keypoint Baseline	0.866	20.002	0.41
	Ours (w/o textured mesh)	0.89	20.112	0.41
	Ours (w/o depth and normal)	0.89	20.127	**0.398**
	Ours	**0.891**	**21.125**	**0.398**
Seth	2D-keypoint Baseline	0.787	14.589	0.399
	Ours (w/o textured mesh)	0.795	14.933	0.328
	Ours (w/o depth and normal)	0.798	14.862	0.334
	Ours	**0.801**	**15.044**	**0.312**

Details. For the user study, each video was assigned to 50 MTurkers. The videos from part of the test sequence are approximately 30 seconds long, whereas the videos generated from input audio are 8 seconds long on average. In their evaluation, we requested the participants to (1) ignore the shifts in the background video because of the aforementioned reason and (2) pay more attention to the quality of generated arms and hands.

4.1 Mesh to Video

As mentioned in Sect. 4, we partition the entire video into training and test data. In this context, we evaluate the quality of video generation conditioned on the ground-truth untextured mesh from the test segments within each of the four speaker's video sequences.

Quantitative Comparison. Since the corresponding ground-truth frame sequences are available, we can assess the similarity and fidelity of the generated video in comparison to the ground-truth video. Table 1 presents a comparison between the frames generated by our method and a baseline method utilizing 2D keypoint as input conditioning (described in Sect. 4). It is noted that the metrics are computed only w.r.t the foreground speaker in the generated frames, and not the background. We observe that videos generated by our method consistently outperform the ones generated by the 2D keypoint-based baseline method on all the metrics and for all four speaker identities, demonstrating that our method generates videos that are more pixel-aligned with the ground-truth video sequence.

Qualitative Comparison. In Fig. 4, we compare the quality of generated video using our method against video generated using a 2D keypoint-based baseline on all actors, Notably, even for actors with a considerably small training set, the videos generated by our method (Fig. 4, 4-th row), employing rendering of textured meshes as conditioning, yields consistent hand regions. In contrast, the video generated by the 2D keypoint-based baseline displays significant artifacts within the hand region. The generated hands appear mostly indistinct and disappear in many frames. We hypothesize that discrepancy may arise from the grounding provided by the textured mesh, which ensures the permanence of the hands, while the 2D keypoint cannot provide a strong prior for the video generator, G_{frame}.

4.2 Audio to Video

In this section, we evaluate our end-to-end method, i.e., the visual quality of the generated video given input audio.

Qualitative Comparison. We show videos generated using our method from input audio during inference, featuring two different actors, "Oliver" (Fig. 3, top row) and "Chemistry" (Fig. 3, bottom row), alongside their respective generated textured meshes. These two actors serve as very distinct settings, as "Oliver" contains a relatively large number of training frames (25K), whereas "Chemistry" is much more challenging (7K) frames. Although the meshes generated by Sect. 3.1 are in very novel poses compared to that seen by the frame generator during training, for both of the actors, the frame generator can generate a high-fidelity video of the actors following the structure in the input textured mesh. This further highlights the robustness of our method, leveraging 3D (textured) intermediate representation to accommodate novel poses and gestures during inference. We also conduct a user study in Table 2. We see that MTurkers prefer our video by a very large margin compared to the ones generated by the 2D-keypoint baseline method. Please check the video results on the website.

4.3 Albation Study

Fig. 4. Qualititative comparison of our method with the 2D keypoint baseline. Our method produces more realistic hands and gesture motion compared to the baseline that relies on 2D keypoint representation. Please refer to the website for more detailed comparisons.

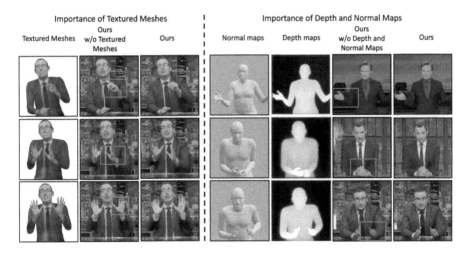

Fig. 5. Ablation study (qualitative comparison). We present an ablation study with qualitative comparisons between our full method and two ablated versions: ours w/o textured mesh (left) and ours w/o depth and normal (right). We observe that our method consistently produces more visually appealing results, with particular improvements noticeable in the hand regions (highlighted in the red bounding boxes). (Color figure online)

In the following section, we study the importance of using textured meshes (compared to untextured meshes) and using depth and normal maps as additional conditioning for video generation. We evaluate these variants on four videos using various automated metrics, including SSIM [47], PSNR, and LPIPS [55] in Table 1 and also conduct user preference study (in Table 2). Please check the video results in the website.

Mesh Texturing. From Table 1, we see that not using textured meshes hurts the performance of the generator the most in terms of PSNR, SSIM, and LPIPS [55]. From Table 2, we see that MTurkers prefer our video by a very large margin compared to our method using untextured meshes. Additionally, we include a visual comparison of results generated by our method with and without using textured meshes. Figure 5 (left) shows the loss of high-frequency details and blurry artifacts around the hands and the face region (highlighted by the red bounding box).

Depth and Normal Maps. As shown in Fig. 5 and Table 1, the performance of our method also decreases slightly when we remove the depth and normal maps from the conditional inputs of the GAN-based generator. From Table 2, we draw a similar conclusion that MTurkers prefer videos generated by our method in most cases, but not by a very large margin compared to the aforementioned ablation study. As highlighted in Fig. 5 (right) red bounding box, ours (w/o depth and normal maps) synthesizes more blurry hand gestures. The blurriness may not be reflected significantly in automated metrics. However, it will reduce the video quality by a large margin.

Table 2. User Study. We perform a user study to evaluate the quality of the generated videos, comparing our method with baseline and ablations for four speakers. The videos are generated directly from an audio input. The numbers show the percentage of users who prefer videos generated by our method compared to ablations. Users prefer our method to the 2D-keypoint baseline, ours (w/o textured meshes), and ours (w/o depth and normal maps).

Method	Speaker			
	Oliver	Chemistry	Conan	Seth
2D-keypoint Baseline	70%	96%	84%	96%
Ours (w/o textured mesh)	68%	88%	86%	68%
Ours (w/o depth and normal)	54%	42%	58%	66%

In essence, from both Tables 1 and 2, we observe that using textured meshes leads to significant improvement in the quality of generated videos, whereas using depth and normal maps as extra conditioning marginally benefits the video generation quality.

5 Discussion and Limitations

In summary, we have introduced a complete system designed to generate co-speech gesture videos from audio inputs. By employing explicit 3D representations of the human body, face, and hands, we have effectively bridged the gap between audio input and video output. We have demonstrated better results on the TalkSHOW dataset compared to a baseline method using 2D skeleton-based representation for gestures in terms of both quantitative metrics and human preference.

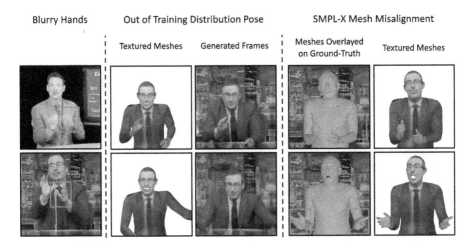

Fig. 6. Limitations. Some of the limitations of our method are that: (left) Our method sometimes synthesizes blurry textures in the arms and hands in the generated frames (highlighted in red). (middle) Our method struggles to generate realistic videos when the mesh pose at inference differs significantly from the speaker's training pose. In this example, the mesh (pose and camera) is taken from the speaker of 'Chemistry,' and the final frames are generated from the model trained on 'Oliver' sequences. (right) Our method does not further optimize the mesh parameters obtained from TalkSHOW training data, leading to misalignment between the ground truth textured mesh and the RGB image (highlighted in red), mainly caused by slight errors in tracking. This also leads to learning blurry textures for the mesh using optimization, especially in the hands region. (Color figure online)

Limitations. Our current method still has several limitations. First, We cannot handle videos with dynamic backgrounds with moving objects, as we only model the 3D geometry and motion of the foreground humans. How to model the dynamics and geometry of more objects will be an interesting future direction.

Second, our method occasionally struggles to synthesize realistic textures for regions occluded by the arms and hands across most frames. Integrating a better inpainting method could further enhance our approach.

Third, we need to perform per-character texture optimization and generator training, which can be time-consuming. Developing encoder-based methods for both appearance and motion modeling will be a meaningful next step.

Fourth, as shown in Fig. 6 (middle), our method also has difficulty generating photorealistic videos in cases where the speaker's pose differs dramatically from that in the training data.

Finally, our method also works best when trained on a relatively large training video sequence of an actor, and sometimes, it struggles to generate proper hand structures, as seen in Fig. 6 (right), especially in cases with less training data.

Acknowledgments. We thank Kangle Deng, Yufei Ye, and Shubham Tulsiani for their helpful discussion. The project is partly supported by Ping An Research.

References

1. Ao, T., Zhang, Z., Liu, L.: GestureDiffuCLIP: gesture diffusion model with CLIP latents. ACM Trans. Graph. (2023). https://doi.org/10.1145/3592097
2. Baevski, A., Zhou, Y., Mohamed, A., Auli, M.: wav2vec 2.0: a framework for self-supervised learning of speech representations. In: NeurIPS, vol. 33 (2020)
3. Boukhayma, A., Bem, R.D., Torr, P.H.: 3D hand shape and pose from images in the wild. In: CVPR (2019)
4. Ceylan, D., Huang, C.H.P., Mitra, N.J.: Pix2Video: video editing using image diffusion. In: ICCV (2023)
5. Chan, C., Ginosar, S., Zhou, T., Efros, A.A.: Everybody dance now. In: IEEE International Conference on Computer Vision (ICCV) (2019)
6. Chen, W., et al.: Control-a-video: controllable text-to-video generation with diffusion models. arXiv:2305.13840 (2023)
7. Cudeiro, D., Bolkart, T., Laidlaw, C., Ranjan, A., Black, M.J.: Capture, learning, and synthesis of 3D speaking styles. In: CVPR (2019)
8. Fan, Y., Lin, Z., Saito, J., Wang, W., Komura, T.: FaceFormer: speech-driven 3D facial animation with transformers. In: CVPR (2022)
9. Geyer, M., Bar-Tal, O., Bagon, S., Dekel, T.: TokenFlow: consistent diffusion features for consistent video editing. arXiv preprint arxiv:2307.10373 (2023)
10. Ginosar, S., Bar, A., Kohavi, G., Chan, C., Owens, A., Malik, J.: Learning individual styles of conversational gesture. In: Computer Vision and Pattern Recognition (CVPR). IEEE (2019)
11. Guan, J., et al.: StyleSync: high-fidelity generalized and personalized lip sync in style-based generator. In: Proceedings of the IEEE/CVF Conference on Computer Vision and Pattern Recognition (CVPR) (2023)
12. Guo, Y., Chen, K., Liang, S., Liu, Y., Bao, H., Zhang, J.: AD-NeRF: audio driven neural radiance fields for talking head synthesis. In: IEEE/CVF International Conference on Computer Vision (ICCV) (2021)
13. Habibie, I., et al.: Learning speech-driven 3D conversational gestures from video. In: Proceedings of the 21st ACM International Conference on Intelligent Virtual Agents, pp. 101–108 (2021)
14. Heusel, M., Ramsauer, H., Unterthiner, T., Nessler, B., Hochreiter, S.: GANs trained by a two time-scale update rule converge to a local nash equilibrium. In: NeurIPS, vol. 30 (2017)

15. Huang, Z., et al.: Make-your-anchor: a diffusion-based 2D avatar generation framework. In: CVPR (2024)
16. Huh, M., Zhang, R., Zhu, J.Y., Paris, S., Hertzmann, A.: Transforming and projecting images to class-conditional generative networks. In: ECCV (2020)
17. Isola, P., Zhu, J.Y., Zhou, T., Efros, A.A.: Image-to-image translation with conditional adversarial networks. In: CVPR (2017)
18. Karras, T., Aila, T., Laine, S., Herva, A., Lehtinen, J.: Audio-driven facial animation by joint end-to-end learning of pose and emotion. ACM Trans. Graph. (TOG) **36**(4), 1–12 (2017)
19. Li, T., Bolkart, T., Black, M.J., Li, H., Romero, J.: Learning a model of facial shape and expression from 4D scans. ACM Trans. Graph. **36**(6), 194 (2017)
20. Lin, S., Yang, L., Saleemi, I., Sengupta, S.: Robust high-resolution video matting with temporal guidance. In: WACV (2022)
21. Liu, X., et al.: Audio-driven co-speech gesture video generation. In: NeurIPS (2022)
22. Loper, M., Mahmood, N., Romero, J., Pons-Moll, G., Black, M.J.: SMPL: a skinned multi-person linear model. In: Seminal Graphics Papers: Pushing the Boundaries, vol. 2, pp. 851–866 (2023)
23. Lu, Y., Chai, J., Cao, X.: Live speech portraits: real-time photorealistic talking-head animation. ACM Trans. Graph. **40**(6), 1–7 (2021). https://doi.org/10.1145/3478513.3480484
24. Ma, Y., et al.: StyleTalk: one-shot talking head generation with controllable speaking styles. In: Proceedings of the AAAI Conference on Artificial Intelligence, vol. 37, pp. 1896–1904 (2023)
25. Mallya, A., Wang, T.-C., Sapra, K., Liu, M.-Y.: World-consistent video-to-video synthesis. In: Vedaldi, A., Bischof, H., Brox, T., Frahm, J.-M. (eds.) ECCV 2020. LNCS, vol. 12353, pp. 359–378. Springer, Cham (2020). https://doi.org/10.1007/978-3-030-58598-3_22
26. Mensah, D., Kim, N.H., Aittala, M., Laine, S., Lehtinen, J.: A hybrid generator architecture for controllable face synthesis. In: ACM SIGGRAPH 2023 Conference Proceedings, pp. 1–10 (2023)
27. Van den Oord, A., Kalchbrenner, N., Espeholt, L., Vinyals, O., Graves, A.: Conditional image generation with PixelCNN decoders. In: NeurIPS, vol. 29 (2016)
28. Ouyang, L., et al.: Training language models to follow instructions with human feedback. In: NeurIPS, vol. 35 (2022)
29. Park, T., Liu, M.Y., Wang, T.C., Zhu, J.Y.: Semantic image synthesis with spatially-adaptive normalization. In: CVPR (2019)
30. Pavlakos, G., et al.: Expressive body capture: 3D hands, face, and body from a single image. In: CVPR (2019)
31. Prajwal, K.R., Mukhopadhyay, R., Namboodiri, V.P., Jawahar, C.: A lip sync expert is all you need for speech to lip generation in the wild. In: ACM MM (2020)
32. Z Qi, C., et al.: FateZero: fusing attentions for zero-shot text-based video editing. arXiv:2303.09535 (2023)
33. Qian, S., Tu, Z., Zhi, Y., Liu, W., Gao, S.: Speech drives templates: co-speech gesture synthesis with learned templates. In: 2021 IEEE/CVF International Conference on Computer Vision (ICCV). IEEE (2021)
34. Ravi, N., et al.: Accelerating 3D deep learning with PyTorch3D. arXiv:2007.08501 (2020) ¡error l="308" c="Invalid
    command: paragraph not started." /¿
35. Richard, A., Zollhöfer, M., Wen, Y., De la Torre, F., Sheikh, Y.: MeshTalk: 3D face animation from speech using cross-modality disentanglement. In: Proceedings of the IEEE/CVF International Conference on Computer Vision (2021)

36. Ronneberger, O., Fischer, P., Brox, T.: U-Net: convolutional networks for biomedical image segmentation. In: Navab, N., Hornegger, J., Wells, W.M., Frangi, A.F. (eds.) MICCAI 2015. LNCS, vol. 9351, pp. 234–241. Springer, Cham (2015). https://doi.org/10.1007/978-3-319-24574-4_28
37. Shen, S., et al.: DiffTalk: crafting diffusion models for generalized audio-driven portraits animation. In: CVPR (2023)
38. Siarohin, A., Lathuilière, S., Tulyakov, S., Ricci, E., Sebe, N.: First order motion model for image animation. In: NeurIPS (2019)
39. Skorokhodov, I., Tulyakov, S., Elhoseiny, M.: StyleGAN-V: a continuous video generator with the price, image quality and perks of styleGAN2. In: CVPR (2022)
40. Tulyakov, S., Liu, M.Y., Yang, X., Kautz, J.: MoCoGAN: decomposing motion and content for video generation. In: CVPR (2018)
41. Van Den Oord, A., Vinyals, O.: Neural discrete representation learning. In: NeurIPS, vol. 30 (2017)
42. Vondrick, C., Pirsiavash, H., Torralba, A.: Generating videos with scene dynamics. In: NeurIPS (2016)
43. Wang, J., Qian, X., Zhang, M., Tan, R.T., Li, H.: Seeing what you said: talking face generation guided by a lip reading expert. In: CVPR (2023)
44. Wang, T.C., et al.: Video-to-video synthesis. In: NeurIPS (2018)
45. Wang, T.C., Liu, M.Y., Zhu, J.Y., Tao, A., Kautz, J., Catanzaro, B.: High-resolution image synthesis and semantic manipulation with conditional GANs. In: CVPR (2018)
46. Wang, T.C., Mallya, A., Liu, M.Y.: One-shot free-view neural talking-head synthesis for video conferencing. In: CVPR (2021)
47. Wang, Z., Bovik, A.C., Sheikh, H.R., Simoncelli, E.P.: Image quality assessment: from error visibility to structural similarity. IEEE Trans. Image Process. **13**(4), 600–612 (2004)
48. Wu, H., Jia, J., Wang, H., Dou, Y., Duan, C., Deng, Q.: Imitating arbitrary talking style for realistic audio-driven talking face synthesis. In: ACM MM (2021)
49. Wu, J.Z., et al.: Tune-a-video: one-shot tuning of image diffusion models for text-to-video generation. arXiv preprint arXiv:2212.11565 (2022)
50. Yan, W., Zhang, Y., Abbeel, P., Srinivas, A.: VideoGPT: Video generation using VQ-VAE and transformers. arXiv preprint arXiv:2104.10157 (2021)
51. Yang, S., Zhou, Y., Liu, Z., , Loy, C.C.: Rerender a video: zero-shot text-guided video-to-video translation. In: ACM SIGGRAPH Asia Conference Proceedings (2023)
52. Yang, S., et al.: DiffuseStyleGesture: stylized audio-driven co-speech gesture generation with diffusion models. In: IJCAI (2023)
53. Yao, X., Fried, O., Fatahalian, K., Agrawala, M.: Iterative text-based editing of talking-heads using neural retargeting. ACM Trans. Graph. (TOG) **40**(3), 1–14 (2021)
54. Yi, H., et al.: Generating holistic 3D human motion from speech. In: CVPR (2023)
55. Zhang, R., Isola, P., Efros, A.A., Shechtman, E., Wang, O.: The unreasonable effectiveness of deep features as a perceptual metric. In: CVPR (2018)
56. Zhang, W., et al.: SadTalker: learning realistic 3D motion coefficients for stylized audio-driven single image talking face animation. arXiv preprint arXiv:2211.12194 (2022)
57. Zhao, J., Zhang, H.: Thin-plate spline motion model for image animation. In: CVPR (2022)

58. Zhu, L., Liu, X., Liu, X., Qian, R., Liu, Z., Yu, L.: Taming diffusion models for audio-driven co-speech gesture generation. In: CVPR (2023)
59. Zielonka, W., Bolkart, T., Thies, J.: Instant volumetric head avatars. In: CVPR (2023)

Understanding the Impact of Negative Prompts: When and How Do They Take Effect?

Yuanhao Ban[1], Ruochen Wang[1], Tianyi Zhou[2], Minhao Cheng[3], Boqing Gong[4], and Cho-Jui Hsieh[1](✉)

[1] University of California, Los Angeles, CA 90025, USA
banyh2000@gmail.com, ruocwang@g.ucla.edu, chohsieh@cs.ucla.edu
[2] University of Maryland, College Park, MD 20742, USA
tianyi@umd.edu
[3] The Pennsylvania State University, University Park, PA 16802, USA
mmc7149@psu.edu
[4] Google, Seattle, WA 98103, USA
bgong@google.com

Abstract. The concept of negative prompts, emerging from conditional generation models like Stable Diffusion, allows users to specify what to exclude from the generated images. Despite the widespread use of negative prompts, their intrinsic mechanisms remain largely unexplored. This paper presents the first comprehensive study to uncover how and when negative prompts take effect. Our extensive empirical analysis identifies two primary behaviors of negative prompts. *Delayed Effect*: The impact of negative prompts is observed after positive prompts render corresponding content. *Deletion Through Neutralization*: Negative prompts delete concepts from the generated image through a mutual cancellation effect in latent space with positive prompts. These insights reveal significant potential real-world applications; for example, we demonstrate that negative prompts can facilitate object inpainting with minimal alterations to the background via a simple adaptive algorithm. We believe our findings will offer valuable insights for the community in capitalizing on the potential of negative prompts.

Keywords: Diffusion Models · Negative prompt · Inpainting

1 Introduction

It has been widely acknowledged that diffusion models have made tremendous breakthroughs in image and video generation [8, 11, 18, 21, 23]. Despite their capabilities, these models sometimes produce images that do not fully align with

---

Supplementary Information The online version contains supplementary material available at https://doi.org/10.1007/978-3-031-73024-5_12.

the intended meaning of their textual prompts, motivating a surge in research aimed at enhancing image fidelity and relevance [6,10,28]. Notable advancements include the development of classifier-free guidance [11], manipulation of cross-attention map [6,13], integration with large language models [14,33], and usage of the semantic information of the prompts [9,10]. Among these innovations, the concept of negative prompts—guiding models by specifying what not to generate—has gained great attention for its effectiveness [3,4,12,19,29]. However, most of the works are merely relying on experimental results and lack a deep understanding on how negative prompts work. Such a lack of analysis of the negative prompts further prevents people from designing more effective negative prompts to obtain better prompts alignment.

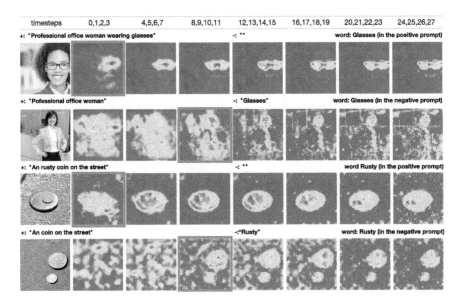

Fig. 1. Illustration on when the negative prompts attend to the "right" place. For example, we consider the face of the person as the "right place" for the "glasses" token. Every row represents an independent diffusion process where the first and the third rows show the tokens in the positive prompt and the second and fourth rows visualize those in the negative prompt. The positive prompt(+), negative prompt(-), and the corresponding token of the attention map are listed on top of each of the rows. Every column denotes the different diffusion steps used to visualize the cross-attention heat maps. We also enclose the feature map which attends to the "right" place for the first time, with a square box □. (Color figure online)

In our work, we perform a systemic study on negative prompts to fill this gap. With a focus on the dynamics of the diffusion steps, our central research question is, *"When and how do negative prompts take effect?"*. Our investigation breaks down the mechanism of negative prompts into noun-based removal and adjective-based alteration tasks, leading to intriguing insights through experimentation.

Specifically, to investigate when negative prompts start to exert their influence, we analyze the model's cross-attention maps that illustrate the likelihood of specific tokens appearing in the image pixels. We identify the **critical step** at which negative prompts begin to influence the generation process, highlighting the dramatic difference in how negative and positive prompts operate. The study reveals a significant delay in the critical step of negative prompts compared to positive ones, as clearly illustrated in Fig. 1.

To figure out how negative prompts take effect, we delve into the architecture of mainstream text-to-image diffusion models to uncover a possible cause: an insufficient exchange of information between the pathways dealing with positive and negative prompts. Analyzing the patterns of estimated noises in object deleting tasks, **we find that negative prompts initially generate a target object at a specific location within the image, which neutralizes the positive noise through a subtractive process, effectively erasing the object.** Furthermore, we observe a counter-intuitive model behavior "Reverse Activation" as shown in Fig. 2. That is, introducing a negative prompt in the early stages of diffusion paradoxically results in the generation of the specified object in the initial generation stage. We give a detailed explanation for this model behavior based on two findings of diffusion dynamics: the Inducing Effect and the Momentum Effect. The former effect reveals that the negative prompt can induce the positive estimated noises to increase in some specific directions, while the latter shows that the estimated noises tend to keep in the same direction in the diffusion process, which means noises exhibit a significant correlation with their preceding segments. We also point out the potential hazards of applying negative prompts too early, that they may distort the original structure of the image.

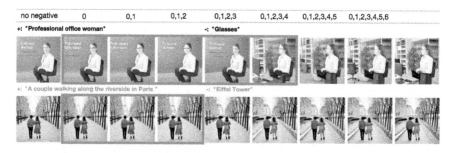

Fig. 2. Illustration: Reverse activation. Each column shows an image generated by applying negative prompts in some specific steps which is shown at the top of the picture. In these two examples, the diffusion process without applying a negative prompt does not produce the object mentioned in the negative prompt. But interestingly, introducing a negative prompt in the early stages results in the generation of the specified object, which is marked with ▫ (Color figure online)

Building on our insights, we introduce a novel controllable inpainting approach aimed at **deleting undesired elements while preserving as much of**

the remaining content as possible. As shown in Fig. 8, applying the negative prompts too early can disrupt the layout of an image that has not yet been fully formed. The best timing for introducing these prompts should be after the critical step. Based on the insights, we propose to involve the negative prompts in the middle of the reverse-diffusion process which shows great success. Note that our method does not need any model retraining and modifications in the sampling step during inference.

Our contributions can be summarized as follows: (1) We have uncovered the critical steps and underlying dynamics that govern the effectiveness of negative prompts (2) We have identified and highlighted the fundamental issue of information lag that occurs between the activation of negative and positive prompts. (3) We provide insights into the strategic design of negative prompts and introduce a novel approach for controllable image inpainting tasks.

2 Related Work

Prompts Analysis: Since the development of text-to-image diffusion models, there has been a surge of interests in understanding its image generation mechanism through the lens of prompts. Tang [24] employed cross-attention maps to analyze prompts through the lens of computational linguistics. Hertz [10] revealed that cross-attention layers are imbued with significant semantic content derived from text prompts. Tumanyan [25] demonstrated that self-attention layers encode layout information. Furthermore, Balaji [5] and Mahajan [16] showed that different stages in the process focus on different kinds of features including color, texture, and shape. In contrast to the extensive focus on positive prompts, research on negative prompts remains unexplored, let alone their dynamics along the temporal dimension. Our research primarily delves into negative prompts, exploring the interplay between negative and positive prompts through the lens of temporal evolution.

Object Removal: Object removal is the process of eliminating undesired objects from an image. Criminisi [7] initially conceptualized object removal as an image inpainting task. Yu [32]proposed a novel deep generative model-based approach that can synthesize novel image structures and utilize surrounding image features to boost performance. Yildirim [31] trained a diffusion model that can remove objects based on the instructions given as text prompts. Yang [30] introduces an attention guidance strategy to constrain the sampling process of diffusion models to enable efficient removal. Existing methods primarily aim to delete a specific object from a given image. In contrast, our approach adopts a different perspective, tailored to the needs of text-to-image model users. Starting with a textual description, we initially create an image using text-to-image diffusion models. Subsequently, we adjust the text to produce a revised image, effectively removing any undesired objects present in the original. There have already been some attempts in this field. For example, Woolf [29], O'Connor [19], Andrew [3] and Kapoor [12] have proposed to utilize negative prompts to do the task. However, their approaches tend to significantly alter the context of the generated image compared to the original. In contrast, our findings suggest that negative prompts can be used to effectively remove the object, while at the same time preserving the background information if they are applied only within a critical period of the generation process. Moreover, this approach is training-free and assumes zero modification to the structure of diffusion models.

3 Preliminary

Denoising Diffusion Probabilistic Models (DDPM). As a new family of powerful generative models, diffusion models could achieve superb performance on high-quality image synthesis. The complete modeling of DDPM consists of 1). *a forward process* and 2). *a reverse process*. Given a sample from data distribution $\mathbf{x}_0 \sim p_{data}(\mathbf{x})$, the forward process gradually injects Gaussian noise to the original data ($\mathbf{x}_0$):

$$q(\mathbf{x}_t|\mathbf{x}_{t-1}) = \mathcal{N}(\mathbf{x}_t; \sqrt{1-\beta_t}\mathbf{x}_{t-1}, \beta_t \mathbf{I}), \tag{1}$$

where β_t is a scheduler designed so that the Markov chain converges to standard Gaussian noise ($\mathbf{x}_T \sim \mathcal{N}$) after T steps. The reverse process then starts with this standard Gaussian noise and repeatedly applies a model (θ) to denoise it back to the real data:

$$p_\theta(\mathbf{x}_{0:T}) = p(\mathbf{x}_T) \prod_{t=1}^{T} p_\theta(\mathbf{x}_{t-1}|\mathbf{x}_t), \tag{2}$$

$$\text{where } p_\theta(\mathbf{x}_{t-1}|\mathbf{x}_t) = \mathcal{N}(\mathbf{x}_{t-1}; \mu_\theta(\mathbf{x}_t, t), \Sigma_\theta(x_t, t)). \tag{3}$$

Classifer-Free Guidance for Conditional Generation. Text-to-image diffusion models introduce the classifier-free context information into the reverse diffusion process through cross attention map. At each sampling step, the predicted error is obtained by subtracting the unconditional error from the conditional error with a guidance strength w:

$$\hat{\epsilon}_\theta((\mathbf{x}_t), c(s), t) = (1+w)\epsilon_\theta(\mathbf{x}_t, c(s), t) - w\epsilon_\theta(\mathbf{x}_t, c(\emptyset), t), \tag{4}$$

where $c(s)$ is the conditional signal of text s, $c(\emptyset)$ is obtained by passing an empty string to the text encoder.

Negative Prompts. Woolf [29] finds that the generative process could be better guided with text prompts that instruct the AI model that it should not include certain elements in its generated images. Specifically, when the empty string $\emptyset$ in the unconditional error is replaced by an actual prompt, it represents what to remove from the (generated) image due to the negative sign. This can be formally written as:

$$\hat{\epsilon}_\theta((\mathbf{x}_t), c(s), t) = (1+w)\epsilon_\theta(\mathbf{x}_t, c(p_+), t) - w\epsilon_\theta(\mathbf{x}_t, c(p_-), t), \tag{5}$$

where p_+ is the regular user prompt (positive prompt) and p_- is the negative prompt.

Stable diffusion [21] is a latent text-to-image diffusion model featuring processing over a lower dimensional latent space to reduce memory and compute complexity. In our experiments, we adopt Stable Diffusion v2 [22] provided by diffusers [20] and set the diffusion steps as 30 in all of the experiments.

4 When Do Negative Prompts Take Effect

4.1 Qualitive Analysis

Visualising Cross-Attention Maps Across Diffusion Steps. In conditional diffusion models, cross-attention layers contextualize text embeddings with coordinate-aware latent representations of the image and output scores for each token-image patch pair. Hence, each element in the cross-attention map can be viewed as the probability that the specific token appears in that position. Following the approach of Daam [24], we gather the scores from various layers for each token we focus on. Then we resize the feature maps to the same size and average them. Notably, different from Daam which averages the scores across all the time steps, we collect and present maps of different steps individually. These heat maps are then organized into sets of four and shown in Fig. 1.

There Exists a Delay in the Effect of Negative Prompts Following the Impact of Positive Prompts for Both Nouns and Adjectives. As shown in Fig. 1, we observe a delayed effect of negative prompts. Take the images of "woman with glasses" as an example (top 2 rows). In the first row, the glasses in the positive prompt are correctly attended to the woman's head from the very beginning, within the first four steps. Conversely, the glasses in the negative prompt cannot attend to the right position until the eighth step. Intuitively, this delay stems from classifier-free guidance in Eq. (5): at every step, both negative and positive prompts attend to the same noise map independently, with their interaction occurring only indirectly after the subtraction step. As a result, the negative prompt has to wait for the target object (woman's face) specified in the positive prompt to appear before it can attend to it. The above analysis also applies to the case where the negative prompt specifies an adjective (bottom two roles): The negative prompt "rusty" can only attend to the coin after the coin is been generated.

4.2 Quantitative Analysis

In this subsection, we quantify the exact step at which the word in the negative prompt aligns accurately with the target objective.

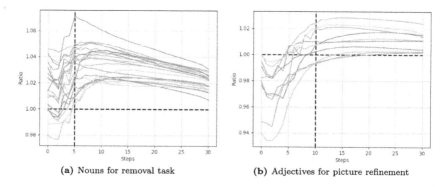

Fig. 3. Illustration of Effectiveness of Negative Prompts Over Time. The x-axis represents the time step. The y-axis denotes the strength of the negative prompt. In the left figure, there is a peak at the 5th step for the noun-based negative prompt, indicating the critical step. Meanwhile, In the right figure, we observe a plateau around the 10th, as the object have been generated and the negative prompt begins to take effect.

New Metric to Measure the Strength of the Negative Prompt. Denoting cross attention map of the k-th layer for the i-th token in the text s at time step t, as $F_{k,s(i)}^{(t)}$, we further define the strength of negative prompt at step t as the ratio of the mean of the squared value of the heat maps of the negative prompt at step t to that of the positive prompt. Notably, we specifically select the token in the positive prompt that is most relevant to the negative prompt and disregard the unrelated ones. For example, we choose the word "woman" in the positive prompt to compare with the negative prompt "glasses". The ratio can be formulated as:

$$r_t = \frac{\Sigma_k \|F_{k,p_-(i)}^{(t)}\|_F}{\Sigma_k \|F_{k,p_+(r(i))}^{(t)}\|_F} \qquad (6)$$

where p_+, and p_- denote the positive and negative prompt respectively. And $r(\cdot)$ represents the mapping function from the negative prompt to its most relevant token in the positive prompt. We categorize the negative prompts into two distinct groups for our examination: nouns, such as 'glasses', which are utilized in object removal tasks, and adjectives, like 'ugly', aimed at refining the visual quality of images. We select a set of 10 corresponding prompt pairs. The experiments are conducted across 10 distinct random seeds to ensure robustness in our findings. We plot the $r_t \sim t$ curves averaged on the seeds in Fig. 3. **Peak at the 5th step for the noun-based negative prompt** as illustrated in Fig. 3a, indicating a critical step here. Initially, the ratio is near 1, possibly due to the Unet framework treating negative and positive prompts with parity, as discussed in Sect. 3. At that time, the negative prompt wants to generate some objects in the middle of the pic regardless of the context of the positive prompt. As we approach the peak, the negative prompt begins to assimilate layout cues from its positive counterpart, trying to remove the object. This results in the peak representing the zenith of its influence. Following this, as the element gradually disappears from the image, the impact of the negative prompt diminishes, with no remaining elements in the image to trigger the neural response. **A plateau around the 10th step for adjective-based negative prompts** is depicted in Fig. 3b, indicating the existence of the critical step. During the initial stages, the absence of the object leads to a subdued response, with the strength below one. Between the 5th and 10th steps, as the object becomes clear, the negative prompt accurately focuses on the intended area and maintains its influence.

5 How Do Negative Prompts Take Effect

To examine the dynamics in the reverse-diffusion process, we focus on analyzing the series of the estimated noises $\{\epsilon^{(t)}\}_{t=0}^{T}$.

5.1 Neutralization Hypothesis

We hypothesize that the negative prompts perform deletion through the canceling effect, where positive and negative noises align and nullify each other post-subtraction in Eq. (5). Supporting this, we observe in the failure cases of object deletion depicted in Fig. 4. A successful case is in the third row, where the attention map starts to target the location of a potted plant between steps 4–7, effectively counteracting the positive noise that would otherwise materialize a potted plant. Conversely, in the bottom row, attention doesn't focus on the lower right corner until step 8. By this stage, with the object nearly fully formed, it's too late for effective cancellation—resulting in only minimal impact on the finer details.

5.2 Reverse Activation

The phenomenon of Reverse Activation is observed when a negative prompt, introduced in the early stages of the diffusion process, unexpectedly leads to the generation of the specified object within the context of that negative prompt. In contrast, omitting negative prompts results in the absence of the object. As demonstrated in Fig. 2, if we apply "Glasses" as negative prompts in the first 3 steps, it will generate the glasses in the final output. In this section, our goal is to shed light on this phenomenon by analyzing the mechanism behind negative prompts. We start by examining the data distribution, highlight two intriguing observations, and ultimately offer an explanation.

Guidance Signals. We borrow the concept of the energy function from Energy-Based Models, as shown in Fig. 5 to represent the data distribution. The function is designed to assign lower energy levels to more 'likely' or 'natural' images according to the model's training data, and higher energy levels to less likely ones. As Real-world distributions often feature elements like a clear blue sky or

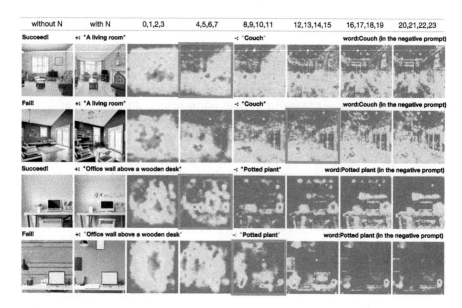

Fig. 4. Illustration: Heat maps showcasing the outcomes of object removal using negative prompts, with both successes and failures. Successful removals are placed in the first and third rows, while the failed attempts occupy the second and fourth rows. The first column shows the pictures **without** applying negative prompts contrasted by the second column, which features images with negative prompts. Notably, the feature map that first targets the relevant location is marked by a red square box ☐ It's evident that the successful cases exhibit earlier attention to the target areas. (Color figure online)

other uniform backgrounds, alongside distinct objects such as the Eiffel Tower, these elements typically possess low energy scores, making the model inclined to generate them. To synthesize a specific object like a tower from scratch, the diffusion process necessarily traverses through an intermediary phase that represents a blurry outline of the object. Given that such blurry representations are atypical in the training data, they present an 'energy barrier' that hinders the seamless generation of the desired object. So the model requires the guidance of prompts to surmount this barrier. We delve into the dynamics of distinct types of guidance as depicted in Fig. 6. To begin, Fig. 6a demonstrates that in the absence of explicit guidance, the model struggles to overcome the energy function's barrier, influenced by the natural data distribution, leading it back to generic backgrounds. Conversely, as depicted in Fig. 6b, when explicit guidance is provided through the inclusion of the object in the positive prompt's context, the model manages to surpass the barrier, with real-world distribution guidance steering it towards the object region.

Inducing Effect. The intriguing part is observed in Fig. 6c. Here, we illustrate an instance where direct negative prompts are applied, yet the context is absent from the positive prompt. As a result, the negative prompt guidance is much stronger than the positive prompt guidance, making this point at a considerable distance in the region opposite to the object. **Consequently, the distribution guidance demonstrates a substantial alignment towards the object and its surrounding area.** Without this, a tower-like structure would emerge at that location. This is because adding or subtracting tower-like features against a uniform background equally contributes to the formation of a tower pattern. As the distribution guidance is encoded into the estimated noises, projecting the positive noise towards the background-to-object direction reveals an enhanced effect in this direction, as opposed to the scenario without a negative prompt in Fig. 6a. We term the phenomenon as the "Inducing Effect", indicating that the negative prompt triggers the positive noise in a direction that represents the context of the negative prompt.

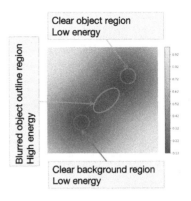

Fig. 5. Illustration: The energy function in the image generation dynamics. The value at the pixel represents the energy of the point in the data distribution. We mark the background region, clear object region, and blurred object outline region by circles. To generate an object from the background, the model should overcome the energy barrier of the blurred object outline region.

Simulation Experiment. To validate our hypothesis on the inducing effect, we conducted a follow-up quantitative experiment employing a variety of prompt pairs. Initially, we generate an image using positive prompt p_+ along with an empty unconditional prompt p_e and record the positive noises $\{\epsilon_{p+}^{(t)}(p_+, p_e)\}_{t=1}^T$. Simultaneously, we calculate a series of negative noises, $\{\epsilon_{p-}^{(t)}(p_+, p_e)\}_{t=1}^T$ but refrain from applying it during the sampling process. Following this, we generate additional noise series, $\{\epsilon_{p+}^{(t)}(p_+, p_-)\}_{t=1}^T$ and $\{\epsilon_{p+}^{(t)}(p_+, p_-)\}_{t=1}^T$ by applying both the positive prompt p_+ and the negative prompt p_- this time. To verify the existence of induction, we project the positive noise onto the negative noise which represents the direction towards the object region, and compute the difference between the two sets. The computation can be formulated as:

$$P_{Ind}^{(t)} = \frac{<\epsilon_{p+}^{(t)}(p_+, p_-), \epsilon_{p-}^{(t)}(p_+, p_-)>}{||\epsilon_{p-}^{(t)}(p_+, p_-)||^2} \epsilon_{p-}^{(t)}(p_+, p_-) \tag{7}$$

$$P_{Ori}^{(t)} = \frac{<\epsilon_{p+}^{(t)}(p_+, p_e), \epsilon_{p-}^{(t)}(p_-, p_e)>}{||\epsilon_{p-}^{(t)}(p_+, p_e)||^2} \epsilon_{p-}^{(t)}(p_+, p_e) \tag{8}$$

$$D^{(t)} = P_{Ind}^{(t)} - P_{Ori}^{(t)} \tag{9}$$

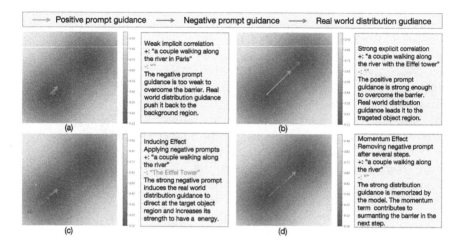

Fig. 6. Illustration: Distinct kinds of guidance. The purple arrow shows the guidance of the data distribution, which is the slope of the energy function. The red, brown arrow shows the guidance of the positive and negative prompts respectively. (Color figure online)

Intuitively, $P_{Ind}^{(t)}$ and $P_{Ori}^{(t)}$ shows the distribution guidance in Fig. 6a and Fig. 6c, respectively. And $D^{(t)}$ suggests the difference in the objection direction if a negative prompt containing the object is applied. **Therefore, a positive difference implies that the presence of the negative prompt effectively induces the inclusion of this component in the positive noise.** We perform experiments on 5 prompt pairs. We run 10 random seeds for each pair, average the results, and plot the $D^{(t)} \sim t$ curve in Fig. 7a. The results demonstrate that the presence of a negative prompt promotes the formation of the object within the positive noise, thereby confirming our hypothesis.

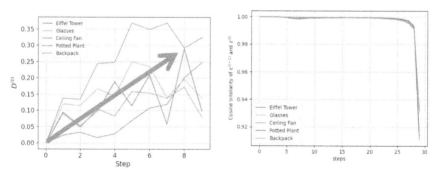

(a) Inducing Effect. The x-axis shows the time step while the y-axis quantifies the amount in the direction towards the object. The upward trajectory of the **red arrow** verifies the Inducing Effect phenomenon.

(b) Momentum Effect. The x-axis denotes the time step. The y-axis measures the cosine similarity between noises at consecutive steps. The diffusion process reveals a strikingly high correlation up to 99.5% in the first 25 steps.

Fig. 7. Illustration: Inducing Effect and Momentum Effect

Momentum Effect. Additionally, we observe a behavior akin to momentum, where the generated noises appear to sustain their trajectory in a specific direction. To confirm this phenomenon, we calculate the cosine similarity between the noise at step t $\epsilon_\theta((\mathbf{x}_t), c(s), t)$ and the noise at the preceding step t-1 $\epsilon_\theta((\mathbf{x}_t - 1), c(s), t - 1)$. As shown in Fig. 7b, there is a notable correlation between each noise and its immediate predecessor, indicating a momentum-like effect.

Explanation. Finally, we can come to an explanation of Reverse Activation in Fig. 2. Figure 6a demonstrates that without the negative prompt, the implicit guidance is insufficient to generate the intended object, explaining why the object fails to appear in the first column of Fig. 2. Conversely, as illustrated in Fig. 6c, the application of a negative prompt intensifies the distribution guidance towards the object, which prevents the object from materializing, clarifying the absence of the object in the last several columns of Fig. 2. Intriguingly, as supported by the Momentum Effect, if we remove the negative prompt after several steps, the real-world distribution guidance will maintain a large component towards the object's direction in the following steps as shown in Fig. 6d. Such a momentum effect finally facilitates the object's emergence, as shown in the middle columns in Fig. 2.

6 Enhanced Controllable Inpainting

In this section, we introduce a novel technique for controllable image inpainting that aims to eliminate undesired objects from generated images while preserving the integrity of the original background. Although Woolf [29] highlights the effectiveness of using negative prompts to remove undesired elements from images, it often leads to substantial modifications to the background, as shown in the second row of Fig. 8.

In Sect. 5, we claim that the negative prompt takes effect by a neutralization effect. But in Sect. 4, we observe a notable delay in the activation of negative prompts compared to their positive counterparts. As a result the negative prompts usually don't attend to the right place until step 5, well after the application of positive prompts. Additionally, as depicted in Fig. 6c, the use of negative prompts in the initial steps can significantly skew the diffusion process, potentially altering the background. This early application throughout the inference process, as practiced by Woolf, could be the reason behind their method's shortcomings. To mitigate these issues, we propose to deploy negative prompts post-'critical step' rather than getting it from the beginning. According to our findings, since negative prompts usually don't attend to the interested region until the critical step, all the neutralization and removal would then happen after the critical step. Meanwhile, the later added negative prompt would focus more on the target area with a reduced effect on surrounding regions in the later phase.

Fig. 8. Illustration of our method's ability to remove unwanted objects in the generated image while preserving the main concept, for various combinations of positive(+) and negative(-) prompts. **From top to bottom**: Initially with solely a positive prompt, followed by the integration of negative prompts throughout all diffusion stages, and finally, applying negative prompts during pivotal stages. Identical seeds were utilized for each column to ensure consistency.

6.1 Experiments

In this section, we conduct large-scale experiments to validate the efficacy of our proposed method.

Finding the Timing for Negative Prompts in Inpainting. We tested various combinations of prompts and negative prompts. The results, depicted in Fig. 9, reveal a U-shaped trend, indicating that employing negative prompts during intermediate steps is most effective. Take the blue line as an example, initiating with negative prompts at the first step necessitates approximately 10 steps for task completion. In contrast, starting at the sixth step significantly reduces this requirement to about 4 steps. Notably, the curves' nadir is around step 5, aligning with earlier insights about the critical step discussed in Sect. 4. Beyond step 11, applying negative prompts appears ineffective in eliminating the desired object. This may be because, in the later stage of diffusion, the shape and structure of the image have been essentially determined. We set steps 5–15 to use negative prompts as default in our later experiments.

Constructing Datasets for Evaluation of Negative Prompts. Due to the absence of pre-existing datasets in our settings, we adapt available datasets for our needs. We begin by selecting text samples from the COCO [15], CC [17], Nocaps [2], Places [34], MSVD [26], and Vatex [27] datasets to serve as prompts for image generation. These images are then analyzed by GPT-4V [1] to identify the contained objects. Then we try to remove these objects when generating the image again. For each dataset, we use 1000 text prompts and every prompt is run with 5 different seeds.

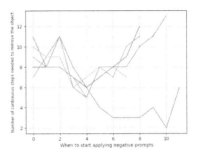

Fig. 9. Experiments on finding the best time to apply negative prompts. Each endpoint on the polyline graph represents a specific scenario: the x-coordinate denotes the starting step for applying negative prompts, while the y-coordinate indicates the number of consecutive steps required to precisely remove the targeted object. When the object cannot be removed, the curve terminates.

Evaluation. We leverage GPT-4V to assess the success of the inpainting process and to determine the relative distance between the original and inpainted images. We also conduct **human evaluation** for further verification. More details on the evaluation protocols can be found in the Appendix.

Metrics. To assess the efficacy of our approach, we employ two key metrics: The Removal Success Rate(RSR) indicates the success rate of the target object removal. The Relative Removal Success Rate (RRSR) gauges the efficiency of our method relative to a baseline by calculating the ratio of our RSR to the baseline's RSR. A higher RRSR suggests that our method remains effective even when negative prompts are applied over fewer steps. Additionally, the Comparison Rate(CR) measures the extent to which our generated images resemble the original images, as judged by GPT-4V or human evaluators. We first ask GPT4V and humans if images generated by our method are more similar to the origin of one. Then we compute the ratio of affirmative responses to the total number of evaluations. The higher the Removal Rate, the better. Details can be found in the Appendix.

Results. Table 1a summarizes the results. As we can see, our method incurs minimal impact on the removal success rate. In fewer than 20% of instances, our method fails to remove the target object where the baseline method succeeds, addressing concerns that applying negative prompts in fewer steps might compromise inpainting effectiveness. Moreover, on average, our method achieves up to 82.64% similarity to the original images, underscoring its efficiency. Meanwhile, the results of the human evaluation can be seen in Table 1b The results show the effectiveness of our method.

Table 1. For both RRSR and CR, the higher the better. The first column shows the RSR of the baseline by applying negative prompts to all steps. The second column shows the RSR of our method.

Dataset	RSR%	RSR%	RRSR%	CR%
COCO	54.41	65.02	83.67	83.32
CC	54.81	64.24	85.31	82.75
MSVD	55.53	66.59	83.39	83.07
Places	49.83	57.21	87.20	84.12
Vatex	61.37	71.00	86.44	82.52
Nocaps	48.88	56.72	86.17	81.26
Avg.	54.16	63.46	85.37	82.84

(a) Main Results of GPT-4V Evaluation

Dataset	RSR%	RSR%	RRSR%	CR%
COCO	87.27	90.90	96.01	92.68
CC	59.32	71.18	83.34	97.67
MSVD	61.11	81.48	75.00	83.67
Places	53.12	76.56	69.38	90.66
Vatex	57.57	73.53	78.29	80.12
Nocaps	72.32	88.46	81.75	87.67
Avg.	65.11	80.35	80.62	88.45

(b) Main Results of Human Evaluation

7 Future Work

In our experiments, we focus primarily on tasks involving the removal of nouns and the attribution based on adjectives, deferring the exploration of other parts of speech and tasks to future research. Our findings highlight the challenge of information lag between pairs of positive and negative prompts. A straightforward remedy could involve increasing interactions during the noise generation phase. Additionally, our method of controllable object removal in image generation presents a novel approach for creating image inpainting datasets. Finally, applying negative prompts in the training process as a form of data augmentation may potentially enhance performance further, which is left as future work.

8 Conclusion

In conclusion, our research provides a comprehensive analysis of negative prompts in diffusion models for image generation. Through systematic experiments, we have identified the critical steps where negative prompts begin to influence the image generation process, uncovering a significant lag in the transition from positive to negative prompt effects. This insight led us to develop a novel approach that strategically applies negative prompts at an optimal stage in the reverse-diffusion process, ensuring the removal of undesired elements while preserving the image's integrity. Our contributions not only shed light on the underlying dynamics of negative prompts but also offer a practical solution for controllable image inpainting tasks, significantly improving upon existing methods without the need for network retraining or modifications during inference.

References

1. Achiam, J., et al.: Gpt-4 technical report. arXiv preprint arXiv:2303.08774 (2023)
2. Agrawal, H., et al.: Nocaps: novel object captioning at scale. In: Proceedings of the IEEE/CVF International Conference on Computer Vision, pp. 8948–8957 (2019)
3. Andrew: How to use negative prompts? (2023). https://stable-diffusion-art.com/how-to-use-negative-prompts/#Why_does_negative_prompt_become_more_important_in_v2
4. Armandpour, M., Zheng, H., Sadeghian, A., Sadeghian, A., Zhou, M.: Re-imagine the negative prompt algorithm: transform 2d diffusion into 3d, alleviate janus problem and beyond. arXiv preprint arXiv:2304.04968 (2023)
5. Balaji, Y., et al.: ediffi: text-to-image diffusion models with an ensemble of expert denoisers. arXiv preprint arXiv:2211.01324 (2022)
6. Chefer, H., Alaluf, Y., Vinker, Y., Wolf, L., Cohen-Or, D.: Attend-and-excite: attention-based semantic guidance for text-to-image diffusion models. ACM Trans. Graph. (TOG) **42**(4), 1–10 (2023)
7. Criminisi, A., Pérez, P., Toyama, K.: Region filling and object removal by exemplar-based image inpainting. IEEE Trans. Image Process. **13**(9), 1200–1212 (2004)
8. Dhariwal, P., Nichol, A.: Diffusion models beat gans on image synthesis. Adv. Neural. Inf. Process. Syst. **34**, 8780–8794 (2021)
9. Feng, W., et al.: Training-free structured diffusion guidance for compositional text-to-image synthesis. arXiv preprint arXiv:2212.05032 (2022)
10. Hertz, A., Mokady, R., Tenenbaum, J., Aberman, K., Pritch, Y., Cohen-Or, D.: Prompt-to-prompt image editing with cross attention control. arXiv preprint arXiv:2208.01626 (2022)
11. Ho, J., Salimans, T.: Classifier-free diffusion guidance. arXiv preprint arXiv:2207.12598 (2022)
12. mukund kapoor: Negative prompts in stable diffusion: a beginner's guide (2023). https://www.greataiprompts.com/imageprompt/what-is-negative-prompt-in-stable-diffusion/
13. Kawar, B., et al.: Imagic: text-based real image editing with diffusion models. In: Proceedings of the IEEE/CVF Conference on Computer Vision and Pattern Recognition, pp. 6007–6017 (2023)
14. Lian, L., Li, B., Yala, A., Darrell, T.: Llm-grounded diffusion: enhancing prompt understanding of text-to-image diffusion models with large language models. arXiv preprint arXiv:2305.13655 (2023)
15. Lin, T.-Y., et al.: Microsoft COCO: common objects in context. In: Fleet, D., Pajdla, T., Schiele, B., Tuytelaars, T. (eds.) ECCV 2014. LNCS, vol. 8693, pp. 740–755. Springer, Cham (2014). https://doi.org/10.1007/978-3-319-10602-1_48
16. Mahajan, S., Rahman, T., Yi, K.M., Sigal, L.: Prompting hard or hardly prompting: prompt inversion for text-to-image diffusion models. arXiv preprint arXiv:2312.12416 (2023)
17. Ng, E.G., Pang, B., Sharma, P., Soricut, R.: Understanding guided image captioning performance across domains. arXiv preprint arXiv:2012.02339 (2020)
18. Nichol, A., et al.: Glide: towards photorealistic image generation and editing with text-guided diffusion models. arXiv preprint arXiv:2112.10741 (2021)
19. O'Connor, R.: Stable diffusion 1 vs 2 - what you need to know (2023). https://www.assemblyai.com/blog/stable-diffusion-1-vs-2-what-you-need-to-know/#negative-prompts

20. von Platen, P., et al.: Diffusers: state-of-the-art diffusion models (2022). https://github.com/huggingface/diffusers
21. Rombach, R., Blattmann, A., Lorenz, D., Esser, P., Ommer, B.: High-resolution image synthesis with latent diffusion models. In: Proceedings of the IEEE/CVF Conference on Computer Vision and Pattern Recognition, pp. 10684–10695 (2022)
22. Rombach, R., Blattmann, A., Lorenz, D., Esser, P., Ommer, B.: High-resolution image synthesis with latent diffusion models. In: Proceedings of the IEEE/CVF Conference on Computer Vision and Pattern Recognition (CVPR), pp. 10684–10695 (2022)
23. Saharia, C., et al.: Photorealistic text-to-image diffusion models with deep language understanding. Adv. Neural. Inf. Process. Syst. **35**, 36479–36494 (2022)
24. Tang, R., et al.: What the daam: interpreting stable diffusion using cross attention. arXiv preprint arXiv:2210.04885 (2022)
25. Tumanyan, N., Geyer, M., Bagon, S., Dekel, T.: Plug-and-play diffusion features for text-driven image-to-image translation. In: Proceedings of the IEEE/CVF Conference on Computer Vision and Pattern Recognition, pp. 1921–1930 (2023)
26. Venugopalan, S., Rohrbach, M., Donahue, J., Mooney, R., Darrell, T., Saenko, K.: Sequence to sequence-video to text. In: Proceedings of the IEEE International Conference on Computer Vision, pp. 4534–4542 (2015)
27. Wang, X., Wu, J., Chen, J., Li, L., Wang, Y.F., Wang, W.Y.: Vatex: a large-scale, high-quality multilingual dataset for video-and-language research. In: Proceedings of the IEEE/CVF International Conference on Computer Vision, pp. 4581–4591 (2019)
28. Wang, Z.J., Montoya, E., Munechika, D., Yang, H., Hoover, B., Chau, D.H.: Diffusiondb: a large-scale prompt gallery dataset for text-to-image generative models. arXiv preprint arXiv:2210.14896 (2022)
29. Woolf, M.: Stable diffusion 2.0 and the importance of negative prompts for good results (2023). https://minimaxir.com/2022/11/stable-diffusion-negative-prompt/
30. Yang, S., Zhang, L., Ma, L., Liu, Y., Fu, J., He, Y.: Magicremover: tuning-free text-guided image inpainting with diffusion models. arXiv preprint arXiv:2310.02848 (2023)
31. Yildirim, A.B., Baday, V., Erdem, E., Erdem, A., Dundar, A.: Inst-inpaint: instructing to remove objects with diffusion models. arXiv preprint arXiv:2304.03246 (2023)
32. Yu, J., Lin, Z., Yang, J., Shen, X., Lu, X., Huang, T.S.: Generative image inpainting with contextual attention. In: Proceedings of the IEEE Conference on Computer Vision and Pattern Recognition, pp. 5505–5514 (2018)
33. Zhong, S., Huang, Z., Wen, W., Qin, J., Lin, L.: Sur-adapter: enhancing text-to-image pre-trained diffusion models with large language models. In: Proceedings of the 31st ACM International Conference on Multimedia, pp. 567–578 (2023)
34. Zhou, B., Lapedriza, A., Khosla, A., Oliva, A., Torralba, A.: Places: a 10 million image database for scene recognition. IEEE Trans. Pattern Anal. Mach. Intell. **40**(6), 1452–1464 (2017)

GS2Mesh: Surface Reconstruction from Gaussian Splatting via Novel Stereo Views

Yaniv Wolf[✉], Amit Bracha, and Ron Kimmel

Technion - Israel Institute of Technology, Haifa, Israel
{yaniv.wolf,amit.bracha,ron}@cs.technion.ac.il
https://gs2mesh.github.io

Abstract. Recently, 3D Gaussian Splatting (3DGS) has emerged as an efficient approach for accurately representing scenes. However, despite its superior novel view synthesis capabilities, extracting the geometry of the scene directly from the Gaussian properties remains a challenge, as those are optimized based on a photometric loss. While some concurrent models have tried adding geometric constraints during the Gaussian optimization process, they still produce noisy, unrealistic surfaces.

We propose a novel approach for bridging the gap between the noisy 3DGS representation and the smooth 3D mesh representation, by injecting real-world knowledge into the depth extraction process. Instead of extracting the geometry of the scene directly from the Gaussian properties, we instead extract the geometry through a pre-trained stereo-matching model. We render stereo-aligned pairs of images corresponding to the original training poses, feed the pairs into a stereo model to get a depth profile, and finally fuse all of the profiles together to get a single mesh.

The resulting reconstruction is smoother, more accurate and shows more intricate details compared to other methods for surface reconstruction from Gaussian Splatting, while only requiring a small overhead on top of the fairly short 3DGS optimization process.

We performed extensive testing of the proposed method on in-the-wild scenes, obtained using a smartphone, showcasing its superior reconstruction abilities. Additionally, we tested the method on the Tanks and Temples and DTU benchmarks, achieving state-of-the-art results.

1 Introduction

The Gaussian Splatting Model for radiance field rendering (3DGS) [19] has recently marked a significant leap forward in the realm of novel view synthesis, surpassing previous neural rendering methods in both speed and accuracy. By optimizing the distribution, size, color, and opacity of a cloud of Gaussian

---

Y. Wolf and A. Bracha—Equal Contribution.

Supplementary Information The online version contains supplementary material available at https://doi.org/10.1007/978-3-031-73024-5_13.

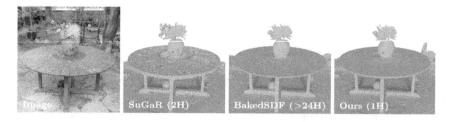

Fig. 1. Qualitative results on Mip-NeRF360 [1] dataset garden scene.

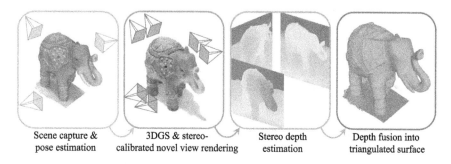

Fig. 2. The proposed pipeline for surface reconstruction. First, we represent the scene by applying a 3DGS model. We then use the 3DGS model to render stereo-aligned pairs of images corresponding to the original views. For each pair, using a shape from stereo algorithm, we reconstruct an RGB-D structure, which is then integrated from all views using TSDF [9] into a triangulated mesh of the scene.

elements, and projecting, or splatting them onto virtual cameras, 3DGS is able to generate realistic images of complex scenes from novel viewing directions in real-time. However, direct reconstruction of surfaces from 3DGS involves significant challenges. The main problem is that the locations of the Gaussian elements in 3D space do not form a geometrically consistent surface, as those are optimized for best matching the input images when projected back onto their image planes. Consequently, reconstructing surfaces based on the centers of the Gaussians yields noisy and inaccurate results. Current state-of-the-art methods attempt to regularize the 3DGS optimization process by adding additional geometric constraints [14], flattening the Gaussian elements, [15], or extracting the geometry using opacity fields [51], but they still rely on the Guassian locations and form noisy, unrealistic surfaces.

We propose an alternative approach for extracting depth from the optimized Gaussian point cloud, which does not rely on the noisy locations of the Gaussians. Instead, we take advantage of a powerful geometric regularizer, trained on real-world data - a pre-trained stereo matching model. Stereo matching models solve a correspondence problem on stereo-aligned pairs of images, from which accurate depth can be extracted. Our main observation is that through 3DGS rendering, we can artificially create stereo-aligned pairs of images corresponding to the original views, feed these pairs into a pre-trained stereo model, and fuse the resulting depths using the Truncated Signed Distance Function (TSDF) algorithm [9]. The result is a smooth, geometrically consistent surface, that is extracted from the noisy 3DGS cloud using real-world regularization.

The proposed method reduces surface reconstruction time dramatically, taking only a small overhead on top of the 3DGS capturing of the scene, which is significantly faster compared to neural surface reconstruction methods. For instance, reconstruction of an in-the-wild scene taken by a standard smartphone camera requires less than five minutes of additional computation time after a 3DGS scene capture. Additionally, since we reconstruct the surface based on the 3DGS capture, it is straightforward to bind the mesh to the original model, as mentioned in [14,37], for mesh-based manipulation of the Gaussian elements. Moreover, since our mesh is more accurate, it does not require any additional refinements [14].

We tested the proposed method on the Tanks and Temples (TnT) benchmark [21] as well as the DTU [16] benchmark, two commonly used 3D reconstruction datasets, and achieved state-of-the-art results. Additionally, we extensively tested our method on in-the-wild scenes captured with a smartphone, showing qualitative results of the proposed method's reconstruction abilities. To summarize, our main contribution,

– We propose a novel method for fast and accurate in-the-wild surface reconstruction, by using a pre-trained stereo matching model as a geometric prior for extracting depth from a 3DGS model.

2 Related Efforts

2.1 Multi-view Stereo and Stereo Matching

Multi-View Stereo (MVS) is a fundamental geometry reconstruction method, where depth maps are extracted for each reference image based on correspondences with neighboring images. In the field of deep MVS methods, the pioneering work of MVSNet [46] introduced an end-to-end framework for MVS learning, which can be divided into three parts: 2D feature extraction, homography, and 3D cost volume with 3D convolutions. Latter methods presented an improvement to this scheme, by improving the 3D cost volume [26,55], improving the architecture for 2D feature extraction [38], using a vision transformer (ViT) architecture for feature extraction [4], and improving 3D convolutions for more efficient computations using a coarse-to-fine method [13,45]. To fuse the extracted depth maps into one point cloud or mesh there are two main methods: Fusibile [12], which has recently been generalized by [44], and TSDF [9]. MVS methods deeply rely on accurate camera poses for the calculation of epipolar lines, and on in-the-wild scenes they struggle to achieve high accuracy as in the controlled environment, since small errors in pose estimation result in a noisy reconstruction, as we show in our ablation study.

Deep stereo matching methods [5,18,22,29,34,53] are related to deep MVS methods, however, since it is guaranteed that matching pixels between two images must lie in the same row, the cost volume layer works on the disparity instead of the depth. Recent state-of-the-art stereo matching models, such as RAFT [25], IGEV [43], and DLNR [56], use iterative refinement using GRU or LSTM layers. Unlike MVS methods, stereo methods require only two images which are typically closer to each other compared to MVS, and share the same image plane, resulting in less occluded regions which are visible only in one of the views.

2.2 Neural Rendering for Novel View Synthesis and Surface Reconstruction

Novel view synthesis methods are trained on a set of images from a scene, and aim to render views of a scene from any given pose. The pioneering work of Neural Radiance Fields (NeRF) [27] presented a major leap forward in accuracy by incorporating importance sampling and positional encoding to enhance rendering quality. However, the use of relatively large Multi-Layer Perceptrons (MLP) to capture the scene resulted in long training times. Later, Mip-NeRF [1] improved the quality of the rendered view with a different sampling method, although training and rendering times remained long. InstantNGP [28] tackled the extended training times of previous efforts, by incorporating a hash grid and an occupancy grid with a small MLP.

Neural surface reconstruction methods [35,39,48,49], in addition to accurately rendering novel views, are also capable of reconstructing the surface of the scene. IDR [49] trained an SDF represented by an MLP for both color and geometry reconstruction. Neus [39] reduced the geometric error by utilizing weighted volume rendering, and HF-Neus [40] enabled coarse-to-fine refinement for high-frequency detail reconstruction by decomposing the implicit SDF into a base function and a displacement function. RegSDF [54] used a point cloud obtained from shape-from-motion (SfM) as regularization, in addition to regularizing the curvature of the zero-level of the SDF function. NeuralWarp [10] suggested refining the geometry by regularizing image consistency between different views through warping based on implicit geometry. Neuralangelo [24], using a 3D hash encoded grid, enabled detailed reconstruction and achieved state-of-the-art results on leading benchmarks. However, reconstruction time of these methods can reach up to several days per scene. In the context of novel stereo views, a recent model [36] has managed to successfully perform unsupervised training of a stereo model using rendered stereo-aligned triplets from a neural scene reconstruction method, showcasing the possibility of novel view synthesis as a data factory.

2.3 Gaussian Splatting for Novel View Synthesis and Surface Reconstruction

Recently, a major leap forward was presented by 3DGS [19], a faster and more accurate method for scene capturing. 3DGS represents the scene as a point cloud of 3D Gaussians, where each Gaussian has the properties of opacity, rotation, scale, location, and spherical harmonics. The scaling of the Gaussians is anisotropic, which allows them to represent thin structures in the scene. The Gaussians are initialized using an SfM algorithm [32,33], which extracts the camera poses and provides an initial guess for the locations of the Gaussians. The capturing time is short compared to other methods based on MLPs, and it is capable of real-time rendering. Concurrent works on GS [6,8,11,42,50] have improved on the vanilla 3DGS in various ways, such as by reducing optimization time, increasing accuracy, reducing aliasing and removing the need for COLMAP [32,33] poses.

As discussed earlier, the Gaussian locations in the vanilla 3DGS do not form a geometrically consistent surface. Recent methods try to manipulate the Gaussian elements to extract more accurate surfaces [7,14,15]. SuGaR [14], the pioneering method in surface reconstruction from Gaussian Splatting, added a regularization term for post-process optimization based on the opacity levels of the Gaussians, forcing the Gaussian element cloud to align with the surface. 2DGS [15] flattens the Gaussians into 2D elements, and GOF [51] extracts the surface by creating an opacity field from the Gaussians. However, since these methods utilize the location and opacity of the Gaussian elements, they reconstruct the surface with noisy undulations.

3 Method

We propose a novel pipeline for surface reconstruction from 3DGS, as illustrated in Fig. 2. In this section, we will explain in detail each step of the pipeline. We note that additionally, we can mask out specific objects by projecting segmentation masks from Segment Anything Model (SAM) [20] between consecutive images using depth maps. Additional information on masking is available in the supplementary material.

3.1 Scene Capture and Pose Estimation

We start with a video or images of a static scene as input. Following the vanilla 3DGS, we employ COLMAP [32,33] for SFM to identify points of interest and deduce camera matrices from the provided images.

3.2 3DGS and Stereo-Aligned Novel View Rendering

The elements extracted from the previous stage are then fed into the 3DGS model to accurately represent the scene. For completeness, we will give a short formulation of the 3DGS process; In 3DGS, 3D Gaussian elements are defined in space by $G(\boldsymbol{x}) = \exp(-\frac{1}{2}(\boldsymbol{x}-\boldsymbol{x}_p)^\top \boldsymbol{\Sigma}^{-1}(\boldsymbol{x}-\boldsymbol{x}_p))$, where $\boldsymbol{x}_p$ is the center of the Gaussian, and $\boldsymbol{\Sigma}$ is its covariance matrix. During optimization, $\boldsymbol{\Sigma}$ is factorized into the rotation $\boldsymbol{R}$ and scale $\boldsymbol{S}$ matrices: $\boldsymbol{\Sigma} = \boldsymbol{R}\boldsymbol{S}\boldsymbol{S}^\top \boldsymbol{R}^\top$. When rendering, the Gaussians are projected onto the image plane: $\boldsymbol{\Sigma}' = \boldsymbol{J}\boldsymbol{W}\boldsymbol{\Sigma}\boldsymbol{W}^\top \boldsymbol{J}^\top$, where

W is the view transformation, and J is the Jacobian of the affine projective transformation onto the image plane. By removing the last row and column of Σ', we remain with 2D Gaussians in the image plane. To calculate the color of a pixel in the image plane, 3DGS employs alpha blending which applies weights to the opacity from front to back, $C = \sum_{i \in N} \alpha_i c_i \prod_{j=1}^{i-1}(1 - \alpha_j)$, where α_i is the product of the ith 2D Gaussian with its opacity parameter, and c_i is the directional appearance component. For more details, see the original 3DGS paper [19].

During the 3DGS process the Gaussians are optimized based on the photometric loss between the given source images and their corresponding rendered images. This creates a representation of the scene that allows rendering novel views which were not present in the original training data. It is important to note that the vanilla 3DGS relies on sufficient coverage of the scene, and in areas lacking sufficient coverage, noisy artifacts might appear, as seen in Fig. 8. Additionally, since the 3DGS is optimized based on the training images, staying close to a training image will result in a cleaner render.

Therefore, when generating novel stereo views of the scene, we input a sufficient amount of images that cover the region of interest. Additionally, we stay as close as possible to the original training poses, by choosing the left image of the stereo pair to be at the same pose R_L, T_L as a training image. Following this choice, the right pose with a horizontal baseline of b is formulated as follows: $R_R = R_L$, $T_R = T_L + (R_L \times [b,0,0])$. This ensures that the resulting left-right cameras are stereo-calibrated.

3.3 Stereo Depth Estimation

With the rendered stereo-aligned image pairs, we can essentially turn a scene captured from a single camera into a scene captured from a pair of stereo-calibrated cameras, using the novel view synthesis capabilities of 3DGS. We then apply a stereo matching algorithm to form depth profiles from every stereo pair. We have tested several stereo matching algorithms in the experimental section, and achieved the best qualitative and quantitative results with DLNR [56], a state-of-the-art neural stereo matching model, with the pre-trained Middlebury [31] weights. To further enhance the resulting reconstructions, we apply several masks to the output of the stereo model. The first mask is an occlusion mask, which is calculated by applying a threshold on the difference between the left-to-right and right-to-left disparities of the same pair of images. This masks out parts of the scene that were only visible in one of the cameras, and therefore the stereo model's output in these areas is unreliable. We justify the use of this mask by the fact that the occluded areas will be filled in from adjacent stereo views. An example of an occlusion mask can be seen in Fig. 3, and we added an experiment in the supplementary material which demonstrates the effectiveness of the occlusion mask.

The second mask is applied based on the depth of the stereo output. The relationship between stereo matching errors can be described as $\epsilon(Z) \approx \dfrac{\epsilon(d)}{f_x \cdot B} Z^2$

[3], where $\epsilon(d)$ represents the disparity output error, Z is the ground-truth depth, $\epsilon(Z)$ is the error of the depth estimate, f_x denotes the camera's horizontal focal length, and B is the baseline. Conversely, the disparity between matching pixels in two images of an object that is positioned at a short distance from the cameras can exceed the maximum disparity limit produced by stereo matching algorithms. Thus, estimating the depth of an object that is too close to the camera can result in an error due to the limitation of the matching algorithms, and estimating the depth of an object that is distant results in a quadratic error. Therefore, we consider depth in the range $4B \leq Z \leq 20B$. This approach enhances the overall accuracy and reliability of the depth estimation process, ensuring more consistent geometric reconstructions. With the above considerations taken into account, we now have two contradicting factors when setting the horizontal baseline of the stereo pair; On the one hand, a larger baseline allows for a wider "sweet spot" for the stereo model. On the other hand, the 3DGS limits how far we can stray from the original training images without producing noisy renders. In the experimental section, we tested different baselines and found that a horizontal baseline of 7% of the scene radius, or 3.5% of the scene diameter, which allows for a "sweet spot" in the range of 14% to 70% of the scene diameter, provides the best results.

Fig. 3. Example of our method's output on DTU [16] scan105. From left to right: The rendered left and right images, segmentation mask, left-right disparity, occlusion mask, and shading - depth gradient.

3.4 Depth Fusion Into Triangulated Surface

To further enhance geometric consistency and smooth out any noise and errors which might have originated from the individual depth profiles, we aggregate all of the extracted depths using the Truncated Signed Distance Function (TSDF) algorithm [9], followed by the Marching-Cubes meshing algorithm [41].

4 Experiments and Results

We present experiments which demonstrate that our method is able to accurately reconstruct surfaces in a more geometrically consistent way than other 3DGS-based or MVS approaches, as well as achieve comparable performance to neural reconstruction methods while taking significantly less time to run. For quantitative results, we tested our method on the Tanks and Temples [21] and DTU

[16]datasets, and compared our results to various neural and 3DGS-based reconstruction methods. We also compared between different versions of our model to justify our design choices. Additionally, we show qualitative reconstruction results from Mip-NeRF360 [1], demonstrating that our method achieves comparable visual quality to neural reconstruction methods, and on in-the-wild videos taken from smartphones, we show our superiority in terms of geometric consistency and smoothness when compared to SuGaR [14]. Finally, we perform an ablation study on the MobileBrick [23] dataset, which validates the contribution of novel-view image generation and stereo, by replacing two different points in our pipeline with a deep MVS model. We note that in the MobileBrick dataset the camera poses are manually refined, and are shown to be more accurate than COLMAP [32,33] poses for reconstruction [23]. The comparison we present thus favors MVS models in that respect.

Fig. 4. Qualitative comparison of mesh reconstruction from in-the-wild videos between our method and SuGaR [14].

4.1 Datasets

DTU [16]. This dataset is an MVS dataset, containing scans of small objects, as well as accurate camera poses and 3D point clouds. We use the dataset and evaluation code from [15], which calculates the Chamfer Distance (CD) between the reconstructed and ground-truth point clouds.

Tanks and Temples (TnT) [21]. This dataset contains videos of large objects such as vehicles, buildings and statues. These objects are scanned with a laser scanner for an accurate ground-truth 3D point cloud. As the videos and the laser scanned objects are difficult to align [30], for evaluation we use the official TnT [21] evaluation alignment method. It first aligns the point clouds using ICP [2], and then calculates the precision, recall, and F1 score.

Mip-NeRF360 [1]. This dataset contains scenes taken from a 360°C view, with emphasis on minimizing photometric variations through controlled capture conditions. Since there is no ground-truth in terms of surface reconstruction, we leave this as a qualitative comparison only.

MobileBrick [23]. This dataset contains videos of LEGO models, with corresponding 3D ground-truth meshes created from the LEGO 3D model. The poses are manually refined and are more accurate than the COLMAP ones [32,33]. This dataset is challenging since most of the videos in the test set are taken from a top view of the model, thus, creating occlusions and leaving areas in the model with little visibility. We use the official evaluation code.

In-the-Wild Videos. For reconstruction of in-the-wild objects, our method presents a favorable balance between accuracy and computation time. To validate this claim, we captured scenes containing various objects such as plants, sculptures, figures and everyday items, with intricate geometries and textures, and reconstructed their surface. Each video contains one or two cycles of moving around the object, depending on the object's size, without any measures to maintain a persistent radius or pose of the camera, and without any control of the lighting in the environment. Since these objects are filmed only with a smartphone camera, there is no ground-truth reconstruction for these objects.

4.2 Baselines

Gaussian Splatting-Based Methods. For the DTU [16] dataset, We compare our method with the vanilla 3DGS [19] and SuGaR [14], as well as additional state-of-the-art methods, namely 2DGS [15] and Gaussian Opacity Fields (GOF) [51]. For the TNT [21] and Mip-NeRF360 [1] datasets, as well as for in-the-wild scenes, we compare with SuGaR [14].

Neural Rendering Methods. For the DTU [16] dataset, we compare our method with Neuralangelo [24], VolSDF [47], and NeuS [39]. For the TnT [21] dataset, we compare with Neuralangelo [24], NeuralWarp [10] and NeuS [39]. For the Mip-NeRF360 [1] dataset, we compare with BakedSDF [48].

Deep MVS. for in-the-wild scenes and the MobileBrick [23] dataset, we compare with MVSformer [4], a state-of-the art deep MVS network.

4.3 Results

DTU [16]. Table 1 presents quantitative results on the DTU [16] dataset. We ran the 3DGS step of our method for 30000 iterations, but as we show in the

Table 1. Quantitative results on the DTU [16] dataset, comparing our method with state-of-the art neural and Gaussian Splatting-based methods. Chamfer distance - lower is better. Red-1st, Orange-2nd, Yellow-3rd. Table adapted from [51].

		24	37	40	55	63	65	69	83	97	105	106	110	114	118	122	Mean	Time
Neural	NeRF [27]	1.90	1.60	1.85	0.58	2.28	1.27	1.47	1.67	2.05	1.07	0.88	2.53	1.06	1.15	0.96	1.49	>12 h
	VolSDF [47]	1.14	1.26	0.81	0.49	1.25	0.70	0.72	1.29	1.18	0.70	0.66	1.08	0.42	0.61	0.55	0.86	>12 h
	NeuS [39]	1.00	1.37	0.93	0.43	1.10	0.65	0.57	1.48	1.09	0.83	0.52	1.20	0.35	0.49	0.54	0.84	>12 h
	Neuralangelo [24]	0.37	0.72	0.35	0.35	0.87	0.54	0.53	1.29	0.97	0.73	0.47	0.74	0.32	0.41	0.43	0.61	>12 h
Gaussian Splatting	3DGS [19]	2.14	1.53	2.08	1.68	3.49	2.21	1.43	2.07	2.22	1.75	1.79	2.55	1.53	1.52	1.50	1.96	11.2 m
	SuGaR [14]	1.47	1.33	1.13	0.61	2.25	1.71	1.15	1.63	1.62	1.07	0.79	2.45	0.98	0.88	0.79	1.33	~1 h
	2DGS [15]	0.48	0.91	0.39	0.39	1.01	0.83	0.81	1.36	1.27	0.76	0.70	1.40	0.40	0.76	0.52	0.80	18.8 m
	GOF [51]	0.50	0.82	0.37	0.37	1.12	0.74	0.73	1.18	1.29	0.68	0.77	0.90	0.42	0.66	0.49	0.74	30 m
	Ours - DLNR Baseline 3.5%	0.61	0.85	0.64	0.39	0.96	1.25	0.80	1.52	1.10	0.68	0.59	0.93	0.45	0.60	0.54	0.79	~20 m
	Ours - DLNR Baseline 10.5%	0.69	0.81	0.95	0.51	0.82	1.06	0.72	1.18	0.93	0.61	0.54	0.66	0.37	0.54	0.50	0.73	~20 m
	Ours - RAFT Baseline 7%	0.59	0.81	0.68	0.40	0.83	1.15	0.73	1.35	1.05	0.62	0.53	0.80	0390	0.55	0.49	0.73	~20 m
	Ours - DLNR Baseline 7%	0.59	0.79	0.70	0.38	0.78	1.00	0.69	1.25	0.96	0.59	0.50	0.68	0.37	0.50	0.46	0.68	~20 m

supplementary material, we can achieve nearly identical results with only 7000 iterations, reaching a mean Chamfer Distance of 0.70 with only ~ 12m of total runtime per scan. We used the same TSDF as in 2DGS [15] and GOF [51], which is based on the Open3D implementation [9], for a fair comparison. We apply the mask supplied with the dataset before inputting the depths into the TSDF algorithm. The table is adapted from GOF [51] for consistency. Within the splatting-based methods, our method achieves the best score, while maintaining a similar runtime. Additionally, when compared to the neural methods, which take more than 12 h to reconstruct a single scene, our method surpasses some of the methods, and is comparable with Neuralangelo [24], the state-of-the-art method. Additionally, we test our method with RAFT [25] as the stereo model with the RVC weights [17] and with DLNR [56] as the stereo model with the Middlebury weights [31], achieving better results with DLNR. Finally, we compare between three different horizontal baselines: 3.5%, 7% and 10.5% of the scene radius. We achieve the best results with 7%, noting that increasing or decreasing the horizontal baseline has a negative effect on the results. Figure 3 shows an example of the intermediate representations of our method on one of DTU [16] scan105, and the full set of reconstructed meshes is available in the supplementary material.

Table 2. Quantitative results on the Tanks and Temples [21] benchmark. F1 score - higher is better.

		Barn	Caterpillar	Ignatius	Truck	Mean F1 ↑	Runtime
Neural	NeuralWarp [10]	0.22	0.18	0.02	0.35	0.19	
	NeuS [39]	0.29	0.29	0.83	0.45	0.47	~16 h–48 h
	Neuralangelo [24]	0.70	0.36	0.89	0.48	0.61	
GS	SuGaR [14]	0.01 (0.08)	0.02 (0.09)	0.06 (0.34)	0.05 (0.17)	0.04 (0.17)	~2 h
	Ours	0.21 (0.22)	0.17 (0.12)	0.64 (0.68)	0.46 (0.40)	0.37 (0.36)	~1 h

Fig. 5. Qualitative results on Tanks and Temples [21]. Top row: Ignatius scene, compared to SuGaR [14]. Bottom row: Barn scene, compared to SuGaR.

Tanks and Temples [21]. Tab. 2 presents a summary of the reconstruction results on the TnT [21] benchmark. Since SuGaR [14] yields a sparse mesh, and thus its recall drops significantly, we include a precision metric that is unaffected by mesh sparsity. However, it is important to note that this metric does not account for missing parts in the reconstruction. The results show that our method outperforms SuGaR [14] in both F1 and precision. Additionally, it is evident from Fig. 5 that our method is able to reconstruct fine details such as in the Barn scene. Moreover, our method has a significant advantage in terms of processing time, requiring less than 60 min of total computation time per TnT [21] scene, compared to the 16–48 h needed by neural reconstruction methods. The reason for our relatively longer computation times for TnT [21] is since each scene containing hundreds of frames, compared to a typical in-the-wild scene which contains less than 100 frames. It is important to note that the TnT [21] dataset predominantly features large scenes, whereas our method is based on 3DGS reconstruction that is designed for accurate reconstruction of smaller ones, and TSDF which is better suited for reconstruction of specific objects. This is particularly evident in the case of the Ignatius and Truck scenes, relatively small scenes, where our method performed on-par with the neural reconstruction methods.

Mip-NeRF360 [1]. As illustrated in Fig. 1 and Fig. 6, we present a qualitative analysis of scenes from the Mip-NeRF360 [1] dataset. This comparison reveals that our approach surpasses SuGaR [14] in terms of reconstruction quality and presents on-par results with BakedSDF [48]. Notably, our method excels in reconstructing fine details; for instance, even the small groves in the garden scene's table are evident in the reconstruction, and there are intricate details in the

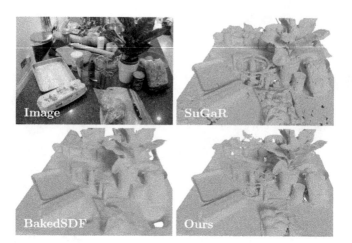

Fig. 6. Qualitative comparison on Mip-NeRF360 [1] dataset with BakedSDF [48] and SuGaR [14].

objects on the countertop scene. Furthermore, while BakedSDF [48] requires 48 h for training, our method achieves comparable results in less than an hour. Compared to SuGaR [14], our method generates smoother and more realistic surfaces, especially in reflective areas; we note that our countertop is smooth and flat, while SuGaR's countertop has many bumps in areas with glare. This is likely due to our model's use of a small baseline for stereo matching, where the reconstruction distortion is relatively small, and additionally, due to our model integrating the reconstructed patches from various viewing directions, which further reduces potential distortions.

In-the-Wild Comparison. Our comprehensive in-the-wild comparisons demonstrate the superior performance of our method across various scenes, as illustrated in Fig. 4, with additional results provided in the supplementary material. Our method surpasses SuGaR [14] in extracting accurate and noise-free meshes from 3DGS.

4.4 Ablation Study

Our main contribution is the use of a pre-trained stereo model to extract depth from a 3DGS scene using novel stereo views. To strengthen our claim of the benefit of using novel stereo views, we perform two ablations, which include replacing steps of our pipeline with deep MVS methods.

MVS on Original Images. Our method extracts depth from each original pose, by creating novel stereo-aligned views from that pose and applying a pre-trained stereo matching model. One obvious comparison would be to take each original pose and extract the depth using a pre-trained deep MVS model which will take as input the original training set of the scene. In the first ablation, we run a pre-trained deep MVS model on the original images, and fuse the resulting depths using TSDF [9].

MVS on Rendered Images. Applying 3DGS to the scene and re-rendering the images from the original poses can reduce distortion and camera noise, which may enhance the quality of the reconstruction regardles of the novel stereo views. In the second ablation, we run a pre-trained deep MVS model on the rendered images from the original poses, which are the left image of each stereo-aligned novel view, and fuse the resulting depths using TSDF [52].

Evaluation. We evaluate on the MobileBrick [23] test set, and compare our method against MVSFormer [4], a state-of-the-art deep MVS model. To ensure a fair comparison, we use TSDF [9] as the fusion method both for our method and the deep MVS model.

Results. Table 3 shows the mean accuracy, recall, F1 and Chamfer Distance of Our method, compared to MVSFormer [4] with the original and rendered images as input. Figure 7 shows a qualitative comparison on one of the scans in the MobileBrick [23] dataset, with the rest of the scans, as well as additional examples from in-the-wild scenes, available in the supplementary material. Qualitative comparison shows that applying deep MVS directly on the original images results in a reconstruction filled with holes. Applying MVS on the rendered images slightly improves the quality of the reconstruction, however, our method still produces a significantly smoother reconstruction. Quantitative comparison confirms that inputting rendered images to the same MVS model results in a smoother reconstructed surface, as evident by the higher recall, with a slight trade-off in accuracy. Overall, our method performs better, as evident by the higher recall and F1 and lower Chamfer distance, even though the manually refined poses given by the MobileBrick [23] dataset should give an advantage to MVSFormer [4].

Table 3. Ablation study results on MobileBrick [23] dataset. We compare our method against MVSFormer [4] with two types of inputs: the original images with the original refined poses, and the rendered images with the same poses.

Method	2.5 mm Radius			5 mm Radius			Chamfer Distance (mm) ↓
	Acc ↑	Recall ↑	F1 ↑	Acc ↑	Recall ↑	F1 ↑	
MVSFormer [4]	**80.77**	55.02	64.60	96.33	71.32	81.14	9.11
MVSFormer + Rendered	80.16	59.92	68.04	**96.84**	77.50	85.59	7.10
Ours	68.77	**69.27**	**68.94**	89.46	**87.37**	**88.28**	**4.94**

5 Limitations

Our pipeline consists of 3DGS, depth extraction via stereo, and TSDF fusion. Each of these steps exhibits limitations that can impact the final reconstruction: 3DGS can produce noisy "floater" Gaussians in areas which aren't sufficiently covered in the original training images, as can be seen in the right side of Fig. 8.

Fig. 7. Example from MobileBrick [23] dataset, on the castle scene. From left to right: the ground truth mesh, reconstruction of MVSFormer [4] with original images, reconstruction of MVSFormer with rendered images, and our reconstruction.

Fig. 8. Examples of limitations of our method. On the left, we show a rendered image from the Caterpillar scene from TnT [21] dataset, highlighting an area with "floater" Gaussians. On the right, the Truck scene from TnT [21] dataset, highlighting the missing windshield.

Additionally, stereo matching models are known to struggle with transparent surfaces, as can be seen in the left side of Fig. 8. Finally, TSDF fusion does not scale well for larger scenes, such as the Meetingroom and Courthouse scenes from TnT [21]. Swapping the 3DGS and stereo with future versions which will have improved accuracy and robustness, as well as adding fusion methods better suited for larger scenes, should help mitigate the effect of these limitations.

6 Conclusion

We introduce a novel approach for bridging the gap between noisy Gaussian point clouds and smooth surfaces in 3D. Instead of applying geometric optimizations directly on the Gaussians and extracting the depth using their locations, we use a pre-trained stereo model as a geometric prior with real-world knowledge to extract the depth. While this approach preserves the inherent properties of the 3DGS representation, it also enhances the accuracy and fidelity of the reconstructed surfaces. Our experimental results on DTU [16], Tanks and Temples [21], Mip-NeRF360 [1], MobileBrick [23] and real-world scenes captured using smartphones - demonstrate the superiority of our method over the current state-of-the-art methods for surface reconstruction from Gaussian splatting models, offering both improved accuracy and significantly shorter computation times compared to neural methods.

References

1. Barron, J.T., Mildenhall, B., Tancik, M., Hedman, P., Martin-Brualla, R., Srinivasan, P.P.: Mip-NeRF: a multiscale representation for anti-aliasing neural radiance fields. In: Proceedings of the IEEE/CVF International Conference on Computer Vision, pp. 5855–5864 (2021)
2. Besl, P.J., McKay, N.D.: Method for registration of 3-D shapes. In: Sensor Fusion IV: Control Paradigms and Data Structures, vol. 1611, pp. 586–606. Spie (1992)
3. Bracha, A., Rotstein, N., Bensaïd, D., Slossberg, R., Kimmel, R.: Depth refinement for improved stereo reconstruction. arXiv preprint arXiv:2112.08070 (2021)
4. Cao, C., Ren, X., Fu, Y.: Mvsformer: learning robust image representations via transformers and temperature-based depth for multi-view stereo. arXiv preprint arXiv:2208.02541 (2022)
5. Chang, J.R., Chen, Y.S.: Pyramid stereo matching network. In: Proceedings of the IEEE Conference on Computer Vision and Pattern Recognition, pp. 5410–5418 (2018)
6. Chen, G., Wang, W.: A survey on 3D Gaussian splatting. arXiv preprint arXiv:2401.03890 (2024)
7. Chen, H., Li, C., Lee, G.H.: Neusg: neural implicit surface reconstruction with 3D Gaussian splatting guidance. arXiv preprint arXiv:2312.00846 (2023)
8. Cheng, K., et al.: GaussianPro: 3D Gaussian splatting with progressive propagation. In: Forty-first International Conference on Machine Learning (2024)
9. Curless, B., Levoy, M.: A volumetric method for building complex models from range images. In: Proceedings of the 23rd Annual Conference on Computer Graphics and Interactive Techniques, pp. 303–312 (1996)
10. Darmon, F., Bascle, B., Devaux, J.C., Monasse, P., Aubry, M.: Improving neural implicit surfaces geometry with patch warping. In: Proceedings of the IEEE/CVF Conference on Computer Vision and Pattern Recognition, pp. 6260–6269 (2022)
11. Fu, Y., Liu, S., Kulkarni, A., Kautz, J., Efros, A.A., Wang, X.: Colmap-free 3D Gaussian splatting. arXiv preprint arXiv:2312.07504 (2023)
12. Galliani, S., Lasinger, K., Schindler, K.: Massively parallel multiview stereopsis by surface normal diffusion (2015)
13. Gu, X., Fan, Z., Zhu, S., Dai, Z., Tan, F., Tan, P.: Cascade cost volume for high-resolution multi-view stereo and stereo matching. In: Proceedings of the IEEE/CVF Conference on Computer Vision and Pattern Recognition, pp. 2495–2504 (2020)
14. Guédon, A., Lepetit, V.: SuGaR: surface-aligned gaussian splatting for efficient 3D mesh reconstruction and high-quality mesh rendering. In: Proceedings of the IEEE/CVF Conference on Computer Vision and Pattern Recognition, pp. 5354–5363 (2024)
15. Huang, B., Yu, Z., Chen, A., Geiger, A., Gao, S.: 2D Gaussian splatting for geometrically accurate radiance fields. In: SIGGRAPH 2024 Conference Papers. Association for Computing Machinery (2024)
16. Jensen, R., Dahl, A., Vogiatzis, G., Tola, E., Aanæs, H.: Large scale multi-view stereopsis evaluation. In: 2014 IEEE Conference on Computer Vision and Pattern Recognition, pp. 406–413. IEEE (2014)
17. Jiang, H., Xu, R., Jiang, W.: An improved RAFTstereo trained with a mixed dataset for the robust vision challenge 2022. arXiv preprint arXiv:2210.12785 (2022)

18. Kendall, A., et al.: End-to-end learning of geometry and context for deep stereo regression. In: Proceedings of the IEEE International Conference on Computer Vision, pp. 66–75 (2017)
19. Kerbl, B., Kopanas, G., Leimkühler, T., Drettakis, G.: 3D Gaussian splatting for real-time radiance field rendering. ACM Trans. Graph. **42**(4) (2023)
20. Kirillov, A., et al.: Segment anything. In: Proceedings of the IEEE/CVF International Conference on Computer Vision, pp. 4015–4026 (2023)
21. Knapitsch, A., Park, J., Zhou, Q.Y., Koltun, V.: Tanks and temples: benchmarking large-scale scene reconstruction. ACM Trans. Graph. **36**(4) (2017)
22. Laga, H., Jospin, L.V., Boussaid, F., Bennamoun, M.: A survey on deep learning techniques for stereo-based depth estimation. IEEE Trans. Pattern Anal. Mach. Intell. **44**(4), 1738–1764 (2020)
23. Li, K., Bian, J.W., Castle, R., Torr, P.H., Prisacariu, V.A.: Mobilebrick: building LEGO for 3D reconstruction on mobile devices. In: Proceedings of the IEEE/CVF Conference on Computer Vision and Pattern Recognition, pp. 4892–4901 (2023)
24. Li, Z., et al.: Neuralangelo: high-fidelity neural surface reconstruction. In: IEEE Conference on Computer Vision and Pattern Recognition (CVPR) (2023)
25. Lipson, L., Teed, Z., Deng, J.: Raft-stereo: multilevel recurrent field transforms for stereo matching. In: 2021 International Conference on 3D Vision (3DV), pp. 218–227. IEEE (2021)
26. Ma, Z., Teed, Z., Deng, J.: Multiview stereo with cascaded epipolar raft. In: European Conference on Computer Vision, pp. 734–750. Springer, Heidelberg (2022). https://doi.org/10.1007/978-3-031-19821-2_42
27. Mildenhall, B., Srinivasan, P.P., Tancik, M., Barron, J.T., Ramamoorthi, R., Ng, R.: NeRF: representing scenes as neural radiance fields for view synthesis. Commun. ACM **65**(1), 99–106 (2021)
28. Müller, T., Evans, A., Schied, C., Keller, A.: Instant neural graphics primitives with a multiresolution hash encoding. ACM Trans. Graph. **41**(4), 102:1–102:15 (2022)
29. Poggi, M., Tosi, F., Batsos, K., Mordohai, P., Mattoccia, S.: On the synergies between machine learning and binocular stereo for depth estimation from images: a survey. IEEE Trans. Pattern Anal. Mach. Intell. **44**(9), 5314–5334 (2021)
30. Rotstein, N., Bracha, A., Kimmel, R.: Multimodal colored point cloud to image alignment. In: Proceedings of the IEEE/CVF Conference on Computer Vision and Pattern Recognition, pp. 6656–6666 (2022)
31. Scharstein, D., Szeliski, R.: A taxonomy and evaluation of dense two-frame stereo correspondence algorithms. Int. J. Comput. Vision **47**, 7–42 (2002)
32. Schönberger, J.L., Frahm, J.M.: Structure-from-motion revisited. In: Conference on Computer Vision and Pattern Recognition (CVPR) (2016)
33. Schönberger, J.L., Zheng, E., Frahm, J.-M., Pollefeys, M.: Pixelwise view selection for unstructured multi-view stereo. In: Leibe, B., Matas, J., Sebe, N., Welling, M. (eds.) ECCV 2016. LNCS, vol. 9907, pp. 501–518. Springer, Cham (2016). https://doi.org/10.1007/978-3-319-46487-9_31
34. Shen, Z., Dai, Y., Rao, Z.: Cfnet: cascade and fused cost volume for robust stereo matching. In: Proceedings of the IEEE/CVF Conference on Computer Vision and Pattern Recognition, pp. 13906–13915 (2021)
35. Sitzmann, V., Martel, J., Bergman, A., Lindell, D., Wetzstein, G.: Implicit neural representations with periodic activation functions. Adv. Neural. Inf. Process. Syst. **33**, 7462–7473 (2020)

36. Tosi, F., Tonioni, A., De Gregorio, D., Poggi, M.: Nerf-supervised deep stereo. In: Proceedings of the IEEE/CVF Conference on Computer Vision and Pattern Recognition, pp. 855–866 (2023)
37. Waczyńska, J., Borycki, P., Tadeja, S., Tabor, J., Spurek, P.: GaMeS: mesh-based adapting and modification of Gaussian splatting. arXiv preprint arXiv:2402.01459 (2024)
38. Wang, F., Galliani, S., Vogel, C., Speciale, P., Pollefeys, M.: Patchmatchnet: learned multi-view patchmatch stereo. In: Proceedings of the IEEE/CVF Conference on Computer Vision and Pattern Recognition, pp. 14194–14203 (2021)
39. Wang, P., Liu, L., Liu, Y., Theobalt, C., Komura, T., Wang, W.: Neus: learning neural implicit surfaces by volume rendering for multi-view reconstruction (2021)
40. Wang, Y., Skorokhodov, I., Wonka, P.: HF-NeuS: improved surface reconstruction using high-frequency details. Adv. Neural. Inf. Process. Syst. **35**, 1966–1978 (2022)
41. We, L.: Marching cubes: a high resolution 3D surface construction algorithm. Comput. Graph. **21**(1), 7–12 (1987)
42. Wu, T., et al.: Recent advances in 3D Gaussian splatting. In: Computational Visual Media, pp. 1–30 (2024)
43. Xu, G., Wang, X., Ding, X., Yang, X.: Iterative geometry encoding volume for stereo matching. In: Proceedings of the IEEE/CVF Conference on Computer Vision and Pattern Recognition, pp. 21919–21928 (2023)
44. Yan, J., et al.: Dense hybrid recurrent multi-view stereo net with dynamic consistency checking. In: Vedaldi, A., Bischof, H., Brox, T., Frahm, J.-M. (eds.) ECCV 2020. LNCS, vol. 12349, pp. 674–689. Springer, Cham (2020). https://doi.org/10.1007/978-3-030-58548-8_39
45. Yang, J., Mao, W., Alvarez, J.M., Liu, M.: Cost volume pyramid based depth inference for multi-view stereo. In: Proceedings of the IEEE/CVF Conference on Computer Vision and Pattern Recognition, pp. 4877–4886 (2020)
46. Yao, Y., Luo, Z., Li, S., Fang, T., Quan, L.: Mvsnet: depth inference for unstructured multi-view stereo. In: Proceedings of the European Conference on Computer Vision (ECCV), pp. 767–783 (2018)
47. Yariv, L., Gu, J., Kasten, Y., Lipman, Y.: Volume rendering of neural implicit surfaces. Adv. Neural. Inf. Process. Syst. **34**, 4805–4815 (2021)
48. Yariv, L., et al.: BakedSDF: meshing neural SDFs for real-time view synthesis. In: ACM SIGGRAPH 2023 Conference Proceedings, pp. 1–9 (2023)
49. Yariv, L., et al.: Multiview neural surface reconstruction by disentangling geometry and appearance. Adv. Neural. Inf. Process. Syst. **33**, 2492–2502 (2020)
50. Yu, Z., Chen, A., Huang, B., Sattler, T., Geiger, A.: Mip-splatting: alias-free 3D Gaussian splatting. In: Proceedings of the IEEE/CVF Conference on Computer Vision and Pattern Recognition, pp. 19447–19456 (2024)
51. Yu, Z., Sattler, T., Geiger, A.: Gaussian opacity fields: efficient high-quality compact surface reconstruction in unbounded scenes. arXiv preprint arXiv:2404.10772 (2024)
52. Zeng, A., Song, S., Nießner, M., Fisher, M., Xiao, J., Funkhouser, T.: 3dmatch: learning local geometric descriptors from rgb-d reconstructions. In: CVPR (2017)
53. Zhang, F., Prisacariu, V., Yang, R., Torr, P.H.: Ga-net: guided aggregation net for end-to-end stereo matching. In: Proceedings of the IEEE/CVF Conference on Computer Vision and Pattern Recognition, pp. 185–194 (2019)
54. Zhang, J., et al.: Critical regularizations for neural surface reconstruction in the wild. In: Proceedings of the IEEE/CVF Conference on Computer Vision and Pattern Recognition, pp. 6270–6279 (2022)

55. Zhang, J., Yao, Y., Li, S., Luo, Z., Fang, T.: Visibility-aware multi-view stereo network. arXiv preprint arXiv:2008.07928 (2020)
56. Zhao, H., Zhou, H., Zhang, Y., Chen, J., Yang, Y., Zhao, Y.: High-frequency stereo matching network. In: Proceedings of the IEEE/CVF Conference on Computer Vision and Pattern Recognition, pp. 1327–1336 (2023)

CARFF: Conditional Auto-Encoded Radiance Field for 3D Scene Forecasting

Jiezhi Yang[1]($\boxtimes$), Khushi Desai[2], Charles Packer[3], Harshil Bhatia[4], Nicholas Rhinehart[3], Rowan McAllister[5], and Joseph E. Gonzalez[3]

[1] Harvard University, Cambridge, MA 02138, USA
jiezhi@berkeley.edu
[2] Columbia University, New York, NY 10025, USA
[3] UC Berkeley, Berkeley, CA 94720, USA
[4] Avataar.ai, Bengaluru 560103, Karnataka, India
[5] Toyota Research Institute, Los Altos, CA 94022, USA

Abstract. We propose CARFF: Conditional Auto-encoded Radiance Field for 3D Scene Forecasting, a method for predicting future 3D scenes given past observations. Our method maps 2D ego-centric images to a distribution over plausible 3D latent scene configurations and predicts the evolution of hypothesized scenes through time. Our latents condition a global Neural Radiance Field (NeRF) to represent a 3D scene model, enabling explainable predictions and straightforward downstream planning. This approach models the world as a POMDP and considers complex scenarios of uncertainty in environmental states and dynamics. Specifically, we employ a two-stage training of Pose-Conditional-VAE and NeRF to learn 3D representations, and auto-regressively predict latent scene representations utilizing a mixture density network. We demonstrate the utility of our method in scenarios using the CARLA driving simulator, where CARFF enables efficient trajectory and contingency planning in complex multi-agent autonomous driving scenarios involving occlusions. Video and code are available at: www.carff.website.

1 Introduction

Humans often imagine what they cannot see given partial visual context. Consider a scenario where reasoning about the unobserved is critical to safe decision-making: for example, a driver navigating a blind intersection. An expert driver will plan according to what they believe may or may not exist in occluded regions of their vision. The driver's belief – defined as the understanding of the world modeled with consideration for inherent environment uncertainties – is informed

---

J. Yang and K. Desai—Core contributors.

Supplementary Information The online version contains supplementary material available at https://doi.org/10.1007/978-3-031-73024-5_14.

by their partial observations (i.e., the presence of other vehicles on the road), as well as their prior knowledge (e.g., past experience navigating this intersection).

When reasoning about the unobserved, humans form complex beliefs about the existence, position, shapes, colors, and textures of occluded scene portions (e.g., an oncoming car). Autonomous systems with high-dimensional sensor data, like video or LiDAR, traditionally reduce this data to low-dimensional state information (e.g., position and velocity of tracked objects) for prediction and planning.

In addition to tracking fully observed objects, this object-centric framework handles partially observed settings by considering potentially dangerous unobserved objects. These systems often plan for worst-case scenarios, such as a "ghost car" at the edge of the visible field of view [45].

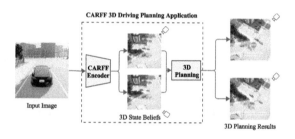

Fig. 1. CARFF 3D planning application for driving. An input image containing a partially observable view of an intersection is processed by CARFF's encoder to establish 3D environment state beliefs, i.e. the predicted possible state of the world: whether or not there could be another vehicle approaching the intersection. These beliefs are used to forecast the future in 3D for planning, generating one among two possible actions for the vehicle to merge into the other lane.

Recent advances in neural rendering, particularly Neural Radiance Fields (NeRF), have significantly improved 3D scene representation learning. NeRF enables novel view synthesis, thus simplifying the process of viewing behind occlusions. NeRF decouples the dependancy of scene representation from traditional object detection and tracking, allowing for the capture of vital visual information that might be missed by detectors, yet is crucial for safe decision-making. NeRF's implicit density representation of explicit geometry also facilitates its direct application in motion planning without the need for rendering. NeRF's ability to represent both visual and geometric information makes them a more general and intuitive 3D representation for autonomous systems.

Despite NeRF's advantages, achieving probabilistic predictions in 3D based on reasoning from occluded views is challenging. For example, discriminative models that yield categorical predictions are unable to capture the underlying 3D structure, impeding their ability to model uncertainty. While prior work on 3D representation captures view-invariant structures, their application is primarily confined to simple scenarios [20]. We present CARFF, which to our knowledge, is the first forecasting approach in scenarios with partial observations that uniquely facilitates stochastic predictions in a partially observable Markov decision process (POMDP) within a 3D representation, effectively integrating visual perception and geometry. Specifically, we make the following contributions:

1. We propose a novel architecture *PC-VAE*: Pose-Conditioned Variational Autoencoder. The encoder maps potentially partially observable ego-centric images to pose-invariant latent scene representations, which hold state beliefs of the POMDP with implicit probability distributions (see Sect. 3.1).
2. We develop the two-stage training pipeline that uniquely enables complex scene modeling with a probabilistic objective. This involves separately training the PC-VAE and a latent conditioned neural radiance field that functions as a 3D decoder, enabling interpretable predictions (see Sect. 3.1).
3. We design a mixture density model to predict the evolution of 3D scenes over time stochastically and regressively in the encoder belief space (see Sect. 3.2). This allows for an effective sampling based-controller to output actions in the POMDP.

We demonstrate how CARFF can be used to enable contingency planning in complex driving scenarios that require reasoning into visual occlusions on CARLA simulated datasets inspired by autonomous driving planning tasks [29, 56–58]. A potential application of CARFF is illustrated in Fig. 1.

2 Related Work

2.1 NeRF and 3D Representations

Neural Radiance Fields. Neural Radiance Fields (NeRF) [2,27,44] for 3D representations generate high-resolution, photorealistic scenes. Instant Neural Graphics Primitive (Instant-NGP) [28] speeds up training and rendering time by introducing a multi-resolution hash encoding. Other works like Plenoxels [12] and DirectVoxGo (DVGO) [42] also provide similar speedups. Recent advancements in volumetric representations such as 3D Gaussian Splatting [18] enhance rendering efficiency while maintaining compatibility with traditional NeRF applications [11]. We utilize Instant-NGP for its accessibility, although our approach is adaptable to alternative rendering methods. NeRFs have also been extended for several tasks such as modeling large-scale unbounded scenes [2,43,50], scene from sparse views [7,39,49] and multiple scenes [20,51]. For an in-depth survey on neural representation learning and its applications we refer the reader to [46].

Generalizable novel view synthesis models, like pixelNeRF and pixelSplat [5, 55], learn a scene prior to render novel views from sparse existing ones. In contrast, CARFF is based on a VAE, encoding a probabilistic objective and decoding to future 3D scenes. Dynamic NeRF models scenes with moving or deforming objects, within which a widely used approach is to construct a canonical space and predict a deformation field [22,33,34,36]. The canonical space is usually a static scene, and the model learns an implicitly represented flow field [33,36]. A

Fig. 2. Novel view planning application. CARFF allows reasoning behind occluded views from the ego car as simple as moving the camera to see the sampled belief predictions, allowing simple downstream planning using, for example, density probing or 2D segmentation models from arbitrary angles.

recent line of work also models dynamic scenes via different representations and decomposition [3,41]. These approaches tend to perform better for spatially bounded and predictable scenes with relatively small variations [3,23,33,55]. Moreover, these methods only solve for changes in the environment but are limited in incorporating stochasticity in the environment.

Multi-scene NeRF: Our approach builds on multi-scene NeRF approaches [20, 48,51,52] that learn a global latent scene representation, which conditions the NeRF, allowing a single NeRF to effectively represent various scenes. A similar method, NeRF-VAE, was introduced by Kosiorek et al. [20] to create a geometrically consistent 3D generative model with generalization to out-of-distribution cameras. However, NeRF-VAE [20] is prone to mode collapse when faced with complex visual information (see Sect. 4.2).

2.2 Scene Forecasting

Planning in 2D Space: Planning in large, continuous state-action spaces is challenging due to exponentially large search spaces [32], leading to various approximation methods for tractability [26,35]. Model-free [13,31,47] and model-based [4] reinforcement learning frameworks, along with other learning-based methods [6,29], have emerged as viable approaches. Additionally, methods forecast for downstream control [16], learn behavior models for contingency planning [38], or predict the existence and intentions of unobserved agents [30]. While these methods operate in 2D, we reason under partial observations and account for these factors in 3D.

NeRF in Robotics: Recent works have applied NeRFs in robotics for localization [54], navigation [1,25], dynamics modeling [10,22], and robotic grasping [15, 19]. Adamkiewicz et al. [1] propose quadcopter motion planning in NeRF models by sampling the learned density function, useful for forecasting and planning. Driess et al. [10] employ a graph neural network to learn dynamics in a multi-object NeRF scene. Li et al. [21] focus on pushing tasks and address grasping and planning with NeRF and a separate latent dynamics model. Prior approaches work in simple, static scenes [1] or uses deterministic dynamics models [21].

CARFF addresses complex, realistic environments with both state and dynamics uncertainty, considering potential object existence and unknown movements.

3 Method

Recent advancements in 3D scene representation allow for modeling environments in a contextually rich and interactive 3D space. This offers analytical benefits, such as spatial analysis with soft occupancy grids and object detection through novel view synthesis. Given these advantages, our primary objective is to develop a model for probabilistic 3D scene forecasting in dynamic environments. However, direct integration of 3D scene representation via NeRF and probabilistic models like VAE often involves non-convex and inter-dependent optimization, which causes unstable training. For instance, NeRF's optimization may rely on the VAE's latent space being structured to provide informative gradients. To navigate these complexities, our method bifurcates the training process into two stages (see Fig. 3). First, we train the PC-VAE to learn view-invariant scene representations. Next, we replace the decoder with a NeRF to learn a 3D scene from the latent representations. The latent scene representations capture the environmental states and dynamics over possible underlying scenes, while NeRF synthesizes novel views within the belief space, giving us the ability to see the unobserved (see Fig. 2 and Sect. 3.1). During prediction, uncertainties can be

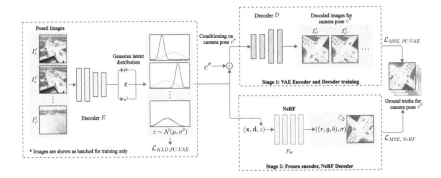

Fig. 3. Visualizing CARFF's two stage training process. Left: The convolutional VIT-based encoder encodes each image I at timestamps t, t' and camera poses c, c' into Gaussian latent distributions. Assuming two timestamps and an overparameterized latent, one Gaussian distribution will have a smaller σ^2, and different μ across timestamps. **Upper Right:** The pose-conditional decoder stochastically decodes the sampled latent z using the camera pose c'' into images $I_{c''}^{t}$ and $I_{c''}^{t'}$. The decoded reconstruction and ground truth images are used for the loss $\mathcal{L}_{\text{MSE, PC-VAE}}$. **Lower Right:** A NeRF is trained by conditioning on the latent variables sampled from the optimized Gaussian parameters. These parameters characterize the distinct timestamp distributions derived from the PC-VAE. An MSE loss is calculated for NeRF as $\mathcal{L}_{\text{MSE, NeRF}}$.

modeled by sampling latents auto-regressively from a predicted Gaussian mixture, allowing for effective decision-making. To this extent, we approach scene forecasting as a POMDP over latent distributions, which enables us to capture multi-modal beliefs for planning amidst perceptual uncertainty (see Sect. 3.2).

3.1 Pose-Conditional VAE (PC-VAE) and NeRF

Architecture: We assume that the model follows a Markovian process, and thus each belief state only depends on the previous. Given a scene S_t at timestamp t, we have an ego-centric observation image I_c^t captured from camera pose c. The objective is to formulate a 3D representation of the image that holds implicit probability distributions of the possible states, where we can perform a forecasting step that evolves the scene forward. Here, the POMDP can be seen as an MDP in belief space [17]. To achieve this, we utilize a radiance field conditioned on latent variable z sampled from the posterior distribution $q_\phi(z|I_c^t)$. Now, to learn the posterior, we utilize PC-VAE. We construct an encoder using convolutional layers and a pre-trained ViT on ImageNet [8]. The encoder learns a mapping from the image space to a Gaussian distributed latent space $q_\phi(z|I_c^t) = \mathcal{N}(\mu, \sigma^2)$ parametrized by mean μ and variance σ^2. The decoder, $p(I|z,c)$, conditioned on camera pose c, maps the latent $z \sim \mathcal{N}(\mu, \sigma^2)$ into the image space I. This helps the encoder to generate latents that are invariant to the camera pose c.

To enable 3D scene modeling, we employ Instant-NGP [28], which incorporates a hash grid and an occupancy grid to enhance computation efficiency. Additionally, a smaller multilayer perceptron (MLP), $F_\theta(z)$ can be utilized to model the density and appearance, given by:

$$F_\theta(z) : (\mathbf{x}, \mathbf{d}, z) \to ((r, g, b), \sigma) \tag{1}$$

Here, $\mathbf{x} \in \mathbb{R}^3$ and $\mathbf{d} \in (\theta, \phi)$ represent the location vector and the viewing direction respectively. The MLP is conditioned on the sampled scene latents $z \sim q_\phi(z|I_c^t)$ (see Appendix B).

Training Methodology: The architecture alone does not enable us to model complex scenarios, as seen through a similar example in NeRF-VAE [20]. A crucial contribution of our work is our two-stage training framework which stabilizes the training. First, we optimize the convolutional ViT based encoder and pose-conditional convolutional decoder in the pixel space for reconstruction. This enables our method to deal with more complex and realistic scenes as the encoding is learned in a semantically rich 2D space. By conditioning the decoder on camera poses, we achieve disentanglement between camera view angles and scene context, making the representation view-invariant and the encoder 3D-aware. Once rich latent representations are learned, we replace the decoder with a latent-conditioned NeRF over the latent space of the frozen encoder. The NeRF reconstructs encoder beliefs in 3D for novel view synthesis.

Loss: PC-VAE is trained using standard VAE loss, with mean square error (MSE) and a Kullback–Leibler (KL) divergence given by evidence lower bound:

$$\mathcal{L}_{PC\text{-}VAE} = \mathcal{L}_{MSE,\ PC\text{-}VAE} + \mathcal{L}_{KLD,\ PC\text{-}VAE} =$$
$$||p(I|z,c'') - I^t_{c''}||^2 + \mathbb{E}_{q(z|I^t_c)}[\log p(I|z)] - w_{KL} D_{KL}(q_\phi(z|I^t_c)\ ||\ p(I|z)) \quad (2)$$

where w_{KL} denotes the KL divergence loss weight and $z \sim q_\phi(z|I^t_c)$. To make our representation 3D-aware, our posterior is encoded using camera c while the decoder is conditioned on a randomly sampled pose c''.

KL divergence regularizes the latent space to balance conditioned reconstruction and stochasticity under occlusion. An elevated KL divergence loss weight w_{KL} pushes the latents closer to a standard normal distribution, $\mathcal{N}(0,1)$, thereby ensuring probabilistic sampling in scenarios under partial observation. However, excessive regularization causes the latents to be less separable, leading to mode collapse. To mitigate this, we adopt delayed linear KL divergence loss weight scheduling to strike a balanced w_{KL}.

Next, we learn a NeRF decoder on the posterior of the VAE to model scenes. At any timestamp t we use a standard photometric loss for training the NeRF, given by the following equation:

$$\mathcal{L}_{MSE,\ NeRF} = ||I^t_c - render(F_\theta(\cdot|q_\phi(z|I^t_c)))||^2 \quad (3)$$

We use a standard rendering algorithm as proposed by Müller *et al.* [28]. Next, we build a forecasting module over the learned latent space of our pose-conditional encoder.

3.2 Scene Forecasting

Formulation: The current formulation allows us to model scenes with different configurations across timestamps. In order to forecast future configurations of a scene given an ego-centric view, we need to predict future latent distributions. We formulate the forecasting as a POMDP over the posterior distribution $q_\phi(z|I^t_c)$ in the PC-VAE's latent space.

During inference, we observe stochastic behaviors under occlusion, which motivates us to learn a mixture of several Gaussian distributions that potentially denote different scene possibilities. Therefore, we model the POMDP using a Mixture Density Network (*MDN*), with multi-headed MLPs, that predicts a mixture of K Gaussians. At any timestamp t the distribution is given as:

$$q'_\phi(z_t|I^{t-1}_c) = MDN(q_\phi(z_{t-1}|I^{t-1}_c)) \quad (4)$$

The model is conditioned on the posterior distribution $q_\phi(z_{t-1})$ to learn a predicted posterior distribution $q'_\phi(z_t|I^{t-1}_c)$ at each timestamp. The predicted posterior distribution is given by the mixture of Gaussian:

$$q'_\phi(z_t) = \sum_{i=1}^{K} \pi_i\ \mathcal{N}(\mu_i, \sigma_i^2) \quad (5)$$

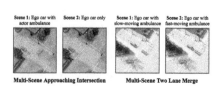
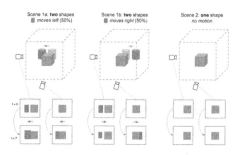

Fig. 4. Multi-scene CARLA datasets. Varying car configurations and scenes for the Multi-Scene Two Lane Merge dataset (**left**) and the Multi-Scene Approaching Intersection dataset (**right**).

Fig. 5. Blender dataset. Blender dataset with a blue cube and a potential red cylinder exhibiting probabilistic temporal movement. The possible occlusions from different camera angles demonstrate how movement needs to be modeled probabilistically.

here, π_i, μ_i, and σ_i^2 denote the mixture weight, mean, and variance of the ith Gaussian distribution within the posterior distribution. Here, K is the total number of Gaussians. For brevity we remove their conditioning on the posterior $q_\phi(z_{t-1})$ and sampled latent z_{t-1}. We sample z_t from the mixture of Gaussians $q'_\phi(z_t)$, where z_t likely falls within one of the Gaussian modes. The configuration corresponding to the mode is reflected in the 3D scene rendered by NeRF.

Loss: To optimize the MDN, we minimize a negative log-likelihood function, given by:

$$\mathcal{L}_{MDN} = -\sum_{j=1}^{N} \log \left(\sum_{i=1}^{K} \pi_i \mathcal{N}(y_j; \mu_i, \sigma_i^2) \right) \qquad (6)$$

where $y_i \sim q_\phi(z_t)$ is sampled from the distribution of latent z_t, learned by the encoder, and N denotes the total number of samples.

Inference: We consider an unseen ego-centric image and retrieve its posterior $q_\phi(z_t)$ through the encoder. Next, we predict the possible future posterior distribution $q'_\phi(z_{t+1})$. From the predicted posterior, we sample a scene latent and perform localization. We achieve this via (a) density probing the NeRF or (b) segmenting the rendered novel views using off-the-shelf methods such as YOLO [37] (see Fig. 2). These allow us to retrieve a corresponding Gaussian distribution $q_\phi(z_{t+1})$ in encoder latent space. This is auto-regressively fed back into the MDN to predict the next timestamp. See Fig. 6 for an overview of the pipeline.

4 Results

Decision-making under perceptual uncertainty is a pervasive challenge faced in robotics and autonomous driving, as the real environment is mostly likely par-

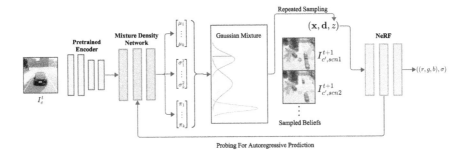

Fig. 6. Auto-regressive inference in scene prediction. The input image at timestamp t, I_c^t, is encoded using the pre-trained encoder from PC-VAE. The corresponding latent distribution is fed into the Mixture Density Network, which predicts a mixture of Gaussians. Each of the K Gaussians is a latent distribution that may correspond to different beliefs at the next timestamp. The mixture of Gaussians is sampled repeatedly for the predicted latent beliefs, visualized as $I_{c',scni}^{t+1}$, representing potentially the ith possible outcome. This is used to condition the NeRF to generate 3D views of the scene. To accomplish autoregressive predictions, we probe the NeRF for the location of the car and feed this information back to the pre-trained encoder to predict the scene at the next timestamp.

tially observable, making it a POMDP. In a partially observable driving scenario, accurate inference regarding the presence of potentially obscured agents is pivotal. We evaluate the effectiveness of CARFF on common driving situations with partial observability and added complexity. We implemented several scenarios in the CARLA driving simulator [9] (see Fig. 4). A single NVIDIA RTX 3090 GPU is used to train PC-VAE, NeRF, and the MDN. All models, trained sequentially, tend to converge within a combined time frame of 24 h. A detailed experimental setup can be found in Appendix B. We show that, given partially observable 2D inputs, CARFF performs well in predicting latent distributions that represent complete 3D scenes. Using these predictions we design a CARFF-based controller for performing downstream planning tasks.

4.1 Data Generation

We conduct experiments on (a) synthetic blender dataset for principle experiments to test the probabilistic modeling capacities in isolation of the vision encoder (it is visually as simple as possible, but requires the full predictive model proposed in CARFF) and (b) CARLA-based driving datasets for more complex driving scenarios [9]. To deliver convincing results, we model these driving scenarios off of related works [29,56–58] that concern planning for driving under difficult situations. We generate the datasets in 3D by programming an ego object and varying actor objects in different configurations.

Blender Synthetic Dataset: This comprises of a stationary blue cube (ego) accompanied by a red cylinder (actor) that may or may not be present (see

Fig. 5). If the actor is present, it exhibits lateral movement as depicted in Fig. 5. This simplistic setting provides an interpretable framework to evaluate our model.

CARLA Dataset: Each dataset is simulated for N timestamps and uses $C = 100$ predefined camera poses to capture images of the environment under full observation, partial observation, and no visibility. These datasets are modeled after common driving scenarios involving state uncertainty that have been proposed in related works such as Active Visual Planning [29].

a) Single-Scene Approaching Intersection: The ego vehicle is positioned at a T-intersection. An actor vehicle traverses the crossing along an evenly spaced, predefined trajectory. We simulate this for $N = 10$ timestamps. We mainly use this dataset to predict the evolution of timestamps under full observation.

b) Multi-Scene Approaching Intersection: We extend the previous scenario to a more complicated setting with state uncertainty, by making the existence of the actor vehicle probabilistic. A similar intersection crossing is simulated for $N = 3$ timestamps for both possibilities. The ego vehicle's view of the actor may be occluded as it approaches the T-intersection over the N timestamps. The ego vehicle either moves forward or halts at the junction (see Fig. 4).

c) Multi-Scene Multi-actor Two Lane Merge: To add more environment dynamics uncertainty, we consider a multi-actor setting at an intersection of two merging lanes. We simulate the scenario at an intersection with partial occlusions, with the second approaching actor having variable speed. Here the ego vehicle can either merge into the left lane before the second actor or after all the actors pass, (see Fig. 4). Each branch is simulated for $N = 3$ timestamps.

4.2 CARFF Evaluation

Table 1. Qualitative comparison of CARFF to related works. CARFF accomplishes all highlighted objectives as opposed to NeRF-VAE [20], NeRF for Visuomotor Control [21], Vision-only NeRF Navigation [1], and AVP [29]. We compare whether methods reason in a 3D environment and perform novel view synthesis; work on complex scenarios; predict probabilistically under state and dynamics uncertainty; forecast into the future; and use model predictions for decision-making.

Method	3D	Complex Scenarios	State Uncertainty	Dynamics Uncertainty	Prediction	Planning	Code Released
CARFF	✓	✓	✓	✓	✓	✓	✓
[20]	✓		✓				
[21]	✓	✓			✓	✓	
[1]	✓	✓				✓	✓
[29]		✓	✓	✓	✓	✓	

A desirable behavior from our model is that it should predict a complete set of possible scenes consistent with the given ego-centric image, which could be partially observable. This is crucial for autonomous driving in unpredictable environments as it ensures strategic decision-making based on potential hazards. To achieve this we require a rich PC-VAE latent space, high-quality novel view synthesis, and auto-regressive probabilistic predictions of latents at future timestamps. We evaluate CARFF on a simple synthetic blender-based dataset and each CARLA-based dataset. Additionally, we extend our model application to a hand-manipulation dataset in Appendix A.

Comparisons with Related Work: We attempt to compare CARFF to existing approaches. NeRF-VAE has the most comparable objective, but during our experiments, it collapse to black using CARLA datasets. We make further qualitative comparisons to other most similar methods in Tab. 1, but none aligns with ours enough to make any possible quantitative comparisons.

Evaluation on Blender Dataset: In Fig. 5, for both Scene 1a and 1b, our model correctly forecasts the lateral movement of the cylinder to be in either position approximately 50% of the time, considering a left viewing angle. In Scene 2, with the absence of the red cylinder in the input camera angle, the model predicts the potential existence of the red cylinder approximately 50% of the time, and predicts lateral movements with roughly equal probability. This validates PC-VAE's ability to predict and infer occlusions in the latent space, aligning with human intuitions. These intuitions, shown in the Blender dataset's simple scenes, can transfer to driving scenarios in our CARLA datasets.

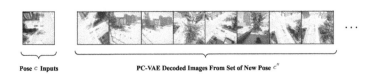

Pose c Inputs   PC-VAE Decoded Images From Set of New Pose c''

Fig. 7. PC-VAE reconstructions. The encoder input, I_c^t, among the other ground truth images I_c viewed from camera pose c at different timestamps, is reconstructed across a new set of poses c'' respecting timestamp t, generating $I_{c''}^t$. A complete grid is in Appendix D.

PC-VAE Performance and Ablations: We evaluate the performance of PC-VAE on CARLA datasets with multiple encoder architectures. We show that PC-VAE effectively reconstructs complex environments involving variable scenes, actor configurations, and environmental noise given potentially partially observable inputs (see Fig. 7). We calculated an average Peak Signal-to-Noise Ratio (PSNR) over the training data, as well as novel view encoder inputs. To evaluate the quality of the latent space generated by the encoder, we utilize t-SNE [24] plots to visualize the distribution of latent samples for each image in a given dataset

Table 2. Averaged PSNR for fully observable 3D predictions. CARFF correctly predicts scene evolution across all timestamps for each dataset. The average PSNR is high for predictions $\hat{I}_{t_i}$ and matching ground truths, I_{t_i}. PSNR values for incorrect correspondences, $\hat{I}_{t_i}, I_{t_j}$, is a result of matching surroundings. See complete table in Appendix D.

Ground Truth Prediction Pair	Avg. PSNR (Scene 1)	Avg. PSNR (Scene 2)
Single-Scene Intersection		
Matching Pairs	**29.06**	N.A
Un-matching P.	24.01	N.A
Multi-Scene Intersection		
Matching Pairs	**28.00**	**28.26**
Un-matching P.	23.27	24.56
Multi-Scene Two Lane Merge		
Matching Pairs	**28.14**	**28.17**
Un-matching P.	22.74	23.32

Table 3. Planning in 3D with controllers with varying sampling numbers n. CARFF-based controllers outperform baselines in success rate over 30 trials. For $n = 10$, the CARFF-based controller consistently chooses the optimal action in potential collision scenarios. To maintain consistency, we use one single image input across 30 trials.

XMulti-Scene Intersection		
Controller Type	Actor	No Actor
Underconfident	30/30	0/30
Overconfident	0/30	30/30
CARFF ($n = 2$)	17/30	30/30
CARFF ($n=10$)	**30/30**	**30/30**
CARFF ($n = 35$)	30/30	19/30
Multi-Scene Two Lane Merge		
Controller Type	Fast	Slow
Underconfident	30/30	0/30
Overconfident	0/30	30/30
CARFF ($n = 2$)	21/30	30/30
CARFF ($n=10$)	**30/30**	**30/30**
CARFF ($n = 35$)	30/30	22/30

(see Appendix D). We introduce a Support Vector Machine (SVM) [14] based metric to measure the visualized clustering quantitatively, where a higher value indicates better clustering based on timestamps. Most latent scene samples are separable by timestamps, which indicates that the latents are view-invariant. Samples that are misclassified or lie on the boundary usually represent partially or fully occluded regions. This is desirable for forecasting, as it enables us to model probabilistic behavior over these samples. In this process, balancing KL divergence weight scheduling maintains the quality of the PC-VAE's latent space and reconstructions (see Appendix B). Additionally, we substantiate the benefits of our PC-VAE encoder architecture through our ablations (see Appendix D.3).

3D Novel View Synthesis: Given an unseen ego-centric view with potentially partial observations, our method maintains all possible current state beliefs in 3D, and faithfully reconstructs novel views from arbitrary camera angles for each belief. Figure 2 illustrates one of the possible 3D beliefs that CARFF holds. This demonstrates our method's ability to generate 3D beliefs that could be used for novel view synthesis in a view-consistent manner. Our model's ability to achieve accurate and complete 3D environmental understanding is important for applications like prediction-based planning.

Inference Under Full and Partial Observations: Under full observation, we use MDN to predict the subsequent car positions in all three datasets. PSNR values are calculated based on bird-eye view NeRF renderings and ground truth bird-eye view images of the scene across different timestamps. In Table 2 we report the PSNR values for rendered images over the predicted posterior with the ground truth images at each timestamp. We also evaluate the efficacy of our prediction model using the accuracy curve given in Fig. 8. This represents CARFF's ability to generate stable beliefs, without producing incorrect predictions, based on actor(s) localization results. For each number of samples between $n = 0$ to $n = 50$, we choose a random subset of 3 fully observable ego images and take an average of the accuracies. In scenarios with partial observable ego-centric images where several plausible scenarios exist, we utilize recall instead of accuracy using a similar setup. This lets us evaluate the encoder's ability to avoid false negative predictions of potential danger.

Figure 8 shows that our model achieves high accuracy and recall in both datasets, demonstrating the ability to model state uncertainty (Approaching Intersection) and dynamic uncertainty (Two Lane Merge). The results indicate CARFF's resilience against randomness in resampling, and completeness in probabilistic modeling of the belief space. Given these observations, we now build a reliable controller to plan and navigate through complex scenarios.

4.3 Planning

In all our experiments, the ego vehicle must make decisions to advance under certain observability. The scenarios are designed such that the ego views contain partial occlusion and the state of the actor(s) is uncertain in some scenarios.

In order to facilitate decision-making using CARFF, we design a controller that takes ego-centric input images and outputs an action. Decisions are made incorporating sample consistency from the mixture density network. For instance, the controller infers occlusion and promotes the ego car to pause when scenes alternate between actor presence and absence in the samples. We use the two multi-scene datasets to assess the performance of the CARFF-based controller as they contain actors with potentially unknown behaviors.

To design an effective controller, we need to find a balance between accuracy and recall (see Fig. 8). A lowered accuracy from excessive sampling means unwanted randomness in the predicted state. However, taking insufficient samples would generate low recall i.e., not recovering all plausible states. This would lead to incorrect predictions as we would be unable to account for the plausible uncertainty present in the environment. To achieve optimal balance, we designed an open-loop planning controller using a sampling strategy that generates $n = 2, 10, 35$ samples. The hyperparameter n is tuned per scene for peak performance but is expected to remain relatively stable across scenes. We demonstrate that $n = 10$ performs well consistently in varying CARLA scenarios (Fig. 8) and do not anticipate this being very different for other experiments.

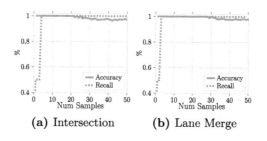

(a) Intersection   (b) Lane Merge

Fig. 8. **Multi-Scene dataset accuracy and recall curves from predicted beliefs.** We test our framework across $n = 1$ and $n = 50$ samples from MDN's predicted latent distributions from ego-centric image input. Across the number of samples n, we achieve an ideal margin of belief state coverage generated under partial observation (recall), and the proportion of correct beliefs sampled under full observation (accuracy). As we significantly increase the number of samples, the accuracy starts to decrease due to randomness in latent distribution resampling.

For sampling values that lie on the borders of the accuracy and recall margin, for example, $n = 2$ and 35, we see that the CARFF-based controller obtains lower success rates, whereas $n = 10$ produces the best result. For actor exists and fast-actor scenes in Table 3, we consider occluded ego-centric inputs to test the controller's ability to avoid collisions. For no-actor and slow-actor scenes, we consider state observability and test the controllers' ability to recognize the optimal action to advance. Across the two datasets, the overconfident controller will inevitably experience collisions in case of a truck approaching, since it does not cautiously account for occlusions. On the other hand, an overly cautious approach results in stasis, inhibiting the controller's ability to advance in the scene. This nuanced decision-making using CARFF-based controller is especially crucial in driving scenarios, as it enhances safety and efficiency by adapting to complex and unpredictable road environments, thereby fostering a more reliable and human-like response in autonomous vehicles.

5 Discussion

Limitations: Like other NeRF-based methods, CARFF currently relies on posed images of specific scenes such as road intersections, limiting its direct applicability to unseen environments. However, we anticipate enhanced generalizability with the increasing deployment of cameras around populated areas, such as traffic cameras at intersections. Additionally, handling very complex dynamics with an extremely large number of actors still poses a challenge for our method, requiring per-scene optimization to balance comprehensive dynamics modeling against accuracy. Potentially stronger models in the near future may offer a promising avenue for further enhancements in this regard.

Conclusion: We present CARFF, a novel method for probabilistic 3D scene forecasting from partial observations. By employing a Pose-Conditional VAE, a NeRF conditioned on the learned posterior, and a mixture density network that forecasts future scenes, we effectively model, predict, and plan in complex environments with state and dynamics uncertainty in a POMDP. We further demonstrate the capabilities of our method in simulated autonomous driving scenarios. Overall, CARFF offers an intuitive framework to perceiving, forecasting, and acting under uncertainty that could prove invaluable for vision-based algorithms in unstructured environments.

References

1. Adamkiewicz, M., et al.: Vision-only robot navigation in a neural radiance world. IEEE Rob. Autom. Lett. **7**(2), 4606–4613 (2022)
2. Barron, J.T., Mildenhall, B., Tancik, M., Hedman, P., Martin-Brualla, R., Srinivasan, P.P.: Mip-nerf: a multiscale representation for anti-aliasing neural radiance fields. In: International Conference on Computer Vision, pp. 5855–5864 (2021)
3. Cao, A., Johnson, J.: Hexplane: Dfor dynamic scenes. In: IEEE Conference on Computer Vision Pattern Recognition, pp. 130–141 (2023)
4. Cao, J., Wang, X., Darrell, T., Yu, F.: Instance-aware predictive navigation in multi-agent environments. In: IEEE International Conference on Robotics and Automation, pp. 5096–5102. IEEE (2021)
5. Charatan, D., Li, S., Tagliasacchi, A., Sitzmann, V.: pixelsplat: 3d gaussian splats from image pairs for scalable generalizable 3d reconstruction (2023)
6. Codevilla, F., Santana, E., López, A.M., Gaidon, A.: Exploring the limitations of behavior cloning for autonomous driving. In: International Conference on Computer Vision, pp. 9329–9338 (2019)
7. Deng, K., Liu, A., Zhu, J.Y., Ramanan, D.: Depth-supervised nerf: fewer views and faster training for free. In: IEEE Conference on Computer Vision Pattern Recognition, pp. 12882–12891 (2022)
8. Dosovitskiy, A., et al.: An image is worth 16×16 words: transformers for image recognition at scale. In: International Conference on Learning Representation (2021)
9. Dosovitskiy, A., Ros, G., Codevilla, F., Lopez, A., Koltun, V.: CARLA: an open urban driving simulator. In: Conference on Robot Learning, pp. 1–16 (2017)

10. Driess, D., Huang, Z., Li, Y., Tedrake, R., Toussaint, M.: Learning multi-object dynamics with compositional neural radiance fields. arXiv preprint arXiv:2202.11855 (2022)
11. Fei, B., Xu, J., Zhang, R., Zhou, Q., Yang, W., He, Y.: 3d gaussian as a new vision era: a survey (2024)
12. Fridovich-Keil, S., Yu, A., Tancik, M., Chen, Q., Recht, B., Kanazawa, A.: Plenoxels: radiance fields without neural networks. In: IEEE Conference on Computer Vision Pattern Recognition, pp. 5501–5510 (2022)
13. Hausknecht, M., Stone, P.: Deep recurrent q-learning for partially observable mdps. In: AAAI (2015)
14. Hearst, M.A., Dumais, S.T., Osuna, E., Platt, J., Scholkopf, B.: Support vector machines. IEEE Intell. Syst. Appl. **13**(4), 18–28 (1998)
15. Ichnowski, J., Avigal, Y., Kerr, J., Goldberg, K.: Dex-nerf: using a neural radiance field to grasp transparent objects. arXiv preprint arXiv:2110.14217 (2021)
16. Ivanovic, B., Elhafsi, A., Rosman, G., Gaidon, A., Pavone, M.: Mats: an interpretable trajectory forecasting representation for planning and control. In: Conference on Robot Learning (2021)
17. Kaelbling, L.P., Littman, M.L., Cassandra, A.R.: Planning and acting in partially observable stochastic domains. Artif. Intell. **101**(1), 99–134 (1998). https://doi.org/10.1016/S0004-3702(98)00023-X. https://www.sciencedirect.com/science/article/pii/S000437029800023X
18. Kerbl, B., Kopanas, G., Leimkühler, T., Drettakis, G.: 3d gaussian splatting for real-time radiance field rendering. ACM Trans. Graph. **42**(4) (2023). https://repo-sam.inria.fr/fungraph/3d-gaussian-splatting/
19. Kerr, J., et al.: Evo-nerf: evolving nerf for sequential robot grasping of transparent objects. In: Conference on Robot Learning (2022)
20. Kosiorek, A.R., et al.: NeRF-VAE: a geometry aware 3d scene generative model. In: ICML, pp. 5742–5752 (2021)
21. Li, Y., Li, S., Sitzmann, V., Agrawal, P., Torralba, A.: 3d neural scene representations for visuomotor control. In: Conference on Robot Learning, pp. 112–123 (2022)
22. Liu, J.W., et al.: Devrf: fast deformable voxel radiance fields for dynamic scenes. Adv. Neural Inform. Process. Syst. **35**, 36762–36775 (2022)
23. Luiten, J., Kopanas, G., Leibe, B., Ramanan, D.: Dynamic 3d gaussians: tracking by persistent dynamic view synthesis (2023)
24. van der Maaten, L., Hinton, G.: Visualizing data using t-SNE. J. Mach. Learn. Res. **9**, 2579–2605 (2008). http://www.jmlr.org/papers/v9/vandermaaten08a.html
25. Marza, P., Matignon, L., Simonin, O., Wolf, C.: Multi-object navigation with dynamically learned neural implicit representations. In: International Conference on Computer Vision, pp. 11004–11015 (2023)
26. McAllister, R., Rasmussen, C.E.: Data-efficient reinforcement learning in continuous state-action gaussian-pomdps. Adv. Neural Inform. Process. Syst. (2017)
27. Mildenhall, B., Srinivasan, P.P., Tancik, M., Barron, J.T., Ramamoorthi, R., Ng, R.: NeRF: representing scenes as neural radiance fields for view synthesis. In: Vedaldi, A., Bischof, H., Brox, T., Frahm, J.-M. (eds) ECCV 2020. LNCS, vol. 12346, pp. 405–421. Springer, Cham (2020). https://doi.org/10.1007/978-3-030-58452-8_24
28. Müller, T., Evans, A., Schied, C., Keller, A.: Instant neural graphics primitives with a multiresolution hash encoding. ACM Trans. Graph. **41**(4), 1–15 (2022)
29. Packer, C., et al.: Is anyone there? learning a planner contingent on perceptual uncertainty. In: Conference on Robot Learning (2022)

30. Packer, C., et al.: Is anyone there? learning a planner contingent on perceptual uncertainty. In: Liu, K., Kulic, D., Ichnowski, J. (eds.) Conference on Robot Learning, pp. 1607–1617 (2023)
31. Pan, X., You, Y., Wang, Z., Lu, C.: Virtual to real reinforcement learning for autonomous driving. arXiv preprint arXiv:1704.03952 (2017)
32. Papadimitriou, C.H., Tsitsiklis, J.N.: The complexity of markov decision processes. Math. Oper. Res. **12**(3), 441–450 (1987)
33. Park, K., et al.: Nerfies: deformable neural radiance fields. In: International Conference on Computer Vision, pp. 5865–5874 (2021)
34. Park, K., et al.: Hypernerf: a higher-dimensional representation for topologically varying neural radiance fields. ACM Trans. Graph (2021)
35. Pineau, J., Gordon, G., Thrun, S., et al.: Point-based value iteration: an anytime algorithm for pomdps. In: IJCAI, vol. 3, pp. 1025–1032 (2003)
36. Pumarola, A., Corona, E., Pons-Moll, G., Moreno-Noguer, F.: D-NeRF: neural radiance fields for dynamic scenes. In: IEEE Conference on Computer Vision Pattern Recognition (2020)
37. Redmon, J., Divvala, S., Girshick, R., Farhadi, A.: You only look once: unified, real-time object detection. In: IEEE Conference on Computer Vision and Pattern Recognition (2016)
38. Rhinehart, N., et al.: Contingencies from observations: tractable contingency planning with learned behavior models. In: IEEE International Conference on Robotics and Automation, pp. 13663–13669 (2021)
39. Roessle, B., Barron, J.T., Mildenhall, B., Srinivasan, P.P., Nießner, M.: Dense depth priors for neural radiance fields from sparse input views. In: IEEE International Conference Computer Vision and Pattern Recognition, pp. 12892–12901 (2022)
40. Russakovsky, O., et al.: Imagenet large scale visual recognition challenge (2015)
41. Shao, R., Zheng, Z., Tu, H., Liu, B., Zhang, H., Liu, Y.: Tensor4d: efficient neural 4d decomposition for high-fidelity dynamic reconstruction and rendering. In: IEEE International Conference Computer Vision and Pattern Recognition, pp. 16632–16642 (2023)
42. Sun, C., Sun, M., Chen, H.T.: Direct voxel grid optimization: super-fast convergence for radiance fields reconstruction. In: IEEE International Conference Computer Vision and Pattern Recognition, pp. 5459–5469 (2022)
43. Tancik, M., et al.: Block-nerf: scalable large scene neural view synthesis. In: IEEE International Conference Computer Vision and Pattern Recognition, pp. 8248–8258 (2022)
44. Tancik, M., et al.: Fourier features let networks learn high frequency functions in low dimensional domains. Adv. Neural Inform. Process. Syst., 7537–7547 (2020)
45. Tas, O.S., Stiller, C.: Limited visibility and uncertainty aware motion planning for automated driving. In: IEEE Intelligent Vehicles Symposium (IV) (2018)
46. Tewari, A., et al.: Advances in neural rendering. Comput. Graph. Forum (2022)
47. Toromanoff, M., Wirbel, E., Moutarde, F.: End-to-end model-free reinforcement learning for urban driving using implicit affordances. In: IEEE International Conference Computer Vision and Pattern Recognition, pp. 7153–7162 (2020)
48. Tretschk, E., Golyanik, V., Zollhoefer, M., Bozic, A., Lassner, C., Theobalt, C.: Scenerflow: time-consistent reconstruction of general dynamic scenes. In: International Conference on 3D Vision (3DV) (2023)
49. Truong, P., Rakotosaona, M.J., Manhardt, F., Tombari, F.: Sparf: neural radiance fields from sparse and noisy poses. In: IEEE International Conference Computer Vision and Pattern Recognition, pp. 4190–4200 (2023)

50. Turki, H., Ramanan, D., Satyanarayanan, M.: Mega-nerf: scalable construction of large-scale nerfs for virtual fly-throughs. In: IEEE International Conference Computer Vision and Pattern Recognition, pp. 12922–12931 (2022)
51. Wang, Q., et al.: Ibrnet: learning multi-view image-based rendering. In: IEEE International Conference Computer Vision and Pattern Recognition, pp. 4690–4699 (2021)
52. Xu, Q., et al.: Point-nerf: point-based neural radiance fields. In: IEEE International Conference Computer Vision and Pattern Recognition, pp. 5438–5448 (2022)
53. Yang, G.: VUER: a 3D visualization and data collection environment for robot learning (2024). https://github.com/vuer-ai/vuer
54. Yen-Chen, L., Florence, P., Barron, J.T., Rodriguez, A., Isola, P., Lin, T.Y.: inerf: inverting neural radiance fields for pose estimation. In: IEEE/RSJ International Conference on Intelligent Robots and Systems (IROS), pp. 1323–1330. IEEE (2021)
55. Yu, A., Ye, V., Tancik, M., Kanazawa, A.: pixelnerf: neural radiance fields from one or few images (2021)
56. Yu, M.Y., Vasudevan, R., Johnson-Roberson, M.: Occlusion-aware risk assessment for autonomous driving in urban environments. IEEE Rob. Autom. Lett. **4**(2), 2235–2241 (2019). https://doi.org/10.1109/lra.2019.2900453
57. Zhang, C., Steinhauser, F., Hinz, G., Knoll, A.: Improved occlusion scenario coverage with a pomdp-based behavior planner for autonomous urban driving. In: 2021 IEEE International Intelligent Transportation Systems Conference (ITSC), pp. 593–600 (2021). https://doi.org/10.1109/ITSC48978.2021.9564424
58. Zhang, C., Steinhauser, F., Hinz, G., Knoll, A.: Occlusion-aware planning for autonomous driving with vehicle-to-everything communication. IEEE Trans. Intell. Veh. **9**(1), 1229–1242 (2024). https://doi.org/10.1109/TIV.2023.3308098

Snuffy: Efficient Whole Slide Image Classifier

Hossein Jafarinia, Alireza Alipanah, Saeed Razavi, Nahal Mirzaie, and Mohammad Hossein Rohban(✉)

Sharif University of Technology, Tehran, Iran
{jafarinia,alireza.alipanah46,saeed.razavi,
nahal.mirzaie,rohban}@sharif.edu
https://www.sharif.edu

Abstract. Whole Slide Image (WSI) classification with multiple instance learning (MIL) in digital pathology faces significant computational challenges. Current methods mostly rely on extensive self-supervised learning (SSL) for satisfactory performance, requiring long training periods and considerable computational resources. At the same time, no pre-training affects performance due to domain shifts from natural images to WSIs. We introduce *Snuffy* architecture, a novel MIL-pooling method based on sparse transformers that mitigates performance loss with limited pre-training and enables continual few-shot pre-training as a competitive option. Our sparsity pattern is tailored for pathology and is theoretically proven to be a universal approximator with the tightest probabilistic sharp bound on the number of layers for sparse transformers, to date. We demonstrate Snuffy's effectiveness on CAMELYON16 and TCGA Lung cancer datasets, achieving superior WSI and patch-level accuracies. The code is available on https://github.com/jafarinia/snuffy.

Keywords: Whole Slide Image (WSI) · Self-Supervised Learning (SSL) · Sparse Transformer · Multiple Instance Learning (MIL)

1 Introduction

The emergence of whole slide images (WSIs) has presented significant opportunities to leverage machine learning techniques for essential tasks of cancer diagnosis and prognosis [21,23,27]. Nevertheless, integrating contemporary deep learning advancements into these areas faces notable challenges. Primarily, the sheer size of WSIs, with usual dimensions of approximately 150,000 × 150,000 pixels, renders them unmanageable for processing and training with existing deep learning frameworks on current hardware [24].

Supplementary Information The online version contains supplementary material available at https://doi.org/10.1007/978-3-031-73024-5_15.

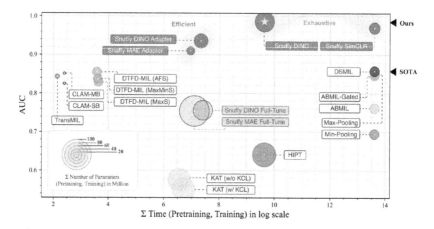

Fig. 1. Performance (AUC) vs. efficiency (size and time) trade off on CAMELYON16.

A common strategy to address this challenge is to divide WSIs into smaller patches followed by the application of Multiple Instance Learning (MIL) [8,24,28,35,38,41]. MIL, a variant of weakly supervised learning, considers instances as elements of sets involving an embedding phase and a pooling operation (MIL-pooling). The embedding phase often employs a pre-trained vision backbone, frequently with a self-supervised learning technique applied on the patches, transforming these patches into embeddings. Subsequently, MIL-pooling aggregates these embeddings, deriving scores at both the patch and WSI levels [20].

Recent advancements in WSI classification have achieved significant outcomes but face challenges like data-hungriness, high computational and memory requirements. These issues hinder the deployment and development of deep learning technologies in clinical and research settings. For example, RNN-MIL [4] and HIPT [6] require tens of terabytes of data, while DSMIL [25] and DGMIL [34] require several months for pre-training phases. DTFD-MIL [46] uses an efficient MIL-pooling strategy but demands over 100 gigabytes of system memory for training, which is only feasible in some settings. Conversely, the absence of pre-training or insufficient pre-training degrades performance because of the domain shift from natural image datasets such as ImageNet-1K to WSIs (CLAM [26] and KAT's [48] low AUC as shown in Fig. 1).

This work presents an approach that significantly reduces the computational demands required for training the embeddings by orders of magnitude. Then, empower its expressivity in a novel MIL-pooling to compensate for performance loss due to limited pre-training. Snuffy makes continual few-shot pre-training possible and a competitive option in this field by balancing efficiency and performance.

Our framework comprises two key components. First, we propose using **self-supervised continual pre-training** with Parameter Efficient Fine Tuning (PEFT) in the pathology domain, specifically utilizing Adaptformer [7] due to its effective and straightforward design. While PEFT has been applied in various fields, its use in pathology imaging is novel. Our results indicate that transitioning from natural images to histopathology is feasible, allowing us to leverage PEFT methods effectively.

Second, inspired by the complex biology of cancer and the importance of the tissue microenvironment in cancer detection, we introduce the **Snuffy MIL-pooling architecture**, which features a new sparsity pattern for sparse transformers. We demonstrate that the Snuffy sparsity pattern acts as a universal approximator, with the number of layers constrained to a linear relationship with the number of patches, denoted as $\mathcal{O}(n)$. This finding represents the tightest probabilistic bound on the number of layers for sparse transformers to date.

We introduce two families within our framework: Efficient Snuffy and Exhaustive Snuffy. The Efficient Snuffy family is trained initially on a natural image dataset and then continues training with PEFT on WSIs. In contrast, the Exhaustive Snuffy family is trained from scratch on WSIs. Both families utilize the Snuffy MIL-pooling architecture. Although the performance of Efficient Snuffy may be slightly inferior to Exhaustive Snuffy, both methods significantly outperform existing benchmarks in Region-of-Interest (ROI) detection and WSI classification, setting a new state-of-the-art (SOTA).

In summary, our main contributions are as follows:

- Using continual SSL pre-training from ImageNet-1K pre-trained models to pathology datasets employing Adapters, substantially reducing the computational time for pre-training by an order of magnitude.
- Introduction of a novel biologically driven sparsity pattern with a new probabilistic sharp bound on the number of layers to guarantee its universal approximation.
- Achieving significant improvements in WSI classification metrics for both the exhaustive and efficient families, and reaching new state-of-the-art scores in WSI classification (AUC of 0.987) and ROI detection (FROC of 0.675) by the exhaustive family.
- Validation of our method on widely recognized datasets, including CAMELYON16 and TCGA Lung Cancer WSI datasets, as well as on three classical multiple instance learning datasets: Musk1, Musk2, and Elephant, demonstrating consistent and superior performance across these varied datasets and establishing our MIL-pooling architecture as a general MIL-pooling architecture.

2 Related Work

Parameter-Efficient Fine-Tuning for Vision. Parameter-Efficient Fine-Tuning (PEFT), initially successful in NLP [18,19], especially with Transformers [16,18,19,32], has recently been applied to Vision Transformers (ViTs) for supervised tasks [7,14,39]. Techniques like Adapters [18] and LoRA [19] help mitigate overfitting during fine-tuning. The high computational demands for self-supervised pre-training, along with SSL pre-training benefits on domain-specific datasets [25], highlight PEFT's potential in SSL contexts. Recent studies [11,47] have proposed continual pre-training from general datasets like ImageNet-1K to domain-specific ones. Our study is the first to apply this approach specifically from ImageNet-1K to pathology datasets, advancing the field.

MIL for WSI Classification. The MIL-pooling operator must be permutation-invariant [20]. Traditional methods like Mean-pooling and Max-pooling have had some success, but parameterized strategies are more effective [20,25]. Recent SOTA MIL-pooling techniques are mainly attention-based, transformer-based, and distribution-based.

Attention-Based Methods: ABMIL [20] uses a trainable weighted average for each patch, incorporating nonlinear functions of tanh(.) and sigm(.) and learnable parameters. DSMIL [25] employs a unique self-attention framework focusing on one patch's interaction with the rest, derived from Max-pooling.

Transformer-Based Approaches: TransMIL [37] uses a ViT variant with the Nyström method [40] for self-attention approximation and introduces the Pyramid Position Encoding Generator (PPEG) for positional information. HIPT [6] uses a three-tier hierarchical pyramid with DINO [5] in the initial stages and a compact ViT in the final stage for WSI-level tasks. KAT [48] introduces kernels reflecting spatial information across various scales for cross-attention with patch embeddings, they described it as a linear memory complexity Transformer.

Distribution-Based Methodologies: They often use a clustering layer to improve embeddings for MIL-Pooling. CLAM [26] enhances embeddings by incorporating clustering atop its attention mechanism, while DGMIL [34] uses patch embedding distribution to create pseudo-labels for patches. This involves training a linear projection head and a linear layer on these pseudo-labels, utilizing the projection head for better embeddings with Mean-pooling for WSI classification. For identifying ROIs, it uses distribution scores from testing patches on these refined embeddings for patch labeling.

While these methods are effective for classification, their ROI detection performance is lacking or absent [6,37,46,48]. Our research aims to improve both WSI classification and localization accuracy while demonstrating that our architecture achieves the universal approximation of self-attention mechanisms with a sharp concentration rate.

3 Background

3.1 Notation

For any positive integer a, we represent the set as $[a] = \{1, 2, ..., a\}$. If we have a matrix $A \in \mathbb{R}^{d \times n}$, we refer to its j-th column as A_j, and A_S denotes the submatrix comprising columns of A indexed by $\subseteq [n]$. The *softmax* operator $\sigma_S[\cdot]$ processes each column of a matrix, resulting in a column stochastic matrix.

In terms of asymptotic behavior, $f = \mathcal{O}(g)$ implies that there exists a constant $C > 0$ such that $f(n) \le Cg(n)$ for all but finitely many n. Conversely, $f = \Omega(g)$ signifies that $g = \mathcal{O}(f)$. We use $\tilde{X}$ to conceal poly-logarithmic factors of X, like $\tilde{\Omega}(n)$.

3.2 MIL Formulation

In a binary classification problem, the dataset $D = \{(X_1, Y_1), ..., (X_n, Y_n)\}$ consists of bags X, where each bag $X = \{x_1, ..., x_k\}$ contains instances x, and $Y_i \in \{0, 1\}$ represents the bag's label. The individual instance labels $\{y_1, ..., y_k\}$ with with $y \in \{0, 1\}$, are unknown during training. This is modeled as:

$$Y = \begin{cases} 0, & \text{iff } \sum_i y_i = 0 \\ 1, & \text{otherwise.} \end{cases} \quad (1)$$

or equivalently:

$$Y = \max_i \{y_i\}. \quad (2)$$

To address the complexities in learning, to navigate this, it is proposed to train the MIL model by optimizing the log-likelihood function:

$$P(Y|X) = \theta(X)^Y (1 - \theta(X))^{1-Y}, \quad (3)$$

where $\theta(X) \in [0, 1]$ is the probability of $Y = 1$ given X.

Given that MIL assumes no ordering or dependency of instances, $\theta(X)$ must be a permutation-invariant (symmetric) function. This is achieved through the Fundamental Theorem of Symmetric Functions, with monomials [45] and a similar Theorem by [33] leading to:

$$\theta(X) = g(\pi(f(x_1), ..., f(x_k))), \quad (4)$$

where f and g are continuous transformations, and π is a permutation-invariant function (MIL-pooling). There are two approaches, the instance-level approach where f is an instance classifier and g is identity function, and the embedding-level approach, where f is a feature extractor and g maps its input to a bag classification score. The embedding-based approach is preferred due to its superior performance [20].

In Deep MIL, f typically uses pre-trained vision backbones to extract features from bag instances [25,26,34,37,46,48]. The aggregation function π ranges from non-parametric methods like max-pooling to parametric ones using attention mechanisms, as detailed in Sect. 2. Finally, g is often implemented as a linear layer with σ to project aggregated instance representations into a bag score.

For multiclass classification, g's output dimension is adjusted to the number of classes, and the *softmax* function is used instead of σ to distribute probabilities across classes.

3.3 Sparse Transformers

In the full-attention mechanism, each attention head considers interactions between every patch in an image. This requires significant memory and computational resources, with complexity scaling quadratically with the number of patches. However, in a sparse attention mechanism, each attention head only attends to a particular subset of patches instead of the entire set of patches. Drawing upon the formalism presented in [42], the ith sparse attention head output for a patch k in the layer l is articulated as follows:

$$SHead^{i,l}(X)_k = W_V^i X_{\mathcal{A}_k^l} \cdot \sigma\left(\left(W_K^i X_{\mathcal{A}_k^l}\right)^T W_Q^i X_k\right) \tag{5}$$

When calculating the attention scores, the query vector of the ith head, $W_Q^i X_k$ of the kth patch interacts exclusively with the key vectors $W_K^i X_{\mathcal{A}_k^l}$ from patches belonging to its specific subset, $\mathcal{A}_k^l$. This means that attention is computed only between the kth patch and the patches in its assigned subset. Consequently, the output of each attention head for the patch k, $SHead^{i,l}(X)_k$ is a result of combining columns from the patch representations $W_V^i X_{\mathcal{A}_k^l}$ within its assigned subset, rather than considering the entire sequence of patches [42]. The collection of these subsets $\mathcal{A}_k^l$, across all layers $l \in [L]$ and patches $k \in [n]$, is termed the *sparsity patterns*.

Sparse attention mechanisms significantly reduce the time and memory complexities of transformers. However, this efficiency comes at the expense of loss of accuracy in attention matrix predictions depending on how sparse the sparsity patterns are. Expanding beyond this limitation, prior studies have identified diverse sparse patterns with $\mathcal{O}(n)$ connections (compared to $\mathcal{O}(n^2)$ in full-attention mechanisms), effectively approximating any full attention matrices [3,9,15,44]. Notably, [42] established adequate conditions to ensure that any collection of sparsity patterns adhering to these criteria, alongside a probability map such as softmax, can serve as a universal approximator for sequence-to-sequence functions (Theorem 1).

Theorem 1. *A sparse transformer with any set of sparsity patterns $\{\mathcal{A}_k^l\}$ satisfying these conditions:*

1. $\forall k \in [n], \forall l \in [L], k \in \{\mathcal{A}_k^l\}$
2. *There is a permutation γ such that $\forall i \in [n-1]$, $\gamma(i) \in \bigcup_{l=1}^{p} \{\mathcal{A}_{\gamma(i+1)}^l\}$.*
3. *Each patch attends to every other patch, directly or indirectly.*

coupled with a probability map generating a column stochastic matrix that closely approximating hardmax operator, is a universal approximator of sequence-to-sequence functions [42].

Put simply, criterion 1 mandates that each patch attends to itself. This condition guarantees that even if contextual attention embeddings are not computed

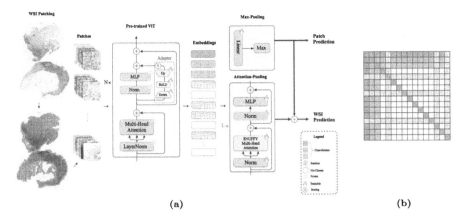

Fig. 2. Overview of the proposed method (a) The WSIs are segmented into 256 × 256 patches at 20X magnification, followed by embedding extraction via a pre-trained ViT [13]. Subsequently, these embeddings are inputted into the Snuffy for patch and WSI classification. (b) The connectivity matrix illustrates the Snuffy attention sparsity patterns, with Class-related Global Attentions, highlighted in darker colors either vertical or horizontal (the darker the more important), Diagonal Attentions depicted with pink, and Random Global Attentions shown in the lightest pink.

within the self-attention framework for a patch, the patch's original embedding remains intact in the output. Additionally, criteria 2 and 3 state that when all patterns in a set of sparsity patterns are combined and an adjacency matrix is created, the resulting graph G, possesses both a Hamiltonian path and strong connectivity. From now on, we refer to any sparsity patterns that meet the conditions outlined in Theorem 1 as *universal approximator sparsity patterns*.

4 Method

4.1 Continual Few-Shot Efficient Self-Supervised Pre-training

To avoid extensive training on large domain-specific datasets, we propose using continual few-shot self-supervised pre-training with AdaptFormer [7] on ViTs pre-trained on ImageNet-1K [47].

AdaptFormer [7], a PEFT variant, adds trainable weights to each layer while freezing the original network's weights. This leverages initial weights while adapting to domain-specific datasets, reducing overfitting [11].

Our experiments show that maintaining a self-supervised methodology during pre-training and adaptation yields a robust backbone with high-quality features, tuning fewer parameters, and reducing memory and computational demands.

For SSL method, we used Masked Autoencoder (MAE) [17], which masks 75% of the input image and trains an encoder-decoder to reconstruct it, making the encoder a feature extractor for downstream tasks. Additionally, we used DINO

[5], known for its effectiveness in pathology [6,22]. DINO [5] employs a student-teacher network setup where the student learns features from the teacher, who has a more global view of the image.

4.2 Snuffy Architecture

Informed by the prerequisites outlined for sparse transformers, pathologists' behavior, and drawing inspiration from DSMIL [25], our Sparse Transformer architecture in MIL-pooling constitutes two main components: Max-pooling component and Attention-pooling (see Fig 2a). Further, we provide detailed descriptions of each component.

Max-Pooling Component. The Max-pooling component serves a pivotal role within our framework. Initially, it functions as an effective instance-level model [46]. Subsequently, it enhances the efficacy of the Attention-pooling module. On the other hand, the Attention-pooling module facilitates more effective optimization of the Max-pooling module. The empirical evidence supporting this assertion is presented in Table 1.

Given these observations, our methodology leverages the top λ_{top} patches from the Max-pooling model as the Class-related Global Attentions.

Attention-Pooling Component. Present sparsity patterns are predominantly designed with a focus on NLP tasks [43,44], often exhibiting a prevalent structure known as the locality windowed pattern. However, in the domain of WSIs, pathologists frequently depend on the identification of non-related tissue within other tissues as a pivotal biomarker for cancer detection, even in instances where the tissue itself does not exhibit overt signs of malignancy [29–31,36].

Inspired by the inherent characteristics of WSI analysis, we propose Snuffy sparsity patterns.

Definition 1. *Snuffy sparsity patterns for patch k in layer l are defined as:*

$$A_k^l = \{k\} \cup \begin{cases} [n] & k \in \Lambda^l \\ \Lambda^l & otherwise \end{cases} \quad (6)$$

where $\Lambda^l := \Lambda_{top} \cup \Lambda_r^l$. Here, Λ^l is set of patches selected in layer l which consists of Λ_{top} representing a set of top λ_{top} patches from the max-pooling component, and Λ_r^l denoting a set of random patches $\in [n]\backslash\Lambda_{top}$.

The Snuffy sparsity concept comprises three key elements: **Class-related Global Attentions** (Λ_{top}), **Random Global Attentions** (Λ_r^l), and **Diagonal Attentions** (k). Class-related Global Attentions are crucial for the final classification task, and we leverage our max-pooling mechanism to identify them. We incorporate Random Global Attentions into our design to integrate insights from the tissue microenvironment, enhancing the capture of critical information. Diagonal Attentions ensure that even if contextual attention embeddings are not computed within the self-attention framework for a patch, the patch's original embedding remains preserved in the output.

After several iterations, the max-pooling component converges, and the scores for patches become fixed, allowing us to assume Λ_{top} remains constant across layers (see illustration of Snuffy patterns in Fig 2b).

5 Universal Approximation of Snuffy

In this section, we demonstrate that our sparse transformer serves as a universal approximator for sequence-to-sequence functions. By applying Theorem 1 and validating the Snuffy sparsity patterns defined in Definition 1, we confirm that our transformer, utilizing softmax as the probability map, satisfies all conditions given in the theorem. Furthermore, we illustrate that our transformer does not necessitate $\tilde{\Omega}(n)$ layers, as previously suggested in studies [44]. Instead, it requires only $\mathcal{O}(\frac{n \log 2}{\lambda_r})$ layers to ensure universal approximation with high probability, achieving the most stringent probabilistic limit of the layer count to our knowledge.

Snuffy sparsity patterns unquestionably satisfy the criteria 1 and 3 outlined in Theorem 1. The first condition is fulfilled through the inclusion of $k \in \mathcal{A}_k^l$, as defined in Definition 1. Moreover, the presence of at least one global attention patch within the patterns ensures connectivity among all patches, with a maximum path length of 2, thus meeting condition 3.

To satisfy the criterion 2, we must demonstrate the existence of a Hamiltonian path in the graph corresponding to the union of patterns in the Snuffy sparsity patterns. Initially, we introduce Proposition 1 from graph theory to facilitate the proof. We employ the proposition and demonstrate that covering half of the patches in all layers, overall satisfies its properties. This leads to the formation of a Hamiltonian path, thus fulfilling the desired proof.

Proposition 1. *Every graph $G(E, V)$ with E and V as the set of edges and nodes, with $|V| \geq 3$ and $\alpha(G) \leq \chi(G)$ has a Hamiltonian cycle. Where $\alpha(G)$ is the maximum independent set, and $\chi(G)$ is the chromatic number of G.*

Proof. see Supplementary Material 1.

Lemma 1. *For G_S, the graph representing Snuffy sparsity patterns, we guarantee that there exists a Hamiltonian path if*

$$|\bigcup_{l \in [L]} \Lambda^l| \geq \frac{n-1}{2} \tag{7}$$

Proof. The maximum independent set of patches in G_S is equal to $[n]\backslash\Lambda$, where $\Lambda := \bigcup_{l \in [L]} \Lambda^l$ is the set of all covered patches. Conversely, the minimum number of colors needed to color the vertices of G_S is $|\Lambda|+1$. Therefore, to satisfy $\alpha(G_S) \leq \chi(G_S)$, we must demonstrate that $|[n]\backslash\Lambda|\geq|\Lambda|+1$, which implies $|\Lambda|\geq \frac{n-1}{2}$. (for more details see supplementary 1)

Considering Lemma 1, we make the keen observation that we only need to ensure that after a finite number of layers, we have covered at least $\lfloor \frac{n}{2} \rfloor$ of the patches. This observation resembles a generalized version of the coupon collector problem, where the goal is to collect half of the coupons in a λ_r-uniform group setting. In this scenario, at each step, we can collect λ_r number of cards simultaneously. This leads us to our main theorem:

Theorem 2. *If $\lambda_r = \mathcal{O}(n)$, where n is number of patches, and $\lambda_r = |\Lambda_r|$, then the number of layers L needed to prove that the Snuffy sparsity pattern (defined in 1) is a universal approximator sparsity pattern is concentrated around $\frac{n \log 2}{\lambda_r}$. More precisely, we have:*

$$\lim_{n \to \infty} \mathbb{P}(|L - \frac{n \log 2}{\lambda_r}| > \frac{c\sqrt{n}}{\lambda_r}) \longrightarrow 1 - \Phi(c) \tag{8}$$

Proof. see Supplementary Material 1.

6 Experiments and Results

6.1 Datasets

For evaluation, we use the CAMELYON16, TCGA Lung Cancer, and MIL Datasets [1,2,10,12]. The CAMELYON16 dataset [2] includes 270 training and 129 testing WSIs, categorized into tumor and normal classes. The TCGA Lung Cancer dataset [10] consists of 1042 WSIs (530 LUAD and 512 LUSC). The MIL Datasets [1,12] are specifically designed for MIL. CAMELYON16 is particularly challenging and referred to as a "needle-in-a-haystack" dataset[1]. For more details on these datasets, refer to the Supplementary Material 2.

6.2 Experimental Setup

For evaluation of the CAMELYON16 dataset [2] we segmented WSIs into 256×256 patches at $20X$ magnification level, excluding patches predominantly featuring background elements and adhered to its official split for testing.

In the analysis of the TCGA Lung Cancer dataset [10], WSIs were similarly segmented into 256×256 patches at a $20X$ magnification, with background patches being discarded. The dataset was divided into training, validation, and test sets, constituting roughly 60%, 15%, and 25% of the total, respectively.

---

[1] https://github.com/mahmoodlab/HIPT/issues/41.

For the MIL datasets [1,12], we implemented a 10-fold cross-validation procedure.

This comprehensive and iterative evaluation approach empowers us to affirm the efficacy of our proposed framework alongside other SOTAs with substantial reliability.

We refer to Snuffy models as Snuffy + SSL pre-training method + Exhaustive for training from scratch, Adapter for fine-tuning with adapter, and Full-tune for fine-tuning all weights without adapter.

For more on the experimental setup and imeplementation details see Supplementary Materials 3, 4 respectively.

6.3 Evaluation Metrics

For WSI classification across the CAMELYON16 [2], TCGA Lung Cancer [10], and MIL datasets [1,12], we utilize Accuracy and Area Under the Receiver Operating Characteristic Curve (AUC) as standard evaluation metrics. Specifically, for the CAMELYON16 dataset [2], recognized for its complexity, we underscore the critical yet mostly overlooked aspect of model calibration evaluation in the high-stakes domain of WSI classification, which directly impacts patient care. In this context, we utilize the evaluation of Expected Calibration Error (ECE), the most widely adopted calibration measure. ECE represents the average discrepancy between the model's prediction confidence and actual performance accuracy. For more details check Supplementary Material 5.

For ROI detection, exclusively applicable to the CAMELYON16 dataset, we employ the Patch classification AUC and the Free Response ROC (FROC) as the benchmark metrics for evaluating ROI detection within WSI classification frameworks. The Patch classification Accuracy is omitted from our reporting due to the unbalanced portion of tumor regions in this dataset.

6.4 Baselines

For baseline we employ traditional approaches, such as Max-pooling and Mean-pooling, and current SOTAs including ABMIL, ABMIL-Gated [20], CLAM-SB, CLAM-MB [26], non-local DSMIL [25], TransMIL [37], DGMIL [34], HIPT [6], DTFD-MIL (MaxS), DTFD-MIL (MaxMinS), DTFD-MIL (AFS) [46], KAT (w/o KCL), KAT (w/ KCL) [48]. If a model does not provide a direct way of ROI detection, we do not report the corresponding metrics for it. Due to computational resource constraints, we could not directly evaluate DGMIL [34] across all datasets, nor could we assess HIPT [6] on the TCGA Lung Cancer dataset [10], as these methods need highly resource-intensive vision backbone pre-training. For Slide and Patch metrics, We have referred to the performance metrics reported in their original publications for these methods.

6.5 Results and Analysis

As shown in Table 1, Snuffy achieves competitive performance in WSI classification and patch classification, using minimal resources. When exhaustive SSL pre-training is applied, Snuffy consistently outperforms other models. Its strong performance in the calibration metric ECE highlights its potential for clinical applications. Snuffy's superiority is further validated by the results in Table 2. Also, Adapter vs Full-tune shows fine-tuning with Adapter is better.

The performance difference between WSI and patch classification in Table 1 for ABMIL [20] and DSMIL [25] is due to their use of attention scores, proxies for actual labels. Mean-pooling shows high ROI detection results because it uses instance-level MIL, making it a strong instance classifier but a weak bag classifier. Mean-pooling's approach, a linear layer followed by a mean function, struggles with the unbalanced nature of CAMELYON16's tumor regions (less than 10%). This imbalance causes normal patches to dominate the final decision, leading to a high false-negative rate in bag classification.

The observed underperformance of HIPT [6] on the challenging CAMELYON16 dataset [2] can be attributed to its dependency on two sequential dimension reduction steps via DINO pre-training [5], which introduces substantial error accumulation detrimental to the classifier's final stage. This issue is combined with the model's reliance on limited input data for a data-hungry ViT.

In ROI detection, our model surpasses existing methods by a significant margin, establishing a new SOTA. The inferior performance of DGMIL [34] in this area is linked to its dependency on noisy labels, which are generated through clustering with a preset and potentially inaccurate number of clusters for embedding enhancement.

Furthermore, our findings across MIL datasets show the versatility of our MIL-pooling strategy, affirming its efficacy as a broadly applicable architecture within the MIL framework which can be seen in Table 3.

For qualitative ROI detection evaluation check Fig. 3.

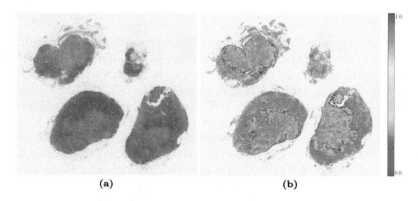

Fig. 3. Qualitative view of ROIs recognized by Suffy through its Patch Classification. (a) An example WSI from the test set of the CAMELYON16 dataset [2]. (b) ROIs are identified by Snuffy with black lines delineating the ground truth ROIs.

7 Ablation Study

In this section, we explore the impact and efficacy of key components of Snuffy on the classification and ROI detection performance on the CAMELYON16 dataset [2]. These experiments are done with Snuffy simCLR Exhaustive.

Effect of Depth: The architecture's depth, evaluated at 1, 2, 4, and 5 layers, is analyzed to determine its influence on the model's performance as a universal approximator. As depicted in Fig. 4, an increase in depth, generally but slightly correlates with enhanced performance and complies with our theoretical insights.

Effect of Number of Random Global Attentions: The examination of the number of Random Global Attentions, at 1, 300, and 700, demonstrates an increase in performance with higher numbers, as illustrated in Fig. 5. This and the observation of diminishing returns beyond 1000 aligns with the theoretical insights.

For more ablation studies check Supplementary Material 6.

Table 1. Results on The CAMELYON16 dataset [2]. Exhaustive is for SSL models trained from scratch, Full-tune for models trained from ImangeNet-1K pre-trained weights, and adapter for models trained with our proposed method. Patch is for patch classification and ROI detection. $\sum$Time shows the sum of pre-training time and MIL training time. $\sum$#Params shows the sum of pre-trained parameters and MIL-trained parameters, and − shows the model is incapable of or has not implemented a quantifiable method for ROI detection. Mem and GPU Mem numbers are not the most accurate due to randomness in CUDA and parallel runs but accurate enough to give a good comparison.

Method	Slide			Patch		System		Resource	
	ACC	AUC	ECE	AUC	FROC	Mem (GB)	GPU Mem (GB)	$\sum$Time (Minute)	$\sum$#Params (Million)
Mean-pooling SimCLR Exhaustive	0.614	0.695	0.070	**0.980**	0.597	8.3	2.4	806,447	11.56
Max-pooling SimCLR Exhaustive	0.829	0.848	0.056	0.790	0.387	8.3	2.4	806,447	11.56
ABMIL SimCLR Exhaustive [20]	0.815	0.762	0.184	0.303	0.182	8.3	1.62	806,422	11.62
ABMIL-Gated SimCLR Exhaustive [20]	0.876	0.847	0.124	0.394	0.245	8.3	1.62	806,423	11.69
DGMIL [34]	0.801	0.836	NA	0.904	0.488	76	11	1,612,920	111.66
DSMIL [25]	0.810	0.858	0.051	0.411	0.328	19.2	3.09	806,448	11.64
TransMIL [37]	0.815	0.843	0.125	−	−	1.51	7.87	8.5	2.7
CLAM-SB [26]	0.820	0.825	0.086	−	−	1.16	1.97	11	0.92
CLAM-MB [26]	0.821	0.851	0.110	−	−	1.15	1.97	11	0.92
HIPT [6]	0.634	0.641	0.122	−	−	0.92	2.40	15,384	69.35
KAT (w/o KCL) [48]	0.579	0.576	0.290	−	−	2.52	16.43	706	61.2
KAT (w/ KCL) [48]	0.576	0.551	0.301	−	−	30	15.22	830	66.4
DTFD-MIL (MaxS) [46]	0.821	0.829	0.107	−	−	29.73	1.38	38	9.86
DTFD-MIL (MaxMinS) [46]	0.818	0.837	0.101	−	−	29.73	1.50	37	9.86
DTFD-MIL (AFS) [46]	0.832	0.855	0.083	−	−	29.72	1.47	36	9.86
Snuffy SimCLR Exhaustive	**0.952**	0.970	0.057	**0.980**	0.622	8.3	4.8	806,496	14.87
Snuffy DINO Adapter	0.876	0.936	0.058	0.911	0.552	6.7	4.8	1,536	23.7
Snuffy DINO Full-tune	0.761	0.756	0.195	0.881	0.381	6.7	4.8	1,658	47.5
Snuffy DINO Exhaustive	0.948	**0.987**	0.083	0.957	**0.675**	6.7	4.8	15,393	47.5
Snuffy MAE Adapter	0.900	0.910	0.078	0.873	0.543	11.7	4.8	1,056	8.1
Snuffy MAE Full-tune	0.782	0.754	0.134	0.875	0.363	11.7	4.8	1,216	113.66

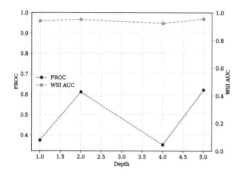

Fig. 4. Ablation on depth.   **Fig. 5.** Random Global Attentions.

Table 2. Results on The TCGA dataset [10].

Method	Slide ACC	AUC
DGMIL [34]	0.920	0.970
TransMIL [37]	0.883	0.949
CLAM-SB [26]	0.875	0.944
CLAM-MB [26]	0.878	0.949
HIPT [6]	NA	0.952
KAT (w/o KCL) [48]	0.849	0.965
KAT (w/ KCL) [48]	0.859	0.971
DTFD-MIL (MaxS) [46]	0.855	0.910
DTFD-MIL (MaxMinS) [46]	0.890	0.938
DTFD-MIL (AFS) [46]	0.898	0.946
Snuffy SimCLR Exhaustive	**0.947**	**0.972**

Table 3. Results on MUSK1, MUSK2 [12], ELEPHANT [1].

Method	MUSK1		MUSK2		ELEPHANT	
	ACC	AUC	ACC	AUC	ACC	AUC
Max-pooling	0.728	0.799	0.676	0.756	0.764	0.861
Mean-pooling	0.800	0.869	0.710	0.855	0.830	0.920
ABMIL [20]	0.826	0.824	**0.812**	0.816	0.849	0.84
ABMIL-Gated [20]	0.831	0.837	0.780	0.784	0.789	0.844
DSMIL [25]	0.786	0.852	0.706	0.813	0.811	0.915
Snuffy	**0.961**	**0.989**	0.789	**0.985**	**0.923**	**0.967**

8 Conclusion and Discussion

We introduced a novel WSI classification framework using PEFT for data efficiency in SSL and a sparse transformer inspired by pathologists. This approach ensures global approximation with high probability and provides a tighter bound for the number of layers in sparse transformers for MIL-pooling, achieving excellent results. However, achieving SOTA results still requires long and exhaustive SSL training. Our Theoretical guarantees need a high number of layers, increasing memory needs (though one layer performs very well empirically). Future work could explore PEFT methods tailored to pathology and develop less resource-intensive MIL-pooling methods. Another avenue is closing the theoretical and practical gap in understanding sparse transformers.

Acknowledgements. We extend our deepest and most special thanks to Danial Hamdi for their efforts. We also thank Mohammad Mosayyebi, Mehrab Moradzadeh, Mohammad Hosein Movasaghinia, Mohammad Azizmalayeri, Hossein Mirzaei, Mohammad Mozafari, Soroush Vafaei Tabar, Mohammad Hassan Alikhani, and Hosein Hasani.

References

1. Andrews, S., Tsochantaridis, I., Hofmann, T.: Support vector machines for multiple-instance learning. Adv. Neural Inf. Process. Syst. **15** (2002)
2. Bejnordi, B.E., et al.: Diagnostic assessment of deep learning algorithms for detection of lymph node metastases in women with breast cancer. JAMA **318**, 2199–2210 (2017). https://api.semanticscholar.org/CorpusID:205086555
3. Beltagy, I., Peters, M.E., Cohan, A.: Longformer: the long-document transformer. arXiv preprint arXiv:2004.05150 (2020)
4. Campanella, G., et al.: Clinical-grade computational pathology using weakly supervised deep learning on whole slide images. Nat. Med. **25**, 1301–1309 (2019). https://api.semanticscholar.org/CorpusID:196814162
5. Caron, M., et al.: Emerging properties in self-supervised vision transformers. In: Proceedings of the IEEE/CVF International Conference on Computer Vision, pp. 9650–9660 (2021)
6. Chen, R.J., et al.: Scaling vision transformers to gigapixel images via hierarchical self-supervised learning. In: IEEE/CVF Conference on Computer Vision and Pattern Recognition, CVPR 2022, New Orleans, LA, USA, 18–24 June 2022, pp. 16123–16134. IEEE (2022). https://doi.org/10.1109/CVPR52688.2022.01567
7. Chen, S., et al.: Adaptformer: adapting vision transformers for scalable visual recognition. In: Koyejo, S., Mohamed, S., Agarwal, A., Belgrave, D., Cho, K., Oh, A. (eds.) Advances in Neural Information Processing Systems, vol. 35, pp. 16664–16678. Curran Associates, Inc. (2022). https://proceedings.neurips.cc/paper_files/paper/2022/file/69e2f49ab0837b71b0e0cb7c555990f8-Paper-Conference.pdf
8. Cheplygina, V., de Bruijne, M., Pluim, J.P.W.: Not-so-supervised: a survey of semi-supervised, multi-instance, and transfer learning in medical image analysis. Medical Image Anal. **54**, 280–296 (2019). https://doi.org/10.1016/J.MEDIA.2019.03.009
9. Child, R., Gray, S., Radford, A., Sutskever, I.: Generating long sequences with sparse transformers (2019). https://openai.com/blog/sparse-transformers
10. Cooper, L.A., Demicco, E.G., Saltz, J.H., Powell, R.T., Rao, A., Lazar, A.J.: Pan-cancer insights from the cancer genome atlas: the pathologist's perspective. J. Pathol. **244**(5), 512–524 (2018)
11. Dadashzadeh, A., Duan, S., Whone, A., Mirmehdi, M.: Pecop: parameter efficient continual pretraining for action quality assessment. In: Proceedings of the IEEE/CVF Winter Conference on Applications of Computer Vision, pp. 42–52 (2024)
12. Dietterich, T.G., Lathrop, R.H., Lozano-Pérez, T.: Solving the multiple instance problem with axis-parallel rectangles. Artif. Intell. **89**(1–2), 31–71 (1997)
13. Dosovitskiy, A., et al.: An image is worth 16×16 words: transformers for image recognition at scale. In: 9th International Conference on Learning Representations, ICLR 2021, Virtual Event, Austria, 3–7 May 2021. OpenReview.net (2021). https://openreview.net/forum?id=YicbFdNTTy
14. Gao, P., et al.: Clip-adapter: better vision-language models with feature adapters. Int. J. Comput. Vis. **132**(2), 581–595 (2024). https://doi.org/10.1007/S11263-023-01891-X
15. Guo, Q., Qiu, X., Liu, P., Shao, Y., Xue, X., Zhang, Z.: Star-transformer. CoRR arxiv:1902.09113 (2019)

16. He, J., Zhou, C., Ma, X., Berg-Kirkpatrick, T., Neubig, G.: Towards a unified view of parameter-efficient transfer learning. In: The Tenth International Conference on Learning Representations, ICLR 2022, Virtual Event, 25–29 April 2022. OpenReview.net (2022). https://openreview.net/forum?id=0RDcd5Axok
17. He, K., Chen, X., Xie, S., Li, Y., Dollár, P., Girshick, R.: Masked autoencoders are scalable vision learners. In: Proceedings of the IEEE/CVF Conference on Computer Vision and Pattern Recognition, pp. 16000–16009 (2022)
18. Houlsby, N., et al.: Parameter-efficient transfer learning for NLP. In: Chaudhuri, K., Salakhutdinov, R. (eds.) Proceedings of the 36th International Conference on Machine Learning, ICML 2019, Long Beach, California, USA, 9–15 June 2019. Proceedings of Machine Learning Research, vol. 97, pp. 2790–2799. PMLR (2019). http://proceedings.mlr.press/v97/houlsby19a.html
19. Hu, E.J., et al.: Lora: low-rank adaptation of large language models. In: The Tenth International Conference on Learning Representations, ICLR 2022, Virtual Event, 25–29 April 2022. OpenReview.net (2022). https://openreview.net/forum?id=nZeVKeeFYf9
20. Ilse, M., Tomczak, J.M., Welling, M.: Attention-based deep multiple instance learning. In: Dy, J.G., Krause, A. (eds.) Proceedings of the 35th International Conference on Machine Learning, ICML 2018, Stockholmsmässan, Stockholm, Sweden, 10–15 July 2018. Proceedings of Machine Learning Research, vol. 80, pp. 2132–2141. PMLR (2018). http://proceedings.mlr.press/v80/ilse18a.html
21. Javed, S., et al.: Cellular community detection for tissue phenotyping in colorectal cancer histology images. Med. Image Anal. **63**, 101696 (2020). https://doi.org/10.1016/J.MEDIA.2020.101696
22. Kang, M., Song, H., Park, S., Yoo, D., Pereira, S.: Benchmarking self-supervised learning on diverse pathology datasets. In: Proceedings of the IEEE/CVF Conference on Computer Vision and Pattern Recognition, pp. 3344–3354 (2023)
23. Kather, J.N., et al.: Multi-class texture analysis in colorectal cancer histology. Sci. Rep. **6** (2016). https://api.semanticscholar.org/CorpusID:4769235
24. van der Laak, J.A., Litjens, G.J.S., Ciompi, F.: Deep learning in histopathology: the path to the clinic. Nat. Med. **27**, 775 – 784 (2021). https://api.semanticscholar.org/CorpusID:234597294
25. Li, B., Li, Y., Eliceiri, K.W.: Dual-stream multiple instance learning network for whole slide image classification with self-supervised contrastive learning. In: IEEE Conference on Computer Vision and Pattern Recognition, CVPR 2021, virtual, 19–25 June 2021, pp. 14318–14328. Computer Vision Foundation/IEEE (2021). https://doi.org/10.1109/CVPR46437.2021.01409. https://openaccess.thecvf.com/content/CVPR2021/html/Li_Dual-Stream_Multiple_Instance_Learning_Network_for_Whole_Slide_Image_Classification_CVPR_2021_paper.html
26. Lu, M.Y., Williamson, D.F.K., Chen, T.Y., Chen, R.J., Barbieri, M., Mahmood, F.: Data efficient and weakly supervised computational pathology on whole slide images. CoRR arxiv:2004.09666 (2020)
27. Ludwig, J.A., Weinstein, J.N.: Biomarkers in cancer staging, prognosis and treatment selection. Nat. Rev. Cancer **5**, 845–856 (2005). https://api.semanticscholar.org/CorpusID:25540232
28. Myronenko, A., Xu, Z., Yang, D., Roth, H.R., Xu, D.: Accounting for dependencies in deep learning based multiple instance learning for whole slide imaging. In: de Bruijne, M., et al. (eds.) MICCAI 2021. LNCS, vol. 12908, pp. 329–338. Springer, Cham (2021). https://doi.org/10.1007/978-3-030-87237-3_32

29. Ng, T.G., Damiris, K., Trivedi, U., George, J.C.: Obstructive jaundice, a rare presentation of lung cancer: a case report. Respir. Med. Case. Rep. **33**, 101425 (2021)
30. Pajaziti, L., Hapçiu, S.R., Dobruna, S., Hoxha, N., Kurshumliu, F., Pajaziti, A.: Skin metastases from lung cancer: a case report. BMC. Res. Notes **8**, 1–6 (2015)
31. Patel, A.M., Vila, D.G.D., Peters, S.G.: Paraneoplastic syndromes associated with lung cancer. Mayo Clin. Proc. **68**(3), 278–287 (1993). https://doi.org/10.1016/S0025-6196(12)60050-0. https://www.sciencedirect.com/science/article/pii/S0025619612600500
32. Pfeiffer, J., Vulic, I., Gurevych, I., Ruder, S.: MAD-X: an adapter-based framework for multi-task cross-lingual transfer. In: Webber, B., Cohn, T., He, Y., Liu, Y. (eds.) Proceedings of the 2020 Conference on Empirical Methods in Natural Language Processing, EMNLP 2020, Online, 16–20 November 2020, pp. 7654–7673. Association for Computational Linguistics (2020).https://doi.org/10.18653/V1/2020.EMNLP-MAIN.617
33. Qi, C.R., Su, H., Mo, K., Guibas, L.J.: Pointnet: deep learning on point sets for 3d classification and segmentation. In: Proceedings of the IEEE Conference on Computer Vision and Pattern Recognition, pp. 652–660 (2017)
34. Qu, L., Luo, X., Liu, S., Wang, M., Song, Z.: DGMIL: distribution guided multiple instance learning for whole slide image classification. In: Wang, L., Dou, Q., Fletcher, P.T., Speidel, S., Li, S. (eds.) Medical Image Computing and Computer Assisted Intervention - MICCAI 2022 - 25th International Conference, Singapore, September 18-22, 2022, Proceedings, Part II. Lecture Notes in Computer Science, vol. 13432, pp. 24–34. Springer, Heidelberg (2022). https://doi.org/10.1007/978-3-031-16434-7_3
35. Rony, J., Belharbi, S., Dolz, J., Ayed, I.B., McCaffrey, L., Granger, E.: Deep weakly-supervised learning methods for classification and localization in histology images: a survey. CoRR arxiv:1909.03354 (2019)
36. Shalata, W., et al.: Dermatomyositis associated with lung cancer: a brief review of the current literature and retrospective single institution experience. Life **13**, 40 (2022). https://doi.org/10.3390/life13010040
37. Shao, Z., et al.: Transmil: transformer based correlated multiple instance learning for whole slide image classification. In: Ranzato, M., Beygelzimer, A., Dauphin, Y.N., Liang, P., Vaughan, J.W. (eds.) Advances in Neural Information Processing Systems 34: Annual Conference on Neural Information Processing Systems 2021, NeurIPS 2021, 6–14 December 2021, Virtual, pp. 2136–2147 (2021). https://proceedings.neurips.cc/paper/2021/hash/10c272d06794d3e5785d5e7c5356e9ff-Abstract.html
38. Srinidhi, C.L., Ciga, O., Martel, A.L.: Deep neural network models for computational histopathology: a survey. Med. Image Anal. **67**, 101813 (2021). https://doi.org/10.1016/J.MEDIA.2020.101813
39. Wu, J., et al.: Medical sam adapter: adapting segment anything model for medical image segmentation. arXiv preprint arXiv:2304.12620 (2023)
40. Xiong, Y., et al.: Nyströmformer: a nyström-based algorithm for approximating self-attention. In: Thirty-Fifth AAAI Conference on Artificial Intelligence, AAAI 2021, Thirty-Third Conference on Innovative Applications of Artificial Intelligence, IAAI 2021, The Eleventh Symposium on Educational Advances in Artificial Intelligence, EAAI 2021, Virtual Event, 2–9 February 2021, pp. 14138–14148. AAAI Press (2021). https://doi.org/10.1609/AAAI.V35I16.17664

41. Xu, Y., Zhu, J., Chang, E.I., Tu, Z.: Multiple clustered instance learning for histopathology cancer image classification, segmentation and clustering. In: 2012 IEEE Conference on Computer Vision and Pattern Recognition, Providence, RI, USA, 16–21 June 2012, pp. 964–971. IEEE Computer Society (2012). https://doi.org/10.1109/CVPR.2012.6247772
42. Yun, C., Chang, Y.W., Bhojanapalli, S., Rawat, A.S., Reddi, S.J., Kumar, S.: O(n) connections are expressive enough: universal approximability of sparse transformers. In: Proceedings of the 34th International Conference on Neural Information Processing Systems. NIPS 2020. Curran Associates Inc., Red Hook (2020)
43. Yun, C., Chang, Y.W., Bhojanapalli, S., Rawat, A.S., Reddi, S.J., Kumar, S.: $o(n)$ connections are expressive enough: universal approximability of sparse transformers. ArXiv arxiv:2006.04862 (2020). https://api.semanticscholar.org/CorpusID:219558319
44. Zaheer, M., et al.: Big bird: transformers for longer sequences. Adv. Neural Inf. Process. Syst. **33** (2020)
45. Zaheer, M., Kottur, S., Ravanbakhsh, S., Poczos, B., Salakhutdinov, R.R., Smola, A.J.: Deep sets. Adv. Neural Inf. Process. Syst. **30** (2017)
46. Zhang, H., et al.: DTFD-MIL: double-tier feature distillation multiple instance learning for histopathology whole slide image classification. In: IEEE/CVF Conference on Computer Vision and Pattern Recognition, CVPR 2022, New Orleans, LA, USA, 18–24 June 2022, pp. 18780–18790. IEEE (2022). https://doi.org/10.1109/CVPR52688.2022.01824
47. Zhang, T., et al.: Pad: self-supervised pre-training with patchwise-scale adapter for infrared images. arXiv preprint arXiv:2312.08192 (2023)
48. Zheng, Y., et al.: Kernel attention transformer for histopathology whole slide image analysis and assistant cancer diagnosis. IEEE Trans. Med. Imaging **42**(9), 2726–2739 (2023). https://doi.org/10.1109/TMI.2023.3264781

Learning to Build by Building Your Own Instructions

Aaron Walsman[1(✉)], Muru Zhang[1], Adam Fishman[1], Ali Farhadi[1], and Dieter Fox[1,2]

[1] University of Washington, Seattle, USA
awalsman@cs.washington.edu
[2] NVIDIA, Santa Clara, USA

Abstract. Structural understanding of complex visual objects is an important unsolved component of artificial intelligence. To study this, we develop a new technique for the recently proposed Break-and-Make problem in LTRON where an agent must learn to build a previously unseen LEGO assembly using a single interactive session to gather information about its components and their structure. We attack this problem by building an agent that we call **InstructioNet** that is able to make its own visual instruction book. By disassembling an unseen assembly and periodically saving images of it, the agent is able to create a set of instructions so that it has the information necessary to rebuild it. These instructions form an explicit memory that allows the model to reason about the assembly process one step at a time, avoiding the need for long-term implicit memory. This in turn allows us to train on much larger LEGO assemblies than has been possible in the past. To demonstrate the power of this model, we release a new dataset of procedurally built LEGO vehicles that contain an average of 31 bricks each and require over one hundred steps to disassemble and reassemble. We train these models using online imitation learning which allows the model to learn from its own mistakes. Finally, we also provide some small improvements to LTRON and the Break-and-Make problem that simplify the learning environment and improve usability. This data and updated environments can be found at github.com/aaronwalsman/ltron/blob/v1.1.0. Additional training code can be found at github.com/aaronwalsman/ltron-torch/tree/eccv-24.

1 Introduction

The ability to understand and execute complex assembly problems is one of the hallmarks of human intelligence. Humans use this ability to construct tools and reverse engineer previously unseen part-based objects.

Supplementary Information The online version contains supplementary material available at https://doi.org/10.1007/978-3-031-73024-5_16.

The recently proposed Break-and-Make problem [41] is designed to train agents to develop these abilities using complex LEGO structures. In this problem, an agent must learn to build a previously unseen assembly by actively inspecting it. To do this the agent is given access to an interactive simulator that allows it to disassemble the structure in order to reveal hidden components and see how everything fits together. Once it is confident that it knows the structure, the agent is presented with an empty scene and must build the model again from scratch. This problem is designed to simulate a reverse engineering problem. The agent must take apart a complex structure in order to learn how to make it. By training agents to effectively reverse engineer these systems, we can discover new tools for understanding and building intelligent systems that can reason about complex structures.

The Break-and-Make problem is quite challenging, as it requires long-term memory, and interaction with a complex visual environment using a 2D cursor-based action space. Baseline approaches that use transformers [39] and LSTMs [16] for long-term memory have struggled to make progress on larger models. We introduce a new model **InstructioNet** for this problem that uses an explicit memory to store a stack of self-curated instruction images. InstructioNet slowly adds to this memory by iteratively disassembling part of the model, then saving its most recent observation to the top of this instruction stack. This stack provides the model with a series of short-term visual targets to use when reassembling the model later on, just like a real-world LEGO instruction book. When rebuilding the model the agent only has to reason about its current observation and the page of the instruction stack that it is currently working on. Once the agent's current assembly matches this instruction, it can turn the page and get a new short-term target to work towards. Figure 1 shows a successful example of this on our new RC-Vehicles dataset.

In addition to this memory-based model, we also detail several practical components that are either necessary or improve training performance in this space. These include online imitation learning, conditional action heads and new loss functions for the dense 2D cursor-based action space.

Our primary contributions are:

1. We introduce a new instruction-stack based model for the Break and Make problem.
2. We provide several modeling and training improvements for complex visual action spaces that use a pixel-based cursor.
3. We provide a new simplified visual interface for the Break and Make problem, along with RC-Vehicles a new dataset of challenging LEGO models for assembly problems.
4. Our new model and training recipe achieve much better performance than previous baselines, especially on larger, more complicated models.

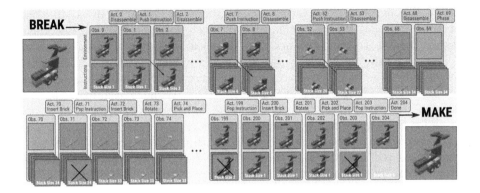

Fig. 1. An example of InstructioNet completing the Break and Make task on a previously unseen example from RC-Vehicles. Our model saves 34 distinct images to the instruction stack over the first 69 steps. It then successfully rebuilds the model from scratch using these images over the course of another 135 steps.

2 Related Work

2.1 Interactive Understanding and Assembly

Despite a proliferation of interactive environments with increasing sophistication [8,33,34,36,43,46], long-term structural understanding and assembly problems remain an open challenge in artificial intelligence. Researchers have explored this in the context of furniture assembly [26,49], Minecraft [11], CAD models [17] and robotic assembly problems [47]. In this work, we build on LTRON [41], a recent LEGO simulator designed to provide a building environment for learning agents. Constructing plans for assembly using a disassembly process has also been explored in the context of multi-part CAD based models [38], however this work is more concerned with finding collision-free paths through free-space, and does not directly reason about connection points.

LEGO bricks have been a popular substrate for learning assembly problems across a variety of subfields in artificial intelligence. These include design problems [29], robotic assembly [14], shape reconstruction [19,20,25], generative modelling [37], and image guided building [6,22]. Most similar to our work, Wang et al. [42] build LEGO structures from existing instructions. Our setting is more challenging because the agent must learn to make its own instructions rather than assuming they are already provided. Furthermore, the action space in LTRON is more difficult as it requires the agent to use a 2D cursor to interact with the scene and contains assemblies with bricks attached to the sides of objects, that cannot be described using simple stacking.

2.2 Memory

Memory structures for interactive problems have been studied for decades. Early attempts at implicit (problem-agnostic) memory structures include simple RNNs

[18,32] and more complex variants such as LSTMs [16] and GRUs [4]. These methods propagate information forward in time using specialized network architecture. Neural Turing Machines [13] use a learned external memory module to read and write to long term storage. Recently attention-based transformers [39] have become one of the most popular ways to build neural networks that rely on past information to make future decisions. While transformers are quite effective, they are computationally expensive for large sequences of observations, which limits their use for very long-term memory.

In contrast to implicit memory structures, explicit memory uses some knowledge of the environment or task in order to build a structure more appropriate to the problem. This can come in the form of geometric or topological map building such as in SLAM [5,15,23], semantic maps in embodied navigation [12] or more complex structures that combine multiple components [2]. Our approach uses an explicit memory structure designed around the intuitive understanding that assembling a structure can be completed in the reverse order of disassembling it. This motivates the use of a stack that allows the agent to sequentially build up a series of experiences, and then pop them off one-by-one in reverse order later.

2.3 Inverse-Graphics

Reconstructing geometry and reasoning about 3D structures from images has been an important issue in research fields such as computer-aided design [10,45] and robotics [3,21,35]. In particular, prior works such as [24,44] use 3D shapes while [27,28,48] use images to guide the inverse inference process. However, when building complex structures such as LEGO models, it is challenging to generate a set of sufficient visual images to predict the reconstruction without dynamically interacting with the object or the environment.

3 Methods

3.1 Environments and Data

The Break and Make environment in LTRON [41] is designed to test an agent's ability to perform complex building tasks on previously unseen target LEGO assemblies. In this, and many other complex construction problems, no single view of the target object is enough to fully describe it. Therefore, in order for the model to complete its objective, it must first interactively discover all the components of the LEGO assembly by disassembling it and remembering where all the individual bricks went during that process. In order to allow for this kind of interactive discovery, the Break and Make problem divides each interactive episode into two distinct phases. In the first Break phase, the agent is presented with a previously unseen LEGO model and is allowed to interactively inspect and disassemble it, while constructing some long-term memory of the object. Once it is done, it takes a dedicated Phase Switch action at which point the agent is presented with an empty scene and must use the memory it built in the Break phase to rebuild the model from scratch.

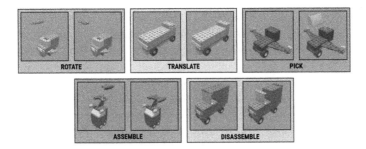

Fig. 2. Our modified LTRON action space without the extra Hand viewport. We only show the manipulation actions here and do not show the camera rotation and done actions which are unchanged.

In this paper, we use a slightly modified version of the original Break and Make environment with a few changes for simplicity of training. The original Break and Make environment provided two images as its observation space, one "table" image representing the current scene, and another "hand" image representing a single brick that had just been removed from the scene, or was about to be added. In order to simplify the observation space of our model, we have removed the "hand" image. In this updated environment, when a new brick is inserted into the scene, it is simply placed in a floating location above the existing assembly computed using the bounding box of existing bricks, or at the origin if no bricks are present. We also add a new Translate action mode which allows the agent to shift a brick by a fixed amount by using the cursor to select a connection point and specifying a direction and discrete offset value. We found this useful for helping the agent recover from small mistakes.

The complete set of action primitives in the new environment are: **Rotate:** rotates a single brick about a connection point specified by clicking a 2D screen location. The rotation angle is selected from a set of discrete values in 90° increments. **Translate:** translates a single brick specified by clicking a 2D screen location. The translation distance and direction is again selected from a set of discrete values corresponding to multiples of the stud spacing and brick heights. **Pick:** inserts a new brick into the scene, by selecting a discrete shape and color index. The new brick is placed in a floating location above the current assembly. **Assemble:** attaches one brick to another by specifying two connection points in the scene. **Disassemble:** removes a brick from the scene by specifying a connection point. All of these actions will only succeed if the operation can be performed without causing a collision. Figure 2 shows the manipulation components of the action space of our updated environment.

Note that despite including collision checking, this environment design does not attempt to simulate real-world physical interaction, as it does not consider forces or other complex dynamics. This means that the LTRON interface is closer to CAD modelling, where an operator must use a set of discrete tools and a 2D mouse in order to manipulate a virtual 3D object, than a physical manipulation setting that might be encountered in robotics or industrial automation.

We also use a reduced resolution of 128 × 128 pixels for training efficiency. In addition, we have made a number of improvements to the original LTRON simulator. These updates provide better collision checking and support for a larger number of connection point styles than was available previously. Note that while both the original and our updated environment allow for camera rotation, we found that rotating the camera was not necessary to successfully reassemble the models in the datasets considered here, so we used a fixed camera angle for all experiments.

In order to focus on construction ability and avoid the confounding issues of long-tailed part distributions, we used the 2, 4 and 8 brick random construction assemblies in LTRON and did not train on the Open Model Repository (OMR) data. Due to the simulator updates mentioned above, we regenerated this random construction data to account for improvements in collision checking. These new assemblies avoid certain rare configurations that resulted in small brick penetrations in the original data.

In addition to the random construction data, we have also developed a new RC-Vehicles dataset of randomly constructed vehicles. These were generated with a series of scripted rules defining distributions over the vehicle dimensions, swappable components such as tires and windshield shapes, and optional features such as wings, headlights and helicopter blades. When combined, these distributions have over 21 bits of entropy defining the shape of the vehicle and 29 bits of entropy defining its color combinations. These models vary in size from 19 to 73 bricks, making them substantially larger and more complex than the previous random construction data. Examples of these vehicles can be seen in Fig. 3.

Note that both the random brick and the RC-Vehicles assemblies are not merely top-down stacks of bricks, but require some parts to be placed on the sides, front and back of the constructed object.

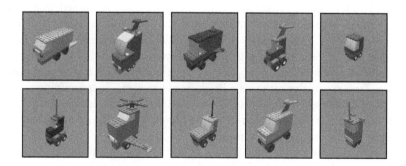

Fig. 3. Examples of the RC-Vehicles dataset.

3.2 Instruction Stack

Our new model **InstructioNet** works by storing an explicit stack of instruction images in order to remember the structure of an assembly at various stages of deconstruction. To do this, we augment the action space discussed in the previous section with two additional **Push Instruction** and **Pop Instruction** actions. Push takes the current image from the simulator and adds it to the top of the instruction stack, while Pop removes the top image from the instruction stack. During training and inference, we do not restrict the size of the instruction stack.

Our learned policy takes in the current image from the simulator as well as the top image of the instruction stack. During the Break phase when the agent is trying to gather more information about the LEGO assembly, the agent can compare these two images and see if they are similar. If they are, then the agent should disassemble the model further. Otherwise, the agent should take the Push action to store the new information that has just been gathered. After completing this process several times, the agent should have an instruction stack with an image of the fully completed assembly on the bottom, and increasingly disassembled images as you move closer to the top. At the start of the Make phase, the agent will be presented with an empty scene, so the current image from the environment will be empty, while the top of the instruction stack will contain the last brick the model saw during the disassembly process. The agent must then build until the assembly in the current image matches the assembly in the top instruction image. When they match, the agent can take the Pop action and will then reveal a new instruction image with slightly more of the original assembly remaining. In this way, the model only needs to reason about two images at a time, which greatly reduces the complexity of the policy.

3.3 Model

The learned policy is implemented using a modified vision transformer [9] with multiple heads that are responsible for different components of the action space. This model tokenizes the workspace and the instruction images into 16×16 pixel patches. Given that the environment produces images that are 128×128 pixels, this results in 64 tokens per image. The patches from both images are passed through a single linear layer and added to a learned positional encoding. The model then concatenates a single decoder token and a binary embedding of the current phase (Break or Make) for 130 total tokens. The transformer consists of 12 blocks with 512 channels and 8 heads each.

To compute an action, the output of the decoder token is then passed through a set of decoder heads to predict distributions for the action mode such as Disassemble, Assemble, or Rotate and mode-specific parameters such as Rotate Direction when performing a Rotate action or Brick Shape and Color when inserting a new brick. In order to predict 2D cursor click and release locations, the 64 output tokens of transformer blocks 3,6,9 and 12 that correspond to the current image are combined using two separate DPT [30] decoders to produce dense feature maps at the resolution of the original input image. We found it beneficial to condition these click locations on the high level action and parameters sampled from the initial decoder heads. This is accomplished by passing the sampled high level actions to an embedding layer and adding the resulting feature to the output of the decoder token. This value is then used to compute a distribution over click locations using dot product attention over the dense features computed by the DPT decoder. This conditional structure is similar to models used in game AI with complex action spaces [40]. Figure 4 shows our policy model and how these components fit together.

3.4 Online Training

We trained our model using online imitation learning similar to DAgger [31] and show ablations using behavior cloning as well. To do this we built a fast expert that can provide online supervision for trajectories generated by the learning model during training. This is in contrast to [41] which used an expert that was too slow for online labeling. Note that while our new expert is much faster, it is not able to construct plans in cases where the agent makes too many mistakes or deviates too far from the target assembly. In these cases, we simply terminate the training episode early. When there are multiple possible best actions that the expert could suggest, one of them is selected at random. To avoid ambiguity, our expert instructs the agent to push an image to the instruction stack each time a brick is removed during the break phase.

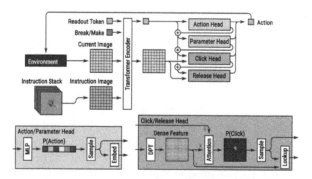

Fig. 4. Architecture of the InstructioNet model. The current image from the environment, and the top image of the instruction stack are tokenized and provided as input to a vision transformer encoder, along with a single readout token and another discrete token that indicates whether the current phase is Break or Make. The readout token's feature decodes a series of discrete action and parameter heads that determine the high level action mode (Rotate/Translate/Pick/Assemble/Disassemble) as well as action parameters such as the rotation angle or translate distance and direction. The cursor click and release locations are sampled from an attention map comparing features from a DPT decoder.

The online training algorithm alternates between generating new data by acting in the environment according to either the expert or the learning model, and then training on a randomized subset of the data generated over the past several iterations. When generating data, a fixed percentage of the environment steps are generated by sampling actions according to the expert, and the rest are sampled by acting according to the agent. We refer to this expert mixture constant as α, and for our main experiments, we found that $\alpha = 0.75$, representing a mix of 75% expert-generated and 25% model-generated data worked well. Note that regardless of which model is controlling the simulator, the expert's actions are always used for supervision. Incorporating trajectories generated by the agent in this way allows the model to learn to recover from its mistakes: when the model takes an inappropriate action it will reach part of the state space that would not have been encountered if acting according to the expert, yet seeing the expert's advice in these states shows the model how to correctly recover from this behavior.

As noted by Czarnecki et al. [7], this method of using the model to generate data with direct supervision from the expert can be unstable. This instability occurs frequently when the data distribution shifts as the model gets better at some parts of the state space and learns to take different actions. This problem is exacerbated in settings with long trajectories where many nearby frames appear similar to each other. If proper care is not taken, the model can quickly overfit to the data it has most recently seen and catastrophically forget prior examples. The original DAgger algorithm addresses this by periodically retraining the entire model on all data collected so far. Unfortunately, this is somewhat impractical for long training runs with millions of steps. To mitigate this issue we maintain a replay buffer and train on data randomly sampled from this experience. See the supplementary material for the pseudo-code we use to train our model.

3.5 Cursor Losses

We experimented with several approaches to computing losses for the cursor click and release locations. During online training, we would like the click-and-release decoders to each produce a distribution over locations that can be sampled in order to generate a variety of training data. To compute the probability of clicking on a particular pixel $p(i, j)$, the dense raw value $x_{i,j}$ predicted at that pixel location is normalized using a standard softmax:

$$p(i,j) = \frac{\exp(x_{i,j})}{\sum_{i',j'} \exp(x_{i',j'})}$$

When the expert's action suggests clicking on a particular LEGO connection point in the scene, there are usually multiple "acceptable" pixels that correspond to the same connection point which complicates the choice of loss function.

One option is to use binary cross-entropy loss using the mask of acceptable pixel locations as a target. This assumes a different probabilistic interpretation of the output, one where multiple pixels can be chosen at the same time instead of just one, but still may be a useful way to encourage the model to put a high probability on the acceptable pixels. This loss function encourages the model to increase the probability of all acceptable pixels, without considering the cross-pixel relationships.

Another option commonly used in keypoint detection is to construct target heatmaps using a small Gaussian distribution around correct locations and supervise the output values using a mean-squared-error loss [1]. We opted against this as it does not lend itself well to softmax sampling, and may add probability mass outside the desired pixel boundaries. However, we can use the mean-squared-error loss to simply push all acceptable pixels toward a large positive constant, and all unacceptable pixels towards a large negative constant. Here we used a constant such that if only one pixel in the image assumed the positive constant, and all others assumed the negative constant, the probability of selecting the single pixel in the softmax would be 0.999. For 128 × 128 pixel images, these constants are ±8.3.

Finally, we consider a loss function in which the probability of acceptable pixels is summed and the probability of unacceptable pixels is summed forming a new two-way distribution. We then supervise this new distribution to maximize the probability of choosing an acceptable pixel using cross-entropy.

$$L = -\log \sum_{i,j} y_{i,j} \frac{\exp x_{i,j}}{\sum_{i',j'} \exp x_{i',j'}}$$

This allows the model to place probability mass on any of the acceptable pixels while decreasing the probability mass of all unacceptable pixels. We find that this loss function outperforms the others and discuss this in Sect. 4.4.

4 Experiments

4.1 Evaluation

We use the metrics from [41] to evaluate the quality of our learned agents. The first $(F1_b)$ is an $F1$ score over the brick shape and color, ignoring pose. The second $(F1_a)$ is over the brick shape, color, and pose after computing the best single rigid transformation to bring the estimated and ground-truth assemblies into alignment with each other. Assembly Edit Distance (AED) measures how many rigid transforms (edits) are required to bring all the bricks in the estimated assembly into alignment with the bricks in the ground-truth assembly, with additional penalties for extra and missing bricks. The final $(F1_e)$ is an $F1$ score over the edges (connections) between bricks after using the alignment computed in AED to construct a mapping between bricks in the estimated assembly and bricks in the ground-truth assembly.

4.2 Break and Make

To evaluate the effectiveness of the InstructioNet model, we trained it on our modified version of the Break and Make task using the randomly constructed assemblies and vehicles discussed in Sect. 3.1. Table 1 shows the performance of our model on these datasets compared to the reported numbers of the LSTM, Studnet-A, and Studnet-B baselines from [41]. The Studnet models are causal transformers that take in a series of deduplicated tiles from the beginning of an episode until the current time step and use a series of decoders to make predictions for the current action. All three of these models require a long horizon to make decisions, while InstructioNet requires only the current frame and the top of the instruction stack. See [41] for details of the Studnet and LSTM models. Note that due to the updated observation space and the small changes to the random construction data due to the improved collision checker, these models should not be considered to have been trained in the same environment, and so the comparisons are only approximate. Regardless of these differences, the InstructioNet model is able to reconstruct large models with much greater accuracy than was previously possible, and the large performance gap clearly demonstrates a new level of capability. Note that we were not able to train the LSTM and Studnet methods from [41] on the new RC-Vehicles data as they require the entire history of past frames to make each decision. The RC-Vehicles assemblies can take over 150 steps to correctly disassemble and reassemble, which are much larger sequences than we could effectively train on available hardware.

Table 1. InstructioNet compared against the LSTM and Studnet baselines from [41]. See Sect. 4.1 for details on these metrics, and 4.2 for an important note on the direct comparability of these methods.

RC-2*	$F1_b \uparrow$	$F1_e \uparrow$	$F1_a \uparrow$	$AED \downarrow$
InstructioNet	**0.98**	**0.95**	**0.93**	**0.18**
LSTM	0.61	0.38	0.43	2.16
Studnet-A	0.90	0.86	0.58	1.11
Studnet-B	0.87	0.77	0.57	1.30
RC-4*	$F1_b \uparrow$	$F1_e \uparrow$	$F1_a \uparrow$	$AED \downarrow$
InstructioNet	**0.80**	**0.69**	**0.71**	**2.39**
LSTM	0.41	0.09	0.13	7.25
Studnet-A	0.56	0.29	0.24	5.80
Studnet-B	0.64	0.34	0.25	5.48
RC-8*	$F1_b \uparrow$	$F1_e \uparrow$	$F1_a \uparrow$	$AED \downarrow$
InstructioNet	**0.68**	**0.62**	**0.63**	**6.30**
LSTM	0.02	0.00	0.02	16.05
Studnet-A	0.02	0.01	0.01	15.87
Studnet-B	0.38	0.14	0.12	13.90
RC-Vehicles	$F1_b \uparrow$	$F1_e \uparrow$	$F1_a \uparrow$	$AED \downarrow$
InstructioNet	**0.59**	**0.51**	**0.53**	**43.36**

4.3 Online Training

In order to show the effectiveness of online training using sequences of actions and observations generated by the learning model, we also train a model on sequences generated only by the expert on the RC-2 and RC-4 datasets. To do this, we set the expert mixture (α) to 1.0, which is equivalent to behavior cloning. The results are shown in the **Online Training** section of Table 2. While these models underperform relative to the default expert mixture ($\alpha = 0.75$) that includes training from an online expert, it still shows that significant progress can be made with offline training on this problem.

Table 2. The results of various ablations, see the sections listed above for details.

Online Training (Sect. 4.3)				
RC-2	$F1_b \uparrow$	$F1_e \uparrow$	$F1_a \uparrow$	$AED \downarrow$
$\alpha = 0.75$	**0.98**	**0.95**	**0.93**	**0.18**
$\alpha = 1.0$	0.97	0.93	0.90	0.29
RC-4	$F1_b \uparrow$	$F1_e \uparrow$	$F1_a \uparrow$	$AED \downarrow$
$\alpha = 0.75$	**0.80**	**0.69**	**0.71**	**2.39**
$\alpha = 1.0$	0.77	0.68	0.66	2.88
Loss Functions (Sect. 4.4)				
RC-2	$F1_b \uparrow$	$F1_e \uparrow$	$F1_a \uparrow$	$AED \downarrow$
Summed CE	**0.98**	**0.95**	**0.93**	**0.18**
BCE	0.91	0.72	0.72	0.92
MSE	0.91	0.50	0.66	1.00
Conditional Actions (Sect. 4.5)				
RC-2	$F1_b \uparrow$	$F1_e \uparrow$	$F1_a \uparrow$	$AED \downarrow$
1.7M	**0.98**	0.94	0.92	0.22
2.6M	**0.98**	**0.95**	**0.93**	**0.18**
Cut 1.7M	0.97	0.89	0.84	0.43
Cut 2.6M	**0.98**	0.00	0.50	1.09
Selective Modification (Sect. 4.6)				
RC-2	$F1_b \uparrow$	$F1_e \uparrow$	$F1_a \uparrow$	$AED \downarrow$
Original	0.98	0.95	0.93	0.18
Altered Color	0.98	0.93	0.95	0.21
RC-4	$F1_b \uparrow$	$F1_e \uparrow$	$F1_a \uparrow$	$AED \downarrow$
Original	0.80	0.69	0.71	2.39
Altered Color	0.78	0.67	0.68	2.63

4.4 Loss Functions

We evaluate the effectiveness of our cursor loss function by comparing it against the binary cross entropy and constant regression methods discussed in Sect. 3.5. We find that even on the relatively easy two-brick models, the summed-probability loss outperforms these other techniques. The results are shown in the **Loss Functions** section of Table 2. Note that due to the difference in magnitude between these losses, we adjusted the learning rate for these methods in an attempt to achieve the best results possible. We found that both benefited from a higher learning rate of 5×10^{-4} rather than the default 5×10^{-5} used for the summed-probability loss.

4.5 Conditional Action Generation

We also test the importance of sequentially conditioning the action heads on one another as discussed in Sect. 3.3 by training a new model where these conditional connections are cut. This corresponds to cutting the magenta connections coming out of the Action Head, Parameter Head and Click Head in Fig. 4. We found that cutting these connections leads to training instability where after a certain point the model loses its ability to effectively use the cursor to connect bricks together. In light of this, we report results after 1.7M frames, right before the instability occurs in addition to the default 2.6M frames after the instability occurs. We also report an evaluation of the default model at 1.7M frames for comparison. Note that even before training became unstable, the model without the conditional connections was significantly underperforming the default model. The **Conditional Actions** section of Table 2 shows these results.

4.6 Selective Modification

We also tested our model on a new task that requires the agent to rebuild the model with one of the brick colors altered. In this setting, the model receives two additional tokens, one which specifies the color to change, and the other that specifies a new color. The agent's objective is to rebuild the model with all bricks using the original color instead built with the new color. The **Selective Modification** section of Table 2 shows the performance on this task for two and four brick models. While, performance degrades slightly on this problem, the model still performs quite well. This demonstrates the model's ability to reason not only about reproducing the exact same model seen during the break phase, but also incorporating new instructions when rebuilding.

4.7 Hyperparameters

Unless otherwise mentioned, we used AdamW with a learning rate of 5×10^{-5}, $\beta_1 = 0.9$, $\beta_2 = 0.95$ and weight decay of 0.1. For RC-8 and RC-V, the learning rate was cut to 1×10^{-5} after 15.8M frames. All models were trained on a single Nvidia 4090 or A40 graphics card. The larger RC-8 and RC-Vehicles runs were trained on 19.7M training steps which took five consecutive days per run. The RC-2 and RC-4 datasets were trained on 2.6M and 7.9M steps respectively which took between one and three consecutive days per run. Table 3 shows the training hyperparameters used to train the models for each dataset.

Table 3. Hyperparameters used to train the different datasets.

Hyperparameters	RC-2	RC-4	RC-8	RC-V
Total Training Steps	2.6M	7.9M	19.7M	19.7M
New Data Steps Per Epoch	8K	8K	8K	32K
Training Steps Per Epoch	16K	16K	16K	65K
Replay Buffer Size	32K	32K	32K	131K
Expert Data Mixture (α)	0.75	0.75	0.75	0.75

4.8 Qualitative Evaluation

Figure 5 shows ten representative failure and success cases of our model on the RC-Vehicles dataset sorted by their F1a score. Example A shows a case where the model fails to complete the Break phase due to a small ornament on top of the car that the model does not realize it needs to remove. In examples B, C, and D the model successfully completes the Break phase, but then struggles to complete the early part of the model. In example E, the model initially placed the wings correctly, but then misclicked as it was placing a later piece and inadvertently moved one of the wings to the wrong location and it was not able to recover. In example F, the model incorrectly built the front grille and fails to either undo its mistakes or move on. In example G, the model misplaced one brick in the back of the car and was also not able to correct its mistake. Examples H and I show cases where the model almost completely reconstructs the assembly, but gets hung up on the small ornamental details on the roof. Finally, example J shows an almost perfect reconstruction where one hidden brick is incorrect.

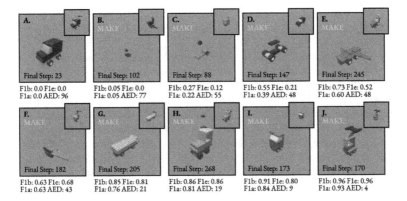

Fig. 5. Examples of InstructioNet reconstructions trained on the RC-Vehicles dataset. The top right overlay shows the target assembly. These examples were chosen to present a diverse array of failure and success cases. See Sect. 4.8 for descriptions of these failures and Sect. 4.1 for an explanation of the evaluation metrics.

5 Conclusion

We have demonstrated substantially improved performance over previous baselines on the Break and Make problem, using a model with explicit instruction memory. The failure modes of this approach suggest that performance could be improved by working on solutions that avoid getting stuck, and that find ways to push forward even if it means making a local mistake. While our approach is successful, it requires an online expert which can provide explicit instructions not only for the inspection and reconstruction process but also for the process of storing and retrieving memory. This limits the utility of this method in real-world settings, where an online expert may not be available. The InstructioNet approach may not be appropriate for problems that do not follow our assumption that assembly can be completed by approximately reversing the disassembly process. Nevertheless the advantage of this method over prior approaches which considered the entire observation history points to the effectiveness of considering only a portion of memory at a time when making decisions.

References

1. Bulat, A., Tzimiropoulos, G.: Human pose estimation via convolutional part heatmap regression. In: Leibe, B., Matas, J., Sebe, N., Welling, M. (eds.) ECCV 2016. LNCS, vol. 9911, pp. 717–732. Springer, Cham (2016). https://doi.org/10.1007/978-3-319-46478-7_44
2. Chang, M., et al.: Goat: go to any thing. arXiv preprint arXiv:2311.06430 (2023)
3. Chen, Q., Memmel, M., Fang, A., Walsman, A., Fox, D., Gupta, A.: Urdformer: constructing interactive realistic scenes from real images via simulation and generative modeling. In: Towards Generalist Robots: Learning Paradigms for Scalable Skill Acquisition@ CoRL2023 (2023)

4. Cho, K., Van Merriënboer, B., Bahdanau, D., Bengio, Y.: On the properties of neural machine translation: encoder-decoder approaches. arXiv preprint arXiv:1409.1259 (2014)
5. Choset, H., Nagatani, K.: Topological simultaneous localization and mapping (slam): toward exact localization without explicit localization. IEEE Trans. Robot. Autom. **17**(2), 125–137 (2001)
6. Chung, H., et al.: Brick-by-brick: combinatorial construction with deep reinforcement learning. Adv. Neural. Inf. Process. Syst. **34**, 5745–5757 (2021)
7. Czarnecki, W.M., Pascanu, R., Osindero, S., Jayakumar, S., Swirszcz, G., Jaderberg, M.: Distilling policy distillation. In: The 22nd International Conference on Artificial Intelligence and Statistics, pp. 1331–1340. PMLR (2019)
8. Deitke, M., et al.: Robothor: an open simulation-to-real embodied ai platform. In: Proceedings of the IEEE/CVF Conference on Computer Vision and Pattern Recognition, pp. 3164–3174 (2020)
9. Dosovitskiy, A., et al.: An image is worth 16×16 words: transformers for image recognition at scale. arXiv preprint arXiv:2010.11929 (2020)
10. Du, T., et al.: Inversecsg: automatic conversion of 3d models to csg trees. ACM Trans. Graph. (TOG) **37**(6), 1–16 (2018)
11. Fan, L., et al.: Minedojo: building open-ended embodied agents with internet-scale knowledge. In: Thirty-sixth Conference on Neural Information Processing Systems Datasets and Benchmarks Track (2022). https://openreview.net/forum?id=rc8o_j8I8PX
12. Gordon, D., Fox, D., Farhadi, A.: What should i do now? marrying reinforcement learning and symbolic planning. arXiv preprint arXiv:1901.01492 (2019)
13. Graves, A., Wayne, G., Danihelka, I.: Neural turing machines. arXiv preprint arXiv:1410.5401 (2014)
14. Gupta, A., Fox, D., Curless, B., Cohen, M.: Duplotrack: a real-time system for authoring and guiding duplo block assembly. In: Proceedings of the 25th Annual ACM Symposium on User Interface Software and Technology, pp. 389–402 (2012)
15. Henry, P., Krainin, M., Herbst, E., Ren, X., Fox, D.: Rgb-d mapping: using kinect-style depth cameras for dense 3d modeling of indoor environments. Int. J. Rob. Res. **31**(5), 647–663 (2012)
16. Hochreiter, S., Schmidhuber, J.: Long short-term memory. Neural Comput. **9**(8), 1735–1780 (1997)
17. Jones, B., Hildreth, D., Chen, D., Baran, I., Kim, V.G., Schulz, A.: Automate: a dataset and learning approach for automatic mating of cad assemblies. ACM Trans. Graph. (TOG) **40**(6), 1–18 (2021)
18. Jordan, M.I.: Serial order: a parallel distributed processing approach. In: Advances in Psychology, vol. 121, pp. 471–495. Elsevier (1997)
19. Kim, J.W., Kang, K.K., Lee, J.H.: Survey on automated lego assembly construction (2014)
20. Lee, S., Kim, J., Kim, J.W., Moon, B.R.: Finding an optimal lego® brick layout of voxelized 3d object using a genetic algorithm. In: Proceedings of the 2015 Annual Conference on Genetic and Evolutionary Computation, pp. 1215–1222 (2015)
21. Lee, Y., Hu, E.S., Lim, J.J.: Ikea furniture assembly environment for long-horizon complex manipulation tasks. In: 2021 IEEE International Conference on Robotics and Automation (ICRA), pp. 6343–6349. IEEE (2021)
22. Lennon, K., et al.: Image2lego: customized lego set generation from images. arXiv preprint arXiv:2108.08477 (2021)
23. Leonard, J.J., Durrant-Whyte, H.F.: Simultaneous map building and localization for an autonomous mobile robot. In: IROS, vol. 3, pp. 1442–1447 (1991)

24. Li, C., Pan, H., Bousseau, A., Mitra, N.J.: Sketch2cad: sequential cad modeling by sketching in context. ACM Trans. Graph. (TOG) **39**(6), 1–14 (2020)
25. Li, K., Bian, J.W., Castle, R., Torr, P.H., Prisacariu, V.A.: Mobilebrick: building lego for 3d reconstruction on mobile devices. In: Proceedings of the IEEE/CVF Conference on Computer Vision and Pattern Recognition, pp. 4892–4901 (2023)
26. Lim, J.J., Pirsiavash, H., Torralba, A.: Parsing ikea objects: fine pose estimation. In: Proceedings of the IEEE International Conference on Computer Vision, pp. 2992–2999 (2013)
27. Mo, K., et al.: Structurenet: hierarchical graph networks for 3d shape generation. arXiv preprint arXiv:1908.00575 (2019)
28. Niu, C., Li, J., Xu, K.: Im2struct: recovering 3d shape structure from a single rgb image. In: Proceedings of the IEEE Conference on Computer Vision and Pattern Recognition, pp. 4521–4529 (2018)
29. Peysakhov, M., Regli, W.C.: Using assembly representations to enable evolutionary design of lego structures. Ai Edam **17**(2), 155–168 (2003)
30. Ranftl, R., Bochkovskiy, A., Koltun, V.: Vision transformers for dense prediction. In: Proceedings of the IEEE/CVF International Conference on Computer Vision, pp. 12179–12188 (2021)
31. Ross, S., Gordon, G., Bagnell, D.: A reduction of imitation learning and structured prediction to no-regret online learning. In: Proceedings of the Fourteenth International Conference on Artificial Intelligence and Statistics, pp. 627–635. JMLR Workshop and Conference Proceedings (2011)
32. Rumelhart, D.E., Hinton, G.E., Williams, R.J., et al.: Learning internal representations by error propagation (1985)
33. Savva, M., et al.: Habitat: a platform for embodied ai research. In: Proceedings of the IEEE/CVF International Conference on Computer Vision, pp. 9339–9347 (2019)
34. Shen, B., et al.: igibson, a simulation environment for interactive tasks in large realisticscenes. arXiv preprint arXiv:2012.02924 (2020)
35. Suárez-Ruiz, F., Zhou, X., Pham, Q.C.: Can robots assemble an ikea chair? Sci. Rob. **3**(17), eaat6385 (2018)
36. Szot, A., et al.: Habitat 2.0: training home assistants to rearrange their habitat. Adv. Neural Inf. Process. Syst. **34**, 251–266 (2021)
37. Thompson, R., Ghalebi, E., DeVries, T., Taylor, G.W.: Building lego using deep generative models of graphs. arXiv preprint arXiv:2012.11543 (2020)
38. Tian, Y., et al.: Assemble them all: physics-based planning for generalizable assembly by disassembly. ACM Trans. Graph. **41**(6) (2022)
39. Vaswani, A., et al.: Attention is all you need. Adv. Neural Inf. Process. Syst. **30** (2017)
40. Vinyals, O., et al.: Grandmaster level in starcraft ii using multi-agent reinforcement learning. Nature **575**(7782), 350–354 (2019)
41. Walsman, A., Zhang, M., Kotar, K., Desingh, K., Farhadi, A., Fox, D.: Break and make: interactive structural understanding using lego bricks. In: European Conference on Computer Vision, pp. 90–107. Springer, Heidelberg (2022). https://doi.org/10.1007/978-3-031-19815-1_6
42. Wang, R., Zhang, Y., Mao, J., Cheng, C.Y., Wu, J.: Translating a visual lego manual to a machine-executable plan. In: European Conference on Computer Vision, pp. 677–694. Springer, Heidelberg (2022). https://doi.org/10.1007/978-3-031-19836-6_38

43. Wani, S., Patel, S., Jain, U., Chang, A.X., Savva, M.: Multi-on: benchmarking semantic map memory using multi-object navigation. In: Neural Information Processing Systems (NeurIPS) (2020)
44. Willis, K.D., et al.: Fusion 360 gallery: a dataset and environment for programmatic cad construction from human design sequences. ACM Trans. Graph. (TOG) **40**(4), 1–24 (2021)
45. Xu, X., Peng, W., Cheng, C.Y., Willis, K.D., Ritchie, D.: Inferring cad modeling sequences using zone graphs. In: Proceedings of the IEEE/CVF Conference on Computer Vision and Pattern Recognition, pp. 6062–6070 (2021)
46. Yan, C., Misra, D., Bennnett, A., Walsman, A., Bisk, Y., Artzi, Y.: Chalet: cornell house agent learning environment. arXiv preprint arXiv:1801.07357 (2018)
47. Zakka, K., Zeng, A., Lee, J., Song, S.: Form2fit: learning shape priors for generalizable assembly from disassembly. In: 2020 IEEE International Conference on Robotics and Automation (ICRA), pp. 9404–9410. IEEE (2020)
48. Zhan, G., et al.: Generative 3d part assembly via dynamic graph learning. Adv. Neural. Inf. Process. Syst. **33**, 6315–6326 (2020)
49. Zhang, J., Cherian, A., Liu, Y., Ben-Shabat, Y., Rodriguez, C., Gould, S.: Aligning step-by-step instructional diagrams to video demonstrations. In: Proceedings of the IEEE/CVF Conference on Computer Vision and Pattern Recognition, pp. 2483–2492 (2023)

Exploring Active Learning in Meta-learning: Enhancing Context Set Labeling

Wonho Bae[1]($\boxtimes$), Jing Wang[1], and Danica J. Sutherland[1,2]

[1] University of British Columbia, Vancouver, Canada
{whbae,dsuth}@cs.ubc.ca, jing@ece.ubc.ca
[2] Alberta Machine Intelligence Institute (Amii), Edmonton, Canada

Abstract. Most meta-learning methods assume that the (very small) context set used to establish a new task at test time is passively provided. In some settings, however, it is feasible to actively select which points to label; the potential gain from a careful choice is substantial, but the setting requires major differences from typical active learning setups. We clarify the ways in which active meta-learning can be used to label a context set, depending on which parts of the meta-learning process use active learning. Within this framework, we propose a natural algorithm based on fitting Gaussian mixtures for selecting which points to label; though simple, the algorithm also has theoretical motivation. The proposed algorithm outperforms state-of-the-art active learning methods when used with various meta-learning algorithms across several benchmark datasets.

Keywords: Meta learning · Active learning · Low budget

1 Introduction

Meta-learning has gained significant prominence as a substitute for traditional "plain" supervised learning tasks, with the aim to adapt or generalize to new tasks given extremely limited data. (Hospedales *et al.* [30] give a recent survey.) There has been enormous success compared to learning "from scratch" on each new problem, but could we do even better, with even less data?

One major way to improve data-efficiency in standard supervised learning settings is to move to an *active* learning paradigm, where typically a model can request a small number of labels from a pool of unlabeled data; these are collected, used to further train the model, and the process is repeated. (Settles *et al.* [59] provides a classic overview, and Ren *et al.* [54] a more recent survey.)

Supplementary Information The online version contains supplementary material available at https://doi.org/10.1007/978-3-031-73024-5_17.

Although each of these lines of research are well-developed, their combination – *active meta-learning* – has seen comparatively little research attention. However, this intersection is not only theoretically appealing but also has numerous practical applications. For example, in medical imaging, there is often a large repository of labeled images such as X-rays or MRIs. However, labeling these images requires manual annotations by radiologists or pathologists, which is time-consuming and costly. In manufacturing, there is often a large amount of sensor data generated during the production process. Annotating them to identify patterns indicating quality issues requires domain experts.

How can a meta-learner leverage an active learning setup to learn the best model possible, using only a few labels in its context sets? We are aware of three previous attempts at active selection of context sets in meta-learning: Müller et al. [43] and Al-Shedivat et al. [3] do so at meta-*training* time for text and image classification, while Boney et al. [9] do it at meta-*test* time in semi-supervised few-shot image classification with ProtoNet [61]. "Active meta-learning" thus means very different things in their procedures; these approaches are also entirely different from work on active selection of *tasks* during meta-training as in [32, 37, 47]. Our first contribution is therefore to clarify the different ways in which active learning can be applied to meta-learning, for differing purposes.[1]

We then confirm in extensive experiments that no active learning method for context set selection seems to significantly help with final predictor quality at meta-training time – aligning with previous observations by Setlur et al. [58] and Ni et al. [46] – but that active learning *can* substantially help at meta-test time. In particular, we propose a natural algorithm based on fitting a Gaussian mixture model to the unlabeled data, using meta-learned feature representations; though the approach is simple, we also give theoretical motivation. We show that our proposed selection algorithm works reliably, and often substantially outperforms competitor methods across many different meta-learning and few-shot learning tasks, across a variety of benchmark datasets and meta-learning algorithms. Our contributions are summarized as follows.

- We explore the concept of "active meta-learning," pointing out that active selection can occur in several places (Sect. 2) and highlighting challenges in traditional meta-learning setups regarding sample stratification (Sect. 2.1).
- We identify that existing active learning algorithms, particularly low-budget active learning methods, perform poorly in active meta-learning (Sect. 4), even if meta-learning is in the low-budget regime.
- We propose a method based on Gaussian mixture model using meta-learning specific features (Sect. 3.2), proven to yield a Bayes-optimal classifier under certain assumptions (or an efficient set cover in a general setting in Sect. 3.3).
- Our experiments show that the simple Gaussian mixture method consistently outperforms more complex active learning methods across various few-shot image classification (Sects. 4.1 and 4.2), cross-domain classification (Sect. 4.3),

---

[1] Note that work on meta-learning an active selection criterion for higher-label-budget problems – *e.g.* [16,36] – is essentially unrelated.

and meta-learning regression tasks (Sect. 4.4), irrespective of the type of meta-learning algorithms employed.

2 Meta-learning: Background and Where to Be "Active"

We aim to learn a learning algorithm f_θ, a function which, given a dataset $\mathcal{C}$ consisting of pairs $(x, y) \in \mathcal{X} \times \mathcal{Y}$, returns $g := f_\theta(\mathcal{C})$. The function $g : \mathcal{X} \to \hat{\mathcal{Y}}$ is a classifier, regressor, or so on. We evaluate the quality of g using a loss function $\ell : \hat{\mathcal{Y}} \times \mathcal{Y} \to \mathbb{R}$, e.g. the cross-entropy or square loss:

$$\text{Empirical risk of } g \text{ on } \mathcal{T} \colon \mathcal{R}_\ell(g, \mathcal{T}) = \frac{1}{|\mathcal{T}|} \sum_{(x,y) \in \mathcal{T}} \ell\left(g(x), y\right).$$

To find the θ which gives the best gs, we assume we have access to distributions $\mathcal{P}^{train}, \mathcal{P}^{eval}$ over tasks $\mathcal{D} \subseteq \mathcal{X} \times \mathcal{Y}$. For each task, we will run f_θ on a *context set* $\mathcal{C}$, then evaluate the quality of the learned predictor on a disjoint *target set* $\mathcal{T}$. We call the distribution over possible $(\mathcal{C}, \mathcal{T})$ pairs $\text{Pick}_\theta(\mathcal{D})$.[2] For instance, the default choice in passive meta-learning chooses, say, five random points per class for $\mathcal{C}$ and assigns the rest to $\mathcal{T}$, ignoring θ and x. Our aim is then,

Meta-training: find $\hat{\theta}$ using

$$\hat{\theta} \approx \arg\min_\theta \mathbb{E}_{\mathcal{D} \sim \mathcal{P}^{train}} \left[\mathbb{E}_{(\mathcal{C}, \mathcal{T}) \sim \text{Pick}_\theta^{train}(\mathcal{D})} \left[\mathcal{R}_{\ell^{train}}\left(f_\theta(\mathcal{C}), \mathcal{T}\right) \right] \right]. \quad (1)$$

Many algorithms have been proposed for meta-training (overviewed in Sect. 2.2).

To compare models based on $\mathcal{P}^{eval}$, we might evaluate with a different loss. For instance, it would be typical to use the 0-1 loss (corresponding to accuracy) for classification problems, despite training with cross-entropy.

$$\text{Meta-testing: eval. } f_{\hat\theta} \text{ using } \mathbb{E}_{\widetilde{\mathcal{D}} \sim \mathcal{P}^{eval}} \left[\mathbb{E}_{(\widetilde{\mathcal{C}}, \widetilde{\mathcal{T}}) \sim \text{Pick}_{\hat\theta}^{eval}(\widetilde{\mathcal{D}})} \left[\mathcal{R}_{\ell^{eval}}\left(f_{\hat\theta}(\widetilde{\mathcal{C}}), \widetilde{\mathcal{T}}\right) \right] \right]. \quad (2)$$

Finally, in practice, we might want to use a different selection scheme at deployment time. For instance, in passive meta-learning, one would typically use all available labeled data for context, not a random subset. Given a task $\check{\mathcal{D}}$,

$$\text{Deployment: find a context set via } (\check{\mathcal{C}}, \_) \sim \text{Pick}_{\hat\theta}^{deploy}(\check{\mathcal{D}}) \text{ and use } f_\theta(\check{\mathcal{C}}). \quad (3)$$

2.1 Active Selection of Context in Meta Learning

There are several places where active learning can be applied during meta-learning. In the meta-training phase (1), we could actively choose tasks $\mathcal{D}$, and/or have $\text{Pick}_\theta^{train}$ actively select points for $\mathcal{C}$ and/or $\mathcal{T}$. At meta-testing time (2),

---

[2] If we pick points by some deterministic process, $\text{Pick}_\theta(\mathcal{D})$ is a point mass.

we could have $\text{Pick}_\theta^{eval}$ actively select points for $\widetilde{\mathcal{C}}$ and/or $\widetilde{\mathcal{T}}$; we might also actively choose $\widetilde{\mathcal{D}}$ to use labels efficiently, similarly to active surveying [22]. At deployment time (3), $\text{Pick}_\theta^{deploy}$ might actively choose a context set $\check{\mathcal{C}}$ to label.

Actively selecting $\mathcal{D}$, $\widetilde{\mathcal{D}}$, $\mathcal{T}$, and/or $\widetilde{\mathcal{T}}$ is interesting to minimize the label burden (or, possibly computational cost) of meta-training [32,37,47]. We assume here, however, that $\mathcal{P}^{train}$ and $\mathcal{P}^{eval}$ are based on already-labeled datasets.

Instead, we are primarily concerned with the labeling burden at deployment time, and so would like to actively select $\check{\mathcal{C}}$ with $\text{Pick}_\theta^{deploy}$ to find the best predictor. To evaluate how well we should expect our algorithms to perform at this task, we choose $\text{Pick}_\theta^{eval} = \text{Pick}_\theta^{deploy}$; thus, we actively select $\widetilde{\mathcal{C}}$.

Should we expect this to help? Efficient approaches for data selection in meta-learning have not yet received much research attention. Setlur *et al.* [58] suggest that context set diversity is empirically not particularly helpful for meta-learning, and Ni *et al.* [46] show that data augmentation on context sets is not very useful either. Pezeshkpour *et al.* [50] further provide some evidence using label information that there is not much room to improve few-shot classification with active learning. Agarwal *et al.* [1], however, argue against previous findings, showing that adversarially selected context sets, at both training and test time, significantly change the classification performance. Their approach is not applicable in practice since it requires full label information, but may suggest there is room to improve meta-learning algorithms with better context sets.

Muller *et al.* [43] and Al-Shedivat *et al.* [3] compare traditional active learning algorithms for few-shot text and image classification at training time, i.e. active $\text{Pick}_\theta^{train}$, passive $\text{Pick}_\theta^{eval}$. Boney *et al.* [9] instead compare active learning algorithms for semi-supervised few-shot image classification (more discussion for the relationship with semi-supervised few-shot learning is provided in Appendix C inside $\text{Pick}_\theta^{eval}$, specifically when f_θ is a ProtoNet, with passive $\text{Pick}_\theta^{train}$. Both are feasible settings, but as argued above if we are concerned with performance of our deployed predictor we should use an active $\text{Pick}_\theta^{eval} = \text{Pick}_\theta^{deploy}$. One can choose $\text{Pick}_\theta^{train}$ to be active or not, depending on which learns better predictors; we show in Appendix N that active $\text{Pick}_\theta^{train}$ does not seem to help.

Stratification. In passive few-shot classification, the Pick functions typically choose context points according to a *stratified* sample: for one-shot classification, $\mathcal{C}$ contains exactly one point per class. This is because, if we take a uniform random sample of size N for an N-way classification problem, $\mathcal{C}$ is unlikely to contain all the classes, making classification very difficult. Assuming "nature" gives a stratified uniform sample, as in nearly all work on few-shot classification, also seems reasonable.

In pool-based active settings, however, it is highly unreasonable to assume that $\text{Pick}_\theta^{deploy}$ can be stratified (as illustrated on the left side of Fig. 1): to do so, we would need to know the label of every point in $\widetilde{\mathcal{D}}$, in which case we should simply use all those labels. As we would like $\text{Pick}_\theta^{eval} = \text{Pick}_\theta^{deploy}$, eval-time stratification is then not particularly reasonable; even so, we do report such results per the standards of meta-learning. When $\text{Pick}_\theta^{deploy}$ is unstratified (as

Fig. 1. Meta-training process. Pick$_\theta$ can be stratified or unstratified, active or passive.

in the right side of Fig. 1), it is particularly important for the selection criterion to find samples from each class.

Train-time stratification with unstratified evaluation does not leak data labels, and is plausible when $\mathcal{P}^{train}$ and $\mathcal{P}^{eval}$ are fully labeled. Since this approach trains f_θ in an "easy" setting and evaluates it in a "hard" one, however, we will see it tends to slightly underperform the fully-unstratified default. Regression tasks are not typically stratified; we do not stratify for regression experiments.

2.2 Related Work: Meta-learning Algorithms

Meta-learning algorithms can be divided into several categories; all will be applicable for our active learning strategies, and we evaluate with at least one representative algorithm per category.

Metric-based methods learn a representation space encoding a "good" similarity, where simple classifiers work well [49,64]. ProtoNet [61] finds features so that points from each class are close to the prototype feature of the class.

Optimization-based methods use f_θ that incorporate optimization, e.g. gradient descent as in MAML [4,17], which seeks parameters θ such that gradient descent quickly finds a useful model on a new task. ANIL [51] freezes most of the network and only updates the last layer, while R2D2 [7] and MetaOptNet [40] replace the last layer with a convex problem whose solution can be differentiated; these approaches can improve both performance and speed.

Model-based methods learn a model that explicitly adapts to new tasks, typically by modeling the distribution of y from $\mathcal{T}$ given its x values and $\mathcal{C}$. The most prominent family of methods is Neural Processes (NPs) [15,21], which encode a context set and estimate task-specific distribution parameters. Conditional NPs can have issues with underfitting [20,21], but AttentiveNPs [35] and ConvNPs [25] are more robust. They are more commonly used for regression.

Pre-training methods, such as SimpleShot [68] and Baseline++ [12], are based on repeated demonstrations [24,71] that simply pre-training a multi-class model can surpass the performance of commonly used meta-learners.

3 Active Learning

In pool-based active learning, a model requests labels for the most "informative" data points from a pool of unlabeled data. The key question is how to estimate which data points will be informative.

3.1 Related Work: Existing Active Learning Methods

Uncertainty-Based Methods. Simple but effective uncertainty-based methods such as maximum entropy [67], least confident [59], and margin sampling [56] are widely used for active learning. Since they only consider current models' uncertainty, active learning strategies that consider expected changes in model parameters [6,60] and model outputs [18,26,33,34,42,55,63,73] have been also been proposed. However, recent analyses have empirically demonstrated that at least in certain experimental settings, most active learning methods are not significantly different from one another [38], and may not even improve over random selection [44]. We consider the following methods from this category:

Random Uniformly randomly samples a context set from unlabeled set $\mathcal{U}$.
Entropy Add a point to the context set based on $x^* = \arg\max_{x \in \mathcal{U}} H(\hat{y}(x) \mid x)$, where $H(\cdot)$ is Shannon entropy [67]. Other than in Appendix K, we apply this in "batch mode," i.e. we do not observe points one-by-one but rather choose the $|\mathcal{C}|$ points with the highest "initial" entropy.[3]
Margin Add a point to $\mathcal{C}$ based on $x^* = \arg\min_{x \in \mathcal{U}} p_1(y|x) - p_2(y|x)$, where p_1 and p_2 denote the first and second highest predicted probabilities, respectively [56]. We also run this method in "batch mode."

Although Entropy and Margin are very simple and fast to evaluate, no uncertainty-based method seems to substantially outperform them on typical active image classification tasks *e.g.* Mohamadi et al. [42], and we will see that other methods are unlikely to be competitive in low-budget regimes.

Low-Budget Active Learning. The limitations of typical active learning approaches may especially apply in very-low-budget cases, such as those considered in few-shot classification and meta-learning. In particular, when the "current" model is quite bad, using it to choose points might be counterproductive.

---
[3] Traditional active learning methods would generally retrain between each step, requiring a back-and-forth labeling process not needed by the methods discussed shortly. In modern deep learning settings, this is almost never done due to the expense of retraining; "batch-mode" entropy is still excellent in those settings [38,42]. Appendix K explores more frequent retraining; the takeaway results are overall similar to the rest of our experiments.

In the one-shot case especially, standard active learning methods simply do not apply.

Recently, several papers have proposed novel active learning algorithms for these settings; none of these papers focused on meta-learning, but should be broadly applicable since meta-learning is also a low-budget setting. Rather than picking e.g. the points about which a model is least certain, these papers propose to label the "most representative" data points independently of a "current" model.

DPP Determinantal Point Processes (DPPs) query diverse samples, by selecting a subset that maximizes the determinant of a similarity matrix [8].

Typiclust Run k-means on the unlabeled data points, where $k = |\mathcal{C}|$ is the annotation budget. Select one data point per cluster such that the distance between a data point and its k' nearest neighbors is minimized: $\arg\min_{x \in \mathcal{U}} \sum_{x' \in \mathrm{NN}_{k'}(x)} \|x - x'\|_2$ [27].

Coreset Select a subset of the unlabeled set $\mathcal{U}$ to approximately minimize the distance from unlabeled data points to their nearest labeled point [57].

ProbCover Select data points that roughly maximize the number of unlabeled points within a distance of δ from any labeled point, where δ is chosen according to a "purity" heuristic [70]; see Appendix G for more details.

3.2 Features for Representative-Selection Methods

Notions of the "most representative" data points are highly dependent on a reasonable metric of data similarity. Prior methods operated either on raw data – typically a poor choice for complex datasets like natural images – or, in semi-supervised settings as in ProbCover and Typiclust, on SimCLR [11] features learned on the unlabeled data.

In metric-based meta-learning, we propose to instead use the current meta-learned representation; choosing points representative for the features we will use downstream is the natural choice. In MAML, the most natural equivalent might be features from the empirical neural tangent kernel (NTK) [39] of the current initialization network; this approximates what will happen when the network is trained on $\mathcal{C}$,[4] and so is perhaps the best simple understanding of "how this network views the data." Even empirical NTKs are often expensive to evaluate, however, and we thus propose to instead use features from the penultimate layer of the initialization neural net $f_\theta(\{\})$, corresponding to the NTK of a model that only retrains its last layer (as in ANIL, R2D2, and MetaOptNet). We also use the penultimate-layer representations of $f_\theta(\{\})$ for NP-based meta-learning.

Experiments in Appendix J show that this proposal outperforms off-the-shelf self-supervised features like SimCLR.

---

[4] Theoretical results about the NTK technically depend on a random initialization, which is not the case here. Mohamadi et al. [42] provide some assurance in that if the initialization were obtained by gradient descent on some dataset, the results would still hold, but MAML finds initial parameters differently.

3.3 Gaussian Mixture Selection for Low-Budget Active Learning

We propose the following very simple algorithm for low-budget active learning: fit a mixture of k Gaussians to the unlabeled data features, where k is the label budget, using EM with a k-means initialization. We use a shared diagonal covariance matrix (more details about EM are provided in Appendix L). Once a mixture is fit, we select the highest-density point from each component:

$$x^* = \arg\min_{x \in \mathcal{U}}(x - \mu_j)^\mathsf{T} \Sigma^{-1}(x - \mu_j) \text{ for each } j \in [k]. \qquad (4)$$

Algorithm 1: GMM-based Active Meta-learning

Input: Selection distribution $\text{Pick}_\theta^{\{train, eval\}}$, a learning algorithm f_θ, empirical risk $\mathcal{R}_\ell$, and the size of context sets k

1 Find $\hat{\theta}$ using Eq. (1) where $\text{Pick}_\theta^{train}$ may be stratified   // Meta-train
2 **while** task for evaluation exists **do**   // Meta-test
3   $\quad \tilde{\mathcal{D}} \sim P^{eval}$ and sample $\tilde{\mathcal{T}}$ from $\tilde{\mathcal{D}}$
4   $\quad$ Fit GMM: $\{(\hat{\pi}_j, \hat{\mu}_j, \hat{\Sigma}_j)\}_{j=1}^k$ using Eq. (10)–(12) in Appendix L
5   $\quad$ Select $\{x_j^*\}_{j=1}^k$ such that $\forall j \in [k]$, $x_j^* = \arg\min_{x \in \mathcal{X}}(x - \hat{\mu}_j)^T \hat{\Sigma}_j (x - \hat{\mu}_j)$
6   $\quad$ Annotate $\{x_j^*\}_{j=1}^k$ to create $\tilde{\mathcal{C}}$, and evaluate $f_{\hat{\theta}}$ using $\mathcal{R}_\ell(f_{\hat{\theta}(\tilde{\mathcal{C}})}, \tilde{\mathcal{T}})$

A comprehensive illustration for the proposed method is provided in Algorithm 1.

For metric-based meta-learning, the motivation of this algorithm is clear: we want labeled points that approximately "cover" the data points. Our notion of a "cover" is somewhat different from that of Coreset [57] or ProbCover [70]; we avoid ProbCover's need for a fixed radius, which we show can lead to poor choices (see Appendix G), and are more concerned with "average" covering (and hence perhaps less sensitive to outliers) than Coreset. The quality of selected data points from those methods are compared for a few metrics in Fig. 5.

On ANIL and MetaOptNet: since $|\mathcal{C}|$ is at most, say, 50 (in 10-way 5-shot) and the feature dimension is typically at least 100, ANIL becomes approximately the same multi-class max-margin separator obtained by (unregularized) MetaOptNet.[5] Intuitively, as $|\mathcal{C}|$ grows, the means of an isotropic Gaussian mixture converge to roughly a covering set for the dataset $\mathcal{U}$, and the max-margin separator of a set cover for $\mathcal{U}$ will be similar to the max-margin separator for all of the data. Even in various cases when $|\mathcal{C}| \ll |\mathcal{U}|$, choosing the means yields a max-margin separator that generalizes well.

---

[5] For reasonable distributions and networks, $\mathcal{C}$ is almost surely linear separable; thus ANIL, which is gradient descent for logistic regression, will converge to the multi-class max-margin separator [62].

Figure 2 illustrates that, if class-conditional data distributions are isotropic Gaussians with the same covariance matrices, labeling the cluster centers can be far preferable to labeling a random point from each cluster. This is backed up by the following theoretical results, which are all proved in Appendix A.

Proposition 1. *Suppose that $\{x_i\}_{i=1}^{N}$ are orthonormal. Then, the solution to (6) with the dataset $\{(x_y, y)\}_{y=1}^{N}$ is given by $w_y = x_y - \frac{1}{N}\sum_{i=1}^{N} x_i$, and hence*

$$\text{for any } x, \quad \arg\max_y w_y^\mathsf{T} x = \arg\min_y \|x - x_y\| \tag{5}$$

Proposition 1 says with orthonormal data points, a N-class support vector machine (one form of max-margin separators) defined in Eq. (6) becomes a nearest-neighbor classifier. While these assumptions will not exactly hold in practice, for high-dimensional normalized features, it is reasonable to expect our selected data points to be *almost* orthonormal. In combination with Lemma 1 (in Appendix A), this leads to the following optimality result.

Corollary 1. *Suppose $Y \sim \text{Uniform}([N])$, and $X \mid (Y = y) \sim \mathcal{N}(\mu_y, \sigma^2 I)$, where the μ_i are orthonormal. Then the max-margin separator (6) on $\{(\mu_i, i)\}_{i=1}^{N}$ is Bayes-optimal for $Y \mid (X = x)$.*

For more general settings, we argue that GMM is still a good method based on being an efficient set cover, as shown in Fig. 4 in Appendix A.

Very-Low-Budget Regime. Active learning based on Gaussian mixtures is not new in and of itself. Closely-related methods such as k-means, k-means^{++} or k-medoids have been employed either as standalone selection algorithms [2,65] or in combination with uncertainty-based methods [6,14,27,45]. Some recent work [9] including DPP [8] and Coreset [57] show significant improvements over k-means baselines. These trends, however, do not seem to hold true in the very-low-budget scenarios typically encountered in meta-learning. As shown in Fig. 8, GMM matches or outperforms other low-budget methods with very small numbers of labels for standard image classification tasks, which has not been known in the community. The following section shows that GMM provides substantial improvements in meta-learning.

4 Active Meta-learning Experiments

We now compare various active learning methods for variants of active meta learning as defined in Sect. 2.1, both for classification tasks (in Sect. 4.1 to 4.2) and regression (in Sect. 4.4).

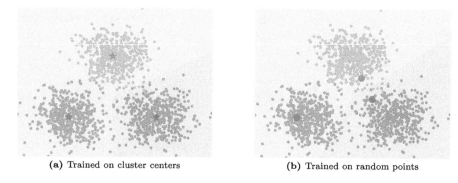

(a) Trained on cluster centers   (b) Trained on random points

Fig. 2. Decision boundaries using a multi-class SVM (6) trained on a one-shot dataset containing (a) cluster centers (stars) and (b) randomly selected points (circles).

4.1 Few-Shot Image Classification

We use four popular few-shot image classification benchmark datasets. **MiniImageNet** [52,64] consists of 60 000 images with 64 training classes, 16 validation, and 20 test. **TieredImageNet** [53] consists of 20 training super-classes, 6 validation, and 8 test; each contains 10 to 30 sub-classes. **FC100** [49] consists of 60 training classes, 20 validation, and 20 test. **CUB** [29,66] consists of 200 classes of bird images, with 140 training classes, 30 validation, and 30 test.

We validate if our active learning method works across various types of meta-learning methods. We run[6] metric-based: ProtoNet [61], optimization-based: MAML [17], ANIL [51], and MetaOpt [40], as well as pre-training-based: Baseline++ [12] and SimpleShot [68].[7] We vary the backbone to demonstrate robustness: for instance, we use 4 convolutional blocks for MAML and ProtoNet, and ResNet10 [28] for Baseline++. As typical in few-shot classification, we report means and 95% confidence intervals for test accuracy on 600 meta-test samples.

We use the meta-learner's features as proposed in Sect. 3.2 for all methods; experiments in Appendix J confirm that they outperform contrastive learning of features on the meta-training set. Additionally, in the main body we only present results where $\text{Pick}_\theta^{train}$ is random; Appendix N demonstrates that, in our setup, active learning at train time is actually mildly *harmful* to overall performance, aligning with observations by Ni et al. [46] and Setlur et al. [58].

For **metric-based** methods, Table 1 shows results for ProtoNet on FC100. The simple GMM method significantly outperforms the other active learning methods on all problems considered here. As reported [27,70], uncertainty-based methods are significantly worse than random selection in this low-budget regime.

For **optimization-based**, Table 2 shows results with MAML on MiniImageNet. GMM again outperforms the other methods in most cases. The perfor-

---

[6] We reproduce ProtoNet, MAML, and ANIL using Learn2Learn [5]; for MetaOptNet, Baseline++, and SimpleShot, we use repositories provided by the authors.

[7] We do not run a model-based method on this case, though we will in Sect. 4.4; most variants do not work well conditioning on images.

mance of ProbCover is sometimes much lower than other methods due to its radius parameter, which is very difficult to tune, with the best choice changing dramatically depending on the sub-task although [70] propose to fix this parameter per dataset (see Appendix G for more). Additional results for ANIL on TieredImageNet and MetaOptNet on FC100 are provided in Appendix H.

For **pre-training-based** methods, we compare active learning strategies with Baseline++ on the CUB dataset in Table 3, seeing that the proposed method is again usually by far the best, though in one five-shot case it essentially ties with DPP. As these methods do not follow the meta-training process in (1), train-time stratification is not applicable. Appendix H shows results for SimpleShot.

Table 1. 5-Way K-Shot on FC100 with ProtoNet, with $\text{Pick}_\theta^{train}$ random. The first, second, **third** best results for each setting are marked in this and all other tables.

$\text{Pick}_\theta^{eval}$	1-Shot			5-Shot		
	Fully strat.	Train strat.	Unstrat.	Fully strat.	Train strat.	Unstrat.
Random	36.73 ± 0.18	31.27 ± 0.21	31.40 ± 0.41	47.98 ± 0.18	42.83 ± 0.20	44.00 ± 0.21
Entropy	33.67 ± 0.16	29.82 ± 0.20	30.01 ± 0.20	44.64 ± 0.17	38.39 ± 0.22	38.36 ± 0.25
Margin	34.28 ± 0.18	29.74 ± 0.20	28.99 ± 0.20	45.31 ± 0.17	39.65 ± 0.21	38.13 ± 0.24
DPP	36.20 ± 0.18	31.34 ± 0.20	31.09 ± 0.20	47.53 ± 0.17	43.69 ± 0.20	44.19 ± 0.20
Coreset	35.79 ± 0.17	30.31 ± 0.20	31.57 ± 0.18	43.08 ± 0.40	41.56 ± 0.20	41.79 ± 0.22
Typiclust	46.01 ± 0.16	30.96 ± 0.19	30.61 ± 0.21	47.54 ± 0.17	43.61 ± 0.18	44.03 ± 0.21
ProbCover	48.66 ± 0.16	32.86 ± 0.22	33.58 ± 0.19	51.11 ± 0.17	44.20 ± 0.23	44.40 ± 0.24
GMM (Ours)	50.22 ± 0.18	34.23 ± 0.23	35.03 ± 0.23	54.76 ± 0.17	46.30 ± 0.21	47.03 ± 0.20

Table 2. 5-Way K-Shot on MiniImageNet with MAML, with $\text{Pick}_\theta^{train}$ random.

$\text{Pick}_\theta^{eval}$	1-Shot			5-Shot		
	Fully strat.	Train strat.	Unstrat.	Fully strat.	Train strat.	Unstrat.
Random	47.93 ± 0.20	28.16 ± 0.17	34.85 ± 0.19	64.16 ± 0.18	53.54 ± 0.20	58.84 ± 0.20
Entropy	48.16 ± 0.20	25.56 ± 0.14	30.44 ± 0.17	61.22 ± 0.20	34.36 ± 0.23	39.57 ± 0.26
Margin	48.31 ± 0.20	28.32 ± 0.16	30.83 ± 0.17	63.73 ± 0.18	49.24 ± 0.22	53.92 ± 0.22
DPP	48.96 ± 0.21	28.90 ± 0.17	36.44 ± 0.19	64.15 ± 0.18	54.18 ± 0.20	57.86 ± 0.19
Coreset	47.74 ± 0.20	29.19 ± 0.18	33.71 ± 0.18	61.28 ± 0.18	30.98 ± 0.19	45.74 ± 0.23
Typiclust	55.65 ± 0.18	27.45 ± 0.17	35.46 ± 0.18	64.16 ± 0.18	46.70 ± 0.21	57.83 ± 0.21
ProbCover	52.07 ± 0.17	23.34 ± 0.11	37.29 ± 0.18	64.66 ± 0.18	40.01 ± 0.21	45.32 ± 0.22
GMM (Ours)	58.82 ± 0.24	33.34 ± 0.24	37.68 ± 0.19	67.18 ± 0.18	54.35 ± 0.20	59.05 ± 0.20

Comparison Between Active Learning Methods. Figure 3 (left) visualizes context set selection using t-SNE [41] for one 5-way, 1-shot, unstratified task. It is vital to select one sample from each class; only GMM does so here. Figure 3 (right) summarizes behavior across many tasks; while not perfect, GMM does a much better job of selecting distinct classes.

Entropy and **Margin** are typically far worse than random. So is **Coreset**, agreeing with prior observations [6, 27, 70]; this may be because of issues with the greedy algorithm and/or sensitivity to outliers. **Typiclust** tends to pick points which, while dense according to its "typicality measure," are far from cluster centers; this may be helpful in traditional active learning, but seems to hurt here. **DPP** is often better than random, but only barely.

ProbCover manages to cover the feature space well, and is usually second-best. However, its "hard" radius causes issues; it may be preferable to use a smoother notion, as in GMM. The "purity" heuristic to choose δ also does not seem to align well with performance for meta-learning, as shown in Appendix G. Appendix M further analyzes the poor performance of other methods.

GMM provides robust performance with few new hyperparameters.[8]

"Soft" k-means would be a special case of GMM with a spherical covariance. For some cases, standard k-means performs about the same as GMM, but GMM is occasionally much better: for Baseline++ on CUB, GMM outperforms k-means by 3.95 points for 5-way 1-shot and 11.79 for 5-shot. We provide a more thorough comparison to k-means in Appendix F.

Table 3. 5-Way K-Shot on CUB with Baseline++, with $\text{Pick}_\theta^{train}$ random.

$\text{Pick}_\theta^{eval}$	1-Shot		5-Shot	
	Test strat.	Test unstrat.	Test strat.	Test unstrat.
Random	68.44 ± 0.92	51.03 ± 0.88	82.66 ± 0.56	79.57 ± 0.67
Entropy	66.33 ± 0.91	45.31 ± 0.89	80.97 ± 0.60	78.33 ± 0.72
Margin	68.65 ± 0.90	50.48 ± 0.94	82.29 ± 0.64	71.07 ± 0.83
DPP	71.53 ± 0.89	54.38 ± 0.92	82.81 ± 0.55	78.62 ± 0.76
Coreset	69.01 ± 0.91	56.22 ± 0.94	82.07 ± 0.55	76.35 ± 0.74
Typiclust	70.58 ± 0.81	29.80 ± 0.32	74.86 ± 0.81	70.00 ± 0.92
ProbCover	78.11 ± 0.69	55.09 ± 0.98	78.59 ± 0.64	65.71 ± 0.97
GMM (Ours)	79.98 ± 0.60	59.55 ± 0.87	82.55 ± 0.58	82.68 ± 0.57

4.2 Comparison with Hybrid Active Learning Methods

We compare the proposed GMM method with "hybrid" methods that select data points for annotation using both uncertainty and representation measures.[9]

---

[8] We did not significantly tune k-means or EM parameters from standard defaults.
[9] We separate the comparison with hybrid to highlight it, because hybrid methods are often considered better than solely uncertainty- or representation-based methods.

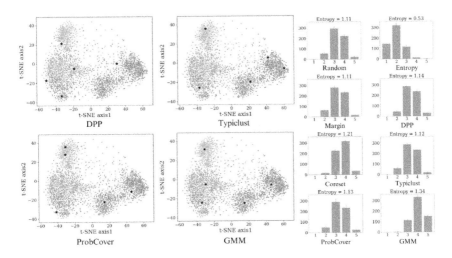

Fig. 3. Left. t-SNE of unlabeled points of one 5-way, 1-shot, unstratified MiniImageNet task. Stars denote selected context points using each method. **Right.** Distributions of the number of classes selected in each $\widetilde{C}$ by ProtoNet on MiniImageNet among 600 meta-test cases, along with the mean empirical entropy of y from $\widetilde{C}$. The higher the value is, the more diverse classes are selected; $\log 5 \approx 1.6$ would be perfect.

Table 4. Comparison of GMM with hybrid active learning methods.

Data & Model	Clustering	1-Shot		5-Shot	
		Train strat.	Unstrat.	Train strat.	Unstrat.
MiniImage. MAML	Weighted Ent.	22.69 ± 0.18	32.27 ± 0.32	23.75 ± 0.25	46.80 ± 0.33
	BADGE	27.71 ± 0.18	34.30 ± 0.21	41.37 ± 0.28	58.79 ± 0.24
	k-means Ent.	30.59 ± 0.28	33.73 ± 0.24	38.24 ± 0.29	54.87 ± 0.26
	GMM (Ours)	33.34 ± 0.24	37.68 ± 0.19	54.35 ± 0.20	59.05 ± 0.20
FC100 ProtoNet	Weighted Ent.	31.80 ± 0.20	28.94 ± 0.19	40.40 ± 0.25	39.95 ± 0.25
	BADGE	30.91 ± 0.23	29.29 ± 0.28	43.85 ± 0.22	44.00 ± 0.29
	k-means Ent.	30.93 ± 0.22	30.43 ± 0.24	41.76 ± 0.27	43.41 ± 0.29
	GMM (Ours)	34.23 ± 0.23	35.03 ± 0.23	46.30 ± 0.21	47.03 ± 0.20

Weighted Entropy Nguyen *et al.* [45] propose weighted expected error for binary classification. For multi-class cases, we derive that it becomes weighted entropy, where weights are likelihood computed using soft k-means.

BADGE This method [6] selects points using k-means^{++} with embeddings derived from the gradients of loss w.r.t the weights of the last layer.

k-means Entropy This approach [3] first clusters unlabeled samples using k-means^{++}, then selects samples per cluster using the classifier's entropy.

Table 4 shows that the proposed GMM-based method significantly outperforms all the hybrid methods. This experiment, along with the poor performance of uncertainty methods such as Entropy, demonstrates that for the very-low-budget regime, diversity is significantly more important than reducing uncertainty.

4.3 Cross-Domain Active Meta-Learning

Cross-domain learning, where $\mathcal{P}^{train}$ is "fundamentally different" from $\mathcal{P}^{eval}$, is typically more difficult than "in-domain" meta-learning. We use a ResNet18 [28] pretrained with standard supervised learning on ImageNet, and meta-test on CUB and **Places** [72], which contains images of "places" such as restaurants. As used for cross-domain meta-learning by [48], it contains 16 classes with an average of 1,715 images each. As the model is not meta-trained, train stratification is not relevant; we show results in Table 5 only for unstratified test sets. GMM is again the clear overall winner; other methods are often worse than random.

Table 5. Cross-domain meta-learning tasks using a ResNet18 pre-trained on ImageNet.

$\text{Pick}_\theta^{eval}$	$\mathcal{P}^{eval}$ on Places		$\mathcal{P}^{eval}$ on CUB	
	1-Shot	5-Shot	1-Shot	5-Shot
Random	44.28 ± 1.93	77.92 ± 1.70	49.93 ± 0.92	84.38 ± 0.72
Entropy	36.12 ± 1.25	57.79 ± 2.93	41.85 ± 0.99	71.15 ± 0.99
Margin	43.31 ± 1.97	73.65 ± 1.94	48.04 ± 0.98	78.84 ± 0.92
DPP	46.76 ± 2.29	78.36 ± 1.89	51.41 ± 0.90	84.19 ± 0.72
Coreset	50.03 ± 0.93	65.20 ± 2.77	50.77 ± 0.95	81.80 ± 0.81
Typiclust	43.76 ± 1.98	77.57 ± 1.84	43.39 ± 1.03	50.69 ± 1.08
ProbCover	47.93 ± 1.08	59.08 ± 2.50	62.13 ± 1.08	69.80 ± 1.16
GMM (Ours)	60.01 ± 0.86	86.45 ± 1.42	59.87 ± 0.86	85.49 ± 0.67

Discussion. Most uncertainty measures tend to be high near decision boundaries. This may be sub-optimal in low-budget settings, as these uncertain points often represent outliers, or are too challenging to generalize.

The primary purpose of context sets in meta-learning is to inform predictions on target samples, necessitating the selection of easily referable points. If the selected context samples are too distant from the target samples, making accurate predictions for the target set becomes difficult. Diversity measures, particularly GMM, ensure that the context set remains close to the target set even in adverse scenarios, such as when target samples are outliers (see Fig. 2). Thus, it is preferable to solely consider diversity for active selection of context sets.

While hybrid methods that incorporate both uncertainty and diversity may be beneficial in mid or high-budget active learning scenarios, they provide limited assistance in extremely low-budget scenarios such as meta-learning.

4.4 Active Meta-Learning for Regression

Each **sinusoidal function** [17] has task $y = a\sin(x+p)$, where $a \sim \text{Unif}(0.1, 5)$ is the amplitude, and $p \sim \text{Unif}(0, \pi)$ is the phase of sine functions; we use MAML for this dataset. **Distractor** and **ShapeNet1D** are vision regression datasets [19]; the task is to predict the position of a specific object in an image ignoring a distractor, or to predict an object's 1D pose (azimuth rotation). **IC** uses objects whose classes were observed during meta-training, while **CC** has novel object classes. We use conditional Neural Processes (NP) for Distractor, and attentive NP for ShapeNet1D. Details are provided in Appendix I.

Table 6 compares active strategies on these datasets; GMM again performs generally the best, followed by Coreset and DPP instead of ProbCover.

Table 6. Meta-learning for regression on a toy dataset and two pose estimation datasets for Intra-Category (IC) and Cross-Category (CC). Sine func. and Distractor use mean squared error, ShapeNet1D uses cosine-sine-distance; lower values are better for each.

Active Strategy	Sine (3-Shots)	Distractor (2-Shots)		ShapeNet1D (2-Shots)	
		IC	CC	IC	CC
Random	24.17 ± 0.43	18.91 ± 2.13	25.79 ± 2.17	16.52 ± 1.08	19.07 ± 1.30
DPP	23.19 ± 0.51	18.08 ± 2.12	19.68 ± 1.92	11.83 ± 0.85	13.68 ± 0.93
Coreset	31.36 ± 0.48	19.58 ± 1.95	24.08 ± 2.19	11.39 ± 0.91	13.05 ± 1.18
Typiclust	21.59 ± 0.40	20.27 ± 2.15	24.96 ± 2.68	12.54 ± 1.08	14.58 ± 1.24
ProbCover	29.36 ± 0.49	21.96 ± 2.45	25.25 ± 2.78	12.31 ± 0.85	13.95 ± 1.08
GMM (Ours)	18.09 ± 0.38	17.95 ± 2.05	22.03 ± 2.42	10.78 ± 0.72	12.35 ± 0.97

5 Conclusion

We clarified the ways in which active learning can be incorporated into meta-learning. While active context set selection does not seem to work at meta-train time (Appendix N), it can be extremely useful at meta-testing/deployment time.

We proposed a surprisingly simple method that substantially outperforms previous proposals. It is intuitive, very easy to implement, and bears theoretical guarantees in a particular "stylized" but informative situation.

Acknowledgements. This work was enabled in part by support provided by the Natural Sciences and Engineering Research Council of Canada, the Canada CIFAR AI Chairs program, Mitacs through the Mitacs Accelerate program, Calcul Québec, the BC DRI Group, and the Digital Research Alliance of Canada.

References

1. Agarwal, M., Yurochkin, M., Sun, Y.: On sensitivity of meta-learning to support data. In: NeurIPS (2021)
2. Aghaee, A., Ghadiri, M., Baghshah, M.S.: Active distance-based clustering using k-medoids. In: PAKDD (2016)
3. Al-Shedivat, M., Li, L., Xing, E., Talwalkar, A.: On data efficiency of meta-learning. In: AISTAT (2021)
4. Antoniou, A., Edwards, H., Storkey, A.: How to train your MAML. In: ICLR (2019)
5. Arnold, S.M.R., Mahajan, P., Datta, D., Bunner, I., Zarkias, K.S.: learn2learn: a library for meta-learning research (2020)
6. Ash, J.T., Zhang, C., Krishnamurthy, A., Langford, J., Agarwal, A.: Deep batch active learning by diverse, uncertain gradient lower bounds. ICLR (2020)
7. Bertinetto, L., Henriques, J.F., Torr, P.H.S., Vedaldi, A.: Meta-learning with differentiable closed-form solvers. In: ICLR (2019)
8. Bıyık, E., Wang, K., Anari, N., Sadigh, D.: Batch active learning using determinantal point processes. In: NeurIPS (2019)
9. Boney, R., Ilin, A.: Semi-supervised and active few-shot learning with prototypical networks. arXiv preprint arXiv:1711.10856 (2017)
10. Chang, A.X., et al.: ShapeNet: an information-rich 3D model repository. arXiv preprint arXiv:1512.03012 (2015)
11. Chen, T., Kornblith, S., Norouzi, M., Hinton, G.: A simple framework for contrastive learning of visual representations. In: ICML (2020)
12. Chen, W.Y., Liu, Y.C., Kira, Z., Wang, Y.C.F., Huang, J.B.: A closer look at few-shot classification. In: ICLR (2019)
13. Crammer, K., Singer, Y.: On the algorithmic implementation of multiclass kernel-based vector machines. JMLR (2001)
14. Donmez, P., Carbonell, J.G., Bennett, P.N.: Dual strategy active learning. In: ECML (2007)
15. Dubois, Y., Gordon, J., Foong, A.Y.: Neural process family (2020). http://yanndubs.github.io/Neural-Process-Family/
16. Fang, M., Li, Y., Cohn, T.: Learning how to active learn: a deep reinforcement learning approach. In: EMNLP (2017)
17. Finn, C., Abbeel, P., Levine, S.: Model-agnostic meta-learning for fast adaptation of deep networks. In: ICML (2017)
18. Freytag, A., Rodner, E., Denzler, J.: Selecting influential examples: active learning with expected model output changes. In: ECCV (2014)
19. Gao, N., Ziesche, H., Vien, N.A., Volpp, M., Neumann, G.: What matters for meta-learning vision regression tasks? In: CVPR (2022)
20. Garnelo, M., et al.: Conditional neural processes. In: ICML (2018)
21. Garnelo, M., et al.: Neural processes. In: ICML Workshop on Theoretical Foundations and Applications of Deep Generative Models (2018)
22. Garnett, R., Krishnamurthy, Y., Xiong, X., Schneider, J., Mann, R.: Bayesian optimal active search and surveying. In: ICML (2012)
23. Gautier, G., Polito, G., Bardenet, R., Valko, M.: DPPy: DPP sampling with Python. JMLR-MLOSS (2019). https://github.com/guilgautier/DPPy/
24. Goldblum, M., Reich, S., Fowl, L., Ni, R., Cherepanova, V., Goldstein, T.: Unraveling meta-learning: understanding feature representations for few-shot tasks. In: ICML (2020)

25. Gordon, J., Bruinsma, W.P., Foong, A.Y., Requeima, J., Dubois, Y., Turner, R.E.: Convolutional conditional neural processes. In: ICLR (2020)
26. Guo, Y., Greiner, R.: Optimistic active-learning using mutual information. In: IJCAI (2007)
27. Hacohen, G., Dekel, A., Weinshall, D.: Active learning on a budget: opposite strategies suit high and low budgets. ICML (2022)
28. He, K., Zhang, X., Ren, S., Sun, J.: Deep residual learning for image recognition. In: CVPR (2016)
29. Hilliard, N., Phillips, L., Howland, S., Yankov, A., Corley, C.D., Hodas, N.O.: Few-shot learning with metric-agnostic conditional embeddings. arXiv preprint arXiv:1802.04376 (2018)
30. Hospedales, T., Antoniou, A., Micaelli, P., Storkey, A.: Meta-learning in neural networks: a survey. PAMI **44**(9), 5149–5169 (2022)
31. Huang, K., Geng, J., Jiang, W., Deng, X., Xu, Z.: Pseudo-loss confidence metric for semi-supervised few-shot learning. In: ICCV (2021)
32. Kaddour, J., Sæmundsson, S., Deisenroth, M.P.: Probabilistic active meta-learning. In: NeurIPS (2020)
33. Käding, C., Rodner, E., Freytag, A., Denzler, J.: Active and continuous exploration with deep neural networks and expected model output changes. In: NeurIPS Workshop on Continual Learning and Deep Networks (2016)
34. Käding, C., et al.: Active learning for regression tasks with expected model output changes. In: BMVC (2018)
35. Kim, H., et al.: Attentive neural processes. In: ICLR (2019)
36. Konyushkova, K., Sznitman, R., Fua, P.: Learning active learning from data. In: NeurIPS (2017)
37. Kumar, R., Deleu, T., Bengio, Y.: The effect of diversity in meta-learning. In: AAAI (2022)
38. Lang, A., Mayer, C., Timofte, R.: Best practices in pool-based active learning for image classification (2021)
39. Lee, J., et al.: Wide neural networks of any depth evolve as linear models under gradient descent. In: NeurIPS (2019)
40. Lee, K., Maji, S., Ravichandran, A., Soatto, S.: Meta-learning with differentiable convex optimization. In: CVPR (2019)
41. van der Maaten, L., Hinton, G.: Visualizing data using t-SNE. JMLR (2008)
42. Mohamadi, M.A., Bae, W., Sutherland, D.J.: Making look-ahead active learning strategies feasible with neural tangent kernels. In: NeurIPS (2022)
43. Müller, T., Pérez-Torró, G., Basile, A., Franco-Salvador, M.: Active few-shot learning with FASL. In: NLDB (2022)
44. Munjal, P., Hayat, N., Hayat, M., Sourati, J., Khan, S.: Towards robust and reproducible active learning using neural networks. In: CVPR (2022)
45. Nguyen, H.T., Smeulders, A.: Active learning using pre-clustering. In: ICML (2004)
46. Ni, R., Goldblum, M., Sharaf, A., Kong, K., Goldstein, T.: Data augmentation for meta-learning. In: ICML (2021)
47. Nikoloska, I., Simeone, O.: BAMLD: Bayesian active meta-learning by disagreement. SPAWC (2022)
48. Oh, J., Kim, S., Ho, N., Kim, J.H., Song, H., Yun, S.Y.: Understanding cross-domain few-shot learning based on domain similarity and few-shot difficulty. In: NeurIPS (2022)
49. Oreshkin, B., Rodríguez López, P., Lacoste, A.: Tadam: task dependent adaptive metric for improved few-shot learning. In: NeurIPS (2018)

50. Pezeshkpour, P., Zhao, Z., Singh, S.: On the utility of active instance selection for few-shot learning. In: HAMLETS Workshop at NeurIPS (2020)
51. Raghu, A., Raghu, M., Bengio, S., Vinyals, O.: Rapid learning or feature reuse? Towards understanding the effectiveness of MAML. In: ICLR (2020)
52. Ravi, S., Larochelle, H.: Optimization as a model for few-shot learning. In: ICLR (2017)
53. Ren, M., et al.: Meta-learning for semi-supervised few-shot classification. In: ICLR (2018)
54. Ren, P., et al.: A survey of deep active learning. ACM Comput. Surv. **54**(9) (2021)
55. Roy, N., McCallum, A.: Toward optimal active learning through Monte Carlo estimation of error reduction. In: ICML (2001)
56. Scheffer, T., Decomain, C., Wrobel, S.: Active hidden Markov models for information extraction. In: ISIDA (2001)
57. Sener, O., Savarese, S.: Active learning for convolutional neural networks: a core-set approach. ICLR (2018)
58. Setlur, A., Li, O., Smith, V.: Is support set diversity necessary for meta-learning? NeurIPS Workshop (2020)
59. Settles, B.: Active learning literature survey (2009)
60. Settles, B., Craven, M., Ray, S.: Multiple-instance active learning. In: NeurIPS (2007)
61. Snell, J., Swersky, K., Zemel, R.S.: Prototypical networks for few-shot learning. In: NeurIPS (2017)
62. Soudry, D., Hoffer, E., Nacson, M.S., Gunasekar, S., Srebro, N.: The implicit bias of gradient descent on separable data. JMLR (2018)
63. Tan, W., Du, L., Buntine, W.: Diversity enhanced active learning with strictly proper scoring rules. In: NeurIPS (2021)
64. Vinyals, O., Blundell, C., Lillicrap, T., Wierstra, D., et al.: Matching networks for one shot learning. In: NeurIPS (2016)
65. Voevodski, K., Balcan, M.F., Röglin, H., Teng, S.H., Xia, Y.: Active clustering of biological sequences. JMLR (2012)
66. Wah, C., Branson, S., Welinder, P., Perona, P., Belongie, S.: The Caltech-UCSD Birds-200-2011 dataset. California Institute of Technology technical report CNS-TR-2011-001, California Institute of Technology (2011)
67. Wang, D., Shang, Y.: A new active labeling method for deep learning. In: IJCNN (2014)
68. Wang, Y., Chao, W.L., Weinberger, K.Q., van der Maaten, L.: Simpleshot: revisiting nearest-neighbor classification for few-shot learning. arXiv preprint arXiv:1911.04623 (2019)
69. Wei, X.S., Xu, H.Y., Zhang, F., Peng, Y., Zhou, W.: An embarrassingly simple approach to semi-supervised few-shot learning. In: NeurIPS (2022)
70. Yehuda, O., Dekel, A., Hacohen, G., Weinshall, D.: Active learning through a covering lens. In: NeurIPS (2022)
71. Zhang, X., Meng, D., Gouk, H., Hospedales, T.M.: Shallow Bayesian meta learning for real-world few-shot recognition. In: ICCV (2021)
72. Zhou, B., Lapedriza, A., Khosla, A., Oliva, A., Torralba, A.: Places: a 10 million image database for scene recognition. PAMI (2017)
73. Zhu, X., Lafferty, J., Ghahramani, Z.: Combining active learning and semi-supervised learning using gaussian fields and harmonic functions. In: ICML Workshops (2003)

BlenderAlchemy: Editing 3D Graphics with Vision-Language Models

Ian Huang(✉), Guandao Yang, and Leonidas Guibas

Stanford University, Stanford, USA
ianhuang@stanford.edu

Abstract. Graphics design is important for various applications, including movie production and game design. To create a high-quality scene, designers usually need to spend hours in software like Blender, in which they might need to interleave and repeat operations, such as connecting material nodes, hundreds of times. Moreover, slightly different design goals may require completely different sequences, making automation difficult. In this paper, we propose a system that leverages Vision-Language Models (VLMs), like GPT-4V, to intelligently search the design action space to arrive at an answer that can satisfy a user's intent. Specifically, we design a vision-based edit generator and state evaluator to work together to find the correct sequence of actions to achieve the goal. Inspired by the role of visual imagination in the human design process, we supplement the visual reasoning capabilities of VLMs with "imagined" reference images from image-generation models, providing visual grounding of abstract language descriptions. In this paper, we provide empirical evidence suggesting our system can produce simple but tedious Blender editing sequences for tasks such as editing procedural materials and geometry from text and/or reference images, as well as adjusting lighting configurations for product renderings in complex scenes (For project website and code, please go to: https://ianhuang0630.github.io/BlenderAlchemyWeb/).

1 Introduction

To produce the compelling graphics content we see in movies or video games, 3D artists usually need to spend hours in software like Blender to find appropriate surface materials, object placements, and lighting arrangements. These operations require the artist to create a mental picture of the target, experiment with different parameters, and visually examine whether their edits get closer to the end goal. One can imagine automating these processes by converting language or visual descriptions of user intent into edits that achieve a design goal. Such a system can improve the productivity of millions of 3D designers and impact various industries that depend on 3D graphic design.

Supplementary Information The online version contains supplementary material available at https://doi.org/10.1007/978-3-031-73024-5_18.

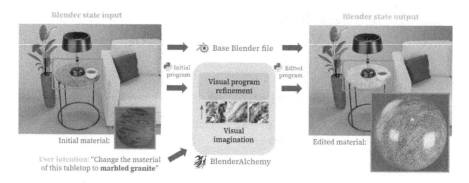

Fig. 1. Overview of BlenderAlchemy. Given an input Blender state and a user intention specified using either language or reference images, BlenderAlchemy edits the Blender state to satisfy that intention by orange *iteratively refining a Blender python program* that executes to produce the final Blender state. Our system additionally leverages text-to-image generation for yellow *visual imagination*, a step that expands a text-only user intention to a concrete visual target to improve program refinement.

Graphic design is very challenging because even a small design goal requires performing a variety of different tasks. For instance, modeling of a game environment requires the 3D artist to cycle between performing modeling, material design, texture painting, animation, lighting, and scene composition. Prior attempts usually focus on specific editing tasks, like material synthesis [8,62]. While these approaches show promising performance in the tasks they are designed for, it is non-trivial to combine these task-specific methods to satisfy an intended design goal. An alternative is to leverage Large Language Models (LLMs) [22,27,45] to digest user intent and suggest design actions by proposing intelligent combination of existing task-specific tools [36] or predicting edits to programs step-by-step [49,57]. While LLMs have excellent abilities to understand user intentions and suggest sequences of actions to satisfy them, applying LLMs to graphical design remains challenging largely because language cannot capture the visual consequences of actions performed in software like Blender.

One promising alternative is to leverage vision language models (VLM), such as LLaVA [23], GPT-4V [28], Gemini [44], and DallE-3 [7]. These VLMs have shown to be highly capable of understanding detailed visual information [13,56, 58,59] and generating compelling images [7]. We posit that these VLMs can be leveraged to complete different kinds of design tasks within the Blender design environment, like editing materials, geometry and lighting setups.

In this paper, we provide a proof-of-concept system using the vision foundation model GPT-4V to generate and edit programs that modify the state of a Blender workspace to satisfy a user intention. Specifically, we first initialize the state of a workspace within Blender. The Blender state is parameterized as a short Python program, and a base Blender file. The user will then input a text description and potentially a reference image to communicate the desirable

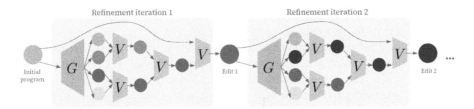

Fig. 2. Iterative visual program editing employs a edit generator G and a state evaluator V in each iteration to explore and prune different potential program edits, where G generates plausible variants of an input program and V picks between two programs based on the consequences they have to the Blender visual state and their alignment to the user intention. Each iteration of the refinement explores variations of the most promising program from the previous iteration. See Algorithm 1 for details.

design outcome. The system is tasked to edit the program so that, when executed on the base Blender file, the rendered image can satisfy the user's intention. Figure 1 provides an illustration of the problem setup.

Naively applying VLMs to this editing setting gives rise to many failure cases, possibly due to the fact that out-of-the-box VLMs have a poor understanding of the visual consequences of Blender program edits. To counter this, we propose a visually-guided program search procedure that combines a vision-aware edit generator and a visual state evaluator to iteratively search for a suitable program edit (Fig. 2). Inspired by the human design process, our system performs guided trial-and-error, capped by some computation budget. Within each iteration, the visual program generator will propose several possible edits on the current program. These edits will be applied and executed in Blender to produce a rendered image. These rendered images will be provided to visual evaluator, which will select the best renders via pairwise comparison by assessing which choice better satisfies the design goal specified by text and reference image. The program achieving the best render will replace the current program as a starting point for the visual program generator. This iterative search process, however, has very low success rate because of the sparsity of correct program edits in the vast program space. To improve the success rate, we further propose two techniques. First, when the proposed program of the new iteration does not contain a viable candidate, we revert to the best candidate of the prior program. This reversion mechanism make sure the search procedure will not diverge when facing a batch of bad edit candidates. Second, to facilitate our visual evaluator and generator to better understand user intent, we leverage the "visual imagination" of text-to-image generative model to imagine a reference image. We show that our method is capable of accomplishing graphical design tasks within Blender, guided by user intention in the form of text and images. We demonstrate the effectiveness of our system on material, geometry and lighting design tasks, all parts of the 3D design process where artists spend a significant amount of time, ranging from 20 hrs to 4–6 workdays *per* model [2,3]. We show that our method can outperform prior works designed for similar problem settings, such as BlenderGPT [1]. In summary, the main contributions of this paper include:

1. We propose BlenderAlchemy, a system that's able to edit visual programs based on user input in the form of text or images.
2. We identify key components that make the system work: a visual state evaluator, a visual edit generator, a search algorithm with an edit reversion mechanism, and a visual imagination module to facilitate the search.
3. We provide evidence showing that our system can outperform prior works in text-based procedural material editing, as well as its applicability to other design tasks including editing geometry and lighting configurations.

2 Related Works

Task-Specific Tools for Material Design. Large bodies of works have been dedicated to using learning-based approaches to generate materials. Prior works exploit 2D diffusion models to generate texture maps either in a zero-shot way [8,21,33,50,54] or through fine-tuning [37,60]. Though these works open the possibility of generating and modifying textures of objects using natural language for 3D meshes, these works do not model the material properties in a way that allows such objects to be relit. Other methods directly predict the physical properties of the surface of a material through learning, using diffusion [46,47] or learning good latent representations from data [10,17,20,25,39,62]. However, for all of the work mentioned so far, their fundamentally image-based or latent-based representations make the output materials difficult to edit in existing 3D graphics pipelines. People have also explored combining learning-based approaches with symbolic representations of materials [34,43]. These works often involve creating differentiable representations of the procedural material graph often used in 3D graphical design pipelines, and backpropagating gradients throughout the graph to produce an image that can match the target [16,19,38]. Other works like Infinigen [32] use rule-based procedural generation over a library of procedural materials. However, no prior works in this direction have demonstrated the ability to edit procedural material graphs using *user intention specified by language* [34], a task that we are particularly interested in. Though the aforementioned works excel at material design, they aren't generalizable to other 3D graphics design task settings. BlenderAlchemy, on the other hand, aims to produce a system that can perform various design tasks according to user intents. This usually requires combining different methods together in a non-trivial way.

LLM as General Problem Solvers. Large language models (LLMs) like GPT-4 [27,28], Llama [45], and Mistral [22] have in recent years demonstrated unprecedented results in a variety of problem settings, like robotics [4,12,18,24,41,52,61], program synthesis [26,35,40], and graphic design [21,50,55]. Other works have shown that by extending such models with an external process like Chain-of-Thought [49], Tree-of-Thought [57] or memory/skill database [11,29,30,48], or by embedding such systems within environments where it can perceive and act [11,29,36,48], a range of new problems that require iterative refinement can be solved. Their application to visual problem settings, however, has mostly been limited due to the nonexistent visual perception capabilities of the LLMs [11,15,48,50,53,55]. While this could be sidestepped by fully condensing the visual state of the environment using text [29] or some symbolic representation [11,48], doing so for 3D graphic design works poorly. For instance, the task of editing a Blender material graph to create a desired material requires many trial-and-error cycles and an accurate understanding of the consequences certain design actions can have on the visual output. Recent works that apply LLMs to graphical design settings [11,15] and ones that more specifically do so within Blender (like BlenderGPT [1], 3D-GPT [42] and L3GO [53]) do not use visual information to inform or refine their system's outputs, leading to unsatisfactory results. BlenderAlchemy borrows ideas from existing LLM literature and tries to address this issue by inputting visual perception into the system.

Vision-Language Models. State-of-the-art vision language models, such as LLaVA [23], GPT-4V [28], and Gemini [44] have demonstrated impressive understanding of the visual world and its connections to language and semantics, enabling many computer vision tasks like scene understanding, visual question answering, and object detection to be one API call away [13,14,56,58,59]. Works such as [6,51] suggest that such models can also be used as a replacement for human evaluators for a lot of tasks, positioning them as tools for guiding planning and search by acting as flexible reward functions. BlenderAlchemy takes the first steps to apply VLMs to solve 3D graphic design tasks, a novel yet challenging application rather unexplored by existing works.

3 Method

The goal of our system is to perform edits within the Blender 3D graphic design environment through iteratively refining programs that define a sequence of edits in Blender. This requires us to (1) decompose the input initial Blender input into a combination of programs and a "base" Blender state (Sect. 3.1) and (2) develop a procedure to edit each program to produce the desired visual state within Blender to match a user intention (Sects. 3.2).

3.1 Representation of the Blender Visual State

The state of the initial Blender design environment can be decomposed into an "base" Blender state S_{base} and a set of initial programs $\{p_0^{(1)}, p_0^{(2)}, ..., p_0^{(k)}\}$ that acts on state S_{base} to produce the initial Blender environment through a dynamics function F that transitions from one state to another based on a set of programmatic actions:

$$S_{\text{init}} = F\left(\left\{p_0^{(i)}\right\}_{i=1...k}, S_{\text{base}}\right)$$

In our problem setting, F is the python code executor within the Blender environment that executes $\{p_0^{(i)}\}_{i=1...k}$ in sequence. We set each initial program $p_0^{(i)}$ to be a program that concerns a single part of the 3D graphical design workflow – for instance, $p_0^{(1)}$ is in charge of the material on one mesh within the scene, and $p_0^{(2)}$ is in charge of the lighting setup of the entire scene. The decomposition of S_{init} into S_{base} and $p_0^1...p_0^k$ can be done using techniques like the "node transpiler" from Infinigen [32], which converts entities within the Blender instance into lines of Python code that can recreate a node graph, like a material shader graph. We develop a suite of tools to do this in our own problem setting.

Though it's possible for *all* edits to be encompassed in a *single* program instead of k programs, this is limiting in practice – either because the VLM's output length isn't large enough for the code necessary or because the VLM has a low success rate, due to the program search space exploding in size. Although it is possible that future VLMs will substantially mitigate this problem, splitting the program into k task-specific programs may still be desirable, given the possibility of querying k task-specific fine-tuned/expert VLMs in parallel.

3.2 Iterative Refinement of Individual Visual Programs

Suppose that to complete a task like material editing for a single object, it suffices to decompose the initial state into S_{base} and a single script, p_0 – that is, $S_{\text{init}} = F(\{p_0\}, S_{\text{base}})$. Then our goal is to discover some edited version of p_0, called p_1, such that $F(\{p_1\}, S_{\text{base}})$ produces a visual state better aligned with some user intention I. Our system assumes that the user intention I is provided in the form of language and/or image references, leveraging the visual understanding of the latest VLM models to understand user intention.

Algorithm 1. Iterative Refinement of Visual Programs

1: **procedure** TOURNAMENT(State candidates $\{S_1, S_2, ...S_k\}$, Visual state evaluator V, User intention I)
2:    **if** $k > 2$ **then**
3:       $w_1 \leftarrow$ TOURNAMENT($\{S_1, S_2, ...S_{k/2}\}, V, I$),
4:       $w_2 \leftarrow$ TOURNAMENT($\{S_{k/2}, S_{k+1}, ...S_k\}, V, I$)
5:    **else**
6:       $w_1 \leftarrow S_1, w_2 \leftarrow S_2$
7:    **end if**
8:    **return** $V(w_1, w_2, I)$
9: **end procedure**
10: **procedure** REFINE(Depth d, Breadth b, Intention I, Edit Generator G, State Evaluator V, Base state S_{base}, Initial program p_0, Dynamics Function F)
11:    $S_0 \leftarrow F(p_0, S_{\text{base}})$, $S_{\text{best}} \leftarrow S_0$, $p_{\text{best}} \leftarrow p_0$
12:    **for** $i \leftarrow 1$ to d **do**
13:       $\mathcal{P}_i = \mathcal{N}(p_{\text{best}})$ if i is odd else $\mathcal{P}$ ▷ "Tweak" or "Leap" edits
14:       $p_i^1, p_i^2...p_i^b \leftarrow G(p_{\text{best}}, S_{\text{best}}, I, b, \mathcal{P}_i)$ ▷ Generate b options
15:       $S_i^1 \leftarrow F(p_i^1, S_0), ..., S_i^b \leftarrow F(p_i^b, S_0)$ ▷ Observe the visual states
16:       $S_i^* \leftarrow$ TOURNAMENT($\{S_i^1, S_i^2...S_i^b\} \cup \{S_{\text{best}}\}, V, I$)   ▷ Choose the best
17:       $S_{\text{best}} \leftarrow S_i^*$, $p_{\text{best}} \leftarrow p_i^*$ ▷ Best visual state and programs so far
18:    **end for**
19:    **return** $S_{\text{best}}, p_{\text{best}}$
20: **end procedure**

To discover a good edit to p_0, we introduce the procedure outlined in Algorithm 1, an iterative refinement loop that repeatedly uses a visual state evaluator V to select among the hypotheses from an edit generator G. A single "agent" for a certain task like procedural material design can be fully described by (G, V).

Inspired by works like [51], we propose a visual state evaluator $V(S_1, S_2, I)$, which is tasked with returning whichever of the two visual states (S_1 or S_2) better matches the user intention I. This evaluator is applied recursively to choose the most suitable visual state candidate among b visual state candidates by making $\mathcal{O}(\log(b))$ queries, as done in TOURNAMENT in Algorithm 1.

Though it seems straightforward to ask the same VLM to edit the code in a single pass, this leads to many failure cases (see more in the Appendix). Due to the VLM's lack of baked-in understanding of the visual consequences of various programs within Blender, a multi-hypothesis and multi-step approach is more appropriate. Extending Tree-of-Thoughts [57] to the visual domain, $G(p, S, I, b, \mathcal{P})$ is a module tasked with generating b different variations of program p, conditioned on the current visual state S and user intention I, constrained such that the output programs fall within some family of programs $\mathcal{P}$, which can be used to instill useful priors to the edit generator.

Below we describe some additional system design decisions that ensured better alignment of the resultant edited program to the user intention, either by improving the stability of the procedure or by supplementing the visual understanding of VLMs. The effect of each is investigated in the Appendix through ablations.

Hypothesis Reversion. To improve the stability of the edit discovery process, we add the visual state of the program being edited at every timestep (S_{best} for p_{best}) as an additional candidate to the selection process, providing the option for the process to revert to an earlier version if the search at a single iteration was unsuccessful. Line 16 in Algorithm 1 shows this.

Tweak and Leap Edits. An important characteristic of visual programs is that continuous values hard-coded within the program can modify the output just as much as structural changes. This is in contrast to non-visual program synthesis tasks based on unit-tests of I/O specs, like [5,9], where foundation models are mostly tasked to produce the right *structure* of the program with minimal hard-coded values. Given a single visual program $p \in \mathcal{P}$, the space of visual outputs achievable through *only* tweaking the numerical values to function parameters and variable assignments can cover a wide range, depending on the fields available in the program. In Algorithm 1, we refer this space as the "neighborhood" of p or "tweak" edits, $\mathcal{N}(p)$. Though this can result in a very small change to the program, this can lead to a large visual difference in the final output, and in a few edits can change the "wrong" program into one more aligned with the user intention. On the other hand, more drastic changes (or "leap" edits) may be needed to accomplish a task. Consider for example the task of changing a perfectly smooth material to have a noised level of roughness scattered sparsely across the material surface. This may require the programmatic addition of the relevant nodes (e.g. color ramps or noise texture nodes), and thus the edited program falls outside of $\mathcal{N}(p)$.

Empirically, we find that the optimal edits are often a mix of tweak and leap edits. As such, our procedure cycles between restricting the edits of G to two different sets: the neighborhood of p_{best}, and the whole program space $\mathcal{P}$ (Line 13 in Algorithm 1). In practice, such restrictions are softly enforced through in-context prompting of VLMs, and though their inputs encourage them to abide by these constraints, the model can still produce more drastic "tweak" edits or conservative "leap" edits as needed.

Visual Imagination. In the case when the user intention is communicated purely textually, their intention may be difficult for the VLM to turn into successful edits. Prior works like [21,50,55] have made similar observations of abstract language for 3D scenes. Consider, for example, the prompt "make me a material that resembles a *celestial nebula*". To do this, the VLM must know what a celestial nebula looks like, *and* how it should change the parameters of the material shader nodes of, say, a wooden material. We find that in such cases, it's hard for the VLM to directly go from abstract descriptions to low-level program edits that affect low-level properties of the Blender visual state.

Instead, we propose supplementing the text-to-program understanding of VLM's with the text-to-image understanding in state-of-the-art image generation models. Intermediary visual artifacts (images generated using the user intention) are created and used to guide the refinement process towards a more plausible program edit to match the desired outcome, as shown in Fig. 1. The generated images act as image references *in addition to the textual intention provided by the user*. This constitutes a simple visual chain of thought [49] for visual program editing, which not only creates a reference image for G and V to guide their low-level visual comparisons (e.g. color schemes, material roughness... *etc.*), but also provides a user-interpretable intermediary step to confirm the desired goals behind an otherwise vague user intention.

4 Experiments

We demonstrate BlenderAlchemy on editing procedural materials, geometry and lighting setups within Blender, three of the most tedious parts of 3D design.

4.1 Procedural Material Editing

Procedural material editing has characteristics that make it difficult for the same reason as a lot of other visual program settings: (1) small edit distances of programs may result in very large visual differences, contributing to potential instabilities in the edit discovery process, (2) it naturally requires trial-and-error when human users are editing them as the desired magnitude of edits depends on the visual output and (3) language descriptions of edit intent typically do not contain low-level information for the editor to know immediately which part of the program to change.

We demonstrate this on two different kinds of editing tasks: (1) turning the same initial material (a synthetic wooden material) into other materials described by a list of **language descriptions** and (2) editing many different initial materials to resemble the same target material described by an **image input**. For the starting materials, we use the synthetic materials from Infinigen [32].

Text-Based Material-Editing. An attractive application of our system is in the modification of preexisting procedural materials using natural language descriptions that communicate user intent, a desireable but thus far undemonstrated capability of neurosymbolic methods [34]. We demonstrate our system's capabilities by asking it to edit a wooden material from Infinigen [32] into many other materials according to diverse language descriptions of target materials that are, importantly, *not wood*. In reality, this is a very challenging task, since this may require a wide range in the size of edits even if the language describing desired target material may be very similar.

Figure 3 shows examples of edits of the same starter wood material using different language descriptions, and Fig. 4 demonstrates the intermediary steps of the problem solving process for a single instance of the problem. Our system is composed of an edit generator that generates 8 hypotheses per iteration, for 4 iterations ($d = 4, b = 8$), cycling between tweak and leap edits. It uses GPT-4V for edit generation and state evaluation, and DallE-3 for visual imagination.

We compare against BlenderGPT, the most recent open-sourced Blender AI agent that use GPT-4 to execute actions within the Blender environment through the Python API.[1] We provide the same target material text prompt to BlenderGPT, as well as the starter code for the initial wood material for reference. We compare the CLIP similarity of their output material to the input text description against our system. BlenderGPT reasons only about how to edit the program using the input text description, doing so in a single pass without state evaluation or multi-hypothesis edit generation. To match the number of edit generator queries we make, we run their method a maximum of 32 times, using the first successful example as its final output. Everything else remains the same, including the starter material program, text description, base Blender state, and lighting setup.

We find that qualitatively, BlenderGPT produces much shallower and more simplistic edits of the input material, resulting in low-quality output materials and poor alignment with the user intention. Examples can be seen in Fig. 5. For instance, observe that for the "digital camouflage" example, BlenderAlchemy is able to produce the "sharper angles" that the original description requests (See Fig. 3) whereas BlenderGPT produces the right colors but fails to create the sharp, digital look. For "metallic swirl" example, our system's visual state selection process would weed out insufficiently swirly examples such as the one given by BlenderGPT, enabling our method to produce a material closer to the prompt.

---

[1] At the time of publishing, works like 3DGPT [42], L3GO [53] have not yet opensourced their code.

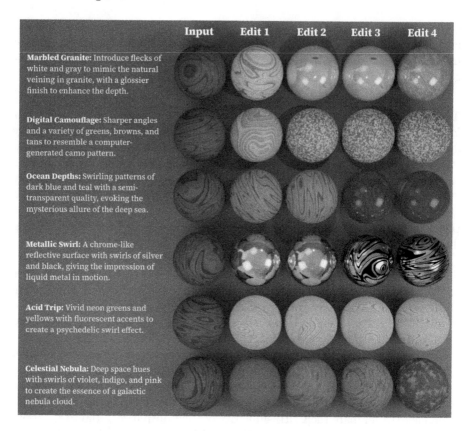

Fig. 3. Text-based Material Editing Results. The step-by-step edits of a 4x8 version of BlenderAlchemy to the same wooden material, given the text description on the left as the input user intention.

Table 1 demonstrates the comparison in terms of the average ViT-B/32 and ViT-L/14 CLIP similarity scores [31] with respect to the language description. Our system's ability to iteratively refine the edits based on multiple guesses at each step gives it the ability to make more substantive edits over the course of the process. Moreover, visual grounding provided both by the visual state evaluator as well as the output of the visual imagination stage guides the program search procedure to better align with the description. For ablations, see the Appendix.

We additionally conducted a user preference study to compare BlenderAlchemy's performance with two material generation baselines that use diffusion: TEXTure [33] and Paint3D [60]. We collect 592 Mechanical Turk comparisons between BlenderAlchemy and the baselines from 24 Turkers on materials created using 32 different text prompts. Users must choose the material that best matches the text-prompt. The results show that BlenderAlchemy is preferred to Paint3D 73% of the time. Users picked BlenderAlchemy over TEXTure 56% of the time. We found that in 71% of comparisons, human users also prefer

Fig. 4. The edit discovery process of turning a wooden material into "marbled granite". Each column shows the hypotheses generated by G, with the most promising candidates chosen by V indicated by the highlights. Note that iteration 3 proved to be unfruitful according to V, and the method reverts to the best candidate from iteration 2, before moving onto iteration 4.

Table 1. CLIP scores of BlenderAlchemy vs. BlenderGPT for the text-based material editing task. We find that a version of our system that has *no visual components* (**-Vision**) still outperforms BlenderGPT. By adding vision to the state evaluator alone (**-Vision G**) and not the edit generator, a further improvement is observed.

Metric	BlenderGPT	-Vision	-Vision G	Ours
ViT-B/32 (↑)	25.2	25.7	27.8	**28.2**
ViT-L/14 (↑)	21.1	21.8	23.4	**24.0**

BlenderAlchemy outputs selected by our visual evaluator. This suggests that our visual evaluator makes decisions that align with users' preferences.

Image-Based Material-Editing. Given an image of a desired material, the task is to convert the code of the starter material into a material that contains many of the visual attributes of the input image, akin to doing a kind of style transfer for procedural materials.

At each step, the edit generator is first asked to textually enumerate a list of obvious visual differences between the current material and the target, then asked to locate lines within the code that may be responsible for these visual differences (*e.g.* "the target material looks more rough" → "line 23 sets a roughness value, which we should try to increase") before finally suggested an edited version of the program. As such, low-level visual differences (*e.g.* color discrepancies) are semantically compressed first into language descriptions (*e.g.* "the target is more red"), before being fed into the editing process, resulting in the behavior that our system produces variations of the input material that resembles the target image along many attributes, even if its outputs don't perfectly match the target image (See Fig. 6). Our system is the same as for text-based material editing, but without the need for visual imagination.

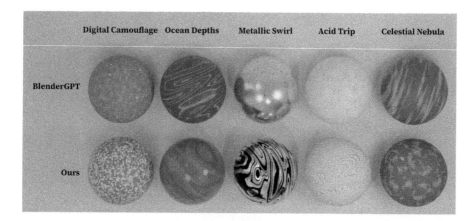

Fig. 5. Comparisons between our method and BlenderGPT for the text-based material editing task setting. Note how our materials align better with the original language prompts (See Fig. 3 for the original prompts). The input material being edited is the same wooden material.

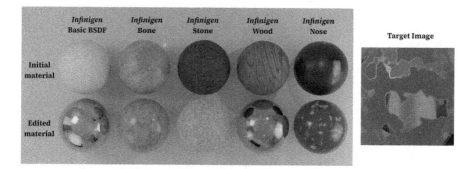

Fig. 6. Material editing based on image inputs. Our edit intention is described by the target image shown on the right. 5 different Infinigen [32] initial materials are shown here, and the final edit. Note how in each case, certain attributes of the target material (metallicness, color, texture) are transferred.

4.2 Geometry Editing

BlenderAlchemy can also be used to manipulate geometry. We show qualitative examples of our system controlling geometry by programmatically 1) interpolating between preset blend shapes, 2) editing of geometry node graphs, and 3) the precise placement of objects within scenes. The blend weights, geometry node graphs and object placements can all be represented programmatically.

As shown in Fig. 7, given an input shape and pre-defined blend shapes, BlenderAlchemy can modify the shape by finding the appropriate blend weights between different shapes to match language, which is non-trivial for human users given the number of blend shapes to interpolate for complex characters and shapes. For instance, the facial expression editing example requires coordinated interpolation between 19 different blend shapes, each which controls a specific

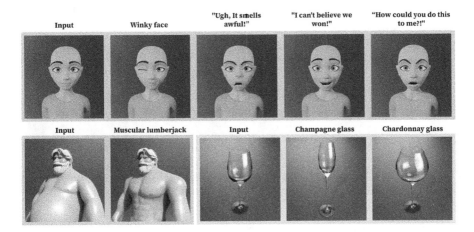

Fig. 7. BlenderAlchemy editing geometry using blend shapes. Edits are made to match a description or a script line. Input shapes from BlenderKit.

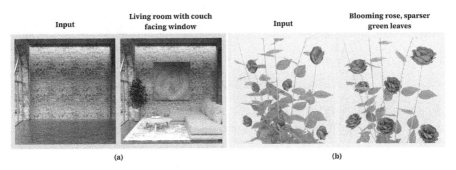

Fig. 8. BlenderAlchemy editing (a) the placement of assets within a living room scene and (b) the procedural geometry nodes of roses, to different text prompts.

kind of motion in one part of the face (e.g. an left eye-brow raise). To aid it in this task, our task description to BlenderAlchemy includes language labels for each blend shape.

In Fig. 8(a), BlenderAlchemy is presented with a scene layout editing task, where assets are initially dropped out of camera view and must be iteratively moved to fit the scene of a living room and the requirement that the couch is "facing the window", through calling a set of object placement helper functions. Figure 8(b) shows examples of BlenderAlchemy editing the procedural geometry of a set of roses to match a nuanced instruction that influences the number and angle of pedals, as well as the density of leaves. The full geometry node graph that BlenderAlchemy edits contains 60 geometry nodes.

4.3 Lighting Setup Editing

We show that BlenderAlchemy can be used to adjust the lighting of scenes according to language instructions as well. Figure 9 shows this being done for an input product visualization designed by a Blender artist.

"Hand lotion under disco lights, like at a nightclub."

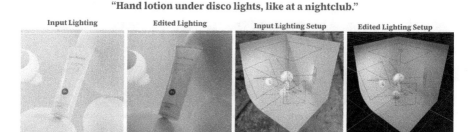

Fig. 9. Optimizing the lighting of the scene setup to the text-based user intention. We base the initial Blender state input based on a product visualization downloaded from BlenderKit.

"Hand cream sitting on a hot comet outside in the night"

Fig. 10. Optimizing the lighting and material iteratively within a product visualization scene, to satisfy the text-based user intention. Note the dimming of the environment lighting for the nighttime lighting, and the glowing-hot material the editing procedure has produced. We base the initial Blender state input based on a product visualization downloaded from BlenderKit.

As mentioned in Sect. 3.1, we can consider iteratively optimizing two separate programs, one controlling the lighting of the whole scene and another controlling the material of an object within the scene. That is, $S_{\text{init}} = F(\{p_0^{(L)}, p_0^{(M)}\}, S_{\text{base}})$ for initial lighting program $p_0^{(L)}$ and material program $p_0^{(M)}$. Figure 10 shows an example of this, where two separate pairs of edit generators and state evaluators, (G_M, V_M) and (G_L, V_L), are used to achieve an edit to the scene that aligns with the user intention. The full algorithm is shown in the Appendix.

5 Conclusion and Discussion

In this paper, we introduce BlenderAlchemy, a system that performs edits within the Blender 3D design environment by leveraging vision-language models to iteratively refining a program to be more aligned with the user intention, by using visual information to both explore and prune possibilities within the program space. We equip our system with visual imagination by providing it access to text-to-image models, a tool it uses to guide itself towards program edits that better align with user intentions. We've demonstrated BlenderAlchemy on editing materials, geometry and lighting, and hope that future works will extend this to other workflows as well.

Acknowledgements. We acknowledge the support of ARL grant W911NF-21-2-0104 and a Vannevar Bush Faculty Fellowship. We'd additionally like to thank Maneesh Agrawala for general discussions, and Purvi Goel, Mika Uy, Vishnu Sarukkai, Fan-yun Sun and Sharon Lee for feedback on paper revisions.

References

1. Blendergpt. https://github.com/gd3kr/BlenderGPT
2. How long does it take to create a 3D model? https://3d-ace.com/blog/how-long-does-it-take-to-create-a-3d-model/
3. How long does it take to make a 3D model? https://pixune.com/blog/how-long-does-it-take-to-create-a-3d-model/
4. Ahn, M., et al.: Do as i can, not as i say: grounding language in robotic affordances. arXiv preprint arXiv:2204.01691 (2022)
5. Austin, J., et al.: Program synthesis with large language models. arXiv preprint arXiv:2108.07732 (2021)
6. Baumli, K., et al.: Vision-language models as a source of rewards. arXiv preprint arXiv:2312.09187 (2023)
7. Betker, J., et al.: Improving image generation with better captions. Comput. Sci. **2**(3), 8 (2023). https://cdn.openai.com/papers/dall-e-3.pdf
8. Chen, D.Z., Siddiqui, Y., Lee, H.Y., Tulyakov, S., Nießner, M.: Text2tex: text-driven texture synthesis via diffusion models. arXiv preprint arXiv:2303.11396 (2023)
9. Chen, M., et al.: Evaluating large language models trained on code (2021)
10. Chen, Y., Chen, R., Lei, J., Zhang, Y., Jia, K.: Tango: text-driven photorealistic and robust 3D stylization via lighting decomposition. Adv. Neural. Inf. Process. Syst. **35**, 30923–30936 (2022)
11. De La Torre, F., Fang, C.M., Huang, H., Banburski-Fahey, A., Fernandez, J.A., Lanier, J.: LLMR: real-time prompting of interactive worlds using large language models. arXiv preprint arXiv:2309.12276 (2023)
12. Firoozi, R., et al.: Foundation models in robotics: applications, challenges, and the future. arXiv preprint arXiv:2312.07843 (2023)
13. Fu, C., et al.: MME: a comprehensive evaluation benchmark for multimodal large language models. arXiv preprint arXiv:2306.13394 (2023)
14. Fu, C., et al.: A challenger to GPT-4V? Early explorations of gemini in visual expertise. arXiv preprint arXiv:2312.12436 (2023)
15. Goel, P., Wang, K.C., Liu, C.K., Fatahalian, K.: Iterative motion editing with natural language. arXiv preprint arXiv:2312.11538 (2023)
16. Guerrero, P., Hašan, M., Sunkavalli, K., Měch, R., Boubekeur, T., Mitra, N.J.: Matformer: a generative model for procedural materials. arXiv preprint arXiv:2207.01044 (2022)
17. Henzler, P., Deschaintre, V., Mitra, N.J., Ritschel, T.: Generative modelling of BRDF textures from flash images. arXiv preprint arXiv:2102.11861 (2021)
18. Hu, Y., et al.: Toward general-purpose robots via foundation models: a survey and meta-analysis. arXiv preprint arXiv:2312.08782 (2023)

19. Hu, Y., Guerrero, P., Hasan, M., Rushmeier, H., Deschaintre, V.: Node graph optimization using differentiable proxies. In: ACM SIGGRAPH 2022 Conference Proceedings, pp. 1–9 (2022)
20. Hu, Y., He, C., Deschaintre, V., Dorsey, J., Rushmeier, H.: An inverse procedural modeling pipeline for SVBRDF maps. ACM Trans. Graph. (TOG) **41**(2), 1–17 (2022)
21. Huang, I., Krishna, V., Atekha, O., Guibas, L.: Aladdin: zero-shot hallucination of stylized 3D assets from abstract scene descriptions. arXiv preprint arXiv:2306.06212 (2023)
22. Jiang, A.Q., et al.: Mistral 7B. arXiv preprint arXiv:2310.06825 (2023)
23. Li, C., et al.: LLaVA-MED: training a large language-and-vision assistant for biomedicine in one day. In: Advances in Neural Information Processing Systems, vol. 36 (2024)
24. Liang, J., et al.: Code as policies: language model programs for embodied control. In: 2023 IEEE International Conference on Robotics and Automation (ICRA), pp. 9493–9500. IEEE (2023)
25. Liu, J., et al.: Perception-driven procedural texture generation from examples. Neurocomputing **291**, 21–34 (2018)
26. Olausson, T.X., Inala, J.P., Wang, C., Gao, J., Solar-Lezama, A.: Is self-repair a silver bullet for code generation? In: The Twelfth International Conference on Learning Representations (2023)
27. OpenAI: GPT-4 system card. OpenAI (2023). https://cdn.openai.com/papers/gpt-4-system-card.pdf
28. OpenAI: GPT-4v(ision) system card. OpenAI (2023). https://api.semanticscholar.org/CorpusID:263218031
29. Park, J.S., O'Brien, J., Cai, C.J., Morris, M.R., Liang, P., Bernstein, M.S.: Generative agents: interactive simulacra of human behavior. In: Proceedings of the 36th Annual ACM Symposium on User Interface Software and Technology, pp. 1–22 (2023)
30. Patil, S.G., Zhang, T., Wang, X., Gonzalez, J.E.: Gorilla: large language model connected with massive APIs. arXiv preprint arXiv:2305.15334 (2023)
31. Radford, A., et al.: Learning transferable visual models from natural language supervision. In: International Conference on Machine Learning, pp. 8748–8763. PMLR (2021)
32. Raistrick, A., et al.: Infinite photorealistic worlds using procedural generation. In: Proceedings of the IEEE/CVF Conference on Computer Vision and Pattern Recognition, pp. 12630–12641 (2023)
33. Richardson, E., Metzer, G., Alaluf, Y., Giryes, R., Cohen-Or, D.: Texture: text-guided texturing of 3D shapes. arXiv preprint arXiv:2302.01721 (2023)
34. Ritchie, D., et al.: Neurosymbolic models for computer graphics. In: Computer Graphics Forum, vol. 42, pp. 545–568. Wiley Online Library (2023)
35. Romera-Paredes, B., et al.: Mathematical discoveries from program search with large language models. Nature **625**(7995), 468–475 (2024)
36. Schick, T., et al.: Toolformer: language models can teach themselves to use tools. In: Advances in Neural Information Processing Systems, vol. 36 (2024)
37. Sharma, P., et al.: Alchemist: parametric control of material properties with diffusion models. arXiv preprint arXiv:2312.02970 (2023)
38. Shi, L., et al.: Match: differentiable material graphs for procedural material capture. ACM Trans. Graph. (TOG) **39**(6), 1–15 (2020)

39. Shimizu, E., Fisher, M., Paris, S., McCann, J., Fatahalian, K.: Design adjectives: a framework for interactive model-guided exploration of parameterized design spaces. In: Proceedings of the 33rd Annual ACM Symposium on User Interface Software and Technology, pp. 261–278 (2020)
40. Shinn, N., Cassano, F., Labash, B., Gopinath, A., Narasimhan, K., Yao, S.: Reflexion: language agents with verbal reinforcement learning (2023). arXiv preprint cs.AI/2303.11366 (2023)
41. Singh, I., et al.: Progprompt: generating situated robot task plans using large language models. In: 2023 IEEE International Conference on Robotics and Automation (ICRA), pp. 11523–11530. IEEE (2023)
42. Sun, C., Han, J., Deng, W., Wang, X., Qin, Z., Gould, S.: 3D-GPT: procedural 3D modeling with large language models. arXiv preprint arXiv:2310.12945 (2023)
43. Tchapmi, L.P., Ray, T., Tchapmi, M., Shen, B., Martin-Martin, R., Savarese, S.: Generating procedural 3D materials from images using neural networks. In: 2022 4th International Conference on Image, Video and Signal Processing, pp. 32–40 (2022)
44. Team, G., et al.: Gemini: a family of highly capable multimodal models. arXiv preprint arXiv:2312.11805 (2023)
45. Touvron, H., et al.: Llama: open and efficient foundation language models. arXiv preprint arXiv:2302.13971 (2023)
46. Vecchio, G., et al.: Controlmat: a controlled generative approach to material capture. arXiv preprint arXiv:2309.01700 (2023)
47. Vecchio, G., Sortino, R., Palazzo, S., Spampinato, C.: Matfuse: controllable material generation with diffusion models. arXiv preprint arXiv:2308.11408 (2023)
48. Wang, G., et al.: Voyager: an open-ended embodied agent with large language models. arXiv preprint arXiv:2305.16291 (2023)
49. Wei, J., et al.: Chain-of-thought prompting elicits reasoning in large language models. Adv. Neural. Inf. Process. Syst. **35**, 24824–24837 (2022)
50. Wen, Z., Liu, Z., Sridhar, S., Fu, R.: Anyhome: open-vocabulary generation of structured and textured 3D homes. arXiv preprint arXiv:2312.06644 (2023)
51. Wu, T., et al.: GPT-4V (ision) is a human-aligned evaluator for text-to-3D generation. arXiv preprint arXiv:2401.04092 (2024)
52. Xiao, X., et al.: Robot learning in the era of foundation models: a survey. arXiv preprint arXiv:2311.14379 (2023)
53. Yamada, Y., Chandu, K., Lin, Y., Hessel, J., Yildirim, I., Choi, Y.: L3go: language agents with chain-of-3D-thoughts for generating unconventional objects. arXiv preprint arXiv:2402.09052 (2024)
54. Yang, H., Chen, Y., Pan, Y., Yao, T., Chen, Z., Mei, T.: 3dstyle-diffusion: pursuing fine-grained text-driven 3D stylization with 2D diffusion models. In: Proceedings of the 31st ACM International Conference on Multimedia, pp. 6860–6868 (2023)
55. Yang, Y., et al.: Holodeck: language guided generation of 3D embodied AI environments. In: The IEEE/CVF Conference on Computer Vision and Pattern Recognition (CVPR 2024), vol. 30, pp. 20–25. IEEE/CVF (2024)
56. Yang, Z., et al.: The dawn of LMMs: preliminary explorations with GPT-4V (ision). arXiv preprint arXiv:2309.17421, vol. 9, no. 1, p. 1 (2023)
57. Yao, S., et al.: Tree of thoughts: deliberate problem solving with large language models. In: Advances in Neural Information Processing Systems, vol. 36 (2024)
58. Yin, S., et al.: A survey on multimodal large language models. arXiv preprint arXiv:2306.13549 (2023)
59. Yin, S., et al.: Woodpecker: hallucination correction for multimodal large language models. arXiv preprint arXiv:2310.16045 (2023)

60. Zeng, X.: Paint3d: paint anything 3D with lighting-less texture diffusion models. arXiv preprint arXiv:2312.13913 (2023)
61. Zhou, H., et al.: Language-conditioned learning for robotic manipulation: a survey. arXiv preprint arXiv:2312.10807 (2023)
62. Zsolnai-Fehér, K., Wonka, P., Wimmer, M.: Gaussian material synthesis. arXiv preprint arXiv:1804.08369 (2018)

DϵpS: Delayed ϵ-Shrinking for Faster Once-for-All Training

Aditya Annavajjala[1]([✉]), Alind Khare[1], Animesh Agrawal[1], Igor Fedorov[3], Hugo Latapie[2], Myungjin Lee[2], and Alexey Tumanov[1]

[1] Georgia Institute of Technology, Atlanta, USA
adityaas@gatech.edu
[2] Cisco Research, San Jose, USA
[3] Meta, Menlo Park, USA

Abstract. CNNs are increasingly deployed across different hardware, dynamic environments, and low-power embedded devices. This has led to the design and training of CNN architectures with the goal of maximizing accuracy subject to such variable deployment constraints. As the number of deployment scenarios grows, there is a need to find scalable solutions to design and train specialized CNNs. Once-for-all training has emerged as a scalable approach that jointly co-trains many models (subnets) at once with a constant training cost and finds specialized CNNs later. The scalability is achieved by training the full model and simultaneously reducing it to smaller subnets that share model weights (weight-shared shrinking). However, existing once-for-all training approaches incur huge training costs reaching 1200 GPU hours. We argue this is because they either start the process of shrinking the full model *too* early or *too* late. Hence, we propose Delayed ϵ-Shrinking (DϵpS) that starts the process of shrinking the full model when it is *partially* trained (∼50%), which leads to training cost improvement and better in-place knowledge distillation to smaller models. The proposed approach also consists of novel heuristics that dynamically adjust subnet learning rates incrementally (ϵ), leading to improved weight-shared knowledge distillation from larger to smaller subnets as well. As a result, DϵpS outperforms state-of-the-art once-for-all training techniques across different datasets including CIFAR10/100, ImageNet-100, and ImageNet-1k on accuracy and cost. It achieves 1.83% higher ImageNet-1k top1 accuracy or the same accuracy with 1.3x reduction in FLOPs and 2.5x drop in training cost (GPU*hrs).

Code is released at https://github.com/gatech-sysml/deps.

Keywords: Scalable vision · NAS · Efficient Inference

1 Introduction

CNNs are pervasive in numerous applications including smart cameras [2], smart surveillance [6], self-driving cars [26], search engines [12], and social media [1].

A. Annavajjala and A. Khare—Authors contributed equally to this research.

Supplementary Information The online version contains supplementary material available at https://doi.org/10.1007/978-3-031-73024-5_19.

As a result, they are increasingly deployed across diverse hardware ranging from server-grade GPUs like V100 [19] to edge-GPUs like Nvidia Jetson [18] and dynamic environments like Autonomous Vehicles [8] that operate under strict latency or power budget constraints. As the diversity in deployment scenarios grows, efficient deployment of CNNs under a myriad of deployment constraints becomes challenging. It calls for developing techniques that find appropriate CNNs suited for different deployment conditions.

Neural Architecture Search (NAS) [4,32] has emerged as a successful technique that finds CNN architectures specialized for a deployment target. It searches for an appropriate CNN architecture and trains it with the goal of maximizing accuracy subject to the target deployment constraints. However, state-of-the-art NAS techniques remain prohibitively expensive, requiring many GPU hours due to the costly operation of the combined search and training of specialized CNNs. The problem is exacerbated when NAS is employed to satisfy multiple deployment targets, as it must be run repeatedly for *each* deployment target. This makes the cost of NAS linear in the number of deployment targets considered ($O(k)$), which is prohibitively expensive and doesn't scale with the growing number of deployment targets. Therefore, there is a need to develop *scalable* NAS solutions able to satisfy multiple deployment targets efficiently.

One such technique is Once-for-all training [3,29,40]—a step towards making NAS computationally feasible to satisfy multiple deployment targets by decoupling training from search. It achieves this decoupling by co-training a family of models (weight-shared subnets with varied shapes and sizes) embedded inside a supernet *once*, incurring a constant training cost. After the supernet is trained, NAS can then be performed for any specific deployment target by simply extracting a specialized subnet from the supernet without retraining (*once-for-all*).

This achieves $O(1)$ training cost w.r.t the number of deployment targets and, therefore, makes NAS scalable. However, the efficiency of this once-for-all training remains limited as it incurs a significant training cost (~1200 GPU hours in [3]). This is primarily due to (a) the large number of training epochs required to overcome training interference (OFA [3] in Fig. 1), and (b) the high average time per-epoch caused by shrinking—defined as sampling and adding smaller

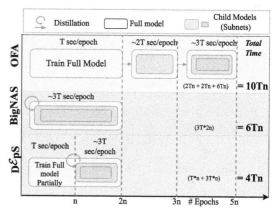

Fig. 1. DϵpS reduces training time compared to existing approaches like OFA [3] & BigNAS [40].

subnets to the training schedule—per minibatch (BigNAS [40] in Fig. 1). Thus, in order to make once-for-all training more efficient, we must reduce its training time without sacrificing state-of-the-art accuracy across the whole operating latency/FLOP range of the supernet.

We propose DϵpS, a technique that increases the scalability of once-for-all training. It consists of three key components designed to meet their respective goals—Full Model warmup (FM-Warmup) provides better supernet initialization, ϵ-Shrinking keeps the accuracy of the full model (largest subnet that contains all the supernet parameters) on par with OFA and BigNAS, and IKD-Warmup boosts the accuracy of small subnets with effective knowledge distillation in once-for-all training. Particularly, with better supernet initialization, FM-Warmup (DϵpS in Fig. 1) reduces both the total number of epochs (compared to OFA) and average time per-epoch (compared to BigNAS). In FM-Warmup, the supernet is initialized with the partially trained full model (~50%) and then subnet sampling (shrinking) is started to train the model family. The partial full model training ensures a lower time per epoch initially. Then, ϵ-Shrinking ensures smooth optimization of the full model. It incrementally warms up the learning rate of subnets using parameter ϵ when the shrinking starts, while keeping the learning rate of the full model higher. Lastly, IKD-Warmup enables knowledge distillation from multiple partially trained full models (that are progressively better) to smaller subnets. The three components, when combined, reduce the training time of once-for-all training and outperform state-of-the-art w.r.t accuracy of subnets across different datasets and neural network architectures. We summarize the contributions of our work as follows:

- FM-Warmup provides better initialization to the weight shared supernet by training the full model only partially and delaying model shrinking. This leads to reduced time per epoch and lower training cost.
- ϵ-Shrinking ensures smooth and fast optimization of the full model by warming up the learning rate of smaller subnets. This enables it to reach optimal accuracy quickly.
- IKD-Warmup provides rich knowledge transfer to subnets, enabling them to quickly learn good representations.

We extensively evaluate DϵpS against existing once-for-all training baselines [3,29,40] on CIFAR10/100 [21], ImageNet-100 [34], and ImageNet-1k [7] datasets. DϵpS outperforms all baselines across all the datasets both w.r.t accuracy (of subnets) and training cost. It achieves 1.83% ImageNet-1k top1 accuracy improvement or the same accuracy with 1.3x FLOPs reduction while reducing training cost by upto 1.8x w.r.t OFA and 2.5x w.r.t BigNAS (in dollars or GPU hours). We also provide a detailed ablation study to demonstrate the benefits of DϵpS components in isolation.

2 Background

Formulation. Let W_o denote the supernet's weights, the objective of once-for-all training is given by—

$$\min_{W_o} \sum_{a \in \mathcal{A}} \mathcal{L}(S(W_o, a)) \tag{1}$$

where $S(W_o, a)$ denotes weights of subnet a selected from the supernet's weight W_o and $\mathcal{A}$ represents the set of all possible neural architectures (subnets). The goal of once-for-all training is to find optimal supernet weights that minimize the loss ($\mathcal{L}$) of all the neural architectures in $\mathcal{A}$ on a given dataset.

Challenges. However, optimizing (1) is non-trivial. On one hand, enumerating gradients of all subnets to optimize the overall objective is computationally infeasible. This is due to the large number of subnets optimized in once-for-all training ($|\mathcal{A}| \approx 22B$ subnets in our case[1]). On the other hand, a naive approximation of objective (1) to make it computationally feasible leads to *interference* (sampling a few subnets in each update step). *Interference* occurs when smaller subnets affect the performance of the larger subnets [3,40]. Hence, interference causes sub-optimal accuracy of the larger subnets. Existing once-for-all training techniques mitigate interference by increasing the training time significantly (Fig. 1). For instance, OFA [3] mitigates interference by first training the full model (largest subnet) and then progressively increasing the size of $|\mathcal{A}|$. This leads to a large number of training epochs and ≈ 1200 GPU hours to perform once-for-all training. Therefore, the following challenges remain in once-for-all training—**(C1)** training supernet at a lesser training cost than SOTA, *and* **(C2)** mitigating interference. We divide challenge **C2** into two sub-challenges—matching existing once-for-all training techniques [3,40] w.r.t.accuracy of **(C2a)** the full model (largest subnet), and **(C2b)** child models (smaller subnets).

3 Related Work

Efficient NN-Architectures in Deep Learning. Efficient deep neural networks (NNs) achieve high accuracy at low FLOPs. These neural nets are easy to deploy as they increase hardware efficiency by operating at low FLOPs. Developing such networks is an active research area. Several efficient neural networks include MobileNets [15], SqueezeNets [17], EfficientNets [33], and TinyNets [10].

Neural Network Compression. Neural network compression reduces the size and computation of neural networks for efficient deployment. The compression occurs after the network is trained. Hence, the performance of compression methods is bounded by the accuracy of the trained neural network. Neural network compression can be broadly divided into two categories—network pruning and quantization. Network pruning removes unimportant units [11,25,30] or channels [22,23,31]. Network quantization converts the representation of neural weights and activations to low bits [16,20,37].

Hardware Aware NAS. Neural architecture search (NAS) automates the design of efficient NN architectures. NAS typically involves searching for and training NN architectures that are more accurate than manually designed NNs [27,42]. Recently, NAS methods are becoming hardware-aware [4,32,38] *i.e.*they find NN architectures suited for deployment at target hardware. These methods

---

[1] 5 stages, 3 depths (2, 3, 4), 3 expands = $(3^2 + 3^3 + 3^4)^5 \approx 22B$ subnets.

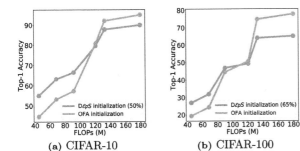
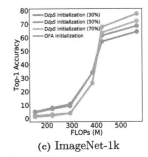

(a) CIFAR-10   (b) CIFAR-100   (c) ImageNet-1k

Fig. 2. Supernetwork initialization. DϵpS provides better initialization for the supernetwork for smaller subnets compared to OFA due to FM-Warmup. This validates the hypothesis that the supernet weights become specialized if the full model is trained to completion (OFA), resulting in poorer accuracy of subnetworks with increased training of the full model.

incorporate deployment constraints of hardware or latency in their search. Then, they find and train efficient NNs that meet the constraints. However, these NAS methods only satisfy a single deployment target. They need to run repeatedly for each deployment target that doesn't scale well.

Once-for-All Training. Once-for-all training is a scalable NAS method that satisfies multiple deployment targets. It co-trains models (subnets) that vary in shape and size embedded inside a single weight-shared supernet. NAS is performed later by extracting specialized subnets from the trained supernet for target hardware. OFA [3], BigNAS [40], and CompOFA [29] exemplify once-for-all training methods. OFA proposes Progressive Shrinking (PS) for once-for-all training that trains the full model first and then progressively introduces smaller subnets into training by dividing the training procedure into multiple training jobs (phases). Compared to OFA, DϵpS performs once-for-all training as a single training job and starts shrinking from a partially trained full model to reduce the training cost. BigNAS starts the process of shrinking early and samples multiple subnets at every minibatch. In contrast, DϵpS initially only trains the full model and delays the shrinking. Finally, CompOFA changes the architecture search space of OFA and performs Progressive Shrinking with reduced phases. DϵpS algorithmically changes the shrinking procedure in once-for-all training and is complementary to architecture space changes proposed in CompOFA.

4 Proposed Approach

We present DϵpS, a once-for-all training technique that trains supernets in less training time. DϵpS consists of three key components that meet the challenges **C1** and **C2**. We describe each component in detail and highlight the core contributions of our work.

4.1 Full-Model Warmup Period (P_{warmup}^{fm}): When to Shrink the Full Model?

Shrinking the full model at an appropriate time is vital for reducing training cost (meet **C1**). Both early or late shrinking isn't sufficient to meet the challenges in once-for-all training. Early shrinking (BigNAS [40] in Fig. 1) doesn't meet the challenge **C1**. It increases the overall training time as multiple subnets are sampled in each update (increasing per-epoch time) to optimize objective (1). Early shrinking also requires a lot of hyper-parameter tuning to meet challenge **C2**. It becomes sensitive to training hyper-parameters due to interference. For instance, training the full model with early shrinking becomes unstable with the standard initialization of the full model [40].

On the other hand, if shrinking happens late after the full model is completely trained (OFA [3] in Fig. 1), the supernet weights become too specialized for the full model architecture and require a large number of training epochs to reduce interference. Hence, late shrinking meets challenge **C2** but not **C1**.

We argue that shrinking should occur after the full model is partially trained (warmed up, trained at least 50%, proposed approach in Fig. 1).

Delayed Shrinking has numerous advantages. It reduces the overall training time to meet challenge **C1**. The initial updates in DϵpS are cheap compared to early shrinking as only the full model gets trained and no subnets are sampled. Moreover, since supernet weights are not specialized for the full model, DϵpS can meet challenge **C2** in less number of epochs. To validate our hypothesis, we ask whether a partially trained full model serves as a good initialization for the supernet. To do this, we compare the accuracy of small subnets (shrinking) on multiple datasets (CIFAR-10, CIFAR-100, ImageNet-1k) in a mobilenet-based supernet [3] when initialized with a partially trained (50%), and completely trained full model (~600 MFLOPs) in Fig. 2.

The takeaway from the experiment in Fig. 2 is that a partially-trained full model-based initialization performs better for smaller subnets than the initialization with the full model completely trained. This validates our hypothesis that supernet weights become too specialized if the full model is trained to completion. Hence, warming up the full model helps in meeting challenge **C1**. DϵpS introduces a hyperparameter P_{warmup}^{fm} denoting the fraction of total epochs used to warm up the full model. P_{warmup}^{fm} is usually kept $\geq 50\%$ in DϵpS.

4.2 ϵ-Shrinking: Learning Rates for Subnets

In addition to the full model warmup, we propose ϵ-Shrinking that enables the full model to reach comparable accuracy with SOTA and meet challenge **C2a**. ϵ-Shrinking ensures that the full model's accuracy doesn't get affected when shrinking is introduced in between its training. When the shrinking starts, the learning rate of subnets is gradually ramped to reach the full model's learning rate (ϵ-Shrinking) as the full model gets sampled with other subnets in each update step.

Without the gradual warmup, the full model becomes prone to an accuracy drop as the supernet weights change rapidly at the start of shrinking. To understand this change, we compare the updates in the supernet with and without shrinking for a minibatch $\mathcal{B}$. Consider supernet weights W_t at iteration t. Without shrinking, the update is given by -

$$W_{t+1}^{noShrink} = W_t - \eta_t \underbrace{\nabla l_\mathcal{B}(S(W_t, a_{full}))}_{=G_{noShrink}^{\mathcal{B},t}} \quad (2)$$

where $l_\mathcal{B}(S(W_t, a_{full}))$ denotes the loss of the full model on minibatch $\mathcal{B}$ and equals $\frac{1}{|\mathcal{B}|}\sum_{x\in\mathcal{B}} l(x, S(W_t, a_{full}))$; x denotes the samples in $\mathcal{B}$. η_t denotes the learning rate at iteration t used to update the weights. Whereas introducing shrinking for the same supernet weights W_t yields the following update -

$$W_{t+1}^{Shrink} = W_t - \eta_t \underbrace{\left(\overbrace{\sum_{a\in\mathcal{U}_k(\mathcal{A})} \nabla l_\mathcal{B}(S(W_t, a))}^{\text{shrinking}} \right)}_{=G_{Shrink}^{\mathcal{B},t}} \quad (3)$$

where $\mathcal{U}_k(\mathcal{A})$ denotes uniformly sampling k subnets from the architecture space $\mathcal{A}$. This update step is the approximation of the objective (1). Clearly, the updates differ, it is *improbable* that $W_{t+1}^{Shrink} = W_{t+1}^{NoShrink}$. This difference in updates causes the supernet weights to change rapidly when shrinking is introduced. The rapid change in supernet weights causes degradation in the full model's accuracy. To avoid rapid changes in weights, a widely adopted technique is to use less aggressive learning rates via learning rate warmup schedules [9,13].

However, applying such principles in the context of weight-sharing is nontrivial but at the same time important. Our key idea is two-fold to a) always sample the full model with other subnets while shrinking, and b) use less aggressive learning rates for subnets at the start of shrinking. Particularly, it is important to ensure $G_{noShrink}^{\mathcal{B},t} \approx G_{Shrink}^{\mathcal{B},t}$ to make $W_{t+1}^{Shrink} \approx W_{t+1}^{NoShrink}$ initially when the shrinking starts. To do this, we introduce a parameter ϵ that controls the effective learning rate of subnets and makes $G_{noShrink}^{\mathcal{B},t} \approx G_{Shrink}^{\mathcal{B},t}$. The gradient in ϵ-Shrinking is given as follows -

$$G_{Shrink}^{\mathcal{B},t}(\epsilon_t) = G_{noShrink}^{\mathcal{B},t} + \epsilon_t * \overbrace{\sum_{a\in\mathcal{U}_{k-1}(\mathcal{A}\setminus\{a_{full}\})} \nabla l_\mathcal{B}(S(W_t, a))}^{\epsilon-\text{shrinking}} \quad (4)$$

where $\epsilon_t \in (0, 1]$. Note that the effective learning rate becomes $\eta_t * \epsilon_t$ for subnets and remains η_t for the full model in ϵ-Shrinking. Hence, slowly increasing ϵ_t warms up the effective learning of subnets. We start with a small value of ϵ_t (=10^{-4}) and increment it by a constant amount to reach 1. Once ϵ_t reaches 1, it stays constant for the rest of the training. We empirically verify if $G_{noShrink}^{\mathcal{B},t}, G_{Shrink}^{\mathcal{B},t}$ differ in magnitude (l_2-norm) and direction (cosine similarity) and whether ϵ-Shrinking is able to reduce the differences with $G_{noShrink}^{\mathcal{B},t}(\epsilon_t)$.

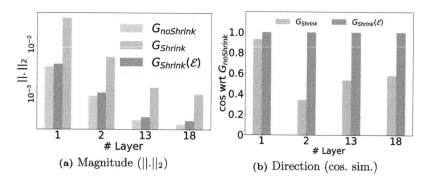

(a) Magnitude ($\|\cdot\|_2$) (b) Direction (cos. sim.)

Fig. 3. Gradients w/ & w/o Shrinking on Mobilenet-Based Supernet. Delayed Shrinking causes gradients (G_{Shrink}) to differ from the full model gradient ($G_{noShrink}$) leading to rapid changes in the supernet's weights. ϵ-Shrinking's gradient ($G_{Shrink}(\mathcal{E})$) reduces such differences and avoids rapid weight changes.

Figure 3 compares the magnitude and direction of the gradients of the full model ($G_{noShrink}$), shrinking (G_{Shrink}) and ϵ-Shrinking ($G_{noShrink}(\epsilon)$) ($\epsilon = 0.001$) on the weights of a mobilenet-based supernet [3] for the ImageNet-1k dataset [28]. $G_{noShrink}$ and G_{Shrink} differ both in magnitude and direction across supernet layers.

The magnitude of G_{Shrink} is an order of magnitude higher than $G_{noShrink}$ for early layers. ϵ-Shrinking maintains the low magnitude of gradient throughout the training as shown in Fig. 4. The magnitude of G_{Shrink} is consistently higher than $G_{Shrink}(\epsilon_t)$ when normalized with the magnitude of $G_{noShrink}$. Such differences cause poor convergence at the start of shrinking and often lead to accuracy drops. Whereas, $G_{noShrink}(\epsilon)$ has minimal differences w.r.t $G_{noShrink}$ enabling healthy convergence and no potential accuracy drops.

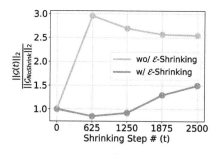

Fig. 4. Gradient Magnitude Over Time. Gradient magnitude with ($\mathcal{G}_{shrink}(\mathcal{E}, t)$) and without ($\mathcal{G}_{shrink}(t)$) $\mathcal{E}$-shrinking is compared w.r.t the initial full model gradient ($\mathcal{G}_{Noshrink}$) over shrinking steps. $\mathcal{E}$-shrinking avoids sudden changes in the supernet parameters by lowering the gradient magnitude.

4.3 IKD-Warmup: In-Place Knowledge Distillation (KD) from Warmed-up Full Model

We now discuss IKD-Warmup that distills knowledge from the full model to subnets and meets challenge **C2b**. Effectively distilling the knowledge from the full model becomes non-trivial due to weight-sharing. On one hand, KD requires the supernet weights biased to the full model to offer meaningful knowledge transfer to subnets. On the other hand, having a large bias in the supernet weights toward the full model may result in subnets' sub-optimal performance since the weights are shared. To tackle this trade-off, OFA [3] biases the supernet weights to a trained full model and then uses it to perform vanilla-KD [14]. However, this results in a long training time during shrinking as the supernet weights are trained to fit subnets' architectures. Another approach like BigNAS [40] doesn't bias the shared weights to the full model by using inplace-KD [39] but lacks in providing rich knowledge transfer to subnets (initially).

This is because inplace-KD distills the knowledge "on the fly" to other subnets as the full model gets trained from randomly initialized weights. Precisely, the full model predictions become ground truth for other subnets. Hence, when the full model is under-trained initially, it doesn't offer rich knowledge transfer.

We believe that the proposed delayed shrinking has an added advantage w.r.tKD for once-for-all training—the partially trained full model (50/60% trained) is rich enough to provide meaningful knowledge transfer to the subnets and doesn't bias the supernet weights to the full model. It has been shown that for vanilla-KD [14], partially trained (intermediate) models provide a comparable or at times better knowledge transfer than the completely trained models [5,36]. This is because they provide more information about non-target classes than the trained models [5]. We use this insight in DϵpS that performs inplace-KD from a partially trained full model (IKD-Warmup).

IKD-Warmup offers two advantages, it—a) distills knowledge from multiple progressively better partially trained models as the full model gets trained (unlike a single partially/fully trained model used in vanilla-KD [36]), and b) provides rich knowledge transfer to the subnets at all times (unlike inplace-KD [39] that distills from an under-trained full model initially).

5 Experiments

We establish that DϵpS **a**) reduces training cost w.r.tSOTA in once-for-all training [3,29,40], **b**) performs at-par or better than SOTA's accuracy across subnets (covering the entire range of architectural space), **c**) generalizes across datasets, **d**) generalizes to different deep neural network (DNN) architecture spaces, and **e**) produces specialized subnets for target hardware without retraining (once-for-all property). We also aim to demonstrate attribution of benefits in DϵpS by providing detailed ablation on **a**) a full model warmup period: empirically demonstrating a sweet spot, **b**) ϵ-Shrinking: showing healthy convergence, and **c**) IKD-Warmup: distilling knowledge better than existing distillation approaches in weight-sharing.

5.1 Setup

Baselines. We first compare DϵpS with the other NAS methods or efficient DNNs [4,15,33,35] w.r.taccuracy. Then, we compare DϵpS with once-for-all training techniques—OFA [3], BigNAS [40], CompOFA [29] w.r.tboth training cost and accuracy of subnets spanned across supernet's FLOP range. Training

Table 1. Comparison of DϵpS with state of the art neural architecture search approaches on ImageNet-1k. DϵpS consistently outperforms the baselines.

Group	Approach	MACs (M)	Top-1 Test Acc (%)
0–100 (M)	OFA [3]	67	70.5
	DϵpS	**67**	**72.3**
100–200 (M)	OFA [3]	141	71.6
	DϵpS	**141**	**73.7**
200–300 (M)	FBNetv2 [35]	238	76.0
	BigNAS [40]	242	76.5
	OFA [3]	230	76
	DϵpS	**230**	**77.3**
300–400 (M)	MNasnet [32]	315	75.2
	ProxylessNAS [4]	320	74.6
	FBNetv2 [35]	325	77.2
	MobileNetV3 [15]	356	76.6
	EfficientNetB0 [33]	390	77.3

time of all the techniques is measured on NVIDIA A40 GPUs. Accuracy of subnets is evaluated without additional finetuning. As once-for-all training trains multiple subnets, the comparison is done by uniformly dividing the entire FLOP range into 6 buckets and picking the most accurate subnet from each bucket for every baseline. All methods are evaluated on the same architecture space.

Success Metrics. DϵpS is compared against the baselines on the following success metrics—a) Training cost measured in GPU hours or dollars (lower is better), b) *Pareto-frontier*: Accuracy of best-performing subnets as a function of FLOPs/latency. To compare Pareto-frontiers obtained from different baselines, we use a metric called *mean pareto accuracy* that is defined as the area under the curve (AUC) of accuracy and normalized FLOPs/latency. The higher the mean pareto accuracy the better.

Datasets. We evaluate all methods on CIFAR10/100 [21], ImageNet-100 [34] and ImageNet-1k [7] datasets. The complexity of datasets progressively increases from CIFAR10 to ImageNet-1k. The datasets vary in the number of classes, image resolution, and number of train/test samples.

DNN Architecture Space. All methods are trained on the supernets derived from two different DNN architecture spaces—MobilenetV3 [15] and ProxylessNAS [4] (same as OFA [3]). The base architecture of ProxylessNAS is derived from ProxylessNAS run for the GPU as a target device. To avoid confounding, we evaluate all baselines on the same DNN architecture space.

Training Hyper-Parameters. The training hyper-parameters of DϵpS are similar to the hyper-parameters of the full model training. The hyper-parameters for MobilenetV3, and ProxylessNAS training are borrowed from [15] and [4]

Table 2. Comparison of DϵpS vs SOTA on ImageNet-1k. Accuracy and Training Cost comparison of DϵpS against SOTA approaches are shown for MobilenetV3-based architecture space. DϵpS outperforms SOTA and achieves 2% better accuracy for the smallest subnet and is at-par with the largest subnet (full model) respectively at 1.8x training cost reduction (in $) compared to OFA. Dollar-cost is calculated based on the on-demand prices for A40 GPUs from exoscale.com

Approach	Smallest Subnet		Largest Subnet		mean pareto acc.	Training Cost			Dollar Cost ($)
	Acc(%)	MACs (M)	Acc (%)	MACs (M)		# Epochs	Avg. GPU min./epoch	Total Time (GPU hours.)	
OFA [3]	71.8	150	77.2	230	75.77	605	125	1256	2675
CompOFA [29]	-*	150	-*	230	-*	330	142	782	1665
BigNAS [40]	70.6	150	74	230	72.51	400	266	1778	3787
DϵpS	**73.8**	150	**77.3**	230	**75.81**	**270**	155	**700**	**1491**

*Since CompOFA reports matching OFA accuracy, we achieve comparison to both by reporting Pareto frontier results for OFA.

respectively. Specifically, we use SGD with Nesterov momentum 0.9, a CosineAnnealing LR [24] schedule, and weight decay $3e^{-5}$. Unless specified, the shrinking is introduced in DϵpS after the full model gets $\approx 50\%$ trained ($P_{warmup}^{fm} \approx 50\%$).

5.2 Evaluation

Comparison with NAS Methods/Efficient Nets on ImageNet-1k. We compare DϵpS with MobilenetV3 [15], FBNet [35], ProxylessNAS [4], BigNAS [40] and efficient nets [33] on the ImageNet-1k dataset.

Takeaway. Table 1 compares accuracy vs MACs of the baselines. DϵpS consistently surpasses the baselines over multiple MAC ranges. Especially in the lower MAC region (0-100M), DϵpS is **1.8%** more accurate. In the larger MAC region (200-300M), DϵpS achieves 77.3% accuracy with upto 1.69x MACs improvement compared to the baselines (efficientNet-B0). DϵpS benefits from supernet initialization and effective knowledge distillation to get superior performance.

Comparison with Once-for-All Training Methods on ImageNet-1k. We now demonstrate the accuracy and training cost benefits of DϵpS on ImageNet-1k dataset [28]. Table 2 compares DϵpS with the baselines[2] on a) the upper-bound (largest subnet) and lower-bound (smallest subnet) top1 accuracy, and b) GPU hours and dollar costs.

Takeaway. DϵpS is atleast **2%** more accurate at 150 MACs (smallest subnet) than baselines and at-par w.r.t accuracy at 230 MACs (largest subnet). DϵpS matches the Pareto-optimality of baselines (with highest mean pareto accuracy) at a reduced training cost (least among all the baselines). It takes 1.8x and 2.5x less dollar cost (or GPU hours) than OFA and BigNAS respectively.

---

[2] Since CompOFA reports matching OFA accuracy, we achieve comparison to both by reporting Pareto frontier results for OFA.

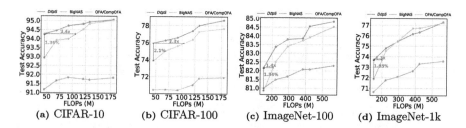

Fig. 5. DϵpS's Accuracy Improvement across Datasets. The comparison of DϵpS with the baselines is shown w.r.t accuracy (of subnets) for CIFAR10/100, ImageNet-100, and ImageNet-1k datasets. DϵpS consistently outperforms the baselines across all the datasets and achieves upto 2.1% better accuracy for the same FLOPs or upto 2.3x FLOP reduction at same accuracy.

The training cost improvement of DϵpS comes due to FM-Warmup. FM-Warmup allows DϵpS to train subnets in less number of total epochs (lowest among the baselines) and a lower average time per epoch than BigNAS (Table 2). The full model's accuracy (largest subnet in Table 2) is improved as ϵ-Shrinking enables its smooth convergence. Finally, DϵpS improves accuracy at lower FLOPs (150 MACs) as IKD-Warmup distills knowledge effectively in once-for-all training.

Generalization Across Datasets. We establish that the accuracy improvements of DϵpS generalize to other vision datasets.

Training Details. DϵpS uses the standard hyper-parameters of the MobileNetV3 for all the datasets using SGD with cosine learning rate decay and nestrov momentum, and shrinking is introduced when the full model is 50% trained. For OFA, we first train the largest network independently. Shrinking occurs after the full model is completely trained and vanilla KD is used for distillation. The depth and expand phases are run for 100 epochs each. The initial learning rate of different phases is set as per OFA [3]. BigNAS uses RMSProp optimizer with its proposed hyper-parameters for ImageNet-1k. However, we use SGD optimizer in BigNAS for CIFAR10/100 and ImageNet-100 datasets as we empirically find that SGD performs better than RMSProp on these datasets. Figure 5 compares the Pareto-frontiers of top1 test accuracy and FLOPs obtained from each baseline across various datasets. The subnets are present in six different FLOP buckets that uniformly divide the supernet's FLOP range. The comparison includes the performance of the smallest and largest subnets to measure the lower-bound and upper-bound test accuracy reached by the baselines.

Takeaway. DϵpS outperforms baselines w.r.t accuracy of smaller subnets (≤ 300 MFLOPs) on all the datasets. It achieves slightly better or at-par accuracy for larger subnets (≥ 300 MFLOPs) than OFA/CompOFA. DϵpS outperforms BigNAS and achieves a better Pareto-Frontier across all the datasets.

Generalization Across DNN-Architecture Spaces. We demonstrate that DϵpS generalizes to other DNN-architecture spaces. We train DϵpS on ImageNet-1k dataset using ProxylessNAS-based supernet (DNN-architecture space) with training-hyperparameters borrowed from [4]. Figure 6 compares Pareto-frontiers obtained from DϵpS and OFA on ImageNet-1k dataset. We reiterate that we don't conduct an additional exhaustive hyperparameter search and instead use training-hyperparameters from [4] and a $P^{fm}_{\text{warmup}} \approx 50\%$.

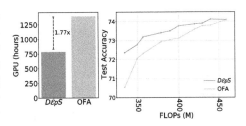

Fig. 6. DϵpS on ProxyLessNAS architecture space: superior Pareto-Frontier with a 1.8% improvement in ImageNet-1k test accuracy on the smallest subnet.

Takeaway. DϵpS outperforms OFA w.r.t ImageNet-1k test accuracy (with 0.5% better mean pareto accuracy). It improves the accuracy of the smallest subnet by **1.8%**. The accuracy improvements come with **1.8x** training cost reduction compared to OFA.

5.3 Ablation Study

We provide detailed ablation on DϵpS components—FM-Warmup, ϵ-Shrinking, and IKD-Warmup to attribute their benefits.

Full Model Warmup Period (P^{fm}_{warmup}). In this ablation, we establish the benefits of delayed shrinking as opposed to early or late shrinking. To do this, we configure DϵpS to run with different full model warmup periods (P^{fm}_{warmup}) – the time at which shrinking starts in DϵpS. Our goal is to empirically demonstrate the existence of a sweet spot in P^{fm}_{warmup} w.r.t accuracy (of subnets). Figure 7a

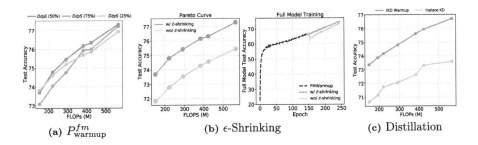

(a) P^{fm}_{warmup} (b) ϵ-Shrinking   (c) Distillation

Fig. 7. DϵpS Ablations. Three ablations are shown for DϵpS—Full model warmup period (P^{fm}_{warmup}), ϵ-Shrinking, and Distillation. a) There exists a sweet spot w.r.t accuracy (of subnets) in P^{fm}_{warmup} (=50%), b) ϵ-Shrinking improves the entire pareto front (left) and prevents drop in accuracy of the full model (right), c) IKD-Warmup performs better than Inplace KD as it uses more information from non-target classes (further details are provided in supplementary material).

compares the accuracy of best-performing subnets in six different FLOP buckets of three P^{fm}_{warmup} periods {25%, 50%, 75%} on ImageNet-1k dataset. P^{fm}_{warmup} =25%, 75% represents early and late shrinking respectively.

Takeaway. DϵpS with P^{fm}_{warmup} = 50% achieves the best test accuracy across subnets compared to DϵpS configured to run with P^{fm}_{warmup} = 25%, 75%. Hence, a sweet spot exists in P^{fm}_{warmup}. The existence of a sweet spot demonstrates that both early (25%) or late (75%) shrinking is sub-optimal in training the model family (discussed in Sect. 4.1). Early shrinking results in sub-optimal accuracy of the larger subnets as training interference occurs very early in the training. While late shrinking causes the specialization of supernet weights to the full model architecture that results in sub-optimal accuracy of smaller subnets ($\approx$ 1% accuracy degradation around 200 MFLOPs for P^{fm}_{warmup} =75% compared to P^{fm}_{warmup} =50%).

$\mathcal{E}$-**Shrinking.** We investigate whether an accuracy drop occurs in the full model's accuracy when shrinking is introduced in DϵpS and if $\mathcal{E}$-Shrinking prevents it. In this ablation, we run DϵpS with and without $\mathcal{E}$-Shrinking and introduce shrinking at 150th epoch while keeping all other training-hyperparameters constant. Figure 7b (right) compares DϵpS with and without $\mathcal{E}$-Shrinking on ImageNet-1k top1 test accuracy of the full model over training epochs. Figure 7b (left) compares subnets for six different FLOP buckets with and without ϵ-Shrinking.

Takeaway. DϵpS without $\mathcal{E}$-Shrinking observes a **2%** drop in full model's accuracy at 150th epoch when the shrinking starts. And, DϵpS with $\mathcal{E}$-Shrinking prevents this huge accuracy drop at the start of shrinking that leads to better full model accuracy overall. The prevention of the drop in full model's accuracy demonstrates that $\mathcal{E}$-Shrinking leads to smooth optimization of the full model. $\mathcal{E}$-Shrinking achieves this by incrementally warming up subnets' learning rate at the start of shrinking to avoid sudden changes in the supernet weight (Fig. 7b, right). $\mathcal{E}$-Shrinking also achieves superior accuracy across the entire FLOP range when compared to the supernet trained without $\mathcal{E}$-Shrinking (Fig. 7b, left).

IKD-Warmup. We assess the benefits of IKD-Warmup in this ablation. IKD-Warmup performs inplace knowledge distillation from a partially trained full model instead of a randomly initialized full model (inplace KD) as proposed in [41]. Hence, to show benefits of IKD-Warmup, we run DϵpS with inplace KD and our proposed IKD-Warmup. Figure 7c compares DϵpS run with IKD-Warmup (blue) and inplace KD (orange) on the ImageNet-1k top1 test accuracy of best-performing subnets in seven different FLOP buckets.

Takeaway. IKD-Warmup outperforms inplace KD across all the subnets that cover the supernet's FLOP range on the ImageNet-1k dataset. It is **3.5%** and **2%** more accurate at 560 MFLOPs and 150 MFLOPs respectively. This shows that IKD-Warmup distills knowledge effectively in once-for-all training as multiple progressively better partially trained full model transfer their knowledge to smaller subnets (Sect. 4.3). Inplace KD is not able to provide meaningful knowledge transfer as the full model is under-trained initially.

6 Conclusion

DϵpS is a training technique that increases the scalability of once-for-all training. DϵpS consists of three key components—FM-Warmup that decreases training costs, ϵ-Shrinking that maintains full model accuracy on par with existing works, and IKD-Warmup that enables effective knowledge distillation in once-for-all training. FM-Warmup's key insight is to delay the process of shrinking (introducing smaller weight-shared subnets) until the full model is partially trained (∼50%) to reduce training cost. ϵ-Shrinking averts accuracy degradation in the full model by avoiding rapid changes in the supernet weights and, instead, enables smooth optimization by progressively warming up subnets' learning rates. IKD-Warmup provides rich knowledge transfer to subnets from multiple partially trained full models that are progressively better w.r.t.accuracy. DϵpS generalizes to different datasets and DNN architecture spaces. It improves the accuracy of smaller subnets, achieves on-par Pareto-optimality, and reduces training cost by upto 2.5x when compared with existing once-for-all weight-shared training techniques.

Acknowledgments. This material is based upon work partially supported by the National Science Foundation under Grant Number CNS-2420977 as well as a sponsored research award from Cisco Research. We would also like to express our sincere gratitude to the reviewers and the AC panel for their insightful comments and thoughtful consideration. We applaud their invaluable and unwavering dedication to the pursuit of scientific truths. **Disclaimer**: Any opinions, findings, and conclusions or recommendations expressed in this material are those of the authors and do not necessarily reflect the views of the National Science Foundation.

References

1. Bai, S., Kolter, J.Z., Koltun, V.: An empirical evaluation of generic convolutional and recurrent networks for sequence modeling. CoRR abs/1803.01271 (2018). http://arxiv.org/abs/1803.01271
2. Bonnard, J., Abdelouahab, K., Pelcat, M., Berry, F.: On building a CNN-based multi-view smart camera for real-time object detection. Microprocess. Microsyst. **77**, 103177 (2020)
3. Cai, H., Gan, C., Wang, T., Zhang, Z., Han, S.: Once-for-all: train one network and specialize it for efficient deployment. In: International Conference on Learning Representations (2020). https://openreview.net/forum?id=HylxE1HKwS
4. Cai, H., Zhu, L., Han, S.: Proxylessnas: direct neural architecture search on target task and hardware. arXiv preprint arXiv:1812.00332 (2018)
5. Cho, J.H., Hariharan, B.: On the efficacy of knowledge distillation. In: Proceedings of the IEEE/CVF International Conference on Computer Vision, pp. 4794–4802 (2019)
6. Cob-Parro, A.C., Losada-Gutiérrez, C., Marrón-Romera, M., Gardel-Vicente, A., Bravo-Muñoz, I.: Smart video surveillance system based on edge computing. Sensors **21**(9), 2958 (2021)

7. Deng, J., Dong, W., Socher, R., Li, L.J., Li, K., Fei-Fei, L.: Imagenet: a large-scale hierarchical image database. In: 2009 IEEE Conference on Computer Vision and Pattern Recognition, pp. 248–255. IEEE (2009)
8. Gog, I., Kalra, S., Schafhalter, P., Wright, M.A., Gonzalez, J.E., Stoica, I.: Pylot: a modular platform for exploring latency-accuracy tradeoffs in autonomous vehicles. In: 2021 IEEE International Conference on Robotics and Automation (ICRA), pp. 8806–8813. IEEE (2021)
9. Goyal, P., et al.: Accurate, large minibatch SGD: training imagenet in 1 hour. arXiv preprint arXiv:1706.02677 (2017)
10. Han, K., Wang, Y., Zhang, Q., Zhang, W., Xu, C., Zhang, T.: Model rubik's cube: twisting resolution, depth and width for tinynets. Adv. Neural. Inf. Process. Syst. **33**, 19353–19364 (2020)
11. Han, S., Mao, H., Dally, W.J.: Deep compression: compressing deep neural networks with pruning, trained quantization and huffman coding. arXiv preprint arXiv:1510.00149 (2015)
12. Hashemi, H.B., Asiaee, A., Kraft, R.: Query intent detection using convolutional neural networks. In: International Conference on Web Search and Data Mining, Workshop on Query Understanding (2016)
13. He, K., Zhang, X., Ren, S., Sun, J.: Deep residual learning for image recognition. In: Proceedings of the IEEE Conference on Computer Vision and Pattern Recognition, pp. 770–778 (2016)
14. Hinton, G., Vinyals, O., Dean, J.: Distilling the knowledge in a neural network. arXiv preprint arXiv:1503.02531 (2015)
15. Howard, A., et al.: Searching for mobilenetv3. In: Proceedings of the IEEE/CVF International Conference on Computer Vision, pp. 1314–1324 (2019)
16. Hubara, I., Courbariaux, M., Soudry, D., El-Yaniv, R., Bengio, Y.: Binarized neural networks. In: Advances in Neural Information Processing Systems, vol. 29 (2016)
17. Iandola, F.N., Han, S., Moskewicz, M.W., Ashraf, K., Dally, W.J., Keutzer, K.: Squeezenet: alexnet-level accuracy with 50x fewer parameters and< 0.5 mb model size. arXiv preprint arXiv:1602.07360 (2016)
18. NVIDIA Inc.: Nvidia Jetson. https://www.nvidia.com/en-in/autonomous-machines/embedded-systems/. Accessed 13 May 2023
19. NVIDIA Inc.: Nvidia v100. https://www.nvidia.com/en-in/data-center/v100/. Accessed 13 May 2023
20. Jacob, B., et al.: Quantization and training of neural networks for efficient integer-arithmetic-only inference. In: Proceedings of the IEEE Conference on Computer Vision and Pattern Recognition, pp. 2704–2713 (2018)
21. Krizhevsky, A., Hinton, G., et al.: Learning multiple layers of features from tiny images (2009)
22. Li, H., Kadav, A., Durdanovic, I., Samet, H., Graf, H.P.: Pruning filters for efficient convnets. arXiv preprint arXiv:1608.08710 (2016)
23. Lin, M., et al.: Hrank: filter pruning using high-rank feature map. In: Proceedings of the IEEE/CVF Conference on Computer Vision and Pattern Recognition, pp. 1529–1538 (2020)
24. Loshchilov, I., Hutter, F.: SGDR: stochastic gradient descent with warm restarts. arXiv preprint arXiv:1608.03983 (2016)
25. Luo, J.H., Wu, J., Lin, W.: Thinet: a filter level pruning method for deep neural network compression. In: Proceedings of the IEEE International Conference on Computer Vision, pp. 5058–5066 (2017)
26. Ouyang, Z., Niu, J., Liu, Y., Guizani, M.: Deep CNN-based real-time traffic light detector for self-driving vehicles. IEEE Trans. Mob. Comput. **19**(2), 300–313 (2019)

27. Real, E., Aggarwal, A., Huang, Y., Le, Q.V.: Regularized evolution for image classifier architecture search. In: Proceedings of the AAAI Conference on Artificial Intelligence, vol. 33, pp. 4780–4789 (2019)
28. Russakovsky, O., et al.: ImageNet large scale visual recognition challenge. Int. J. Comput. Vision (IJCV) **115**(3), 211–252 (2015). https://doi.org/10.1007/s11263-015-0816-y
29. Sahni, M., Varshini, S., Khare, A., Tumanov, A.: Compofa – compound once-for-all networks for faster multi-platform deployment. In: International Conference on Learning Representations (2021). https://openreview.net/forum?id=IgIk8RRT-Z
30. Sanh, V., Wolf, T., Rush, A.: Movement pruning: adaptive sparsity by fine-tuning. Adv. Neural. Inf. Process. Syst. **33**, 20378–20389 (2020)
31. Sun, W., Zhou, A., Stuijk, S., Wijnhoven, R., Nelson, A.O., Corporaal, H., et al.: Dominosearch: find layer-wise fine-grained n: M sparse schemes from dense neural networks. Adv. Neural. Inf. Process. Syst. **34**, 20721–20732 (2021)
32. Tan, M., et al.: Mnasnet: platform-aware neural architecture search for mobile. In: Proceedings of the IEEE/CVF Conference on Computer Vision and Pattern Recognition, pp. 2820–2828 (2019)
33. Tan, M., Le, Q.: Efficientnet: rethinking model scaling for convolutional neural networks. In: International Conference on Machine Learning, pp. 6105–6114. PMLR (2019)
34. Tian, Y., Krishnan, D., Isola, P.: Contrastive multiview coding. In: Vedaldi, A., Bischof, H., Brox, T., Frahm, J.-M. (eds.) ECCV 2020. LNCS, vol. 12356, pp. 776–794. Springer, Cham (2020). https://doi.org/10.1007/978-3-030-58621-8_45
35. Wan, A., et al.: Fbnetv2: differentiable neural architecture search for spatial and channel dimensions. In: Proceedings of the IEEE/CVF Conference on Computer Vision and Pattern Recognition, pp. 12965–12974 (2020)
36. Wang, C., Yang, Q., Huang, R., Song, S., Huang, G.: Efficient knowledge distillation from model checkpoints. In: Oh, A.H., Agarwal, A., Belgrave, D., Cho, K. (eds.) Advances in Neural Information Processing Systems (2022). https://openreview.net/forum?id=0ltDq6SjrfW
37. Wang, L., Dong, X., Wang, Y., Liu, L., An, W., Guo, Y.: Learnable lookup table for neural network quantization. In: Proceedings of the IEEE/CVF Conference on Computer Vision and Pattern Recognition, pp. 12423–12433 (2022)
38. Wu, B., et al.: Fbnet: hardware-aware efficient convnet design via differentiable neural architecture search. In: Proceedings of the IEEE/CVF Conference on Computer Vision and Pattern Recognition, pp. 10734–10742 (2019)
39. Yu, J., Huang, T.S.: Universally slimmable networks and improved training techniques. In: Proceedings of the IEEE/CVF International Conference on Computer Vision, pp. 1803–1811 (2019)
40. Yu, J., et al.: BigNAS: scaling up neural architecture search with big single-stage models. In: Vedaldi, A., Bischof, H., Brox, T., Frahm, J.-M. (eds.) ECCV 2020. LNCS, vol. 12352, pp. 702–717. Springer, Cham (2020). https://doi.org/10.1007/978-3-030-58571-6_41
41. Yu, J., Yang, L., Xu, N., Yang, J., Huang, T.: Slimmable neural networks. arXiv preprint arXiv:1812.08928 (2018)
42. Zoph, B., Vasudevan, V., Shlens, J., Le, Q.V.: Learning transferable architectures for scalable image recognition. In: Proceedings of the IEEE Conference on Computer Vision and Pattern Recognition, pp. 8697–8710 (2018)

Customize-A-Video: One-Shot Motion Customization of Text-to-Video Diffusion Models

Yixuan Ren[1](✉), Yang Zhou[2], Jimei Yang[2], Jing Shi[2], Difan Liu[2], Feng Liu[2], Mingi Kwon[3], and Abhinav Shrivastava[1]

[1] University of Maryland, College Park, MD 20770, USA
{yxren,abhinav}@cs.umd.edu
[2] Adobe Research, San Jose, CA 95110, USA
{yazhou,jimyang,jingshi,diliu,fengl}@adobe.com
[3] Yonsei University, Seoul 03722, Republic of Korea
kwonmingi@yonsei.ac.kr

Abstract. Image customization has been extensively studied in text-to-image (T2I) diffusion models, leading to impressive outcomes and applications. With the emergence of text-to-video (T2V) diffusion models, its temporal counterpart, motion customization, has not yet been well investigated. To address the challenge of one-shot video motion customization, we propose Customize-A-Video that models the motion from a single reference video and adapts it to new subjects and scenes with both spatial and temporal varieties. It leverages low-rank adaptation (LoRA) on temporal attention layers to tailor the pre-trained T2V diffusion model for specific motion modeling. To disentangle the spatial and temporal information during training, we introduce a novel concept of appearance absorbers that detach the original appearance from the reference video prior to motion learning. The proposed modules are trained in a staged pipeline and inferred in a plug-and-play fashion, enabling easy extensions to various downstream tasks such as custom video generation and editing, video appearance customization and multiple motion combination. Our project page can be found at https://customize-a-video.github.io.

Keywords: Video Motion Customization · Text-to-Video Diffusion Models · Low-Rank Adaptation

1 Introduction

Replicating an iconic motion in novel scenes is highly desirable for video creation. Recent large-scale diffusion-based text-to-video (T2V) generation models [5,45] demonstrate impressive outcomes in generating imaginative videos

Y. Ren and M. Kwon—Major work was done during an internship at Adobe.

Supplementary Information The online version contains supplementary material available at https://doi.org/10.1007/978-3-031-73024-5_20.

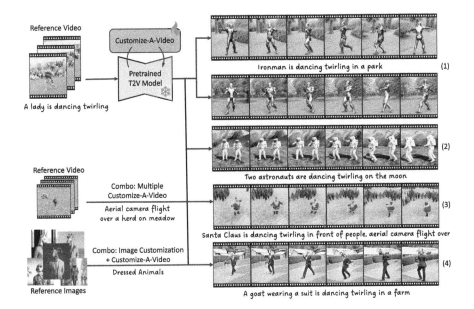

Fig. 1. Customize-A-Video takes as input a single reference video (top left) and transfers its motion onto new generated videos with plausible variance. (**1**) Transferring the dancing twirling from the lady onto Ironman with two random output variants. (**2**) Transferring the motion onto multiple subjects. (**3**) Combining multiple motion customization together, i.e., both *dancing twirling* and with *aerial camera flight over*. (**4**) Combining proposed motion customization and existing image customization methods ([21] in the example) to support both appearance and motion customization.

based on text depictions. However, they struggle with precise motion control and often demand extensive prompt engineering. Another thread of work on video editing [4,9,47,50] leverages large image generative models for appearance alteration, and introduces frame-wise precise controls via DDIM inversion [32,42] or ControlNet [54]. While achieving promising motion transfer results with variations in appearance and texture, these methods rigidly adhere to the reference frame structure and layout and fail to provide variability in the motion itself, such as new positions, intensities, camera views, or quantity of subjects.

Image customization of T2I models has been widely explored [10,39] where a specific unique appearance is modeled and composed into novel roles and scenes. These modules are trained on a small set of images that share the same concept. They are then able to reproduce the desired concept needless of complex prompt engineering, while also allowing for diversity in poses, views, lighting, etc. compared to direct stitching and editing approaches. Inspired by this, we introduce a new task of video motion customization and present a novel one-shot method named *Customize-A-Video* (Fig. 1) built upon T2V diffusion models. It customizes the pre-trained model with the motion learned from the reference video, enabling it to be easily adapted to new subjects and scenes. This includes not only precise transfer but also variations in motion intensities, positions, quantity of subjects, and camera views. These variations make the output videos more dynamic and engaging, as opposed to the robotic rhythm or unnatural appearance of frame-wise tampering.

Specifically, we start from utilizing a common customization technique, Low-Rank Adaptation (LoRA) [19], applied on a pre-trained T2V diffusion model [45] to capture the motion signature in the reference video. Applying LoRA directly to the entire T2V models proves less effective in motion preservation, as spatial and temporal characteristics are intricately entangled and both will be learned simultaneously. Therefore, we apply LoRA only on temporal cross-frame attention layers, creating Temporal LoRA (T-LoRA), which is more concentrated on capturing motion dynamics from the video. In comparison to other popular customization algorithms, LoRA also offers a portable model size and requires minimal training data, as well as the simplicity of plug-and-play for easy extensibility to collaborate with additional customization modules.

While LoRA works well on few-shot customization tasks through the residual module weights, a portion of spatial features still leak into it when trained on a single reference video. Concurrent efforts attempt to address this challenging yet significant issue by either demanding a small dataset with diverse appearances and the consistent motion [31,48], or stopping training early and supplementing the underfit temporal modules with direct control signals from the reference video [20]. To tackle this issue and facilitate one-shot video customization, we introduce an innovative Appearance Absorber module to further decompose static signals from dynamics. The key idea of this module is to *absorb* the appearance out of the reference video, leaving only the desired motion information for the Temporal LoRA to model.

We introduce a staged training and inference pipeline as illustrated in Fig. 2 to connect all the components we have proposed while keeping them independent. In the first stage, we build and train the appearance absorber on unordered reference video frames to capture frame-wise spatial information, such as the subject appearance and the background scene. In the second stage, we load the trained appearance absorber in frozen state, and construct the Temporal LoRA on the temporal layers of the T2V model to train. The appearance absorber has encoded the static frames and therefore helps the Temporal LoRA focus primarily on temporal signals, minimizing the spatial information leakage into motion customization modules. During the inference stage, we remove the appearance absorber and load solely the trained Temporal LoRA. Given a text prompt containing novel subjects and scenes, our model not only accurately transfers the learned motion signature to the new appearance, but also produces diverse motions in terms of their intensities, positions, and camera views.

To summarize, our contributions involve:

– We present a novel one-shot motion customization method for single reference video based on pre-trained text-to-video diffusion models;
– We introduce Temporal LoRA to learn the motion from a single reference video, facilitating motion transfer with not only accuracy but also variety;

- We propose the general class of Appearance Absorbers to dedicatedly decompose the spatial information out of the reference video, effectively excluding it from the motion customization process;
- Our modules feature the plug-and-play and staged fashion and can be smoothly extended to various downstream applications.

2 Related Work

Text-to-Video Generation Models. Text-to-video (T2V) generation task generates videos from given text prompts specifying the expected appearances and motions. It has been widely explored previously using GANs [23,25,35] and transformers [11,18,44,49,52]. With the boost of T2I diffusion models, T2V diffusion models become subsequently under fast development. [22] reprogram the 2D spatial attentions into 3D temporal attentions to handle the new temporal dimension. [3,6,14,16,17,27,41,45,62] insert spatio-temporal 3D convolutions and/or cross-frame attentions to regulate the output temporal consistency from the random input noise. [12,22,27,29] design explicitly disentangled noise prior between key frames and residues to enforce temporal coherency. T2V models designate the generated content through text prompts, demanding significant engineering effort to prompt it to produce desired motions in details.

T2I-Based Video Editing. Leveraging the control signal directly from a reference video by editing it into new appearances is an efficient practice to precisely transfer the motion and has been studied by various methods. [13,28,37,40,46,50,58] leverage the inverse denoising process or degradation of the reference video frames to maintain the desired motion while altering its appearance through T2I generation. [1,8,9,26,30,47,56] adopt controllable image generation approaches [34,54] and extract the low-level reference signals such as their depth or edge maps to guide the generation process. [4,7,53,59,61] make use of the combination of above techniques. However, such methods fall short as they focus more on adopting novel appearances, and merely duplicate the original motion exactly but with no temporal diversity to vary in the motion intensity and velocity, subject position and quantity etc.

Video Motion Customization. Model customization is the task of adapting the original output to a new specific domain by adjusting the pre-trained model weights. It was first introduced for T2I models to personalize in spatial aspects such as identity, art style or pose [10,24,39]. Recently, the idea of customizing the motion given reference videos has also been emerging and evolving rapidly. [28,40,50,58] add temporal attentions from scratch on pre-trained T2I models and finetune them on a single video. Concurrent work [20,31] tunes the temporal layers in place in a pre-trained T2V model with either a regularization set or a frame residual loss to reduce the impact of training videos' appearances. Instead, our method appends residual weights to the original model using LoRA [19] and enables ligthweight training strategy and flexible inference utility.

[33] represents the first attempt to finetune the spatial and temporal attentions independently for the appearance and motion of a reference video, which

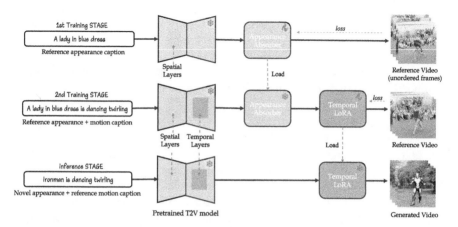

Fig. 2. Our Temporal LoRA, Appearance Absorbers and their training and inference processes. All noise and denoising schedules are omitted for simplicity. **(1)** We bypass all temporal layers in a base T2V diffusion model and apply appearance absorber such as S-LoRA or Textual Inversion on its spatial attention layers. The module is trained on unordered video frames. **(2)** We apply T-LoRA on all temporal attentions in the full base T2V model. The trained appearance absorber is also loaded and frozen. The module is trained on the target video. **(3)** During inference, only the trained T-LoRA is loaded. A new video with the customized motion is generated by a prompt describing the new appearance and the target motion.

have however entangled inference. Concurrent work [57] employs two parallel UNets and tune one of them with an appearance normalization loss to disentangle the motion as well as maintain the generalization ability on new appearances. Another concurrent work [48] adds adapters conditioned on one frame to decompose pure motion from its appearance, requiring additional input of an image when inference while ours asks for the minimal input of text prompt only. Concurrent work [60] applies dual-path LoRAs and trains them jointly with an appearance-debiased loss. In contrast, our approach adopts a staged training pipeline with independent tuning configurations, where our appearance absorber class can be easily extended to more candidates than LoRA, or reusing third-party modules pre-trained on in-the-wild images or videos.

3 Method

We present a novel motion customization method based on pre-trained T2V diffusion models for a single reference video. We suggest learning the motion concept from the reference video through a LoRA module designed for temporal layers of the T2V model. Given the challenging nature of working with a single datum, we develop a staged training strategy with an appearance absorber module to disentangle spatial information from motion. Figure 2 shows an illustration of each proposed module and its connection to the base T2V model.

3.1 Preliminary

Text-to-Video Diffusion Models. A text-to-video (T2V) diffusion model trains a 3D UNet to generate videos in a series denoising steps conditioned on a input text prompt. The 3D UNet usually consists of spatial self- and cross-attentions, 2D and 3D convolutions, and temporal cross-frame attentions. Given the F frames $x^{1...F}$ of a video, the 3D UNet θ is trained by minimizing

$$L_\theta = \mathbb{E}_{x^{1...F},\epsilon,t}[\|\epsilon - \epsilon_\theta(x_t^{1...F}, t, \tau_v(y))\|] \tag{1}$$

at every denoising step $t = T,...,0$. ϵ is Gaussian noise and ϵ_θ is the UNet prediction, τ is the text encoder with token sets v, and y is the text prompt.

Low-Rank Adaptation. Low-Rank Adaptation (LoRA) [19] was proposed for adapting pre-trained large language models to downstream tasks. It has also been widely developed for image customization models. LoRA applies a residue path of two low-rank matrices $\theta_B \in \mathbb{R}^{d \times r}, \theta_A \in \mathbb{R}^{r \times k}$ on an attention layer, whose original weight is $\theta_0 \in \mathbb{R}^{d \times k}$, $r \ll \min(d,k)$. The new forward path is

$$\theta = \theta_0 + \alpha \Delta \theta = \theta_0 + \alpha \theta_B \theta_A. \tag{2}$$

where α is a coefficient adjusting the strength of the added LoRA.

3.2 Customize-A-Video

We proposed two critical modules to customize the pre-trained T2V model for a single reference video. *Temporal LoRA* is introduced to learn motion from a reference video, whereas *Appearance Absorbers* are crafted to improve the separation of spatial and temporal information within the single reference video.

Temporal LoRA. Inspired by [19], we introduce Temporal LoRA (T-LoRA), a technique for capturing motion characteristics from input videos and enabling motion customization for new appearance via text prompts. We apply LoRAs on all temporal cross-frame attention layers of the base T2V model [45] to maximize modeling motion signals. Our ablation studies reveal that T-LoRA outperforms applying LoRA to other non-temporal attention layers, as T-LoRA targets at preserving motion while discarding unnecessary input appearance (see Sect. 4.2).

Appearance Absorbers. To separate spatial signals from temporal signals within a single video, we abstract the general class of Appearance Absorbers. Its objective is to absorb the spatial information, including the identity, texture, scene, etc., out of the training video, such that the reference motion can be exclusively modeled by our T-LoRA. To achieve this, we construct the absorbers leveraging a set of image customization modules, and use them in an *inverse* manner compared to their original role:

Image Model + *ImageCustomizationModule* → Image Custom Model;
Video Model − $\underbrace{AppearanceAbsorber}_{removes\,spatial\,signals\,from\,training\,video}$ + T-LoRA → Video Custom Model.

Our appearance absorbers can be built upon including but not limited to:

- *Spatial LoRA.* We apply LoRA on only the spatial attention layers in a T2V model to adopt solely the spatial information out of the video frames. LoRA modules are injected in all self-attention layers of the frames and cross-attention layers between frames and the text prompt. We call it spatial LoRA (S-LoRA) to distinguish from our T-LoRA for temporal modeling.
- *Textual Inversion.* We utilize textual inversion [10] as another approach to collect spatial features from the training video. It creates learnable placeholder tokens, initialized with briefly depicting words of the video appearance, to assimilate relevant spatial information through the text tokenizer.

These image customization modules are adept at modeling appearance signals from limited number of frames of a single video in a few-shot manner, and thus we prefer less finetuning-based customization methods such as [39] since they require a considerable amount of training and regularization data. All types of appearance absorbers can be employed individually or jointly.

Training and Inference Pipelines. Our motion customization pipeline in Fig. 2 consists of two training stages for appearance absorbers and T-LoRA respectively, and one inference stage to finally generate output videos with novel text prompts. Our configurable pipeline has dedicated stage for each module and is universal for extensive types of its components.

First Training Stage. We train appearance absorber modules first. Since they are originated from T2I models, we propose to specially train them by bypassing all temporal layers in the T2V model, including temporal attention layers and 3D convolution layers in the denoising UNet. We train them with the appearance description y_S cut out of the full caption so that they focus on learning the spatial information. The training images are unordered frames of the reference video. We follow their native loss as in [10,19] to train each type of appearance absorber. Formally, for S-LoRA $\Delta\theta_S$:

$$L_{\Delta\theta_S} = \mathbb{E}_{x,\epsilon,t}[\|\epsilon - \epsilon_{\theta_0+\Delta\theta_S}(x_t^f, t, \tau_{v_0}(y_S))\|], \tag{3}$$

and for textual inversion Δv:

$$L_{\Delta v} = \mathbb{E}_{x,\epsilon,t}[\|\epsilon - \epsilon_{\theta_0}(x_t^f, t, \tau_{v_0+\Delta v}(y_S))\|]. \tag{4}$$

Second Training Stage. We inject above trained appearance absorbers into the T2V model and maintain their frozen state. Our T-LoRA is meanwhile injected into the temporal attention layers of the T2V model. It is trained with the reference video and full ground truth caption consisting of both motion verbs and appearance nouns, by which the appearance absorber is also triggered to yield spatially customized content in static frames. We train T-LoRA $\Delta\theta_T$ using the standard reconstruction loss as in diffusion models [38]:

$$L_{\Delta\theta_T} = \mathbb{E}_{x^{1...F},\epsilon,t}[\|\epsilon - (\epsilon_{\theta'+\Delta\theta_T}(x_t^{1...F}, t, \tau_{v'}(y)))\|] \tag{5}$$

where $\theta' = \theta_0 + \Delta\theta_S$ and $v' = v_0 + \Delta v$ if respective AAs are employed.

Inference Stage. During the final inference stage, solely the trained T-LoRA is loaded onto the base T2V model. Given a new text prompt depicting the reference motion with new appearances and scenes, the customized model generates novel videos animated by the desired motion following the standard denoising process. As a result of the customized weights in T-LoRA, our output video transfers the reference motion faithfully as well as with diversity in motion intensities, positions, and camera views etc.

4 Experiments

Base T2V Models. Our methods are applicable to general T2V diffusion models. In the following experiments, we hire the ModelScope T2V model [45] as the pre-trained base model. All videos are pre-processed and generated for 2 s, 8 FPS and 256×256 resolution. Training hyperparameters, model size statistics and time consumption analyses are detailed in the supplementary material.

Datasets. We select videos from mixed sources, including LOVEU-TGVE-2023 [51], WebVid-10M [2] and DAVIS [36] datasets to evaluate our method. [51] provides ground truth captions and target editing prompts, while we create those for videos from other sources. We also apply our method on in-the-wild videos and demonstrate its generalization ability.

Comparison Methods. As of a new task of one-shot video motion customization, we mainly compare to Tune-A-Video [50] and Video-P2P [28], which append raw temporal layers to pre-trained T2I models and finetune them on a single reference video. It is worth noting that they additionally rely on DDIM inverted reference video latent as the input during inference and thus only produce temporally deterministic videos with fixed frame layout and view angle, so we also evaluate their variants removing this condition. We also compare our method against the pre-trained T2V model, i.e. ModelScope, to prove that our method enhances the base foundation model to produce faithful motions following the reference video that are not trivial to depict via prompt engineering. Besides, we run the concurrent work MotionDirector [60] using their released training code with the same configuration as ours, including the same base T2V model and LoRA hyperparameters for fair comparison.

Quantitative Metrics. We measure the performance quantitatively over a subset containing 53 videos out of [51] of 2–3 seconds with standard original and editing captions. We consider comparisons in terms of three metrics: **text alignment** between the generated video frames and the inference prompt gauges both generated appearance and motion accuracy, in the form of CLIPScore [15] that associates text and image in a unified space; **temporal consistency** between consecutive frames of the generated video indicates the generated motion quality, in the form of LPIPS [55] that measures deep feature distance; **diversity** among multiple generated videos with the same prompt and different random

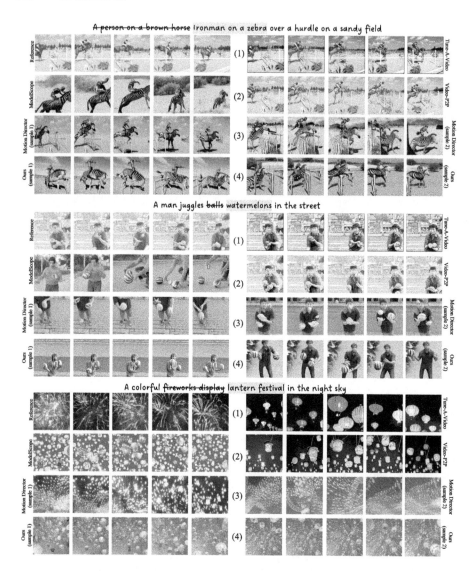

Fig. 3. Results of one-shot motion customization. **(1-left)** Reference video. **(2-left)** ModelScope [45] fails to transfer the reference motion faithfully with only text guidance. **(1-right & 2-right)** Tune-A-Video [50] and Video-P2P [28] rely on DDIM inverted latent input and duplicate the original frame structure deterministically. **(3)** Concurrent work MotionDirector [60] also generates various output following the reference motion while there exist some appearance and motion artifacts especially for hard examples with complex or intensive movements. **(4)** Our methods generate motion with both accuracy and variety in details such as view perspective and frame layout. Two variants generated with random noise are shown for MotionDirector and Ours.

Table 1. Quantitative comparisons on [51] dataset. ~ *w/o DDIM Inversion* represents the above method without DDIM inverted latent input. Video-P2P outputs video clips of 4 FPS with 512 × 512 resolution. MotionDirector is a concurrent work to ours and is tested with either the same LoRA rank or comparable amount of parameters to ours.

Method	Text Alignment↑	Temporal Consistency↓	Diversity↑
ModelScope [45]	31.705	0.175	**0.636**
Tune-A-Video [50]	31.149	0.185	-
~ w/o DDIM Inversion	30.304	0.206	0.348
Video-P2P [28]	31.001	0.162	-
~ w/o DDIM Inversion	30.876	0.251	0.469
MotionDirector [60]	32.500	0.163	0.606
~ w/ comparable #params	31.842	0.166	0.595
Ours No AA	31.687	0.166	0.613
Ours S-LoRA AA	31.913	0.163	0.618
Ours TextInv AA	**32.632**	**0.160**	0.621
Ours Dual AA	32.193	0.164	0.631

noise involves both spatial and temporal diversity by collating aligned frames at the same timestamp, in the form of LPIPS. It is calculated on 4 random samples per reference video.

User Study. We conduct a human user study among five algorithms with stochastic output: ModelScope, Tune-A-Video and Video-P2P without DDIM inverted latent, our method with both S-LoRA and textual inversion as the appearance absorbers, and MotionDirector. Every participant is presented 10 random reference videos from [51] and their output videos. Each algorithm outputs two videos, and participants are asked to assess their motion fidelity and motion diversity respectively, from 1 (worst) to 5 (best) stars. Details of the questionnaire design is provided in the supplementary materials.

4.1 Motion Customization from Single Video

Qualitative Results. Figure 3 illustrates the comparative visual results of one-shot motion customization. The base ModelScope T2V model fails to accurately replicate the specific motions as reference guided by simply the text prompt. On the other hand, Tune-A-Video [50] and Video-P2P [28] leverage DDIM inverted latents extracted from reference videos and produce temporally deterministic output with structural constraints by the reference frame layouts. In contrast to both of them, our approach demonstrates the capability to transfer the reference motion to new scenarios and subjects while introducing temporal variations via random noise input. Our outcomes not only exhibit diverse subject appearances and background scenes but also showcase variability in motion attributes such as action range, intensity, velocity, and camera perspective. Concurrent MotionDirector [60] is also able to generate adapted output with variety, while for the hard examples with complex or intensive motion, such as the juggling and firework in Fig. 3, it yields less competitive visual quality than ours. Our T-LoRA is always trained with well-optimized AAs, while MotionDirector might have appearances leaked into the temporal module when their spatial path is jointly being tuned. We also notice that real-world videos usually have divergent spatial and temporal complexity, and our dedicated tuning procedures with independent steps and other schedules for each module reach their individual optimal.

Quantitative Results. The quantitative results and comparisons are listed in Table 1. Our methods outperform the base ModelScope [45], Tune-A-Video [50], Video-P2P [28] and concurrent MotionDirector [60] on both text alignment and temporal consistency. ModelScope [45] provides the highest comprehensive diversity as a foundation model and loses in motion fidelity with text guidance only. Our methods sacrifice it subtly to gain significant improvements in faithfully customizing the exemplar motion, as well as retaining rich varieties in motion details. Tune-A-Video [50] and Video-P2P [28] have the minimal diversity as strictly constrained by frame structures, although at the cost of which they can achieve acceptable temporal consistency. MotionDirector [60] shows comparable text alignment and temporal consistency to ours but its diversity falls behind.

User Study. We involved 20 evaluators participating our user study and collected 153 and 156 valid ratings per algorithm on each benchmark. The averaged scores are listed in Table 2. Our method leads on both motion fidelity and motion diversity. We asked users to rate pure motion fidelity and diversity which are hard to assess by automatic metrics in need of generic motion representations irrespective to spatial structures. These results are complementary to the text alignment and diversity measured in Table 1 that mix spatial and temporal signals.

Table 2. Human user study results on [51] dataset. Methods are evaluated from 1 (worst) to 5 (best) stars on each benchmark.

Method	Motion Fidelity↑	Motion Diversity↑
ModelScope [45]	2.03	2.97
Tune-A-Video [50] w/o DDIM Inversion	2.29	2.23
Video-P2P [28] w/o DDIM Inversion	2.29	2.01
MotionDirector [60]	3.33	3.50
Ours (w/ dual AA)	**3.72**	**3.72**

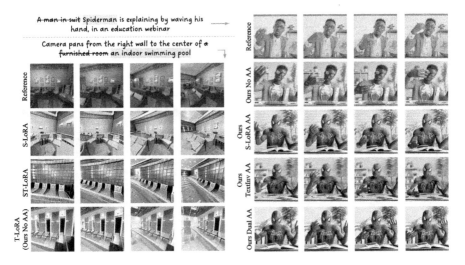

Fig. 4. Left: Ablations on applying LoRAs on different attention layers. S-LoRA memorizes the indoor furniture and wall decorations and T-LoRA converts paintings to entrances and sofas to pool benches. **Right**: Ablations on training T-LoRAs with different types of appearance absorbers. No AA adds stylish glasses and the logo but remains most the original appearance. S-LoRA AA and TextInv AA significantly boost the quality while resulting in the strips on the wall and the partially white sleeves. Dual AA reaches best spatial clearance with clear costume and background.

4.2 Ablations Studies

LoRAs on Non-temporal Attentions. While it is intuitive to apply LoRA on only temporal attention layers to learn video motions without original appearance, we also validate the effects of applying LoRA on the spatial attentions only (*S-LoRA*), or on both spatial and temporal attentions (*ST-LoRA*) in the base T2V model. Figure 4 left displays the visualizations in which adding LoRA to spatial attentions significantly impairs the motion modeling. The models with spatial elements primarily memorize the video by its spatial layout, resulting in a substantial degradation of both appearance and motion adaptation.

Comparisons Among Appearance Absorber Types. We explore four different configurations of Appearance Absorbers (AA). **No AA**: no appearance absorber is used, **S-LoRA AA**: a spatial LoRA based appearance absorber is used, **TextInv AA**: a textual inversion based appearance absorber is used, and **Dual AA**: two appearance absorbers of both above types are used. The comparison results are unveiled by Fig. 4. No AA remains some original appearance in addition to modeling the motion. S-LoRA AA and TextInv AA are both able to capture the pure action with minimal appearance leakage. We notice that S-LoRA AA is easier to overfit and sometimes causes spatial artifacts while TextInv AA might tend to underfit and leave spatial residues on the other hand. We attribute these properties to the spatial structure of S-LoRA weights inside U-Net blocks while

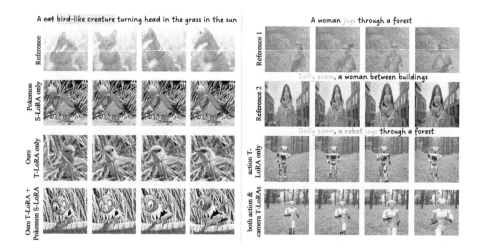

Fig. 5. Left: Video appearance customization with both T-LoRA and existing pre-trained S-LoRA (from [43]). **Right**: Multiple motion combination with two T-LoRAs loaded at the same time. When the robot is slowly jogging, it fast zooms in while the background trees rapidly zoom out (dolly zoom).

textual inversion works via a 1D learnable embedding as new tokens. Dual AA unites their advantages and leads to a comprehensive result with both the reference motion and new appearance clearly reflected.

5 Applications

With the plug-and-play nature of LoRA and our staged training pipeline, we present four downstream applications that demonstrate the collaborative potential of our proposed modules.

Video Appearance Customization. Our motion customization module works on temporal layers and thus can cooperate with image customization approaches to manipulate both the temporal and spatial layers in the base T2V model at the same time. In Fig. 5 left, we inject a T-LoRA to present the reference action as well as an image spatial LoRA to reflect the comic style in one comprehensive output.

Multiple Motion Combination. T-LoRA is applied to the original layers with residual connections. Therefore we can customize the base model with multiple T-LoRA modules trained on different source videos to integrate assorted motions into one outcome. Figure 5 right demonstrates that our method merges the human action of jogging and camera movement of dolly zoom into one target scenario using two T-LoRA modules.

Fig. 6. Left: Precise frame-wise video editing with DDIM inverted latent as the inference input noise. **Right**: A third-party Dreambooth UNet pre-tuned on in-the-wild images of the same subject serves as the appearance absorber to train the T-LoRA.

DDIM Inverted Latent Input. Our method can easily incorporate additional deterministic controls to perform precise video editing. The comparison results to Tuna-A-Video [50] are shown as Fig. 6 left. Our models prove to be also able to benefit from the DDIM inverted latent of the reference video and yield output that reproduces the exact original frame structures.

Third-Party Appearance Absorbers. Our staged training pipeline enables reusing appearance absorbers across videos or loading third-party image customization modules as ready appearance absorbers when they share the similar appearance. This skips the first training stage and extends appearance absorber categories to those demand more training data. In Fig. 6 right, we finetune the spatial layers of the UNet following Dreambooth [39] and Eq. 1 on other photos of the same dog. Our T-LoRA trained with it avoids the leakage of the dog color to the white wolf, compared to that trained with the original base model.

6 Conclusion

We introduce the one-shot motion customization task that learns the motion signature from a single reference video and transfers it to new scenes and subjects with variety in both appearance and motion. We propose Temporal LoRA to model the target motion by adding LoRA residual weights on the temporal attention layers of a pre-trained text-to-video diffusion model. We further propose Appearance Absorbers to decouple the spatial information from the reference video so that Temporal LoRA can focus on motion modeling. Extensive experiments demonstrate that our methods yield faithful and diverse videos compared to both per-frame video editing approaches and the base T2V model. Our method's is plug-and-play nature supports various downstream tasks including precise video editing, video appearance customization, multiple motion combination as well as third-party appearance absorbers.

References

1. AILab-CVC: Ailab-cvc/videocrafter at 1f46314b6609712eea89b67f41d612557eec5b8e. https://github.com/AILab-CVC/VideoCrafter/tree/1f46314b6609712eea89b67f41d612557eec5b8e
2. Bain, M., Nagrani, A., Varol, G., Zisserman, A.: Frozen in time: a joint video and image encoder for end-to-end retrieval. In: IEEE International Conference on Computer Vision (2021)
3. Blattmann, A., et al.: Align your latents: high-resolution video synthesis with latent diffusion models. In: Proceedings of the IEEE/CVF Conference on Computer Vision and Pattern Recognition, pp. 22563–22575 (2023)
4. Ceylan, D., Huang, C.H.P., Mitra, N.J.: Pix2video: video editing using image diffusion. In: Proceedings of the IEEE/CVF International Conference on Computer Vision, pp. 23206–23217 (2023)
5. Chai, W., Guo, X., Wang, G., Lu, Y.: Stablevideo: text-driven consistency-aware diffusion video editing. In: Proceedings of the IEEE/CVF International Conference on Computer Vision, pp. 23040–23050 (2023)
6. Chen, H., et al.: Videocrafter1: open diffusion models for high-quality video generation. arXiv preprint arXiv:2310.19512 (2023)
7. Chen, W., et al.: Control-a-video: controllable text-to-video generation with diffusion models. arXiv preprint arXiv:2305.13840 (2023)
8. Chu, E., Lin, S.Y., Chen, J.C.: Video controlnet: towards temporally consistent synthetic-to-real video translation using conditional image diffusion models. arXiv preprint arXiv:2305.19193 (2023)
9. Esser, P., Chiu, J., Atighehchian, P., Granskog, J., Germanidis, A.: Structure and content-guided video synthesis with diffusion models. In: Proceedings of the IEEE/CVF International Conference on Computer Vision, pp. 7346–7356 (2023)
10. Gal, R., et al.: An image is worth one word: personalizing text-to-image generation using textual inversion. arXiv preprint arXiv:2208.01618 (2022)
11. Ge, S., et al.: Long video generation with time-agnostic VQGAN and time-sensitive transformer. In: Avidan, S., Brostow, G., Cissé, M., Farinella, G.M., Hassner, T. (eds.) ECCV 2022. LNCS, vol. 13677, pp. 102–118. Springer, Cham (2022). https://doi.org/10.1007/978-3-031-19790-1_7
12. Ge, S., et al.: Preserve your own correlation: a noise prior for video diffusion models. In: Proceedings of the IEEE/CVF International Conference on Computer Vision, pp. 22930–22941 (2023)
13. Geyer, M., Bar-Tal, O., Bagon, S., Dekel, T.: Tokenflow: consistent diffusion features for consistent video editing. arXiv preprint arXiv:2307.10373 (2023)
14. Guo, Y., et al.: Animatediff: animate your personalized text-to-image diffusion models without specific tuning. In: The Twelfth International Conference on Learning Representations (2024). https://openreview.net/forum?id=Fx2SbBgcte
15. Hessel, J., Holtzman, A., Forbes, M., Bras, R.L., Choi, Y.: Clipscore: a reference-free evaluation metric for image captioning. arXiv preprint arXiv:2104.08718 (2021)
16. Ho, J., et al.: Imagen video: high definition video generation with diffusion models. arXiv preprint arXiv:2210.02303 (2022)
17. Ho, J., Salimans, T., Gritsenko, A., Chan, W., Norouzi, M., Fleet, D.J.: Video diffusion models. Adv. Neural. Inf. Process. Syst. **35**, 8633–8646 (2022)
18. Hong, W., Ding, M., Zheng, W., Liu, X., Tang, J.: Cogvideo: large-scale pretraining for text-to-video generation via transformers. arXiv preprint arXiv:2205.15868 (2022)

19. Hu, E.J., et al.: Lora: low-rank adaptation of large language models. arXiv preprint arXiv:2106.09685 (2021)
20. Jeong, H., Park, G.Y., Ye, J.C.: VMC: video motion customization using temporal attention adaption for text-to-video diffusion models. arXiv preprint arXiv:2312.00845 (2023)
21. Kappa_Neuro: Dressed animals - SD 2.1: Stable diffusion lora. https://civitai.com/models/35733?modelVersionId=41939
22. Khachatryan, L., et al.: Text2video-zero: text-to-image diffusion models are zero-shot video generators. arXiv preprint arXiv:2303.13439 (2023)
23. Kim, D., Joo, D., Kim, J.: Tivgan: text to image to video generation with step-by-step evolutionary generator. IEEE Access **8**, 153113–153122 (2020)
24. Kumari, N., Zhang, B., Zhang, R., Shechtman, E., Zhu, J.Y.: Multi-concept customization of text-to-image diffusion. In: Proceedings of the IEEE/CVF Conference on Computer Vision and Pattern Recognition, pp. 1931–1941 (2023)
25. Li, Y., Min, M., Shen, D., Carlson, D., Carin, L.: Video generation from text. In: Proceedings of the AAAI Conference on Artificial Intelligence, vol. 32 (2018)
26. Liew, J.H., Yan, H., Zhang, J., Xu, Z., Feng, J.: Magicedit: high-fidelity and temporally coherent video editing. arXiv preprint arXiv:2308.14749 (2023)
27. Liu, B., Liu, X., Dai, A., Zeng, Z., Cui, Z., Yang, J.: Dual-stream diffusion net for text-to-video generation. arXiv preprint arXiv:2308.08316 (2023)
28. Liu, S., Zhang, Y., Li, W., Lin, Z., Jia, J.: Video-p2p: video editing with cross-attention control. arXiv preprint arXiv:2303.04761 (2023)
29. Luo, Z., et al.: Videofusion: decomposed diffusion models for high-quality video generation. In: Proceedings of the IEEE/CVF Conference on Computer Vision and Pattern Recognition, pp. 10209–10218 (2023)
30. Ma, Y., et al.: Follow your pose: pose-guided text-to-video generation using pose-free videos. arXiv preprint arXiv:2304.01186 (2023)
31. Materzyńska, J., Sivic, J., Shechtman, E., Torralba, A., Zhang, R., Russell, B.: Customizing motion in text-to-video diffusion models. arXiv preprint arXiv:2312.04966 (2023)
32. Mokady, R., Hertz, A., Aberman, K., Pritch, Y., Cohen-Or, D.: Null-text inversion for editing real images using guided diffusion models. In: Proceedings of the IEEE/CVF Conference on Computer Vision and Pattern Recognition, pp. 6038–6047 (2023)
33. Molad, E., et al.: Dreamix: video diffusion models are general video editors. arXiv preprint arXiv:2302.01329 (2023)
34. Mou, C., et al.: T2i-adapter: learning adapters to dig out more controllable ability for text-to-image diffusion models. arXiv preprint arXiv:2302.08453 (2023)
35. Pan, Y., Qiu, Z., Yao, T., Li, H., Mei, T.: To create what you tell: generating videos from captions. In: Proceedings of the 25th ACM International Conference on Multimedia, pp. 1789–1798 (2017)
36. Perazzi, F., Pont-Tuset, J., McWilliams, B., Van Gool, L., Gross, M., Sorkine-Hornung, A.: A benchmark dataset and evaluation methodology for video object segmentation. In: Computer Vision and Pattern Recognition (2016)
37. Qi, C., et al.: Fatezero: fusing attentions for zero-shot text-based video editing. arXiv preprint arXiv:2303.09535 (2023)
38. Rombach, R., Blattmann, A., Lorenz, D., Esser, P., Ommer, B.: High-resolution image synthesis with latent diffusion models. In: Proceedings of the IEEE/CVF Conference on Computer Vision and Pattern Recognition, pp. 10684–10695 (2022)

39. Ruiz, N., Li, Y., Jampani, V., Pritch, Y., Rubinstein, M., Aberman, K.: Dreambooth: fine tuning text-to-image diffusion models for subject-driven generation. In: Proceedings of the IEEE/CVF Conference on Computer Vision and Pattern Recognition, pp. 22500–22510 (2023)
40. Shin, C., Kim, H., Lee, C.H., Lee, S.g., Yoon, S.: Edit-a-video: single video editing with object-aware consistency. arXiv preprint arXiv:2303.07945 (2023)
41. Singer, U., et al.: Make-a-video: text-to-video generation without text-video data. arXiv preprint arXiv:2209.14792 (2022)
42. Song, J., Meng, C., Ermon, S.: Denoising diffusion implicit models. arXiv preprint arXiv:2010.02502 (2020)
43. tungdop2: tungdop2/pokemon-lora_sd2.1. https://huggingface.co/tungdop2/pokemon-lora_sd2.1
44. Villegas, R., et al.: Phenaki: variable length video generation from open domain textual description. arXiv preprint arXiv:2210.02399 (2022)
45. Wang, J., Yuan, H., Chen, D., Zhang, Y., Wang, X., Zhang, S.: Modelscope text-to-video technical report. arXiv preprint arXiv:2308.06571 (2023)
46. Wang, W., et al.: Zero-shot video editing using off-the-shelf image diffusion models. arXiv preprint arXiv:2303.17599 (2023)
47. Wang, X., et al.: Videocomposer: compositional video synthesis with motion controllability. arXiv preprint arXiv:2306.02018 (2023)
48. Wei, Y., et al.: Dreamvideo: composing your dream videos with customized subject and motion. arXiv preprint arXiv:2312.04433 (2023)
49. Wu, C., et al.: NÜWA: visual synthesis pre-training for neural visual world creation. In: Avidan, S., Brostow, G., Cissé, M., Farinella, G.M., Hassner, T. (eds.) ECCV 2022. LNCS, vol. 13676, pp. 720–736. Springer, Cham (2022). https://doi.org/10.1007/978-3-031-19787-1_41
50. Wu, J.Z., et al.: Tune-a-video: one-shot tuning of image diffusion models for text-to-video generation. In: Proceedings of the IEEE/CVF International Conference on Computer Vision, pp. 7623–7633 (2023)
51. Wu, J.Z., et al.: CVPR 2023 text guided video editing competition. arXiv preprint arXiv:2310.16003 (2023)
52. Yan, W., Zhang, Y., Abbeel, P., Srinivas, A.: Videogpt: video generation using VQ-VAE and transformers. arXiv preprint arXiv:2104.10157 (2021)
53. Yang, S., Zhou, Y., Liu, Z., Loy, C.C.: Rerender a video: zero-shot text-guided video-to-video translation. arXiv preprint arXiv:2306.07954 (2023)
54. Zhang, L., Rao, A., Agrawala, M.: Adding conditional control to text-to-image diffusion models. In: Proceedings of the IEEE/CVF International Conference on Computer Vision, pp. 3836–3847 (2023)
55. Zhang, R., Isola, P., Efros, A.A., Shechtman, E., Wang, O.: The unreasonable effectiveness of deep features as a perceptual metric. In: Proceedings of the IEEE Conference on Computer Vision and Pattern Recognition, pp. 586–595 (2018)
56. Zhang, Y., Wei, Y., Jiang, D., Zhang, X., Zuo, W., Tian, Q.: Controlvideo: training-free controllable text-to-video generation. arXiv preprint arXiv:2305.13077 (2023)
57. Zhang, Y., et al.: Motioncrafter: one-shot motion customization of diffusion models. arXiv preprint arXiv:2312.05288 (2023)
58. Zhang, Z., Li, B., Nie, X., Han, C., Guo, T., Liu, L.: Towards consistent video editing with text-to-image diffusion models. arXiv preprint arXiv:2305.17431 (2023)
59. Zhao, M., Wang, R., Bao, F., Li, C., Zhu, J.: Controlvideo: adding conditional control for one shot text-to-video editing. arXiv preprint arXiv:2305.17098 (2023)

60. Zhao, R., et al.: Motiondirector: motion customization of text-to-video diffusion models. arXiv preprint arXiv:2310.08465 (2023)
61. Zhao, Y., Xie, E., Hong, L., Li, Z., Lee, G.H.: Make-a-protagonist: generic video editing with an ensemble of experts. arXiv preprint arXiv:2305.08850 (2023)
62. Zhou, D., Wang, W., Yan, H., Lv, W., Zhu, Y., Feng, J.: Magicvideo: efficient video generation with latent diffusion models. arXiv preprint arXiv:2211.11018 (2022)

Author Index

A
Abbasi, Reza 35
Agrawal, Animesh 315
Akbari, Mohammad 118
Alipanah, Alireza 243
Alvar, Saeed Ranjbar 118
Aminbeidokhti, Masih 51
Annavajjala, Aditya 315

B
Bae, Wonho 279
Baghshah, Mahdieh Soleymani 35
Bakış, Yasin 137
Balhoff, James P. 137
Ban, Yuanhao 190
Bart Jr., Henry L. 137
Berger-Wolf, Tanya 137
Bhatia, Harshil 225
Bracha, Amit 207

C
Chang, Peng 172
Chao, Wei-Lun 137
Charpentier, Caleb 137
Chen, Ziyi 172
Cheng, Minhao 190
Cui, Jiali 86

D
Dahdul, Wasila 137
Daw, Arka 137
Deng, Xizhen 1
Desai, Khushi 225
Dhakal, Aayush 69
Ding, Boyang 172

F
Farhadi, Ali 261
Fatemi, Arezou 118

Fedorov, Igor 315
Fishman, Adam 261
Fox, Dieter 261

G
Gong, Boqing 190
Gonzalez, Joseph E. 225
Gorbunov, Mikhail 19
Granger, Eric 51
Grishina, Ekaterina 19
Guibas, Leonidas 297

H
Hamidi, Shayan Mohajer 1
Han, Mei 172
Han, Tian 86
Hsieh, Cho-Jui 190
Huang, Ian 297

J
Jacobs, Nathan 69
Jafarinia, Hossein 243

K
Karpatne, Anuj 137
Khare, Alind 315
Khurana, Mridul 137
Kimmel, Ron 207
Kwon, Mingi 332

L
Lapp, Hilmar 137
Latapie, Hugo 315
Latortue, David 51
Lee, Myungjin 315
Li, Hanao 86
Li, Peng 101
Li, Renda 172
Liu, Difan 332

© The Editor(s) (if applicable) and The Author(s), under exclusive license to Springer Nature Switzerland AG 2025
A. Leonardis et al. (Eds.): ECCV 2024, LNCS 15147, pp. 351–352, 2025.
https://doi.org/10.1007/978-3-031-73024-5

Liu, Feng 332
Liu, Lingqiao 101

M
Mabee, Paula M. 137
Mahapatra, Aniruddha 172
Maruf, M. 137
McAllister, Rowan 225
Medeiros, Heitor Rapela 51
Mirzaie, Nahal 243
Mishra, Richa 172

P
Packer, Charles 225
Pedersoli, Marco 51
Peña, Fidel Alejandro Guerrero 51

R
Rakhuba, Maxim 19
Razavi, Saeed 243
Ren, Yixuan 332
Rezaei, Ahmad 118
Rhinehart, Nicholas 225
Rohban, Mohammad Hossein 35, 243

S
Salamah, Ahmed Hussein 1
Sanchez Giraldo, Luis Gonzalo 69
Shi, Jing 332
Shrivastava, Abhinav 332
Skean, Oscar 69
Stewart, Charles 137
Sutherland, Danica J. 279

T
Tan, Renhao 1
Tian, Samuel 101
Tumanov, Alexey 315

U
Uyeda, Josef C. 137

W
Walsman, Aaron 261
Wang, Jing 279
Wang, Ruochen 190
Wang, Shoulei 172
Wang, Zihu 101
Weston, Scott Ricardo Figueroa 101
Wolf, Yaniv 207

X
Xiao, Jing 172

Y
Yang, En-hui 154
Yang, Guandao 297
Yang, Jiezhi 225
Yang, Jimei 332
Ye, Linfeng 1, 154
Yuan, Shiyu 86

Z
Zhang, Muru 261
Zhang, Yong 118
Zhou, Tianyi 190
Zhou, Yang 332
Zhu, Jun-Yan 172

Printed in the USA
CPSIA information can be obtained
at www.ICGtesting.com
CBHW081448011224
18277CB00005B/95